Geometric Gradient

Geometric Series Present Worth:

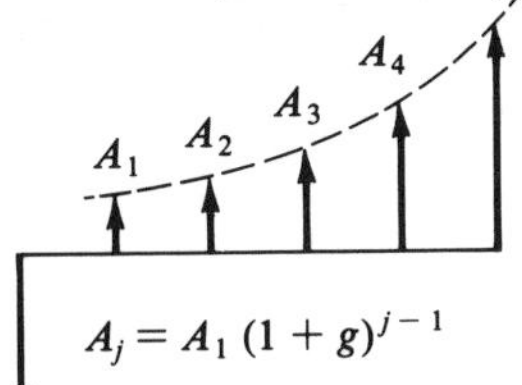

To Find P Given A_1, g	$(P/A,g,i,n)$ When $i = g$	$P = A_1[n(1+i)^{-1}]$
To Find P Given A_1, g	$(P/A,g,i,n)$ When $i \neq g$	$P = A_1\left[\dfrac{1-(1+g)^n(1+i)^{-n}}{i-g}\right]$

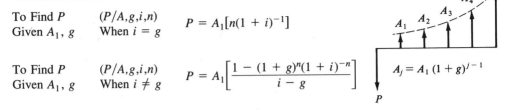

Continuous Compounding at Nominal Rate r

Single Payment: $F = P[e^{rn}]$ $\quad P = F[e^{-rn}]$

Uniform Series:

$$A = F\left[\frac{e^r - 1}{e^{rn} - 1}\right] \qquad A = P\left[\frac{e^{rn}(e^r - 1)}{e^{rn} - 1}\right]$$

$$F = A\left[\frac{e^{rn} - 1}{e^r - 1}\right] \qquad P = A\left[\frac{e^{rn} - 1}{e^{rn}(e^r - 1)}\right]$$

Continuous, Uniform Cash Flow (One Period) With Continuous Compounding at Nominal Rate r

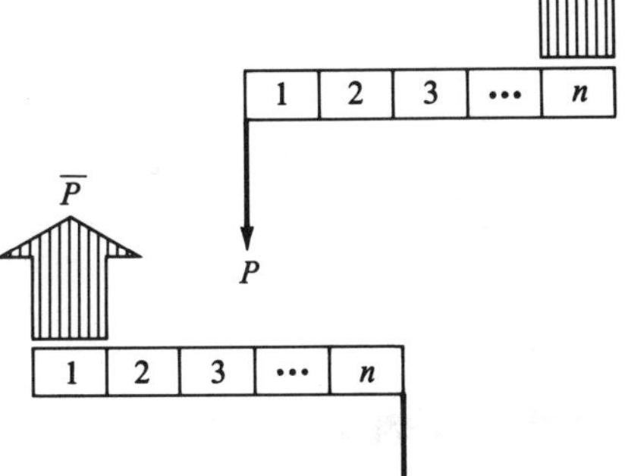

Present Worth:

To Find P Given $\overline{F}$	$(P/\overline{F},r,n)$	$P = \overline{F}\left[\dfrac{e^r - 1}{re^{rn}}\right]$

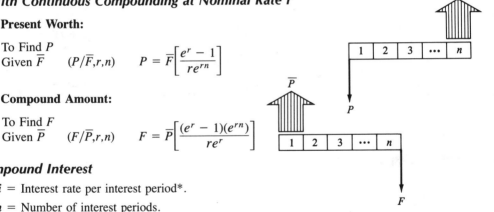

Compound Amount:

To Find F Given $\overline{P}$	$(F/\overline{P},r,n)$	$F = \overline{P}\left[\dfrac{(e^r - 1)(e^{rn})}{re^r}\right]$

Compound Interest

i = Interest rate per interest period*.

n = Number of interest periods.

P = A present sum of money.

F = A future sum of money. The future sum F is an amount, n interest periods from the present, that is equivalent to P with interest rate i.

A = An end-of-period cash receipt or disbursement in a uniform series continuing for n periods, the entire series equivalent to P or F at interest rate i.

G = Uniform period-by-period increase or decrease in cash receipts or disbursements; the arithmetic gradient.

g = Uniform *rate* of cash flow increase or decrease from period to period; the geometric gradient.

r = Nominal interest rate per interest period*.

m = Number of compounding subperiods per period*.

$\overline{P}$, $\overline{F}$ = Amount of money flowing continuously and uniformly during one given period.

*Normally the interest period is one year, but it could be something else.

Engineering Economic Analysis

Engineering Economic Analysis

Fourth Edition

Donald G. Newnan

Engineering Press, Inc. San Jose, California

Library of Congress Cataloging-in-Publication Data

Newnan, Donald G.
Engineering economic analysis / Donald G. Newnan.—4th ed.
p. cm.
Includes bibliographical references and index.
ISBN 0-910554-82-X
1. Engineering economy. I. Title.
TA177.4.N48 1991
658.1'5—dc20 90-40779
CIP

Engineering economic analysis.
Fourth Edition: ISBN 0-910554-82-X
Exam File: ISBN 0-910554-81-1
The two volumes are not sold separately.
They are shrinkwrapped and sold only as a set.
Order ISBN 0-910554-83-8.

The microcomputer programs were written by
Dr. Jan Wolski, New Mexico Institute of Mining and Technology.

Cover illustration courtesy of Monex International Ltd.

Printed in the United States of America 5 4 3 2

Engineering Press, Inc. **P.O. Box 1** **San Jose, California 95103-0001**

Contents

Preface

This book is designed to teach the fundamental concepts of engineering economy to engineers. By limiting the intended audience to engineers, it is possible to provide a more rigorous presentation of engineering economic analysis—and do it more concisely—than if the book were written for a wider audience. Because this is a textbook, the order of subjects and the methods of presentation are those which have proved most satisfactory in the classroom.

We begin by looking at the process of decision making in the first few chapters. ***Engineering decision making***—where problems are resolved into their economic consequences—is a specialized examination of the more general ***decision-making process***. Chapter 4—one of the most important in the book—presents two central subjects. Cash receipts and disbursements may be transformed by ***compound interest*** calculations into equivalent sums at different points in time: most economic analyses are based on this concept of ***equivalence***. This chapter's discussion and calculations are illustrated and developed in many example problems.

Chapters 5 through 9 elaborate ***present worth***, ***uniform annual cost***, ***rate of return***, and a number of other analysis methods for comparing alternatives in economic problem solving. ***Depreciation*** methods are described in Chapter 10, followed by the effects of ***income taxes*** on decision making in Chapter 11.

Equipment replacement, a rather special kind of engineering economic analysis, is described in Chapter 12, followed by the ever-present problems of ***inflation*** and ***deflation***. Chapter 14 discusses methods and difficulties in ***estimating the future consequences*** of alternatives. The next three chapters cover the diverse topics of choosing an ***interest rate*** for economic analysis, ***economic analysis in government***, and rationing money among ***competing projects*** when there are more good projects than money. Chapter 18 is a re-examination of ***rate of return computations*** when there are multiple sign changes or other complexities. The final two chapters explain the ***microcomputer programs*** (contained on the diskette inside the back cover) and provide additional ***homework problems***.

The first three editions of *Engineering Economic Analysis* introduced: the consolidation of the eight principal ***compound interest factors*** on a single page; summaries of important information on the inside front and back covers; and computer programs on a PC-***diskette***.

This fourth edition contains significant improvements. Dr. Jan Wolski has integrated six computer programs into a single highly usable program. *Engineering Economic Analysis Exam File* is included with every textbook. Finally, the number of homework problems in the book has increased to over 570. Most problems are at the end of the individual chapters, but some are in the new Chapter 20 entitled Additional Homework Problems.

A book is a portion of an author's total experience in a particular field, acting as a sort of sieve: it must retain the best ideas and concepts from the author's own education, his colleagues, and his students. I am most grateful to the many people who have indirectly contributed to the content of this book. For their assistance and suggestions, I am particularly grateful to Professors:

William Baron, Clemson University;
Dick Bernhard, North Carolina State University;
James Diegel, Lawrence Institute of Technology;
Lou Freund, San Jose State University;
Bruce Hartsough, University of California, Davis;
Bruce Johnson, U.S. Naval Academy;
Robert E. Keith, The University of Texas;
Jack Lohmann, University of Michigan;
Robert Michel, Old Dominion University;
Nic Nigro, Cogswell College North;
Susan Richards, GMI Engineering & Management Institute;
Ralph E. Smith, University of Georgia;
Spencer B. Smith, Illinois Institute of Technology;
Charles Stevens, Old Dominion University;
Tom Ward, University of Louisville;
Martin Wohl, Carnegie-Mellon University;
Jan Wolski, New Mexico Institute of Mining and Technology.

The transformation of the manuscript into a book was the combined effort of Landis Gwynn, editor; John Foster, technical illustrator; and the staff of Key Strokes and Engineering Press, Inc. I would appreciate being informed of errors and receiving other comments about the book.

Donald G. Newnan

Engineering Economic Analysis

CHAPTER 1

Introduction

This book is about making decisions. ***Decision making***, however, is quite a broad topic, for it is a major aspect of everyday human existence. This book will isolate those problems that are commonly faced by engineers and develop the tools to properly grasp, analyze, and solve them. Even very complex situations can be broken down into components from which sensible solutions are produced. If one understands the decision-making process and has tools for obtaining realistic comparisons between alternatives, one can expect to make better decisions.

Although we will focus on solving problems that confront firms in the marketplace, many techniques we will use may be applied to the problems one faces in daily life, as will be evident in the problems at the end of each chapter. Since problem solving (which is possibly a less glamorous name for decision making) is our objective, let us start by looking at some problems.

People Are Surrounded By A Sea Of Problems

A careful look at the world around us clearly demonstrates that we are surrounded by a sea of problems. Just in a single day, problems begin with the sound of an alarm clock:

> Should Joe shut off the alarm? Yes, by all means, the darn thing is far too noisy to let it ring on. Wow! Awake for three seconds and there's the first decision of the day. Should Joe get up and go to work today? That's really two more questions which require, at least, two more decisions; for early-morning decision making, it might be better to split the questions and take them one at a time: Shall Joe get up? It's obvious that unless Joe plans to

just lie there forever, the answer at some point in time has to be "yes." Shall Joe go to work? This all depends on what day of the week it is and—assuming Joe *has* a job—whether he's needed at work, if he needs the income, and whether he wants to keep his job or not!

About thirty or forty decisions later, Joe finds himself dressed and at the breakfast table. A question is posed: Would Joe like toast with the rest of his breakfast? That kind of decision making is pretty straightforward. And even if Joe later decides he made the wrong decision, the consequences are not of much importance. But there are harder decisions to come.

A look at the morning newspaper poses more serious questions. There is a lengthy article concerning world disarmament. That doesn't look easy to solve, so Joe had better turn to another page. The financial pages offer possibly 2000 different stocks and bonds in which he can invest his money. Some of these are bound to be excellent. In fact, it wouldn't take more than a couple of hundred dollars and two or three years to become a millionaire—if only Joe could make the right decisions. But those kinds of decisions are not very easy to make at all.

It has been a decision-a-minute situation for Joe so far, and it probably will remain that way for the rest of the day. One could ask, however, "Did he or did he not make a decision to shut off the alarm clock?" Although Joe might point out that he did not make a *conscious* decision to shut off the alarm—he *always* shuts it off—the answer actually depends on whether or not he had an alternative. Did he? Yes, of course he did: he could have let the alarm clock keep ringing. If it were a spring-wound clock, it would eventually run down, or an electric clock might just ring for hours.

It is clear, then, that Joe *did* have an alternative but he didn't even consider it. As a result of not recognizing that there was an alternative (and therefore not considering its merits), Joe thought he was taking the *only* course of action. By ignoring alternatives, Joe slipped onto the well-worn path that he was accustomed to taking each workday morning. There is a second, equally important, fact about Joe's morning. Although he may not actually recognize it, he is surrounded by a great variety of ***problems***, or events that call for ***decisions***. Some are relatively easy problems to solve, but many are of increasing complexity and have far-ranging consequences to Joe's life.

Yet, despite the similarity of Joe's situation to all of ours, there does not seem to be any exact way of classifying problems, simply because problems are so diverse in complexity and "personality." One approach to classification would be to arrange problems by their *difficulty*. A simple problem could be, "Shall I have toast for breakfast?" or, "Shall we replace a burned-out motor?" or even, "Shall I stop smoking?" An example of an intermediate problem might be, "Should a manual or semi-automatic machine be purchased for the factory?" or, "Shall I buy or lease my next car?" A complex problem might be, "How should the U.S. Government budget its money?" or, "Should a firm build an assembly plant in a foreign country?" Problems appear to group themselves into three broad categories.

Simple Problems

On the lower end of our classification of problems are less difficult situations. While deciding on the components of your breakfast may not be *easy*, the problem does not seem of great importance. In fact, when compared to complex problems, it may not be much of a problem at all. However, it is not clear that *all* apparently straightforward decisions are truly *simple*. After all, consider how many people smoke cigarettes even though they fully realize that smoking may be harmful to health. That problem, while seemingly simple, has all sorts of social and psychological complexities.

Intermediate Problems

At this level of complexity we find problems that are primarily ***economic***. The selection between a manual or a semi-automatic machine is an instance in which the economics of the situation will be the primary basis of decision making. (There will, of course, be other aspects as well.) A semi-automatic machine implies that less labor would be needed than if the manual machine were selected. Moreover, in this kind of problem, whether or not an extra machine-tender would need to be hired is still more an economic consideration than, say, a social one.

Complex Problems

On the upper end of our classification system, we discover problems that are indeed complex. They represent a mixture of ***economic***, ***political***, and ***humanistic***—that is, lively but unpredictable—elements.

For example, the preparation of the annual budget of the United States is *economic* in that it attempts to allocate available money to all the various federal agencies so that the best possible use is made of the money. But the allocation of money clearly has *political* consequences (for example, to the President, the Congress, and individual members of Congress). The closing of a large military installation in a particular congressional district may create unemployment and, consequently, dissatisfaction among the people with their representatives. No member of Congress wants to create such an unfavorable climate in his district; thus, he can be expected to oppose government actions that are *economically* sound but that create political problems.

Meanwhile, what might be a truly economic level of defense spending? Economics, in fact, may have little or nothing to do with this problem! Rather, defense spending appears to be based primarily on certain critical decisions made by the President as part of a complex world strategy. Other problems at the upper end of the classification scale are *humanistic* and involve relationships between people, with economics playing a subordinate role.

The Role Of Engineering Economic Analysis

What kinds of problems can be solved by engineering economic analysis? Our classification of problems suggests that those we consider simple can be solved quickly in one's head; there does not seem to be much need for analytical techniques to aid in their solution. Since these problems are of little consequence (after all, if you don't have toast for breakfast today, you can have it tomorrow), one can hardly justify making any calculations, even meaningful calculations.

At the other end of the scale are complex problems, which are a mixture of "people problems," with economics being only one of the multiple elements. We will not expect engineering economic analysis to be *too* helpful in solving complex problems.

It is the intermediate level of problems that appears best-suited for solution by engineering economic analysis. In this classification, the economics of the problem are a major component in decision making. There may well be a great many other aspects of the problem to consider before making a decision,* but the economic aspects dominate the problem and are therefore dominant in determining its best solution. Also, these intermediate-level problems are of sufficient importance that we can afford to sit down and spend some time in trying to solve them. And time is important! We certainly would not want to spend ten dollars' worth of time and effort to solve a fifty-cent problem! Such a tiny problem simply would not justify that amount of effort.

We may generalize by saying that the problems most suitable for solution by engineering economic analysis have these qualities:

1. The problem is *sufficiently important* that we are justified in giving it some serious thought and effort.
2. The problem can't be worked in one's head—that is, a careful analysis *requires that we organize* the problem and all the various consequences, and this is just too much to be done all at once.
3. The problem has *economic aspects* that are sufficiently important to be a significant component of the analysis leading to a decision.

When problems have these three criteria, engineering economic analysis is an appropriate technique for seeking a solution. Since there are vast numbers of problems that one will encounter in the business world (and in one's personal life) that meet these criteria, engineering economic analysis can be a valuable tool in a great many situations.

---

*For example, in deciding to buy a $10,000 car, one would probably spend some time choosing exterior and interior colors as well as quite a few other options (questions of preference), yet none of these choices would be as significant as the decision whether or not to buy the car itself (an economic problem).

Problems

1-1 Think back to your first hour after awakening this morning. List fifteen decision-making opportunities that existed during that one hour. After you have done that, mark the decision-making opportunities that you actually recognized this morning and upon which you made a conscious decision.

1-2 Some of the problems listed below would be suitable for solution by engineering economic analysis. Which ones are they?

a. Would it be better to buy an automobile with a diesel engine or a gasoline engine?

b. Should an automatic machine be purchased to replace three workers now doing a task by hand?

c. Would it be wise to enroll for an early morning class so you could avoid travelling during the morning traffic rush hours?

d. Would you be better off if you changed your major to Electrical Engineering?

e. One of the people you might marry has a job that pays very little money, while another one has a professional job with an excellent salary. Which one should you marry?

1-3 Which one of the following problems is *most* suitable for analysis by engineering economic analysis?

a. Some 35¢ candy bars are on sale for twelve bars for $3.00. Sandy eats a couple of candy bars a week, and must decide whether or not to buy a dozen.

b. A woman has $150,000 in a bank checking account that pays no interest. She can either invest it immediately at a desirable interest rate, or wait one week and know that she will be able to obtain an interest rate that is 0.15% higher.

c. Joe backed his car into a tree, damaging the fender. He has automobile insurance that will pay for the fender repair. But if he files a claim for payment, they may change his "good driver" rating downward, and charge him more for car insurance in the future.

1-4 If you have $300 and could make the right decisions, how long would it take you to become a millionaire? Explain briefly what you would do.

1-5 Many people write books explaining how to make money in the stock market. Apparently the authors plan to make *their* money selling books telling other people how to profit from the stock market. Why don't these authors forget about the books, and make their money in the stock market?

1-6 The owner of a small machine shop has just lost one of his larger customers. The solution to his problem, he says, is to fire three machinists to balance his workforce with his current level of business. The owner says it is a simple problem with a simple solution. The three machinists disagree. Why?

1-7 Every college student had the problem of selecting the college or university to attend. Was this a simple, intermediate, or complex problem for you? Explain.

1-8 Recently the U. S. Government wanted to save money by closing a small portion of all its military installations throughout the United States. While many people agreed it was a desirable goal, areas potentially affected by selection to close soon reacted negatively. The Congress finally selected a panel of people whose task was to develop a list of installations to close, with the legislation specifying that Congress could not alter the list. Since the goal was to save money, why was this problem so hard to solve?

1-9 The college bookstore has put pads of engineering computation paper on sale at half price. What is the minimum and maximum number of pads you might buy during the sale? Explain.

1-10 Consider the seven situations described. Which one situation seems most suitable for solution by engineering economic analysis?

a. Jane has met two college students that interest her. Bill is a music major who is lots of fun to be with. Alex, on the other hand, is a fellow engineering student, but does not like to dance. Jane wonders what to do.

b. You drive periodically to the post office to pick up your mail. The parking meters require 10¢ for six minutes—about twice the time required to get from your car to the post office and back. If parking tickets cost $8.00, do you put money in the meter or not?

c. At the local market, candy bars are 40¢ each or three for $1.00.

d. The cost of automobile insurance varies widely from insurance company to insurance company. Should you check with several companies when your insurance comes up for renewal?

e. There is a special local sales tax ("sin tax") on a variety of things that the town council would like to remove from local distribution. As a result a store has opened up just outside the town and offers an abundance of these specific items at prices about 30% less than is charged in town.

f. Your mother reminds you she wants you to attend the annual family picnic. That same Saturday you already have a date with a girl you have been trying to date for months.

g. One of your professors mentioned that you have a poor attendance record in his class. You wonder whether to drop the course now or wait to see how you do on the first midterm exam. Unfortunately, the course is required for graduation.

1-11 An automobile manufacturer is considering locating an automobile assembly plant in Tennessee. List two simple, two intermediate, and two complex problems associated with this proposal.

CHAPTER 2

The Decision Making Process

Decision making may take place by default, that is, without consciously recognizing that an opportunity for decision making exists. This fact leads us to a first element in a definition of decision making. To have a decision-making situation, there must be at least two alternatives available. If only one course of action is available, there can be no decision making, for there is nothing to decide. We would have no alternative but to proceed with the single available course of action. (One might argue that it is a rather unusual situation when there are no alternative courses of action. More frequently, alternatives simply are not recognized.)

At this point we might conclude that the decision-making process consists of choosing from among alternative courses of action. But this is an inadequate definition. Consider the following:

> At a horse race, a bettor was uncertain which of the five horses to bet on in the next race. He closed his eyes and pointed his finger at the list of horses printed in the racing program. Upon opening his eyes, he saw that he was pointing to horse number four. He hurried off to place his bet on that horse.

Does the racehorse selection represent the process of decision making? Yes, it clearly was a process of choosing among alternatives (assuming the bettor had already ruled out the "do-nothing" alternative of placing no bet). But the particular method of deciding seems inadequate and irrational. We want to deal with rational decision making.

Rational Decision Making

Rational decision making is a complex process that contains a number of essential elements; although somewhat arbitrary, we define the rational decision-making process in terms of eight steps:

1. Recognition of a problem;
2. Definition of the goal or objective;
3. Assembly of relevant data;
4. Identification of feasible alternatives;
5. Selection of the criterion for judging which is the best alternative;
6. Construction of the interrelationships between the objective, alternatives, data, and the criterion;
7. Prediction of the outcomes for each alternative; and,
8. Choice of the best alternative to achieve the objective.

The following sections will describe these elements in greater detail.

Recognition of a Problem

The starting point in any conscious attempt at rational decision making must be recognition that a problem exists. Only when a problem has been recognized can the work toward its solution begin in a logical manner.

Some years ago, for example, it was discovered that a number of species of ocean fish contained substantial concentrations of mercury. The decision-making process began with this recognition of a problem, and the rush was on to determine what should be done. Research into the problem revealed that fish taken from the ocean decades before, and preserved in laboratories, also contained similar concentrations of mercury. Thus, the problem had existed for a long time but had not been recognized.

In typical situations, recognition is obvious and immediate. An auto accident, an overdrawn check, a burned-out motor, an exhausted supply of parts all produce the recognition of a problem. Once we are aware of the problem, we can take action to solve it as best we can.

Definition of the Goal or Objective

In a sense, every problem is a situation that prevents us from achieving previously determined goals. If a personal goal is to lead a pleasant and meaningful life, then any situation that would prevent it is viewed as a problem. Similarly, in a business situation, if a company objective is to operate profitably, then problems are those occurrences which prevent the company from achieving its previously defined profit objective.

But an objective need not be a grand, overall goal of a business or an individual. It may be quite narrow and specific: "I want to pay off the loan on my car by May," or "The plant must produce 300 golf carts in the next two weeks," are more limited objectives. Thus, defining the objective is the act of exactly describing the task or goal.

Assembly of Relevant Data

To make a good decision, one must first assemble good information. It has been said that 100 years ago an individual could assemble all the published knowledge of the world in his library at home. Today it is doubtful that all published knowledge could even be assembled. Sheer volume alone dictates that, even if we were able to gather it, we probably could not organize it in any very meaningful way. Pertinent published data thus have become both more voluminous, and more difficult to assemble.

In addition to published information, there is a vast quantity of information that is not written down anywhere, but is stored as part of the knowledge and experience of individuals. And finally, there is information that remains ungathered. A question like "How many people in Lafayette, Indiana, would be interested in buying a pair of left-handed scissors?" cannot be answered by examining published data or by asking any one person. Market research or other data gathering would be required to obtain the desired information.

From all of this information, which of it is relevant in a specific decision-making process? It may be a complex task to decide which data are important and which data are not. The availability of data further complicates this task. Some data are available immediately at little or no cost in published form; other data are available by consulting with specific knowledgeable people; still other data require surveys or research to assemble the information. Information that can be gathered only by the two latter means may be both expensive and time consuming to collect.

In developing and selecting relevant data, the analyst must often decide whether the value of certain information justifies the cost to obtain it. This constitutes in itself yet another problem in rational decision making. In decision making, the assembly of relevant data is generally one of the more difficult parts of the process.

Identification of Feasible Alternatives

For decision making to take place, alternative courses of action must be available. With some thought, we can usually devise a variety of ways to achieve most objectives. But there is an ever-present danger that, in devising alternatives, we may overlook the best alternative of all. If this happens, we now have a situation where the best of the *identified* alternatives has been selected, but the result is not the *best possible* solution.*

There is no way, however, to ensure that the best alternative *is* among the alternatives being considered. One should try to be certain that all conventional alternatives have been listed, then make a serious effort to suggest innovative solutions. Sometimes a group of people considering alternatives in an innovative atmosphere—***brainstorming***—can be helpful.

Any good listing of alternatives will produce both practical and impractical alternatives. It would be of little use, however, to seriously consider an alternative that cannot be adopted. An alternative may be infeasible for a variety of reasons, such as, it violates fundamental laws of science, or it requires resources or materials that cannot be obtained, or it cannot be available in the time specified in the problem objective. After elimination, the feasible alternatives are retained for further analysis.

Selection of a Criterion to Determine the Best Alternative

The central task of decision making is choosing from among alternatives. How is the choice made? Logically, one wants to choose the best alternative. This can only be done, however, if we can define what we mean by *best*. There must be a ***criterion***, or set of ***criteria***, for judging which alternative is best. Now, we recognize that *best* is a relative adjective. It is on one end of a spectrum that might read:

Worst Bad Fair Good Better Best

Since we are dealing in *relative terms*, rather than *absolute values*, the selection will be the alternative that is relatively the most desirable. Consider a driver found guilty of speeding by a judge and given the alternatives of a $175 fine or three days in jail. In absolute terms, neither alternative is desirable. But on a relative basis, one simply chooses the better of the undesirable alternatives. In this case we would be following the old adage to "make the best of a bad situation."

There must be an unlimited number of ways that one might judge the various alternatives. Several possible criteria are:

- Create the least disturbance to the ecology;
- Improve the distribution of wealth among people;
- Use money in ways that are economically efficient;
- Minimize the expenditure of money;

*A group of techniques called ***value analysis*** are sometimes used to examine past decisions. Where the decision made was somehow inadequate in results, value analysis re-examines the entire decision-making process with the goal of identifying a better solution and, hence, improving future decision making.

- Ensure that the benefits to those who gain from the decision are greater than the losses of those who are harmed by the decision;*
- Minimize the time to accomplish the goal or objective;
- Minimize unemployment.

The selection of the criterion for choosing the best alternative may not be easy. If one were to apply the seven criteria above to some situation in which there were a number of alternatives, it seems likely that the different criteria would result in different decisions. It may be impossible, for example, to minimize unemployment without at the same time increasing the expenditure of money. The disagreement between management and labor in collective bargaining (concerning wages and conditions of employment) reflects a disagreement over the criterion for selecting the best alternative. Management's idea of the best alternative based on its criterion is seldom the best alternative using organized labor's criterion!

Construction of the Model

At some point in the decision-making process, the various elements must be brought together. The *objective*, the *relevant data*, the *feasible alternatives*, and the *selection criterion* must be merged. The relationships may be obscure and complex, as in trying to measure the impact of a domestic decision on world peace (a complex problem). They may be impossible to define on paper in any meaningful way. On the other hand, if one were considering borrowing money to pay for an automobile (an intermediate problem), there is a readily defined mathematical relationship between the following variables: amount of the loan, loan interest rate, duration of the loan, and monthly payment.

The construction of the interrelationships between the decision-making elements is frequently called ***model building*** or ***construction of the model***. To an engineer, modeling may be of two forms: a scaled *physical representation* of the real thing or system; or a *mathematical equation*, or set of equations, that describe the desired interrelationships. In a laboratory there may be a physical model, but in economic decision making, the model is usually mathematical.

In modeling, it is helpful to represent only that part of the real system that is important to the problem at hand. Thus, the mathematical model of the student capacity of a classroom might be,

$$\text{Capacity} = \frac{lw}{k}, \quad \text{where } l = \text{length of classroom in meters,}$$
$$w = \text{width of classroom in meters, and}$$
$$k = \text{classroom arrangement factor.}$$

*Kaldor Criterion.

The equation for student capacity of a classroom is a very simple model; yet it may be adequate for the problem being solved. Other situations have much more elaborate mathematical models, as we will soon discover.

Prediction of the Outcome for Each Alternative

A model is used to predict the outcome for each of the feasible alternatives. As was suggested earlier, each alternative might produce a variety of outcomes. Selecting a motorcycle, rather than a bicycle, for example, may make the fuel supplier happy, the neighbors unhappy, the environment more polluted, and one's savings account smaller. But, to avoid unnecessary complications, we assume that decision making is based on a single criterion for measuring the relative attractiveness of the various alternatives. The other outcomes or consequences are ignored, and this single criterion* is used to judge the alternatives. Using the model, the magnitude of the selected criterion is computed and recorded for each alternative.

Choice of the Best Alternative

When the seven prior elements of the rational decision-making process have been completed, the final step is choosing the best alternative. If the other elements of decision making have been done carefully, we may select the alternative that best meets the chosen criterion with some confidence in having found the best solution to the particular problem.

Decision-Process System

We know that decision making cannot begin until the existence of a problem is recognized. But from that point on, there is no fixed path to choosing the best alternative. Problems seldom can be solved by the sequential approach of Figure 2-1. This is because it is usually difficult, or impossible, to complete one element in the process without considering the effect on other elements in decision making. The gathering of relevant data may suggest feasible alternatives. But it could just as easily be that in identifying feasible alternatives, one will need additional data not yet assembled. Thus, decision making cannot be seen as an eight-step process that proceeds sequentially from Step 1 to Step 8.

A somewhat better diagram of the decision process is illustrated in Fig. 2-2. This diagram groups the elements in a more flexible, and therefore more realistic, manner. There is no attempt to dictate which comes first—the objective or goal, the feasible alternatives, or the relevant data. In fact, the implication is that once

*If necessary, one could devise a single composite criterion that is the weighted average of several different choice criteria.

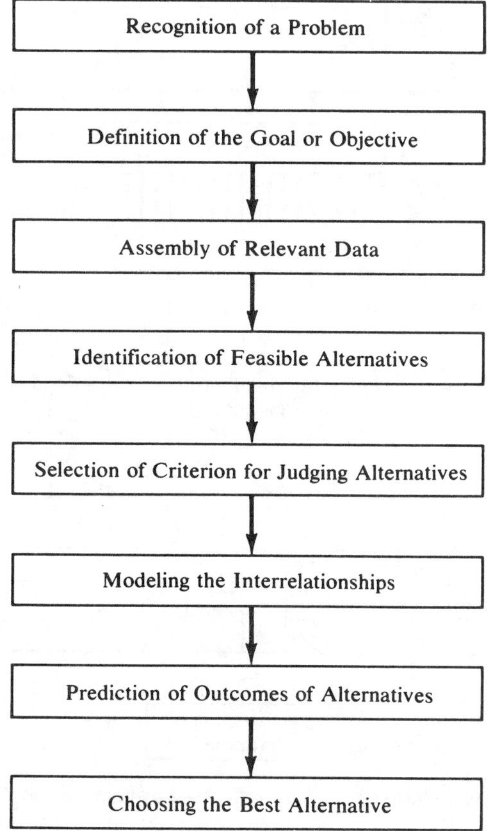

Figure 2-1 A sequential approach to decision making.

one has recognized the problem, several elements of the decision-making process may be considered concurrently. We mentioned earlier that the eight elements of Figures 2-1 and 2-2 are somewhat arbitrary and artificial, so it is not too surprising that we have difficulty drawing a diagram that properly represents the interrelationship between the elements.

Even Fig. 2-2 seems to suggest that once the relevant data, for example, are determined, that element of the decision process has been concluded. We objected to that concept in the linear relationship in Fig. 2-1, and so we are equally critical of Fig. 2-2 in this respect. The missing aspect of both Figures 2-1 and 2-2 is ***feedback***. No matter where one is in the decision-making process, there will frequently be a need to go back and redo or extend the work on some other element in the process. In other words, one may pass through a particular element several times while in the decision process system. This feedback, where subsequent elements influence previously determined elements, is difficult to show

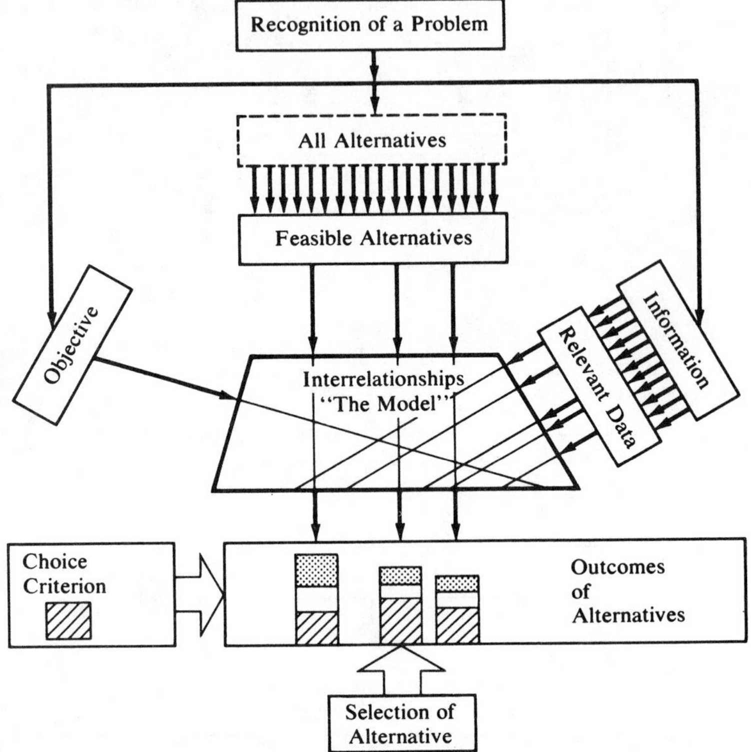

Figure 2-2 A more realistic—but still imperfect—flowchart of the decision process.

in a diagram, for there would likely be prospective paths from any element to most other elements. To redraw Fig. 2-2 with these additional pathways would obscure the decision-process mechanism, so we'll leave it as an imperfect but useful schematic diagram of the decision process.

When Is A Decision Made?

Having examined the structure of the decision-making process, it is appropriate to ask, "When is a decision made and who makes it?" If one person performs *all* the steps in decision making, then he is the decision maker. *When* he makes the decision is less clear. The selection of the feasible alternatives may be the key item, with the rest of the analysis a methodical process leading to the inevitable decision. We can see that the decision may be drastically affected, or even predetermined, by the way in which the decision-making process is carried out. This is illustrated by the following example.

> Dave, a young engineer, was assigned to make an analysis of what additional equipment to add to the machine shop. The criterion for selection was that the equipment selected should be the most economical, considering both initial costs and future operating costs. A little investigation by Dave revealed three practical alternatives:
>
> 1. A new specialized lathe;
> 2. A new general-purpose lathe;
> 3. A rebuilt lathe available from a used equipment dealer.
>
> A preliminary analysis indicated that the rebuilt lathe would be the most economical. Dave did not like the idea of buying a rebuilt lathe so he decided to discard that alternative. He prepared a two-alternative analysis which showed the general-purpose lathe was more economical than the specialized lathe. Dave presented his completed analysis to his manager. The manager assumed that the two alternatives presented were the best of all feasible alternatives, and he approved Dave's recommendation.

At this point we should ask: who was the decision maker, Dave or his manager? Although the manager signed his name at the bottom of the economic analysis worksheets to authorize purchasing the general-purpose lathe, he was merely authorizing what already had been made inevitable, and thus he was not the decision maker. Rather, Dave had made the key decision when he decided to discard the most economical alternative from further consideration. The result was a decision to buy the better of the two *less economically desirable* alternatives.

SUMMARY

For rational decision making to take place, there must be an effort to select, by a logical method of analysis, the best alternative from among the feasible alternatives. While difficult to isolate into discrete items, the analysis can be thought of as including eight elements.

1. Recognition of a problem: the realization that a problem exists is the first step in problem solving.
2. Definition of the goal or objective to be accomplished: what is the task?
3. Assembly of relevant data: what are the facts? Do we need to gather additional data? Is the additional information worth at least what it costs us to obtain it?
4. Identification of feasible alternatives: what are the practical alternative ways of accomplishing our objective or task?
5. Selection of the criterion for judging the best alternative: what is the single criterion most important to the solution of the problem? There are many possible criteria from which to choose. They may be political,

economic, ecological, humanitarian, or whatever. A single criterion may be selected, or it may be a composite of several different criteria.

6. Construction of the various interrelationships: this phase is frequently called *mathematical modeling*.
7. Prediction of outcomes for each alternative.
8. Choice of the best alternative to achieve the objective.

The decision process system is not a matter of proceeding from the first element to the last one, for there is no mandatory sequence that must be followed. In fact, as one proceeds it is often necessary to go back and re-examine earlier elements in a feedback process. Finally, we note that the actual decision maker is more likely to be the person who performs the analysis than the person who selects the resulting alternative to be adopted.

Problems

2-1 A college student determines that he will have only $50 per month available for his housing for the coming year. He is determined to continue in the university, so he has decided to list all feasible alternatives for his housing. To help him, list five feasible alternatives.

2-2 An electric motor on a conveyor burned out. The foreman told the plant manager the motor had to be replaced. The foreman indicated that "there are no alternatives," and asked for authorization to order the replacement. In this situation, is there any decision making taking place? By whom?

2-3 Bill Jones' parents insisted that Bill buy himself a new sportshirt. Bill's father gave specific instructions, saying the shirt must be in "good taste," that is, neither too wildly colored nor too extreme in tailoring. Bill found in the local department store there were three types of sportshirts available:

- rather somber shirts that Bill's father would want him to buy;
- good looking shirts that appealed to Bill; and
- weird shirts that were even too much for Bill.

He wanted a good looking shirt but wondered how to convince his father to let him keep it. The clerk suggested that Bill take home two shirts for his father to see and return the one he did not like. Bill selected a good looking blue shirt he liked, and also a weird lavender shirt. His father took one look and insisted that Bill keep the blue shirt and return the lavender one. Bill did as his father instructed. What was the key decision in this decision process, and who made it?

2-4 Bob Johnson decided to purchase a new home. After looking at tracts of new homes, he decided a custom-built home was preferable. He hired an architect to prepare the drawings. In due time, the architect completed the drawings and submitted them.

Bob liked the plans; he was less pleased that he had to pay the architect a fee of $4000 to design the house. Bob asked a building contractor to provide a bid to construct the home on a lot Bob already owned. While the contractor was working to assemble the bid, Bob came across a book of standard house plans. In the book was a home that he and his wife liked better than the one designed for them by the architect. Bob paid $75 and obtained a complete set of plans for this other house. Bob then asked the contractor to provide a bid to construct this "stock plan" home. In this way Bob felt he could compare the costs and make a decision. The building contractor submitted the following bids:

Custom designed home	$98,000
Stock-plan home	98,500

Both Bob and his wife decided they were willing to pay the extra $500 for it. Bob's wife, however, told Bob they would have to go ahead with the custom-designed home, for, as she put it, "We can't afford to throw away a set of plans that cost $4000." Bob agrees, but he dislikes the thought of building a home that is less desirable than the stock-plan home. He asks your advice. Which house would you advise him to build? Explain.

2-5 Willie Lohmann travels from city to city in the conduct of his business. Every other year he buys a new car for about $12,000. The auto dealer allows about $8000 as a trade-in allowance with the result that the salesman spends $4000 every other year for a car. Willie keeps accurate records which show that all other expenses on his car amount to 12.3¢ per mile for each mile he drives. Willie's employer has two plans by which salesmen are reimbursed for their car expenses:

1. Willie will receive all his operating expenses, and in addition will receive $2000 each year for the decline in value of the automobile.
2. Instead of Plan 1, Willie will receive 22¢ per mile.

If Willie travels 18,000 miles per year, which method of computation gives him the larger reimbursement? At what annual mileage do the two methods give the same reimbursement?

2-6 Jeff Martin, a college student, is getting ready for three final examinations at the end of the school year. Between now and the start of exams, he has 15 hours of study time available. He would like to get as high a grade average as possible in his Math, Physics, and Engineering Economy classes. He feels he must study at least two hours for each course and, if necessary, will settle for the low grade that the limited study would yield. How much time should Jeff devote to each class if he estimates his grade in each subject as follows:

Mathematics		*Physics*		*Engineering Economy*	
Study hours	*Grade*	*Study hours*	*Grade*	*Study hours*	*Grade*
2	25	2	35	2	50
3	35	3	41	3	61
4	44	4	49	4	71
5	52	5	59	5	79
6	59	6	68	6	86
7	65	7	77	7	92
8	70	8	85	8	96

2-7 A grower estimates that if he picks his apple crop now, he will obtain 1000 boxes of apples, which he can sell at $3 per box. However, he thinks his crop will increase an additional 120 boxes of apples for each week he delays picking, but that the price will drop at a rate of 15¢ per box per week; in addition, he estimates approximately 20 boxes per week will spoil for each week he delays picking. When should he pick his crop to obtain the largest total cash return? How much will he receive for his crop at that time?

2-8 In the Fall, Jay Thompson decided to live in a university dormitory. He signed a dorm contract under which he was obligated to pay the room rent for the full college year. One clause stated that if he moved out during the year, he could sell his dorm contract to another student who would move into the dormitory as his replacement. The dorm cost was $600 for the two semesters, which Jay already has paid.

A month after he moved into the dorm, he decided he would prefer to live in an apartment. That week, after some searching for a replacement to fulfill his dorm contract, Jay had two offers. One student offered to move in immediately and to pay Jay $30 per month for the eight remaining months of the school year. A second student offered to move in the second semester and pay $190 to Jay.

Jay now has $1050 left (after paying the $600 dorm bill and food for a month) which must provide for all his room and board expenses for the balance of the year. He estimates his food cost per month is $120 if he lives in the dorm and $100 if he lives in an apartment with three other students. His share of the apartment rent and utilities will be $80 per month. Assume each semester is 4½ months long. Disregard the small differences in the timing of the disbursements or receipts.

- ***a.*** What are the three alternatives available to Jay?
- ***b.*** Evaluate the cost for each of the alternatives.
- ***c.*** What do you recommend that Jay do?

2-9 Seven criteria are given in the chapter for judging which is the best alternative. After studying the list, devise three additional criteria that might be used.

2-10 The local garbage company charges $6.00 a month for garbage collection. It had been their practice to send out bills to their 100,000 customers at the end of each two-month period. Thus, they would send a bill to each customer for $12 at the end of February for garbage collection during January and February.

Recently the firm changed its billing date so now they send out the two-month bills after one month's service has been performed. Bills for January–February, for example, are sent out at the end of January. The local newspaper claims the firm is receiving half their money before they do the garbage collection. This unearned money, the newspaper says, could be temporarily invested for one month at 1% per month interest by the garbage company to earn extra income.

Compute how much extra income the garbage company could earn each year if it invests the money as described by the newspaper. (*Answer:* $36,000)

2-11 Consider the three situations below. Which ones appear to represent rational decision making? Explain.

- ***a.*** Joe's best friend has decided to become a civil engineer, so Joe has decided that he, too, will become a civil engineer.

b. Jill needs to get to the university from her home. She bought a car and now drives to the university each day. When Jim asks her why she didn't buy a bicycle instead, she replies, "Gee, I never thought of that."

c. Don needed a wrench to replace the sparkplugs in his car. He went to the local automobile supply store and bought the cheapest one they had. It broke before he finished replacing all the sparkplugs in his car.

2-12 Suppose you have just two hours to answer the question, "How many people in your home town would be interested in buying a pair of left-handed scissors?" Give a step-by-step outline of how you would seek to answer this question within two hours.

2-13 Suppose you are assigned the task of determining the route of a new highway through an older section of town. The highway will require that many older homes must be either relocated or torn down. Two possible criteria that might be used in deciding exactly where to locate the highway are:

1. Ensure that there are benefits to those who gain from the decision and no one is harmed by the decision.

2. Ensure that the benefits to those who gain from the decision are greater than the losses of those who are harmed by the decision.

Which criterion will you select to use in determining the route of the highway? Explain.

2-14 Consider a situation where there are only two alternatives available and both are unpleasant and undesirable. What should you do?

2-15 In decision making we talk about the construction of a model. What kind of model is meant?

2-16 A firm must decide which of three alternatives to adopt to expand its capacity. The firm wishes a minimum annual profit of 20% of the initial cost of each separable increment of investment. Any money not invested in capacity expansion can be invested elsewhere for an annual yield of 20% of initial cost.

Alternative	*Initial cost*	*Annual profit*	*Profit rate*
A	$100,000	$30,000	30%
B	300,000	66,000	22
C	500,000	80,000	16

Which alternative should be selected?

2-17 A farmer must decide what combination of seed, water, fertilizer, and pest control will be most profitable for the coming year. The local agricultural college did a study of this farmer's situation and prepared the following table.

Plan	*Cost/acre*	*Income/acre*
A	$ 600	$ 800
B	1500	1900
C	1800	2250
D	2100	2500

The last page of the college's study was torn off, and hence the farmer is not sure which plan the agricultural college recommends. Which plan should the farmer adopt? Explain.

2-18 A firm believes the sales volume (S) of its product depends on its unit selling price (P), and can be determined from the equation P = \$100 − S. The cost (C) of producing the product is \$1000 + 10S.

a. Draw a graph with the sales volume (S) from 0 to 100 on the *x*-axis, and Total Cost and Total Income from 0 to 2500 on the *y*-axis. On the graph draw the line C = \$1000 + 10S. Then plot the curve of Total Income [which is sales volume (S) × Unit Selling Price (\$100 − S)]. Mark the breakeven points on the graph.

b. Determine the breakeven point (lowest sales volume where total sales income just equals total production cost). *Hint:* This may be done by trial and error, or by using the quadratic equation to locate the point where profit is zero.

c. Determine the sales volume (S) at which the firm's profit is a maximum. *Hint:* Write an equation for profit and solve it by trial and error, or as a minima-maxima calculus problem.

2-19 (20-2) Two manufacturing companies, located in cities 90 miles apart, have discovered that they both send their trucks four times a week to the other city full of cargo and return empty. Each company pays its driver \$185.00 a day (the round trip takes all day) and have truck operating costs (excluding the driver) of 60 cents per mile. How much could each company save each week if they shared the task, with each sending their truck twice a week and hauling the other company's cargo on the return trip?

2-20 (20-3) A painting operation is performed by a production worker at a labor cost of \$1.40 per unit. A robot spray-painting machine, costing \$15,000, would reduce the labor cost to \$0.20 per unit. If the paint machine would be valueless at the end of three years, what would be the minimum number of units that would have to be painted each year to justify the purchase of the paint machine?

2-21 (20-4) Venus Computer Co. can produce 23,000 personal computers a year on its daytime shift. The fixed manufacturing costs per year are \$2,000,000 and the total labor cost is \$9,109,000. To increase its production to 46,000 computers per year, Venus is considering adding a second shift. The unit labor cost for the second shift would be 25% higher than the day shift, but the total fixed manufacturing costs would increase only to \$2,400,000 from \$2,000,000.

a. Compute the unit manufacturing cost for the daytime shift.

b. Would adding a second shift increase or decrease the unit manufacturing cost at the plant?

CHAPTER 3

Engineering Decision Making

Not every problem can be solved by the decision-making process. One need only pick up a daily newspaper to read of situations where decision makers do not seem to know even what the desired objective or task is! The problems of real life do not often lend themselves to an orderly presentation as described in Chapter 2.

There are many problems, however, that *are* more readily solvable. It is from this set of problems, to which we may apply the decision-making process, that we narrow our objectives even further. In this book we are interested in *engineering* decision making and *engineering* economic analysis. We will examine substantial problems to be solved by engineers where economic factors dominate and economic efficiency is the most significant criterion for choosing among alternatives.

Engineering decision making is based on the same eight elements presented in the last chapter, plus one final element—the ***post audit of results***. We first examine the individual elements as they apply to engineering decision making, then use the engineering decision-making techniques to solve some design problems and other short-range economic problems. In the last part of the chapter, we will convert problems into a series of money receipts or disbursements, or what is called a cash flow.

The Decision-Process System For Engineers

The nine elements of engineering decision making are:

1. Recognition of a problem.
2. Definition of the goal or objective.

3. Assembly of relevant data.
4. Identification of feasible alternatives.
5. Selection of the criterion for judging the alternatives.
6. Construction of the model.
7. Prediction of the outcome for each alternative.
8. Choice of the best alternative to achieve the objective.
9. Post audit of results.

Everything that we've stated about the decision-making process generally applies in engineering decision making. There are, however, some details specific to engineering decision making which are discussed in the sections below.

1. Recognition of the Problem

Some problems arise from circumstances outside a business organization and beyond its control. A newly enacted law, for example, may have a serious impact on a firm. Problems also occur within companies, such as faulty manufacturing practices. But the fact that a problem exists is not enough, it must be recognized by people who can do something about it. Frequently a problem is well known to a group of workers in a particular area of the plant, but not to those who might initiate the decision-making process. Employee suggestion boxes and similar programs are sometimes used to encourage internal communication about problems.

2. Definition of the Goal or Objective

We need to understand what we seek to accomplish. Activity without a well-defined goal is rather fruitless. Consider the story of an airline pilot informing the passengers, "The good news is that we are proceeding at 1000 kilometers per hour; the bad news is that we are lost."

3. Assembly of Data

Obtaining the relevant data for decision making is always a challenge. In engineering decision making, an important source of data is a firm's own accounting system. These data must be examined quite carefully.

Financial- and cost-accounting is designed to show accounting values and the flow of money—specifically ***costs*** and ***benefits***—in a company's operations. Where costs are directly related to specific operations, there is no difficulty; but there are other costs that are not related to specific operations. These indirect costs, or ***overhead***, are usually allocated to a company's operations and products by some arbitrary method. The results are generally satisfactory for cost-accounting purposes, but may be unreliable for use in economic analysis.

To create a meaningful economic analysis, we must determine the *true* differences between alternatives, which might require some adjustment of cost-accounting data. The following Example illustrates this situation.

EXAMPLE 3-1

The cost-accounting records of a large company show the following average monthly costs for the three-person printing department:

Direct labor and salaries (including employee benefits)	\$ 6,000
Materials and supplies consumed	7,000
Allocated overhead costs 200 m^2 of floor area at \$25/m^2	5,000
	\$18,000

The printing department charges the other departments for its services to recover its \$18,000 monthly cost. For example, the charge to run 1000 copies of an announcement is:

Direct labor	\$7.60
Materials and supplies	9.80
Overhead costs	9.05
Cost to other departments	\$26.45

The shipping department checks with a commercial printer and finds they could have the same 1000 copies printed for \$22.95. Although the shipping department only has about 30,000 copies printed a month, they decide to stop using the printing department and have their printing done by the outside printer. The printing department objects to this. As a result, the general manager has asked you to study the situation and recommend what should be done.

Solution: Much of the printing department's work reveals the company's costs, prices, and other financial information. The company president considers the printing department necessary to prevent disclosing such information to people outside the company.

A review of the cost-accounting charges reveals nothing unusual. The charges made by the printing department cover direct labor, materials and supplies, and overhead. (*Note:* The company's indirect costs—such as heat, electricity, employee insurance, and so forth—must be distributed to its various departments in *some* manner and, like many other firms, it uses *floor space* as the basis for its allocations. The printing department, in turn, must distribute its costs into the charges for the work that it does.)

	Printing department		*Outside printer*	
	1000 copies	*30,000 copies*	*1000 copies*	*30,000 copies*
Direct labor	$ 7.60	$228.00		
Materials and supplies	9.80	294.00	$22.95	$688.50
Overhead costs	9.05	271.50		
	$26.45	$793.50	$22.95	$688.50

The shipping department would reduce its cost from $793.50 to $688.50 by using the outside printer. In that case, how much would the printing department's costs decline? We will examine each of the cost components:

1. *Direct labor*. If the printing department had been working overtime, then the overtime could be reduced or eliminated. But, assuming no overtime, how much would the saving be? It seems unlikely that a printer could be fired or even put on less than a 40-hour work week. Thus, although there might be a $228.00 saving, it is much more likely that there will be *no* reduction in direct labor.
2. *Materials and supplies*. There would be a $294.00 saving in materials and supplies.
3. *Allocated overhead costs*. There will be *no* reduction in the printing department's monthly $5000 overhead, for there will be no reduction in department floor space. (Actually, of course, there may be a slight reduction in the firm's power costs if the printing department does less work.)

The firm will save $294.00 in materials and supplies and may or may not save $228.00 in direct labor if the printing department no longer does the shipping department work. The maximum saving would be $294.00 + 228.00 = $522.00. But if the shipping department is permitted to obtain its printing from the outside printer, the firm must pay $688.50 a month. The saving from not doing the shipping department work in the printing department would not exceed $522.00, and it probably would be only $294.00. The result would be a net increase in cost to the firm. For this reason, the shipping department should be discouraged from sending its printing to the outside commercial printer. ◀

Gathering cost data presents other difficulties. One way to look at the financial consequences—costs and benefits—of various alternatives is as follows.

- *Market Consequences*. These consequences have an established price in the marketplace. We can quickly determine raw material prices, machinery costs, labor costs, and so forth, in this manner.
- *Extra-Market Consequences*. There are other items that are not directly priced in the marketplace. But by indirect means, a price may be assigned

to these items. (Economists call these prices ***shadow prices***.) Examples might be the cost of an employee injury or the value to employees of going from a five-day to a four-day, forty-hour week.

- *Intangible Consequences.* Numerical economic analysis probably never *fully* describes the real differences between a group of alternatives. The tendency to leave out those consequences that do not have a significant impact on the analysis itself, or on the conversion of the final decision into actual money, is difficult to resolve or eliminate. How does one evaluate the potential loss of workers' jobs due to automation? What is the value of landscaping around a factory? These and a variety of other consequences may be left out of the numerical calculations, but they should be considered in conjunction with the numerical results in reaching a decision on the particular problem.

4. Identification of Feasible Alternatives

One must keep in mind that unless the best alternative is considered, the result will always be suboptimal. Two types of alternatives are sometimes ignored. First, in many situations a do-nothing alternative is feasible. This may be the "let's keep doing what we are now doing" alternative, or it may be the "let's not spend any money on that problem" alternative. Second, there are often feasible (but unglamorous) alternatives, such as "patch it up and keep it running for another year before replacing it."

5. Criteria for Judging Alternatives

All economic analysis problems fall into one of three categories:

1. ***Fixed input.*** The amount of money or other input resources (like labor, materials, or equipment) are fixed. The objective is to effectively utilize them. *Examples:*
   - A project engineer has a budget of $350,000 to overhaul a portion of a petroleum refinery.
   - You have $100 to buy clothes for the start of school.

   For economic efficiency, the appropriate criterion is to maximize the benefits or other outputs.
2. ***Fixed output.*** There is a fixed task (or other output objectives or results) to be accomplished. *Examples:*
   - A civil engineering firm has been given the job to survey a tract of land and prepare a "Record of Survey" map.
   - You wish to purchase a new car with no optional equipment.

   The economically efficient criterion for a situation of fixed output is to minimize the costs or other inputs.

3. ***Neither input not output fixed.*** The third category is the general situation where neither the amount of money or other inputs, nor the amount of benefits or other outputs are fixed. *Examples:*
   - A consulting engineering firm has more work available than it can handle. It is considering paying the staff for working evenings to increase the amount of design work it can perform.
   - One might wish to invest in the stock market, but neither the total cost of the investment nor the benefits are fixed.
   - An automobile battery is needed. Batteries are available at different prices, and although each will provide the energy to start the vehicle, their useful lives are different.

   What should be the criterion in this category? Obviously, we want to be as economically efficient as possible. This will occur when we maximize the difference between the return from the investment (benefits) and the cost of the investment. Since the difference between the benefits and the costs is simply profit, a businessperson would define this criterion as ***maximizing profit***.

For the three categories, the proper economic criteria are:

Category	*Economic criterion*
Fixed input	**Maximize the benefits or other outputs.**
Fixed output	**Minimize the costs or other inputs.**
Neither input nor output fixed	**Maximize (benefits–costs) or, stated another way, maximize profit.**

6. Construction of the Model

In engineering economic analysis, the model is often some simple computations to resolve alternatives into comparable values at a selected point in time.

7. Obtaining Comparable Outcomes

To obtain a meaningful basis for choosing the best alternative, the outcomes for each alternative must be arranged in a *comparable* way. Since we wish to choose the best alternative, we must arrange the mathematical calculations to provide a meaningful comparison among the alternatives. The initial step in that direction is the decision to state the consequences of each alternative in terms of money, that is, in the form of costs and benefits. This ***resolution of consequences*** is done with all market and extra-market consequences. Most intangible consequences are unable to be included in the numerical calculations.

In the initial problems we will examine, the costs and benefits occur over a short time period and can be considered as occurring at the same time. In other situations the various costs and benefits take place in a longer time period. The result may be costs at one point in time followed by periodic benefits. We will resolve these into a *cash flow table* to show the timing of the various costs and benefits. A number of methods—for example, present worth, annual cost, rate of return—will be used to resolve the cash flow table of each alternative into comparable values.

8. Choosing the Best Alternative

Earlier we indicated that choosing the best alternative may be simply a matter of determining which alternative best meets the selection criterion. But the solutions to most economics problems have market consequences, extra-market consequences, and intangible consequences. Since the intangible consequences of possible alternatives are left out of the numerical calculations, they should be introduced into the decision-making process at this point. We said that engineering economic analysis techniques are generally used where the economic consequences dominate. That statement implies that the intangible consequences will be of lesser importance than the (numerical) economic consequences. The alternative to be chosen is the one that best meets the choice criterion after looking at both the numerical consequences and the consequences not included in the monetary analysis.

9. Post Audit of Results

In any operating system, it is important to see that the results of a decision analysis are in reasonable agreement with its projections. If a new machine tool was purchased because of labor savings and improvements in quality, it is only logical to see if both those savings are being realized. If they are, the economic analysis projections would seem to be accurate. If the savings are not being obtained, we need to see what has been overlooked. The post audit review may help ensure that projected operating advantages are ultimately obtained. On the other hand, the economic analysis projections may have been unduly optimistic. We want to know this, too, so that these mistakes may be avoided in the future. Finally, an effective way to promote *realistic* economic analysis calculations is for all people involved to know that there *will* be an audit of the results!

Elementary Engineering Decision Making

Some of the easiest forms of engineering decision making deal with problems of alternate *designs, methods,* or *materials*. Since results of the decision occur in a very short period of time, one can quickly add up the costs and benefits for each alternative. Then, using the suitable economic criterion, the best alternative can be identified.

Example 3-2 concerns the economic selection of alternate materials.

EXAMPLE 3-2

A concrete aggregate mix is required to contain at least 31% sand by volume for proper batching. One source of material, which has 25% sand and 75% coarse aggregate, sells for \$3 per cubic meter. Another source, which has 40% sand and 60% coarse aggregate, sells for \$4.40 per cubic meter. Determine the least cost per cubic meter of blended aggregates.

Solution: The least cost of blended aggregates will result from maximum use of the lower cost material. The higher cost material will be used to increase the proportion of sand up to the minimum level (31%) specified.

$$\begin{aligned} \text{Let } x &= \text{Portion of blended aggregates from \$3.00/m}^3 \text{ source} \\ 1 - x &= \text{Portion of blended aggregates from \$4.40/m}^3 \text{ source} \end{aligned}$$

Sand balance:

$$\begin{aligned} x(0.25) + (1 - x)(0.40) &= 0.31 \\ 0.25x + 0.40 - 0.40x &= 0.31 \\ x &= \frac{0.31 - 0.40}{0.25 - 0.40} = \frac{-0.09}{-0.15} \\ &= 0.60 \end{aligned}$$

Thus the blended aggregates will contain:

60% of \$3.00/m^3 material

40% of \$4.40/m^3 material

The least cost per cubic meter of blended aggregates:

$$\begin{aligned} &= 0.60(\$3.00) + 0.40(\$4.40) = 1.80 + 1.76 \\ &= \$3.56/\text{m}^3 \quad \blacktriangleleft \end{aligned}$$

Example 3-3 describes a situation of selecting between alternate methods.

EXAMPLE 3-3

A machine part is manufactured at a unit cost of 40¢ for material and 15¢ for direct labor. An investment of \$500,000 in tooling is required. The order calls for three million pieces. Half-way through the order, a new method of manufacture can be put into effect which will reduce the unit costs to 34¢ for material and 10¢ for direct labor—but it will require \$100,000 for additional tooling. If all tooling costs are to be amortized during the production of the order, and other costs are 250% of direct labor cost, would it be profitable to make the change?

Solution:
Alternative A: Continue with present method.

Material cost	1,500,000 pieces × 0.40 =	$ 600,000
Direct labor cost	1,500,000 pieces × 0.15 =	225,000
Other costs	2.50 × Direct labor cost =	562,500
Cost for remaining 1,500,000 pieces		$1,387,500

Alternative B: Change the manufacturing method.

Additional tooling cost	=	$ 100,000
Material cost	1,500,000 pieces × 0.34 =	510,000
Direct labor cost	1,500,000 pieces × 0.10 =	150,000
Other costs	2.50 × Direct labor cost =	375,000
Cost for remaining 1,500,000 pieces		$1,135,000

Before making a final decision, one should closely examine the *Other costs* to see that they do, in fact, vary as the *Direct labor cost* varies. Assuming they do, the decision would be to change the manufacturing method. ◀
Example 3-4 illustrates selection between alternate designs.

EXAMPLE 3-4
In the design of a cold-storage warehouse, the specifications call for a maximum heat transfer through the warehouse walls of 30,000 joules/hr/sq meter of wall when there is a 30°C temperature difference between the inside surface and the outside surface of the insulation. The two insulation materials being considered are as follows:

Insulation material	*Cost/cubic meter*	*Conductivity J-m/m²-°C-hr*
Rock wool	$12.50	140
Foamed insulation	14.00	110

The basic equation for heat conduction through a wall is:

$$Q = \frac{K(\Delta T)}{L}$$

where Q = Heat transfer in J/hr/m² of wall
K = Conductivity in J-m/m²-°C-hr
ΔT = Difference in temperature between the two surfaces in °C
L = Thickness of insulating material in meters

Which insulation material should be selected?

Solution: There are two steps required to solve the problem. First, the required thickness of each of the alternate materials must be calculated. Then, since the problem is one of providing a fixed output (heat transfer through the wall limited to a fixed maximum amount), the criterion is to minimize the input (cost).

Required insulation thickness:

Rock wool $30{,}000 = \frac{140(30)}{L}$ $L = 0.14$ m

Foamed insulation $30{,}000 = \frac{110(30)}{L}$ $L = 0.11$ m

Cost of insulation per square meter of wall:

Unit cost = Cost/m^3 × Insulation thickness in meters

Rock wool: Unit cost = \$12.50 × 0.14 m = \$1.75/m^2

Foamed insulation: Unit cost = \$14.00 × 0.11 m = \$1.54/m^2

The foamed insulation is the lesser cost alternative. ◀

Computing Cash Flows

In the examples presented so far, we have selected the least-cost alternative to meet a specification or requirement (Examples 3-2 and 3-4) or the savings have been obtained in a short period of time (Ex. 3-3). There are other situations where the alternatives have different consequences (costs and benefits) that continue for an extended period of time. In these circumstances, we do not add up the various consequences; instead, we describe each alternative as cash ***receipts*** or ***disbursements*** at different points in ***time***. In this way, each alternative is resolved into a ***cash flow***. This is illustrated by Examples 3-5 and 3-6.

EXAMPLE 3-5

The manager has decided to purchase a new \$30,000 mixing machine. The machine may be paid for by one of two ways:

1. Pay the full price now *minus* a 3% discount.
2. Pay \$5000 now; at the end of one year, pay \$8000; at the end of four subsequent years, pay \$6000 per year.

List the alternatives in the form of a table of cash flows.

Solution: In this problem the two alternatives represent different ways to pay for the mixing machine. While the first plan represents a lump sum of $29,100 now, the second one calls for payments continuing until the end of the fifth year. (The next chapter will focus on how to treat a series of payments extending over a significant period of time.) The problem is to convert an alternative into cash receipts or disbursements and show the timing of each receipt or disbursement. The result is called a ***cash flow table*** or, more simply, a *cash flow*.

The cash flows for both the alternatives in this problem are very simple. The cash flow table, with disbursements given negative signs, is as follows:

End of Year	*Pay in full now*	*Pay over 5 years*
0 (now)	−$29,100	−$5000
1	0	−8000
2	0	−6000
3	0	−6000
4	0	−6000
5	0	−6000

◀

EXAMPLE 3-6

A man borrowed $1000 from a bank at 8% interest. He agreed to repay the loan in two end-of-year payments. At the end of the first year, he will repay half of the $1000 principal amount plus the interest that is due. At the end of the second year, he will repay the remaining half of the principal amount plus the interest for the second year. Compute the borrower's cash flow.

Solution: In engineering economic analysis, we normally refer to the *beginning* of the first year as "Time 0." At this point the man receives $1000 from the bank. (A positive sign represents a receipt of money and a negative sign, a disbursement.) Thus, at Time 0, the cash flow is +$1000.

At the end of the first year, the man pays 8% interest for the use of $1000 for one year. The interest is $0.08 \times \$1000 = \$80$. In addition, he repays half the $1000 loan, or $500. Therefore, the end-of-Year 1 cash flow is −$580.

At the end of the second year, the payment is 8% for the use of the balance of the principal ($500) for the one-year period, or $0.08 \times 500 = \$40$. The $500 principal is also repaid for a total end-of-Year 2 cash flow of −$540. The cash flow is:

End of year	*Cash flow*
0 (now)	+$1000
1	−580
2	−540

◀

Chapter 4 will examine techniques for resolving cash flows into comparable forms so that decisions can be made on the relative desirability of alternatives.

SUMMARY

Engineering economic analysis refers to the solution of substantial engineering problems where economic aspects dominate and economic efficiency is the criterion for choosing from possible alternatives. It is a particular case of the general decision-making process. Some of the unusual aspects of engineering economic analysis are as follows:

1. Cost-accounting systems, while an important source of cost data, contain allocations of indirect costs that may be inappropriate for use in economic analysis.
2. The various consequences—costs and benefits—of an alternative may be of three types:
   a. Market consequences—where there are established market prices available;
   b. Extra-market consequences—there are no direct market prices, but prices can be assigned by indirect means;
   c. Intangible consequences—the consequences cannot be valued in any practical way, so they are not included in the monetary analysis.
3. The economic criteria for judging alternatives can be reduced to three cases:
   a. For fixed input: maximize benefits or other outputs.
   b. For fixed output: minimize costs or other inputs.
   c. When neither input nor output is fixed: maximize the difference between output and input or, more simply stated, maximize profit.

   The third case states the general rule from which both the first and second cases may be derived.
4. To choose among the alternatives, the market consequences and extra-market consequences are organized into a cash flow. We will see in the next chapter that differing cash flows can be resolved into comparable values of outcomes. These outcomes are compared against the selection criterion. From this comparison *plus* the consequences not included in the monetary analysis, the best alternative is selected.
5. An essential part of engineering economic analysis is the post audit of results. This step helps to ensure that projected benefits are obtained and to encourage realistic estimates in analyses.

In elementary engineering decision making, the problem is often one of selecting appropriate designs, methods, or materials. In some cases all the consequences occur in such a short time period that it is reasonable to add together the various consequences. When the consequences of alternatives occur over a longer time period (say, over one year), an intermediate step in the analysis is to resolve the alternatives into a table of cash flows. Chapter 4 will deal with the manipulation of cash flows.

Problems

3-1 Bill's father read that, at the end of each year, an automobile is worth 25% less than it was at the beginning of the year. After a car is three years old, the rate of decline reduces to 15%. Maintenance and operating costs, on the other hand, increase as the age of the car increases. Because of the manufacturer's warranty, the first year maintenance is very low.

Age of car, in years	*Maintenance expense*
1	$ 50
2	150
3	180
4	200
5	300
6	390
7	500

Bill decided this is a good economic analysis problem. Bill's dad wants to keep his annual cost of automobile ownership low. The car Bill's dad prefers costs $11,200 new. Should he buy a new or a used car and, if used, when would you suggest he buy it, and how long should it be kept? Give a practical, rather than a theoretical, solution.

(*Answer:* Buy a three-year-old car and keep it three years.)

3-2 A city is in need of increasing its rubbish disposal facilities. There is a choice of two rubbish disposal areas, as follows:

Area A: A gravel pit with a capacity of 16 million cubic meters. Due to the possibility of high ground water, however, the Regional Water Pollution Control Board has restricted the lower 2 million cubic meters of fill to inert material only; for example, earth, concrete, asphalt, paving, brick, and so forth. The inert material, principally clean earth, must be purchased and hauled to this area for the bottom fill.

Area B: Capacity is 14 million cubic meters. The entire capacity may be used for general rubbish disposal. This area will require an average increase in a round-trip haul of five miles for 60% of the city, a decreased haul of two miles for 20% of the city. For the remaining 20% of the city, the haul is the same distance as for Area *A*.

Assume the following conditions:

- Cost of inert material placed in Area *A* will be $2.35 per cubic meter.
- Average speed of trucks from last pickup to disposal site is 15 miles per hour.
- The rubbish truck and a two-man crew will cost $35 per hour.

- Truck capacity of $4\frac{1}{2}$ tons per load or 20 cubic meters.
- Sufficient cover material is available at all areas; however, inert material for the bottom fill in Area *A* must be hauled in.

Which of the sites do you recommend? (*Answer:* Area *B*)

3-3 The three economic criteria for choosing the best alternative are: minimize input; maximize output; or maximize the difference between output and input. For each of the following situations, what is the appropriate economic criterion?

a. A manufacturer of plastic drafting triangles can sell all the triangles he can produce at a fixed price. His unit costs increase as he increases production due to overtime pay, and so forth. The manufacturer's criterion should be ______.

b. An architectural and engineering firm has been awarded the contract to design a wharf for a petroleum company for a fixed sum of money. The engineering firm's criterion should be ______.

c. A book publisher is about to set the list price (retail price) on a textbook. If they choose a low list price, they plan on less advertising than if they select a higher list price. The amount of advertising will affect the number of copies sold. The publisher's criterion should be ______.

d. At an auction of antiques, a bidder for a particular porcelain statue would be trying to ______.

3-4 See Problem 3-3. For each of the following situations, what is the appropriate economic criterion?

a. The engineering school held a raffle of an automobile with tickets selling for 50¢ each or three for $1. When the students were selling tickets, they noted that many people were undecided whether to buy one or three tickets. This indicates the buyers' criterion was ______.

b. A student organization bought a soft-drink machine for use in a student area. There was considerable discussion as to whether they should set the machine to charge 30¢, 35¢, or 40¢ per drink. The organization recognized that the number of soft drinks sold would depend on the price charged. Eventually the decision was made to charge 35¢. Their criterion was ______.

c. In many cities, grocery stores find that their sales are much greater on days when they have advertised their special bargains. The advertised special prices do not appear to increase the total physical volume of groceries sold by a store. This leads us to conclude that many shoppers' criterion is ______.

d. A recently graduated engineer has decided to return to school in the evenings to obtain a Master's degree. He feels it should be accomplished in a manner that will allow him the maximum amount of time for his regular day job plus time for recreation. In working for the degree, he will ______.

3-5 A small machine shop, with thirty horsepower of connected load, purchases electricity under the following monthly rates (assume any demand charge is included in this schedule):

First 50 kw-hr per HP of connected load at 8.6¢ per kw-hr;
Next 50 kw-hr per HP of connected load at 6.6¢ per kw-hr;
Next 150 kw-hr per HP of connected load at 4.0¢ per kw-hr;
All electricity over 250 kw-hr per HP of connected load at 3.7¢ per kw-hr.

The shop uses 2800 kw-hr per month.

a. Calculate the monthly bill for this shop.

b. Suppose the proprietor of the shop has the chance to secure additional business that will require him to operate his existing equipment more hours per day. This will use an extra 1200 kw-hr per month. What is the lowest figure that he might reasonably consider to be the "cost" of this additional energy? What is this per kw-hr?

c. He contemplates installing certain new machines that will reduce the labor time required on certain operations. These will increase the connected load by 10 HP but, as they will operate only on certain special jobs, will add only 100 kw-hr per month. In a study to determine the economy of installing these new machines, what should be considered as the "cost" of this energy? What is this per kw-hr?

3-6 On his first engineering job, Jim Hayes was given the responsibility of determining the production rate for a new product. He has assembled data as indicated on the two graphs:

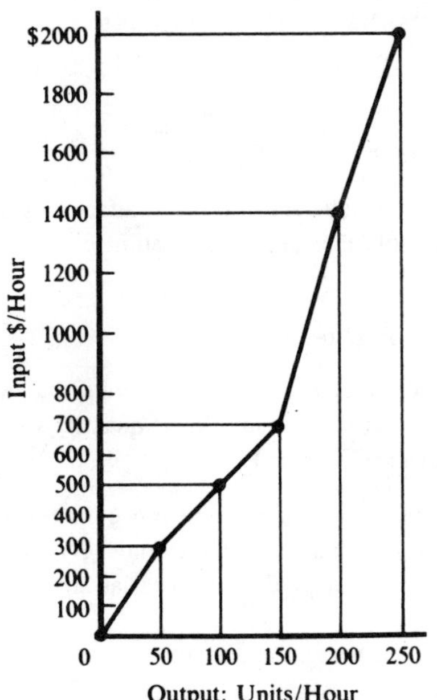

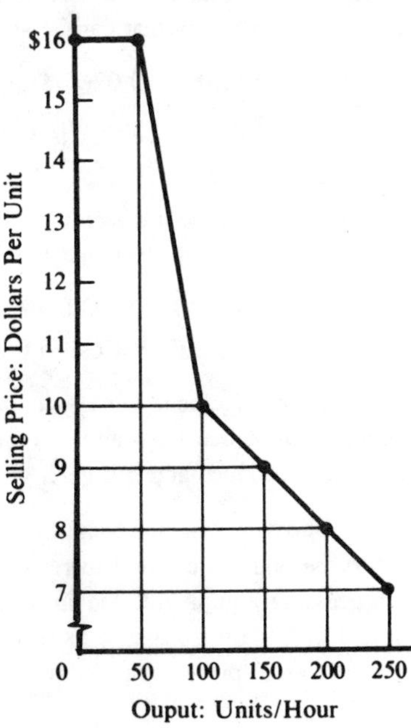

a. Select an appropriate economic criterion and estimate the production rate based upon it.

b. Jim's boss told Jim: "I want you to maximize output with minimum input." Jim wonders if it is possible to achieve his boss's criterion. He asks your advice. What would you tell him?

3-7 An oil company is considering adding an additional grade of fuel at its service stations. To do this, an additional 3000-gallon tank must be buried at each station. Discussions with tank fabricators indicate that the least expensive tank would be cylindrical with minimum surface area. What size tank should be ordered?
(*Answer:* 8-ft diameter by 8-ft length.)

3-8 Mr. Sam Spade, the president of Ajax, recently read in a report that a competitor named Bendix has the following cost–production quantity relationship:

$$C = \$3{,}000{,}000 - \$18{,}000Q + \$75Q^2$$

where C = Total manufacturing cost per year, and
Q = Number of units produced per year

A newly hired employee, who previously worked for Bendix, told Mr. Spade that Bendix is now producing 110 units per year. If the selling price remains unchanged, Sam wonders if Bendix is likely to increase the number of units produced per year, in the near future. He asks you to look at the information and tell him what you are able to deduce from it.

3-9 A firm is planning to manufacture a new product. The sales department estimates that the quantity that can be sold depends on the selling price. As the selling price is increased, the quantity that can be sold decreases. Numerically they estimate:

$$P = \$35.00 - 0.02Q$$

where P = Selling price per unit
Q = Quantity sold per year

On the other hand, the management estimates that the average cost of manufacturing and selling the product will decrease as the quantity sold increases. They estimate

$$C = \$4.00Q + \$8000$$

where C = Cost to produce and sell Q per year

The firm's management wishes to produce and sell the product at the rate that will maximize profit, that is, where income minus cost is a maximum. What quantity should they plan to produce and sell each year? (*Answer:* 775 units)

3-10 The New England Soap Company is considering adding some processing equipment to the plant to allow it to remove impurities from some raw materials. By adding the processing equipment, the firm can purchase lower grade raw material at reduced cost and upgrade it for use in its products.

Four different pieces of processing equipment are being considered:

	A	*B*	*C*	*D*
Initial investment	\$10,000	\$18,000	\$25,000	\$30,000
Annual saving in materials costs	4,000	6,000	7,500	9,000
Annual operating cost	2,000	3,000	3,000	4,000

The company can obtain a 15% annual return on its investment in other projects and is willing to invest money on the processing equipment only so long as it can obtain 15% annual return on each increment of money invested. Which one, if any, of the alternatives should be selected?

3-11 On December 1st, Al Smith purchased a car for $18,500. He paid $5000 immediately and agreed to pay three additional payments of $6000 each (which includes principal and interest) at the end of one, two, and three years. Maintenance for the car is projected at $1000 at the end of the first year, and $2000 at the end of each subsequent year. Al expects to sell the car at the end of the fourth year (after paying for the maintenance work) for $7000. Using these facts, prepare a table of cash flows.

3-12 Sally Stanford is buying an automobile that costs $12,000. She will pay $2000 immediately and the remaining $10,000 in four annual end-of-year principal payments of $2500 each. In addition to the $2500, she must pay 15% interest on the unpaid balance of the loan each year. Prepare a cash flow table to represent this situation.

3-13 The vegetable buyer for a group of grocery stores has decided to sell packages of sprouted grain in the vegetable section of the stores. The product is perishable and any remaining unsold after one week in the store is discarded. The supplier will deliver the packages to the stores, arrange them in the display space, and remove and dispose of any old packages.

The price the supplier will charge the stores depends on the size of the total weekly order for all the stores.

Weekly order	*Price per package*
Less than 1000 packages	35¢
1000–1499	28
1500–1999	25
2000 or more	20

The vegetable buyer estimates the quantity that can be sold per week, at various selling prices, as follows:

Selling price	*Quantity sold per week*
60¢	300 packages
45	600
40	1200
33	1700
26	2300

The sprouted grain will be sold at the same price in all the grocery stores. How many packages should be purchased per week, and at which of the five prices listed above should they be sold?

3-14 Bill Gwynn, a recently graduated engineer, decided to invest some of his money in a "Quick Shop" grocery store. The store emphasizes quick service, a limited assortment of grocery items, and rather high prices. Bill wants to study the business to see if the store hours (currently 0600 to 0100) can be changed to make the store more profitable. Bill assembled the following information.

Time period	*Daily sales in the time period*
0600–0700	$ 20
0700–0800	40
0800–0900	60
0900–1200	200
1200–1500	180
1500–1800	300
1800–2100	400
2100–2200	100
2200–2300	30
2300–2400	60
2400–0100	20

The cost of the groceries sold averages 70% of sales. The incremental cost to keep the store open, including the clerk's wage and other incremental operating costs, is $10 per hour. To maximize profit, when should the store be opened, and when should it be closed?

3-15 Jim Jones, a motel owner, noticed that just down the street the "Motel 36" advertises their $36-per-night room rental rate on their sign. As a result, they rent all of their eighty rooms every day by late afternoon. Jim, on the other hand, does not advertise his rate, which is $54 per night, and averages only a 68% occupancy of his fifty rooms.

There are a lot of other motels nearby and, except for Motel 36, none of the others advertises their rate on their sign. (Their rates vary from $48 to $80 per night.) Jim estimates that his actual incremental cost per night for each room rented, rather than remaining vacant, is $12. This $12 pays for all the cleaning, laundering, maintenance, utilities, and so on. Jim believes his eight alternatives are:

Alternative	*Advertise and charge*	*Resulting occupancy rate*
1	$36 per night	100%
2	42 per night	94
3	48 per night	80
4	54 per night	66
	Do not advertise and charge	
5	$48 per night	70%
6	54 per night	68
7	62 per night	66
8	68 per night	56

What should Jim do? Show how you reached your conclusion.

3-16 A small computer system is to be purchased for the sales department. There are four alternatives available.

Computer	*Pet*	*Pear*	*Pal*	*Pearl*
Initial cost	$6000	$8000	$9000	$10,000
Total annual operating cost	3500	3200	2800	2,650

Each computer system will perform the desired tasks, but the more expensive models will take less effort to use and, hence, will have a lower total annual operating cost.

The company requires a 20% Annual Percentage Rate (APR) on all investments or non-essential increments of investment.

$$\text{APR} = \frac{\text{Annual benefit}}{\text{Initial cost}} \times 100$$

Which computer system should the company purchase? Show your computations.

3-17 One strategy for solving a complex problem is to break the problem into a group of less complex problems, and then find solutions to the less complex problems. The result is the solution of the complex problem.

Give an example where this strategy will work. Then give another example where this strategy will not work.

3-18 (20-5) A woman borrowed $2000 and agreed to repay it at the end of three years, together with 10% simple interest per year. How much will she pay three years hence?

3-19 (20-6) A $5000 loan was to be repaid with 8% simple annual interest. A total of $5350 was paid. How long was the loan outstanding?

3-20 (20-7) A manufacturing firm has received a contract to assemble 1000 units of test equipment in the next year. The firm must decide how to organize its assembly operation. Skilled workers, at $22.00 per hour each, could be assigned to individually assemble the test equipment. Each worker would do all the assembly steps and the task would take 2.6 hours per unit.

An alternate approach would be to set up teams of four less skilled workers (at $13.00 per hour each) and organize the assembly tasks so each worker does his share of the assembly. The four-man team would be able to assemble a unit in one hour. Which approach would result in more economical assembly?

3-21 (20-8) A company uses 8000 wheels per year in its manufacture of golf carts. The wheels cost $15 each and are purchased from an outside supplier. The money invested in the inventory costs 10% per year, and the warehousing cost amounts to an additional 2% per year. It costs $150 to process each purchase order. An order can be placed for any number of wheels. How many orders per year should be placed for wheels?

CHAPTER **4**

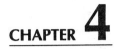

Equivalence And Compound Interest

In the last chapter we saw the full range of the engineering decision-making process. Part of that process includes the prediction of outcomes for each alternative. For many of the situations we examined, the economic consequences of an alternative were immediate, that is, took place either right away or in a very short period of time, as in Example 3-2 (the decision on the design of a concrete aggregate mix) or Ex. 3-3 (the change of manufacturing method). In such relatively simple situations, we total the various positive and negative aspects and quickly reach a decision. But can we do the same if the economic consequences occur over a considerable period of time?

The installation of an expensive piece of machinery in a plant obviously has economic consequences that occur over an extended period of time. If the machinery were bought on credit, then the simple process of paying for it is one that may take several years. What about the usefulness of the machinery? Certainly it must have been purchased because it would be a beneficial addition to the plant. These favorable consequences may last as long as the equipment performs its useful function. Anyone who has visited an industrial plant probably has seen equipment that was ten, twenty, even thirty years old or older still in use. There may then be economic consequences that continue over a substantial time period.

There were a number of situations in Chapter 3 where the outcome from selecting a particular alternative continued over a considerable length of time. For those situations, we created a cash flow table to show the various receipts and disbursements of money and their timing. In this chapter, we will examine the value of money at different points in time. We will be able to compare the value of money at different dates, an ability that is essential to engineering economic analysis. We must be able to compare, for example, a low-cost motor with a higher-cost motor. If there were no other consequences, we would obviously prefer the low-cost one. But if the higher-cost motor were more efficient and thereby reduced the annual electric power cost, we would be faced with the question of whether to spend more money now on the motor to reduce power costs in the future. Through the equivalence relationship, we will be able to relate a present sum of money to future sums. This chapter develops the basic tools for engineering economic analysis.

Time Value Of Money

We often find that the money consequences of any alternative occur over a substantial period of time—say, a year or more. When money consequences occur in a short period of time, we simply add up the various sums of money and obtain a net result. But can we treat money this same way when the time span is greater?

Which would you prefer, $100 cash today or the assurance of receiving $100 a year from now? You might decide you would prefer the $100 now because that is one way to be certain of receiving it. But suppose you were convinced that you would receive the $100 one year hence. Now what would be your answer? A little thought should convince you that it *still* would be more desirable to receive the $100 now. If you had the money now, rather than a year hence, you would have the use of it for an extra year. And if you had no current use for $100, you could let someone else use it.

Money is quite a valuable asset—so valuable that people are willing to pay to have money available for their use. Money can be rented in roughly the same way one rents an apartment, only with money, the charge for its use is called ***interest*** instead of rent. The importance of interest is demonstrated by banks and savings institutions continuously offering to pay for the use of people's money, to pay interest.

If the current interest rate is 9% per year, and you put $100 into the bank for one year, how much will you receive back at the end of the year? You will receive your original $100 together with $9 interest, for a total of $109. This example demonstrates the time preference for money: we would rather have $100 today than the assured promise of $100 one year hence; but we might well consider leaving the $100 in a bank if we knew it would be worth $109 one year hence. This is because there is a ***time value of money*** in the form of the willingness of banks, businesses, and people to pay interest for the use of money.

Repaying a Debt

To better understand the mechanics of interest, let us consider a situation where $5000 is owed and is to be repaid in five years, together with 8% annual interest. There are a great many ways in which debts are repaid; for simplicity, we have selected four specific ways for our example. Table 4-1 tabulates the four plans.

Table 4-1 FOUR PLANS FOR REPAYMENT OF $5000 IN FIVE YEARS WITH INTEREST AT 8%

(a) Year	(b) *Amount owed at beginning of year*	(c) *Interest owed for that year* [8% × (b)]	(d) *Total owed at end of year* [(b) + (c)]	(e) *Principal payment*	(f) *Total end-of-year payment*
Plan 1: At end of each year pay $1000 principal *plus* interest due.					
1	$5000	$ 400	$5400	$1000	$1400
2	4000	320	4340	1000	1320
3	3000	240	3240	1000	1240
4	2000	160	2160	1000	1160
5	1000	80	1080	1000	1080
		$1200		$5000	$6200
Plan 2: Pay interest due at end of each year and principal at end of five years.					
1	$5000	$ 400	$5400	$ 0	$ 400
2	5000	400	5400	0	400
3	5000	400	5400	0	400
4	5000	400	5400	0	400
5	5000	400	5400	5000	5400
		$2000		$5000	$7000
Plan 3: Pay in five equal end-of-year payments.					
1	$5000	$ 400	$5400	$ 852	$1252*
2	4148	331	4479	921	1252
3	3227	258	3485	994	1252
4	2233	178	2411	1074	1252
5	1159	93	1252	1159	1252
		$1260		$5000	$6260
Plan 4: Pay principal and interest in one payment at end of five years.					
1	$5000	$ 400	$5400	$ 0	$ 0
2	5400	432	5832	0	0
3	5832	467	6299	0	0
4	6299	504	6803	0	0
5	6803	544	7347	5000	7347
		$2347		$5000	$7347

*The exact value is $1252.28, which has been rounded to an even dollar amount.

In Plan 1, \$1000 will be paid at the end of each year plus the interest due at the end of the year for the use of money to that point. Thus, at the end of the first year, we will have had the use of \$5000. The interest owed is 8% × \$5000 = \$400. The end-of-year payment is, therefore, \$1000 principal *plus* \$400 interest, for a total payment of \$1400. At the end of the second year, another \$1000 principal plus interest will be repaid on the money owed during the year. This time the amount owed has declined from \$5000 to \$4000 because of the \$1000 principal payment at the end of the first year. The interest payment is 8% × \$4000 = \$320, making the end-of-year payment a total of \$1320. As indicated in Table 4-1, the series of payments continues each year until the loan is fully repaid at the end of the fifth year.

Plan 2 is another way to repay \$5000 in five years with interest at 8%. This time the end-of-year payment is limited to the interest due, with no principal payment. Instead, the \$5000 owed is repaid in a lump sum at the end of the fifth year. The end-of-year payment in each of the first four years of Plan 2 is 8% × \$5000 = \$400. The fifth year, the payment is \$400 interest *plus* the \$5000 principal, for a total of \$5400.

Plan 3 calls for five equal end-of-year payments of \$1252 each. At this point, we have not shown how the figure of \$1252 was computed (see Ex. 4-8). However, it is clear that there is some equal end-of-year amount that would repay the loan. By following the computations in Table 4-1, we see that this series of five payments of \$1252 repays a \$5000 debt in five years with interest at 8%.

Plan 4 is still another method of repaying the \$5000 debt. In this plan, no payment is made until the end of the fifth year when the loan is completely repaid. Note what happens at the end of the first year: the interest due for the first year—8% × \$5000 = \$400—is not paid; instead, it is added to the debt. At the second year, then, the debt has increased to \$5400. The second year interest is thus 8% × \$5400 = \$432. This amount, again unpaid, is added to the debt, increasing it further to \$5832. At the end of the fifth year, the total sum due has grown to \$7347 and is paid at that time.

Note that when the \$400 interest was not paid at the end of the first year, it was added to the debt and, in the second year, there was interest charged on this unpaid interest. That is, the \$400 of unpaid interest resulted in 8% × \$400 = \$32 of additional interest charge in the second year. That \$32, together with 8% × \$5000 = \$400 interest on the \$5000 original debt, brought the total interest charge at the end of the second year to \$432. Charging interest on unpaid interest is called ***compound interest***. We will deal extensively with compound interest calculations later in this chapter.

With Table 4-1 we have illustrated four different ways of accomplishing the same task, that is, to repay a debt of \$5000 in five years with interest at 8%. Having described the alternatives, we will now use them to present the important concept of *equivalence*.

EQUIVALENCE

When we are indifferent as to whether we have a quantity of money now or the assurance of some other sum of money in the future, or series of future sums of money, we say that the present sum of money is ***equivalent*** to the future sum or series of future sums.

If an industrial firm believed 8% was an appropriate interest rate, it would have no particular preference whether it received \$5000 now or was repaid by Plan 1 of Table 4-1. Thus \$5000 today is equivalent to the series of five end-of-year payments. In the same fashion, the industrial firm would accept repayment Plan 2 as equivalent to \$5000 now. Logic tells us that if Plan 1 is equivalent to \$5000 now and Plan 2 is also equivalent to \$5000 now, it must follow that Plan 1 is equivalent to Plan 2. In fact, *all four repayment plans must be equivalent to each other and to \$5000 now.*

Equivalence is an essential factor in engineering economic analysis. In Chapter 3, we saw how an alternative could be represented by a cash flow table. How might two alternatives with different cash flows be compared? For example, consider the cash flows for Plans 1 and 2:

Year	*Plan 1*	*Plan 2*
1	−\$1400	−\$400
2	−1320	−400
3	−1240	−400
4	−1160	−400
5	−1080	−5400
	−\$6200	−\$7000

If you were given your choice between the two alternatives, which one would you choose? Obviously the two plans have cash flows that are different. Plan 1 requires that there be larger payments in the first four years, but the total payments are smaller than the sum of Plan 2's payments. To make a decision, the cash flows must be altered so that they can be compared. The ***technique of equivalence*** is the way we accomplish this.

Using mathematical manipulation, we can determine an equivalent value at some point in time for Plan 1 and a ***comparable equivalent value*** for Plan 2, based on a selected interest rate. Then we can judge the relative attractiveness of the two alternatives, not from their cash flows, but from comparable equivalent values. Since Plan 1 and Plan 2 each repay a *present* sum of \$5000 with interest at 8%, they both are equivalent to \$5000 *now*; therefore, the alternatives are equally attractive. This cannot be deduced from the given cash flows alone. It is necessary to learn this by determining the equivalent values for each alternative at some point in time, which in this case is "the present."

Difference in Repayment Plans

The four plans computed in Table 4-1 are equivalent in nature but different in structure. Table 4-2 repeats the end-of-year payment schedule from the previous table. In addition, each plan is graphed to show the debt still owed at any point in time. Since $5000 was borrowed at the beginning of the first year, all the graphs begin at that point. We see, however, that the four plans result in quite different situations on the amount of money owed at any other point in time. In Plans 1 and 3, the money owed declines as time passes. With Plan 2 the debt remains constant, while Plan 4 increases the debt until the end of the fifth year. These graphs show an important difference among the repayment plans—the areas under the curves differ greatly. Since the axes are *Money Owed* and *Time in Years*, the area is their product: Money owed *times* Time in years.

In the discussion of the time value of money, we saw that the use of money over a time period was valuable, that people are willing to pay interest to have the use of money for periods of time. When people borrow money, they are acquiring the use of money as represented by the area under the Money owed *vs*. Time in years curve. It follows that, at a given interest rate, the amount of interest to be paid will be proportional to the area under the curve. Since in each case the $5000 loan is repaid, the interest for each plan is the Total *minus* the $5000 principal:

Plan	*Total interest paid*
1	$1200
2	2000
3	1260
4	2347

Using Table 4-2 and the data from Table 4-1, we can compute the area under each of the four curves, that is, the area bounded by the abscissa, the ordinate, and the curve itself. We multiply the ordinate (Money owed) *times* the abscissa (1 year) for each of the five years, then *add*:

$$\begin{aligned}\text{Area under curve} = &\ (\text{Money owed in Year 1})(1\text{ year})\\ &+ (\text{Money owed in Year 2})(1\text{ year})\\ &+ \cdots\\ &+ (\text{Money owed in Year 5})(1\text{ year})\end{aligned}$$

or,

Area under curve [(Money owed)(Time)] = ***Dollar-Years***

Table 4-2 END-OF-YEAR PAYMENT SCHEDULES AND THEIR GRAPHS

From Table 4-1: *"Four Plans for Repayment of $5000 in Five Years at 8% Interest"*

Plan 1: At end of each year pay $1000 principal *plus* interest due.

Year	*End-of-year payment*
1	$1400
2	1320
3	1240
4	1160
5	1080
	$6200

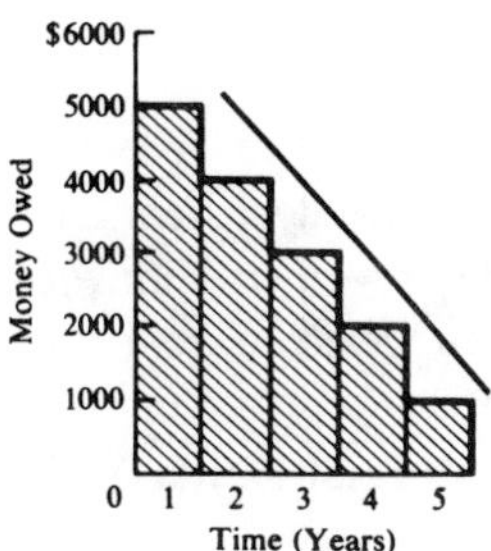

Plan 2: Pay interest due at end of each year and principal at end of five years.

Year	*End-of-year payment*
1	$ 400
2	400
3	400
4	400
5	5400
	$7000

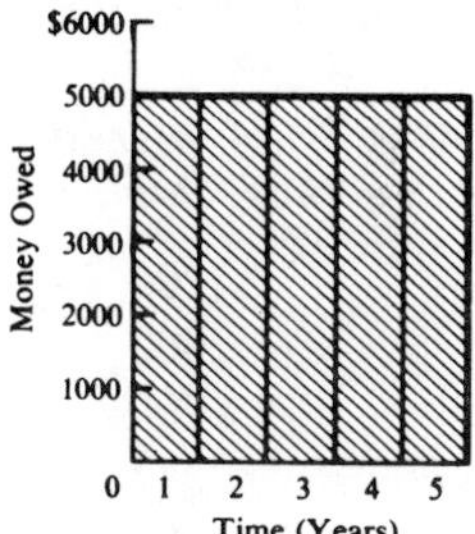

Plan 3: Pay in five equal end-of-year payments.

Year	*End-of-year payment*
1	$1252
2	1252
3	1252
4	1252
5	1252
	$6260

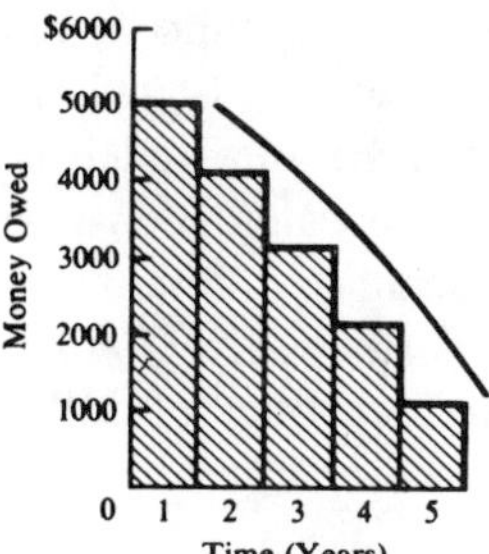

Plan 4: Pay principal and interest in one payment at end of five years.

Year	*End-of-year payment*
1	$ 0
2	0
3	0
4	0
5	7347
	$7347

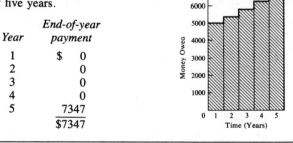

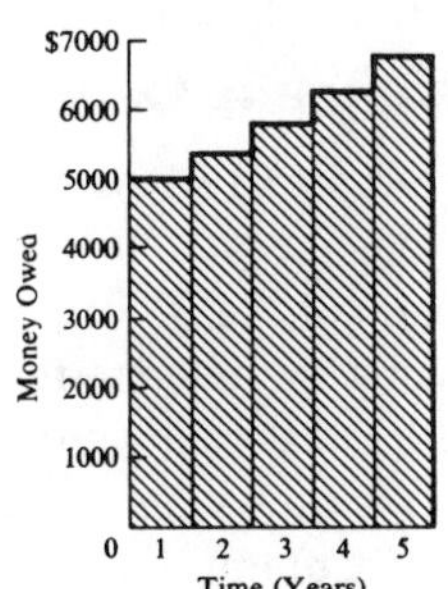

The dollar-years for the four plans would be as follows:

	Dollar-Years			
	Plan 1	*Plan 2*	*Plan 3*	*Plan 4*
(Money owed in Year 1)(1 year)	$ 5,000	$ 5,000	$ 5,000	$ 5,000
(Money owed in Year 2)(1 year)	4,000	5,000	4,148	5,400
(Money owed in Year 3)(1 year)	3,000	5,000	3,227	5,832
(Money owed in Year 4)(1 year)	2,000	5,000	2,233	6,299
(Money owed in Year 5)(1 year)	1,000	5,000	1,159	6,803
Total Dollar-Years	$15,000	$25,000	$15,767	$29,334

With the area under each curve computed in dollar-years, the ratio of total interest paid to area under the curve may be obtained:

Plan	*Total interest paid, dollars*	*Area under curve, dollar-years*	*Ratio:* $\frac{\text{Total interest paid}}{\text{Area under curve}}$
1	$1200	$15,000	0.08
2	2000	25,000	0.08
3	1260	15,767	0.08
4	2347	29,334	0.08

We see that the ratio of total interest paid to area under the curve is constant and equal to 8%. Stated another way, the total interest paid equals the interest rate *times* the area under the curve.

From our calculations, we more easily see why the repayment plans require the payment of different total sums of money, yet are actually equivalent to each other. The key factor is that the four repayment plans provide the borrower with different quantities of dollar-years. Since dollar-years *times* interest rate equals the interest charge, the four plans result in different total interest charges.

Equivalence is Dependent on Interest Rate

In the example of Plans 1–4, all calculations were made at an 8% interest rate. At this interest rate, it has been shown that all four plans are equivalent to a present sum of $5000. But what would happen if we were to change the problem by changing the interest rate?

If the interest rate were increased to 9%, we know that the required interest payment for each plan would increase, and the calculated repayment schedules—Table 4-1, Column **(f)**—could no longer repay the $5000 debt with the higher interest. Instead, each plan would repay a sum *less* than the principal of $5000, because more money would have to be used to repay the higher interest rate.

By some calculations (that will be explained in this chapter), the equivalent present sum that each plan will repay at 9% interest is:

Plan	*Repay a present sum of*
1	$4877
2	4806
3	4670
4	4775

As predicted, at the higher 9% interest the repayment plans of Table 4-1 each repay a present sum less than $5000. But they do not repay the *same* present sum. Plan 1 would repay $4877 with 9% interest, while Plan 2 would repay $4806. Thus, with interest at 9%, Plans 1 and 2 are no longer equivalent, for they will not repay the same present sum. The two series of payments (Plan 1 and Plan 2) were equivalent at 8%, but not at 9%. This leads to the conclusion that ***equivalence is dependent on the interest rate***. Changing the interest rate destroys the equivalence between two series of payments.

Could we create revised repayment schemes that would be equivalent to $5000 now with interest at 9%? Yes, of course we could: to revise Plan 1 of Table 4-1, we need to increase the total end-of-year payment in order to pay 9% interest on the outstanding debt.

Year	*Amount owed at beginning of year*	*9% interest for year*	*Total end-of-year payment* (*$1000 plus interest*)
1	$5000	$450	$1450
2	4000	360	1360
3	3000	270	1270
4	2000	180	1180
5	1000	90	1090

Plan 2 of Table 4-1 is revised for 9% interest by increasing the first four payments to 9% × $5000 = $450 and the final payment to $5450. Two plans that repay $5000 in five years with interest at 9% are:

	Revised end-of-year payments	
Year	*Plan 1*	*Plan 2*
1	$1450	$ 450
2	1360	450
3	1270	450
4	1180	450
5	1090	5450

We have determined that Revised Plan 1 is equivalent to a present sum of $5000 and Revised Plan 2 is equivalent to $5000 now; it follows that at 9% interest, Revised Plan 1 is equivalent to Revised Plan 2.

Application of Equivalence Calculations

To understand the usefulness of equivalence calculations, consider the following:

Year	*Alternative A:* *lower initial cost, higher operating cost*	*Alternative B:* *higher initial cost, lower operating cost*
0 (now)	−$600	−$850
1	−115	−80
2	−115	−80
3	−115	−80
.	.	.
.	.	.
.	.	.
10	−115	−80

Is the least cost alternative the one that has the lower initial cost and higher operating costs or the one with higher initial cost and lower continuing costs? Because of the time value of money, one cannot add up sums of money at different points in time directly. This means that a comparison between alternatives cannot be made in actual dollars at different points in time, but must be made in some equivalent comparable sums of money.

It is not sufficient to compare the initial $600 against $850. Instead, we must compute a value that represents the entire stream of payments. In other words, we want to determine a sum that is equivalent to Alternative *A*'s cash flow; similarly, we need to compute the equivalent present sum for Alternative *B*. By computing equivalent sums at the same point in time ("now"), we will have values that may be validly compared. The methods for accomplishing this will be presented later in this chapter.

Thus far we have discussed computing equivalent present sums for a cash flow. But the technique of equivalence is not limited to a present computation. Instead, we could compute the equivalent sum for a cash flow at any point in time. We could compare alternatives in "Equivalent Year 10" dollars rather than "now" (Year 0) dollars. Further, the equivalence need not be a single sum, but could be a series of payments or receipts. In Plan 3 of Table 4-1, we had a situation where the series of equal payments was equivalent to $5000 now. But the equivalency works both ways: if we ask the question, "What is the equivalent equal annual payment continuing for five years, given a present sum of $5000 and interest at 8%?" the answer is "$1252."

COMPOUND INTEREST

To facilitate equivalence computations, a series of ***interest formulas*** will be derived. To simplify the presentation, we'll use the following notation:

- i = *Interest rate per interest period.* In the equations the interest rate is stated as a decimal (that is, 9% interest is 0.09).
- n = *Number of interest periods.*
- P = *A present sum of money.*
- F = *A future sum of money.* The future sum F is an amount, n interest periods from the present, that is equivalent to P with interest rate i.
- A = *An end-of-period cash receipt or disbursement in a uniform series,* continuing for n periods, the entire series equivalent to P or F at interest rate i.

Single Payment Formulas

Suppose a present sum of money P is invested for one year* at interest rate i. At the end of the year, we should receive back our initial investment P, together with interest equal to iP, or a total amount $P + iP$. Factoring P, the sum at the end of one year is $P(1 + i)$.

Let us assume that, instead of removing our investment at the end of one year, we agree to let it remain for another year. How much would our investment be worth at the end of the second year? The end-of-first-year sum $P(1 + i)$ will draw interest in the second year of $iP(1 + i)$. This means that, at the end of the second year, the total investment will become

$$P(1 + i) + iP(1 + i)$$

This may be rearranged by factoring $P(1 + i)$, which gives us

$$P(1 + i)(1 + i)$$

or $P(1 + i)^2$.

*A more general statement is to specify "one interest period" rather than "one year." It is easier to visualize one year so the derivation will assume that one year is the interest period.

If the process is continued for a third year, the end of the third year total amount will be $P(1 + i)^3$; at the end of n years, it will be $P(1 + i)^n$. The progression looks like this:

	Amount at beginning of interest period	+ *Interest for period*	=	*Amount at end of interest period*
First year	P	$+ iP$	$=$	$P(1 + i)$
Second year	$P(1 + i)$	$+ iP(1 + i)$	$=$	$P(1 + i)^2$
Third year	$P(1 + i)^2$	$+ iP(1 + i)^2$	$=$	$P(1 + i)^3$
nth year	$P(1 + i)^{n-1}$	$+ iP(1 + i)^{n-1}$	$=$	$P(1 + i)^n$

In other words, a present sum P increases in n periods to $P(1 + i)^n$. We therefore have a relationship between a present sum P and its equivalent future sum, F.

$$\text{Future sum} = (\text{present sum})(1 + i)^n$$

$$F = P(1 + i)^n \tag{4-1}$$

This is the ***single payment compound amount formula*** and is written in functional notation as

$$F = P(F/P,i,n)$$

Functional notation is designed so the compound interest factors may be written in an equation in an algebraically correct form. In the equation above, for example, the functional notation is interpreted as

$$F = P\left(\frac{F}{P}\right)$$

which is dimensionally correct. Without proceeding further, we can see that, if we were to derive a compound interest factor to find a present sum P, given a future sum F, the factor would be $(P/F,i,n)$; so, the resulting equation would be

$$P = F(P/F,i,n)$$

which is dimensionally correct.

EXAMPLE 4-1

If \$500 were deposited in a bank savings account, how much would be in the account three years hence if the bank paid 6% interest compounded annually?

We can draw a diagram of the problem. *Note:* To have a consistent notation, we will represent *receipts* by upward arrows (and positive signs), and *disbursements* (or *payments*) will have downward arrows (and negative signs).

Solution: From the viewpoint of the person depositing the \$500, the diagram is:

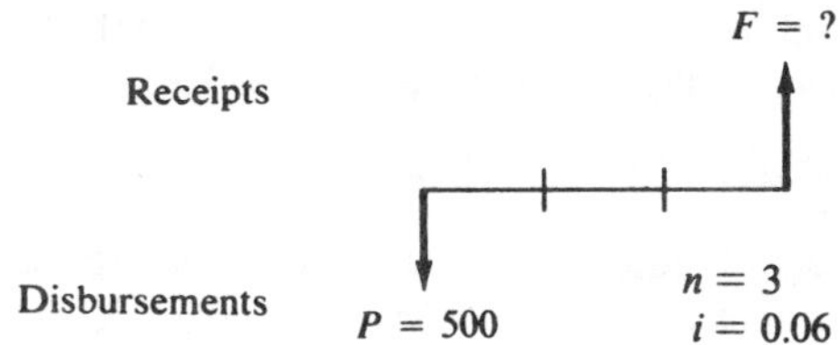

We need to identify the various elements of the equation. The present sum P is \$500. The interest rate per interest period is 6%, and in three years there are three interest periods. The future sum F is to be computed.

$$P = \$500 \qquad i = 0.06 \qquad n = 3 \qquad F = \text{unknown}$$

$$F = P(1 + i)^n = 500(1 + 0.06)^3 = \$595.50$$

If we deposit \$500 in the bank now at 6% interest, there will be \$595.50 in the account in three years.

Alternate Solution: The equation $F = P(1 + i)^n$ need not be solved with a hand calculator. Instead, the *single payment compound amount factor,* $(1 + i)^n$, is readily determined from computed tables. The factor is written in convenient notation as

$$(1 + i)^n = (F/P,i,n)$$

and in functional notation as

$$(F/P,6\%,3)$$

Knowing $n = 3$, locate the proper row in the 6% table. (*Note:* ***Compound Interest Tables*** appear in the tinted pages in this volume. Each table is computed for a particular value of i.) Read in the first column, which is headed "Single Payment, Compound Amount Factor," 1.191. Thus,

$$F = 500(F/P,6\%,3) = 500(1.191) = \$595.50$$

Before leaving this problem, let's draw another diagram of it, this time from the bank's point of view.

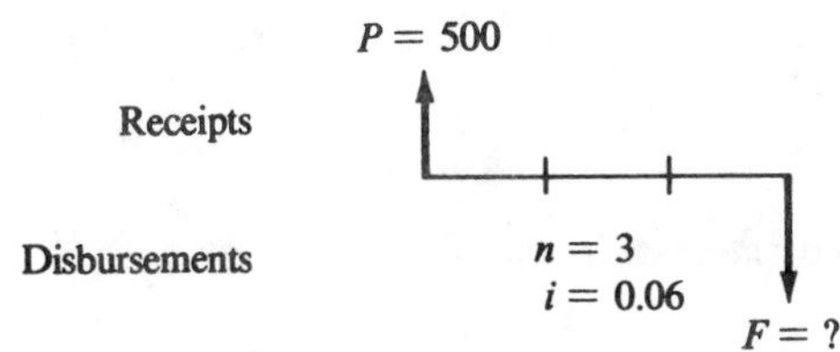

This indicates the bank receives \$500 now and must make a disbursement of F at the end of three years. The computation, from the bank's point of view, is

$$F = 500(F/P,6\%,3) = 500(1.191) = \$595.50$$

This shows that the bank's future disbursement is equal to the depositor's future receipt, and confirms that the correct answer is \$595.50. ◀

If we take $F = P(1 + i)^n$ and solve for P, then

$$P = F\frac{1}{(1 + i)^n} = F(1 + i)^{-n}$$

This is the ***single payment present worth formula***. The equation

$$\mathbf{P = F(1 + i)^{-n}} \qquad (4\text{-}2)$$

in our notation becomes

$$\mathbf{P = F(P/F,i,n)}$$

EXAMPLE 4-2

If you wished to have \$800 in a savings account at the end of four years, and 5% interest was paid annually, how much should you put into the savings account now?

Solution: $F = \$800 \qquad i = 0.05 \qquad n = 4 \qquad P = \text{unknown}$

$$P = F(1 + i)^{-n} = 800(1 + 0.05)^{-4} = 800(0.8227) = \$658.16$$

To have \$800 in the savings account at the end of four years, we must deposit \$658.16 now.

Alternate Solution:

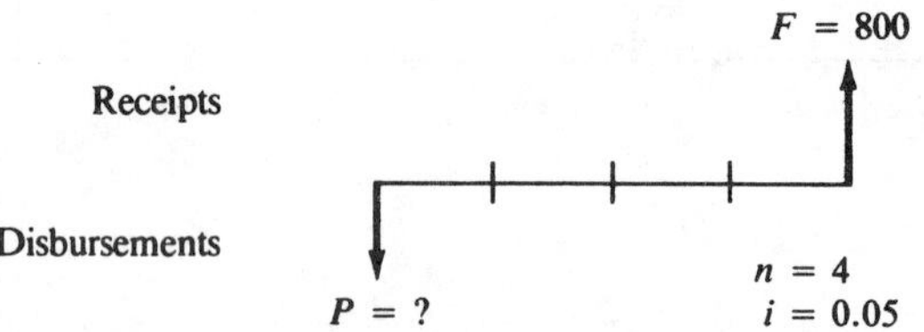

$$P = F(P/F,i,n) = \$800(P/F,5\%,4)$$

From the Compound Interest Tables,

$$(P/F,5\%,4) = 0.8227$$
$$P = \$800(0.8227) = \$658.16$$ ◀

Here the problem has an exact answer. In many situations, however, the answer is rounded off, recognizing that it can only be as accurate as the input information upon which it is based.

EXAMPLE 4-3

Suppose the bank changed their interest policy in Ex. 4-1 to "6% interest, compounded quarterly." For this situation a $500 deposit now would result in how much money in the account at the end of three years?

Solution: First, we must be certain to understand the meaning of *6% interest, compounded quarterly*. There are two elements:

1. *6% interest:* Unless otherwise described, it is customary to assume the stated interest is for a one-year period. *If the stated interest is for other than a one-year period, then it must be explained.*
2. *Compounded quarterly:* This indicates there are four interest periods per year; that is, an interest period is three months long.

We know that the 6% interest is an annual rate, because if it were anything different, it would have been stated. Since we are dealing with four interest periods per year, it follows that the interest rate per interest period is 1½%. For the total three-year duration, there are twelve interest periods.

$$P = \$500 \qquad i = 0.015 \qquad n = (4 \times 3) = 12 \qquad F = \text{unknown}$$

$$\begin{aligned} F &= P(1 + i)^n = P(F/P,i,n) \\ &= \$500(1 + 0.015)^{12} = \$500(F/P,1\tfrac{1}{2}\%,12) \\ &= \$500(1.196) = \$598.00 \end{aligned}$$

A $500 deposit now would be $598.00 in three years. ◀

EXAMPLE 4-4

Consider the following situation:

Year	*Cash flow*
0	$+P$
1	0
2	0
3	−400
4	0
5	−600

Solve for P assuming a 12% interest rate and using the Compound Interest Tables. Recall that receipts have a plus sign and disbursements or payments have a negative sign. Thus, the diagram is:

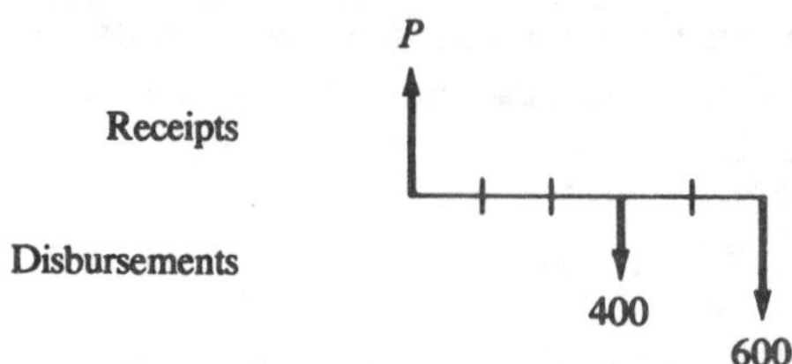

Solution:

$$P = 400(P/F,12\%,3) + 600(P/F,12\%,5)$$
$$= 400(0.7118) + 600(0.5674)$$
$$= 625.16$$

It is important to understand just what the solution, \$625.16, represents. We can say that \$625.16 is the amount of money that would need to be invested at 12% annual interest to allow for the withdrawal of \$400 at the end of three years and \$600 at the end of five years. Let's examine the computations further.

If \$625.16 is invested for one year at 12% interest, it will increase to [625.16 + 0.12(625.16)] = \$700.18. If for the second year the \$700.18 is invested at 12%, it will increase to [700.18 + 0.12(700.18)] = \$784.20. And if this is repeated for another year, [784.20 + 0.12(784.20)] = \$878.30.

We are now at the end of Year 3. The original \$625.16 has increased through the addition of interest to \$878.30. It is at this point that the \$400 is paid out. Deducting \$400 from \$878.30 leaves \$478.30.

The \$478.30 can be invested at 12% for the fourth year and will increase to [478.30 + 0.12(478.30)] = \$535.70. And if left at interest for another year, it will increase to [535.70 + 0.12(535.70)] = \$600. We are now at the end of Year 5; with a \$600 payout, there is no money remaining in the account.

In other words, the \$625.16 was just enough money, at a 12% interest rate, to exactly provide for a \$400 disbursement at the end of Year 3 and also a \$600 disbursement at the end of Year 5. We neither end up short of money, nor with money left over: this is an illustration of equivalence. The initial \$625.16 is *equivalent* to the combination of a \$400 disbursement at the end of Year 3 and a \$600 disbursement at the end of Year 5.

Alternate Formation of Example 4-4: There is another way to see what the \$625.16 value of P represents.

Suppose at Year 0 you were offered a piece of paper that guaranteed you would be paid \$400 at the end of three years and \$600 at the end of five years. How much would you be willing to pay for this piece of paper if you wanted your money to produce a 12% interest rate?

This alternate statement of the problem changes the signs in the cash flow and the diagram:

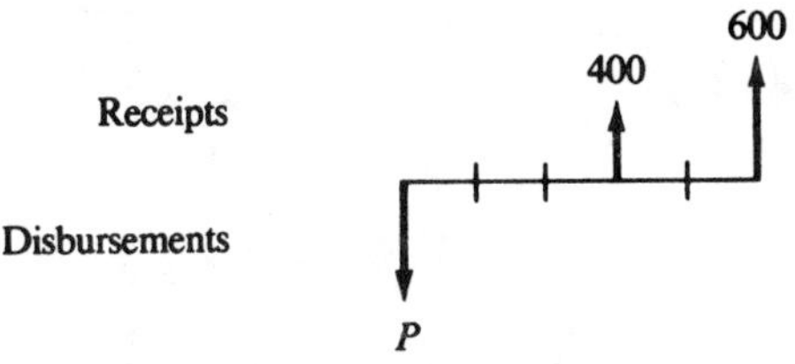

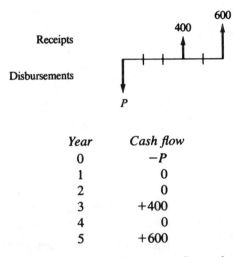

Year	*Cash flow*
0	$-P$
1	0
2	0
3	+400
4	0
5	+600

Since the goal is to recover our initial investment P together with 12% interest per year, we can see that P must be *less* than the total amount to be received in the future (that is, \$400 + 600 = \$1000). We must calculate the present sum P that is *equivalent*, at 12% interest, to an aggregate of \$400 three years hence and \$600 five years hence.

Since we previously derived the relationship

$$P = (1 + i)^{-n},$$

then

$$P = 400(1 + 0.12)^{-3} + 600(1 + 0.12)^{-5}$$
$$= \$625.17$$

This is virtually the same figure as was computed from the first statement of this example. [The slight difference is due to the rounding in the Compound Interest Tables. For example, $(1 + 0.12)^{-5} = 0.567427$, but the Compound Interest Table shows 0.5674.] ◀

Both problems in Example 4-4 have been solved by computing the value of P that is equivalent to \$400 at the end of Year 3 and \$600 at the end of Year 5. In the first problem, we received $+P$ at Year 0 and were obligated to pay out the \$400 and \$600 in later years. In the second (alternate formation) problem, the reverse was true. We paid $-P$ at Year 0 and would receive the \$400 and \$600 sums in later years. In fact, the two problems could represent the buyer and seller of the same piece of paper. The seller would receive $+P$ at Year 0 while the buyer would pay $-P$. Thus, while the problems looked different, they could have been one situation examined from the viewpoint first of the seller and subsequently of that of the buyer. Either way, the solution is based on an equivalence computation.

EXAMPLE 4-5

One of the cash flows in Example 4-4 was

Year	*Cash flow*
0	$-P$
1	0
2	0
3	+400
4	0
5	+600

At a 12% interest rate, P was computed to be \$625.17. Suppose the interest rate is increased to 15%. Will the value of P be larger or smaller?

Solution: One can consider P a sum of money invested at 15% from which one is to obtain \$400 at the end of 3 years and \$600 at the end of 5 years. At 12%, the required P is \$627.17. At 15%, P will earn more interest each year, indicating that we can begin with a *smaller* P and still accumulate enough money for the subsequent cash flows. The computation is:

$$\begin{aligned} P &= 400(P/F,15\%,3) + 600(P/F,15\%,5) \\ &= 400(0.6575) + 600(0.4972) \\ &= \$561.32 \end{aligned}$$

The value of P is smaller at 15% than at 12% interest. ◀

Uniform Series Formulas

Many times we will find situations where there are a uniform series of receipts or disbursements. Automobile loans, house payments, and many other loans are based on a ***uniform payment series***. It will often be convenient to use tables based on a uniform series of receipts or disbursements. The series A is defined:

A = An end-of-period* cash receipt or disbursement in a uniform series, continuing for n periods, the entire series equivalent to P or F at interest rate i.

The horizontal line in Figure 4-1 is a representation of time with four interest periods illustrated. Uniform payments A have been placed at the end of each

*In textbooks on economic analysis, it is customary to define A as an end-of-period event rather than beginning-of-period or, possibly, middle-of-period. The derivations that follow are based on this end-of-period assumption. One could, of course, derive other equations based on beginning-of-period or mid-period assumptions.

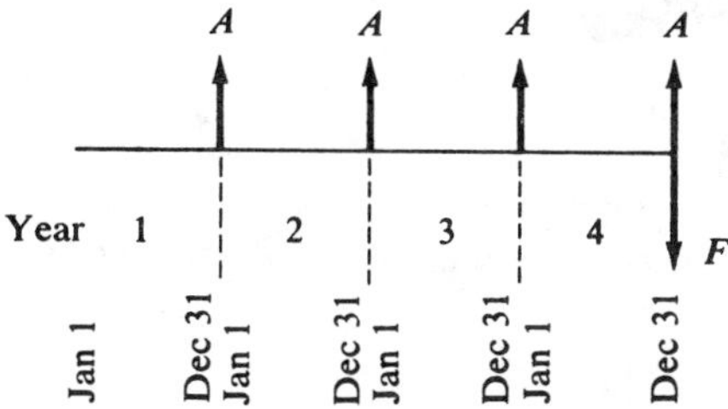

Figure 4-1 The general relationship between A and F.

interest period, and there are as many A's as there are interest periods n. (Both of these conditions are specified in the definition of A.)

In the previous section on single payment formulas, we saw that a sum P at one point in time would increase to a sum F in n periods, according to the equation

$$F = P(1 + i)^n$$

We will use this relationship in our uniform series derivation.

Looking at Fig. 4-1, we see that if an amount A is invested at the end of each year for n years, the total amount F at the end of n years will be the sum of the compound amounts of the individual investments.

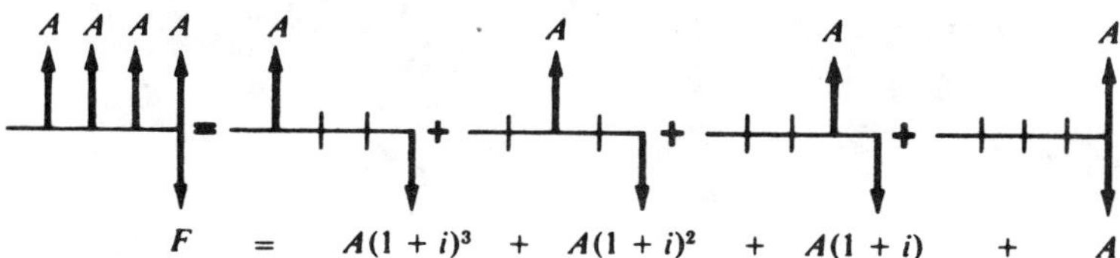

A A A A
A
A
A
A
F = A(1 + i)³ + A(1 + i)² + A(1 + i) + A

In the general case for n years,

$$F = A(1 + i)^{n-1} + \cdots + A(1 + i)^3 + A(1 + i)^2 + A(1 + i) + A \quad (4\text{-}3)$$

Multiplying Eq. 4-3 by $(1 + i)$,

$$(1 + i)F = A(1 + i)^n + \cdots + A(1 + i)^4 + A(1 + i)^3 + A(1 + i)^2 + A(1 + i) \quad (4\text{-}4)$$

Factoring out A and subtracting Eq. 4-3 gives

$$(1 + i)F = A[(1 + i)^n + \cdots + (1 + i)^4 + (1 + i)^3 + (1 + i)^2 + (1 + i)] \quad (4\text{-}5)$$

$$-\quad F = A[(1 + i)^{n-1} + \cdots + (1 + i)^3 + (1 + i)^2 + (1 + i) + 1]$$

$$iF = A[(1 + i)^n - 1] \quad (4\text{-}6)$$

Solving Equation 4-6 for F,

$$F = A\left[\frac{(1+i)^n - 1}{i}\right] \tag{4-7}$$

Thus we have an equation for F when A is known. The term within the brackets

$$\left[\frac{(1+i)^n - 1}{i}\right]$$

is called the ***uniform series compound amount factor*** and is referred to by the notation

$(F/A,i,n)$

EXAMPLE 4-6

A man deposits \$500 in a credit union at the end of each year for five years. The credit union pays 5% interest, compounded annually. At the end of five years, immediately following his fifth deposit, how much will he have in his account?

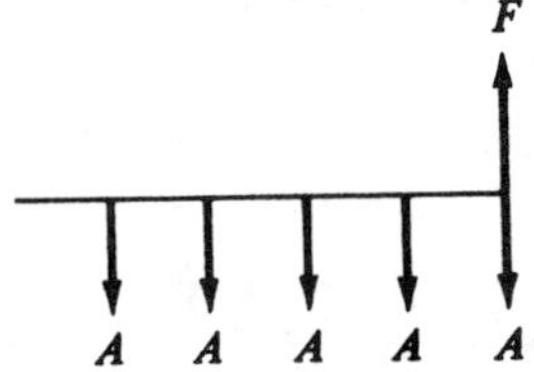

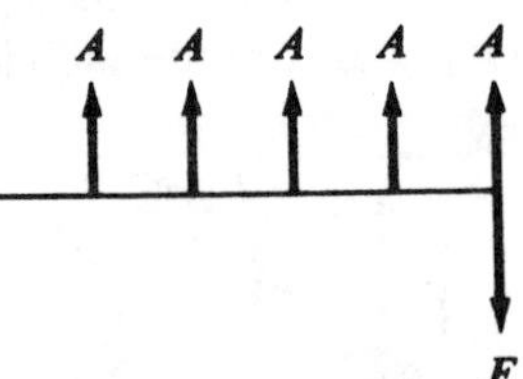

Solution: The diagram on the left shows the situation from the man's point of view; the one on the right, from the credit union's point of view. Either way, the diagram of the five deposits and the desired computation of the future sum F duplicates the situation for the uniform series compound amount formula

$$F = A\left[\frac{(1+i)^n - 1}{i}\right] = A(F/A,i,n)$$

where A = \$500, $n = 5$, $i = 0.05$, F = unknown. Filling in the known variables,

$$F = \$500(F/A,5\%,5) = \$500(5.526) = \$2763$$

He will have \$2763 in his account following the fifth deposit. ◀

If Eq. 4-7 is solved for A, we have

$$A = F\left[\frac{i}{(1+i)^n - 1}\right] \tag{4-8}$$

where

$$\left[\frac{i}{(1+i)^n - 1}\right]$$

is called the ***uniform series sinking fund* factor*** and is written as **$(A/F,i,n)$**.

EXAMPLE 4-7

Jim Hayes read that in the western United States, a ten-acre parcel of land could be purchased for $1000 cash. Jim decided to save a uniform amount at the end of each month so that he would have the required $1000 at the end of one year. The local credit union pays 6% interest, compounded monthly. How much would Jim have to deposit each month?

Solution: In this example,

$$F = \$1000 \qquad n = 12 \qquad i = \tfrac{1}{2}\% \qquad A = \text{unknown}$$

$$A = 1000(A/F,\tfrac{1}{2}\%,12) = 1000(0.0811) = \$81.10$$

Jim would have to deposit $81.10 each month. ◀

If we use the sinking fund formula (Eq. 4-8) and substitute for F the single payment compound formula (Eq. 4-1), we obtain

$$A = F\left[\frac{i}{(1+i)^n - 1}\right] = P(1+i)^n\left[\frac{i}{(1+i)^n - 1}\right]$$

$$A = P\left[\frac{i(1+i)^n}{(1+i)^n - 1}\right] \tag{4-9}$$

We now have an equation for determining the value of a series of end-of-period payments—or disbursements—A when the present sum P is known.

The portion within the brackets,

$$\left[\frac{i(1+i)^n}{(1+i)^n - 1}\right]$$

is called the ***uniform series capital recovery factor*** and has the notation **$(A/P,i,n)$**.

---

*A *sinking fund* is a separate fund into which one makes a uniform series of money deposits (A) with the goal of accumulating some desired future sum (F) at a given future point in time.

EXAMPLE 4-8

On January 1 a man deposits \$5000 in a credit union that pays 8% interest, compounded annually. He wishes to withdraw all the money in five equal end-of-year sums, beginning December 31st of the first year. How much should he withdraw each year?

Solution:

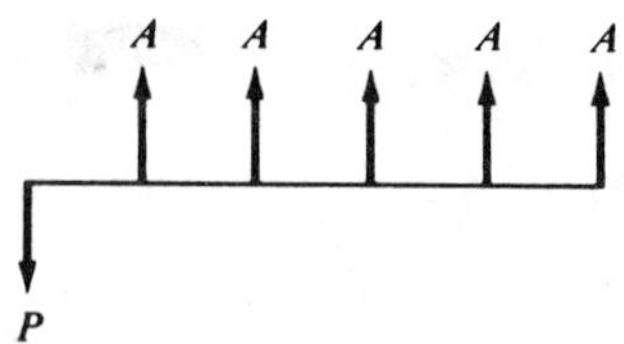

$P = \$5000 \qquad n = 5 \qquad i = 8\% \qquad A = \text{unknown}$

$A = P(A/P,8\%,5) = 5000(0.2505) = \$1252$

The annual withdrawal is \$1252. ◀

In the example, with interest at 8%, a present sum of \$5000 is equivalent to five equal end-of-period disbursements of \$1252. This is another way of stating Plan 3 of Table 4-1. The method for determining the annual payment that would repay \$5000 in five years with 8% interest has now been explained. The calculation is simply

$$A = 5000(A/P,8\%,5) = 5000(0.2505) = \$1252$$

If the capital recovery formula (Eq. 4-9) is solved for the Present sum P, we obtain the uniform series present worth formula.

$$\mathbf{P = A\left[\frac{(1 + i)^n - 1}{i(1 + i)^n}\right]} \qquad (4\text{-}10)$$

$$(P/A,i,n) = \left[\frac{(1 + i)^n - 1}{i(1 + i)^n}\right]$$

which is the ***uniform series present worth factor***.

EXAMPLE 4-9

An investor holds a time payment purchase contract on some machine tools. The contract calls for the payment of \$140 at the end of each month for a five-year period. The first payment is due in one month. He offers to sell you the contract for \$6800 cash today. If you otherwise can make 1% per month on your money, would you accept or reject the investor's offer?

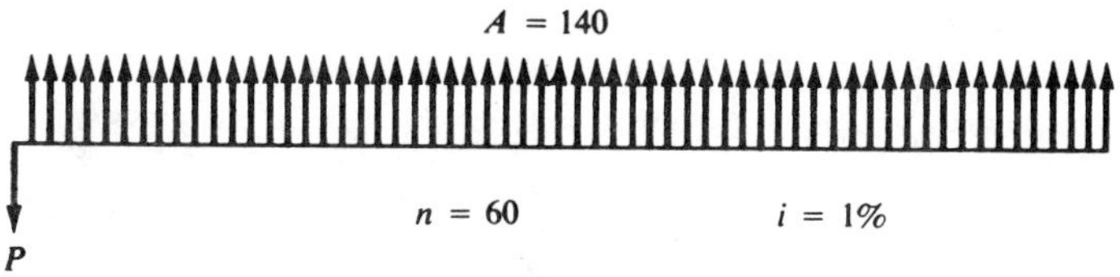

Solution: In this problem we are being offered a contract that will pay \$140 per month for 60 months. We must determine whether the contract is worth \$6800, if we consider 1% per month to be a suitable interest rate. Using the uniform series present worth formula, we will compute the present worth of the contract.

$$P = A(P/A,i,n) = 140(P/A,1\%,60) = 140(44.955)$$
$$= \$6293.70$$

It is clear that if we pay the \$6800 asking-price for the contract, we will receive something less than the 1% per month interest we desire. We will, therefore, reject the investor's offer. ◀

EXAMPLE 4-10

Suppose we decided to pay the \$6800 for the time purchase contract in Example 4-9. What monthly rate of return would we obtain on our investment?

Solution: In this situation, we know *P*, *A*, and *n*, but we do not know *i*. The problem may be solved using either the uniform series present worth formula,

$$P = A(P/A,i,n)$$

or the uniform series capital recovery formula,

$$A = P(A/P,i,n)$$

Either way, we have one equation with one unknown.

$$P = \$6800 \qquad A = \$140 \qquad n = 60 \qquad i = \text{unknown}$$

$$P = A(P/A,i,n)$$
$$\$6800 = \$140(P/A,i,60)$$
$$(P/A,i,60) = \frac{6800}{140} = 48.571$$

We know the value of the uniform series present worth factor, but we do not know the interest rate *i*. As a result, we need to look through several Compound Interest Tables and compute the rate of return *i* by interpolation. From the tables in the back of this book, we find

Interest rate	$(P/A,i,60)$
½%	51.726
¾%	48.174
1%	44.955

The rate of return is between ½% and ¾%, and may be computed by a linear interpolation:

$$\text{Rate of return } i = 0.50\% + 0.25\%\left(\frac{51.726 - 48.571}{51.726 - 48.174}\right)$$

$$= 0.50\% + 0.25\%\left(\frac{3.155}{3.552}\right) = 0.50\% + 0.22\%$$

$$= 0.72\% \text{ per month}$$

The monthly rate of return on our investment would be 0.72% per month. ◀

EXAMPLE 4-11

Using a 15% interest rate, compute the value of F in the following cash flow:

Year	*Cash flow*
1	+100
2	+100
3	+100
4	0
5	$-F$

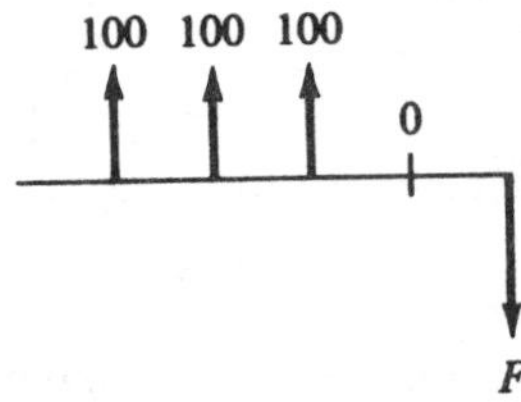

Solution: We see that the cash flow diagram is not the same as the sinking fund factor diagram (see Ex. 4-6):

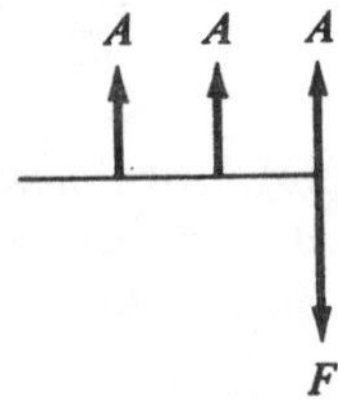

Since the diagrams do not agree, the problem is more difficult than the previous ones we've discussed. The general approach to use in this situation is to convert the cash flow from its present form into standard forms, for which we have compound interest factors and Compound Interest Tables.

One way to solve this problem is to consider the cash flow as a series of single payments P and then to compute their sum F. In other words, the cash flow is broken into three parts, each one of which we can solve.

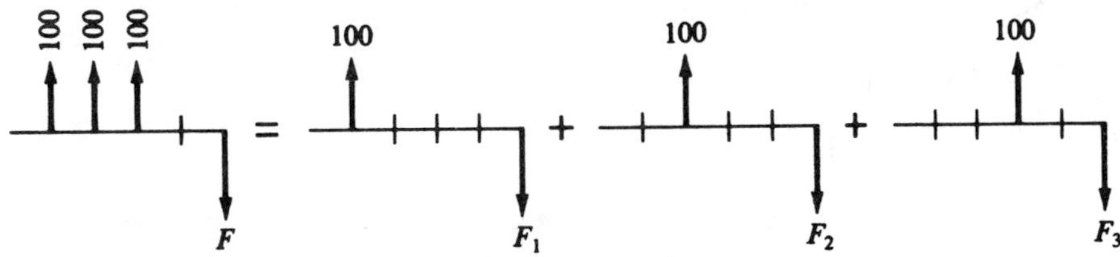

$$F = F_1 + F_2 + F_3 = 100(F/P,15\%,4) + 100(F/P,15\%,3) + 100(F/P15\%,2)$$
$$= 100(1.749) + 100(1.521) + 100(1.322)$$
$$= \$459.20$$

The value of F in the illustrated cash flow is \$459.20.

Alternate Solution:

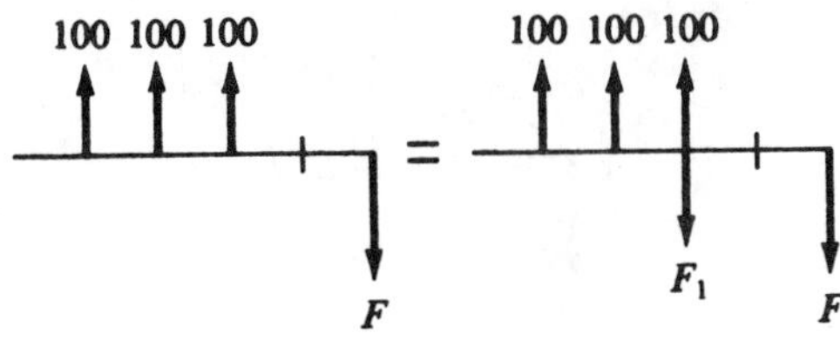

Looked at this way, we first solve for F_1.

$$F_1 = 100(F/A,15\%,3) = 100(3.472) = \$347.20$$

Now F_1 can be considered a present sum P in the diagram:

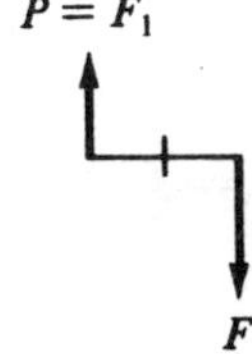

So, $F = F_1(F/P,15\%,2)$
$= 347.20(1.322)$
$= \$459.00$

The slightly different value from the previous computation is due to rounding in the Compound Interest Tables.

This has been a two-step solution:

$$F_1 = 100(F/A,15\%,3)$$
$$F = F_1(F/P,15\%,2)$$

One could substitute the value of F_1 from the first equation into the second equation and solve for F, without computing F_1.

$$F = 100(F/A,15\%,3)(F/P,15\%,2)$$
$$= 100(3.472)(1.322)$$
$$= \$459.00$$ ◀

EXAMPLE 4-12

Consider the following situation:

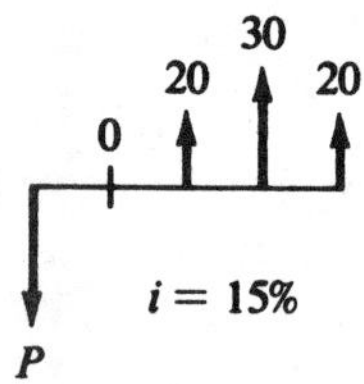

The diagram is not in a standard form, indicating there will be a multiple-step solution. There are at least three different ways of computing the answer. (It is important that you understand how the three computations are made, so please study all three solutions.)

Ex.4-12, Solution One:

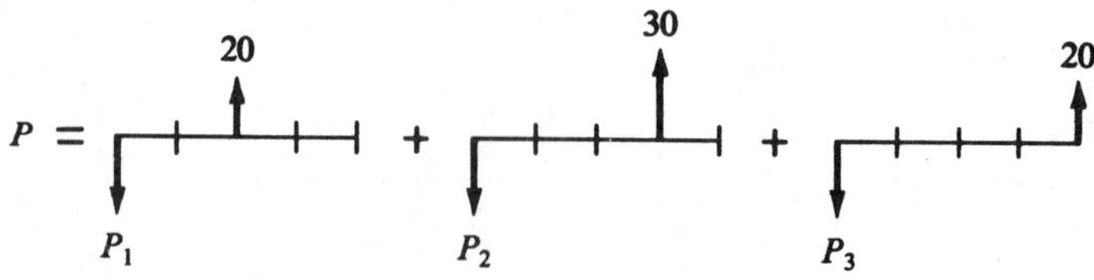

$$\begin{aligned} P &= P_1 + P_2 + P_3 \\ &= 20(P/F,15\%,2) + 30(P/F,15\%,3) + 20(P/F,15\%,4) \\ &= 20(0.7561) + 30(0.6575) + 20(0.5718) \\ &= \$46.28 \end{aligned}$$

Ex. 4-12, Solution Two:

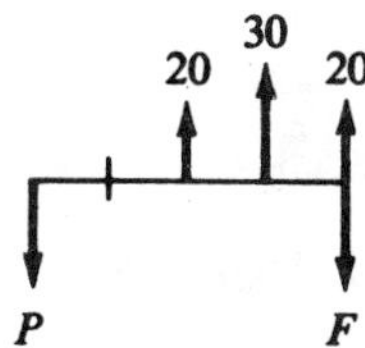

The relationship between P and F in the diagram is

$$P = F(P/F,15\%,4)$$

Next we compute the future sums of the three payments, as follows:

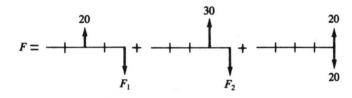

$$\begin{aligned} F &= F_1 + F_2 + 20 \\ &= 20(F/P,15\%,2) + 30(F/P,15\%,1) + 20 \end{aligned}$$

Combining the two equations,

$$\begin{aligned} P &= [F_1 + F_2 + 20](P/F,15\%,4) \\ &= [20(F/P,15\%,2) + 30(F/P,15\%,1) + 20](P/F,15\%,4) \\ &= [20(1.322) + 30(1.150) + 20](0.5718) \\ &= \$46.28 \end{aligned}$$

Ex. 4-12, Solution Three:

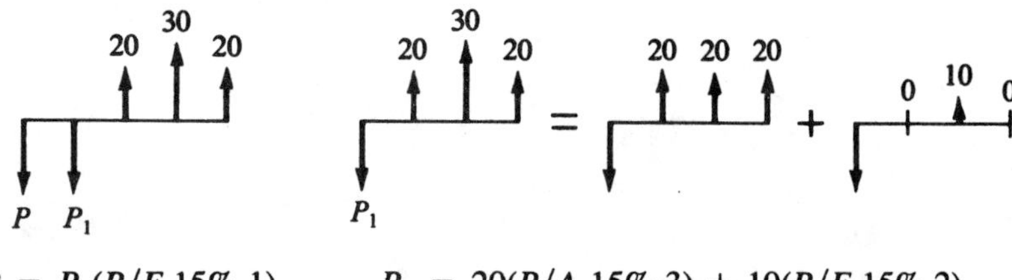

$$P = P_1(P/F,15\%,1) \qquad P_1 = 20(P/A,15\%,3) + 10(P/F,15\%,2)$$

Combining,

$$
\begin{aligned}
P &= [20(P/A,15\%,3) + 10(P/F,15\%,2)](P/F,15\%,1) \\
&= [20(2.283) + 10(0.7561)](0.8696) \\
&= \$46.28 \blacktriangleleft
\end{aligned}
$$

Relationships Between Compound Interest Factors

From the derivations, we see there are several simple relationships between the compound interest factors. They are summarized here.

Single Payment

$$\text{Compound amount factor} = \frac{1}{\text{Present worth factor}}$$

$$(F/P,i,n) = \frac{1}{(P/F,i,n)}$$

Uniform Series

$$\text{Capital recovery factor} = \frac{1}{\text{Present worth factor}}$$

$$(A/P,i,n) = \frac{1}{(P/A,i,n)}$$

$$\text{Compound amount factor} = \frac{1}{\text{Sinking fund factor}}$$

$$(F/A,i,n) = \frac{1}{(A/F,i,n)}$$

The uniform series present worth factor is simply the sum of the n terms of the single payment present worth factor

$$(P/A,i,n) = \sum_{J=1}^{n} (P/F,i,J)$$

For example:

$$
\begin{aligned}
(P/A,5\%,4) &= (P/F,5\%,1) + (P/F,5\%,2) + (P/F,5\%,3) + (P/F,5\%,4) \\
3.546 &= 0.9524 + 0.9070 + 0.8638 + 0.8227
\end{aligned}
$$

The uniform series compound amount factor equals 1 *plus* the sum of $(n - 1)$ terms of the single payment compound amount factor

$$(F/A,i,n) = 1 + \sum_{J=1}^{n-1} (F/P,i,J)$$

For example,

$$(F/A,5\%,4) = 1 + (F/P,5\%,1) + (F/P,5\%,2) + (F/P,5\%,3)$$
$$4.310 = 1 + 1.050 + 1.102 + 1.158$$

The uniform series capital recovery factor equals the uniform series sinking fund factor *plus* i.

$$(A/P,i,n) = (A/F,i,n) + i$$

For example,

$$(A/P,5\%,4) = (A/F,5\%,4) + 0.05$$
$$0.2820 = 0.2320 + 0.05$$

This may be proved as follows:

$$(A/P,i,n) = (A/F,i,n) + i$$

$$\left[\frac{i(1 + i)^n}{(1 + i)^n - 1}\right] = \left[\frac{i}{(1 + i)^n - 1}\right] + i$$

Multiply by $(1 + i)^n - 1$ to get

$$i(1 + i)^n = i + i(1 + i)^n - i = i(1 + i)^n$$

Arithmetic Gradient

We frequently encounter the situation where the cash flow series is not of constant amount A. Instead, there is a uniformly increasing series as shown:

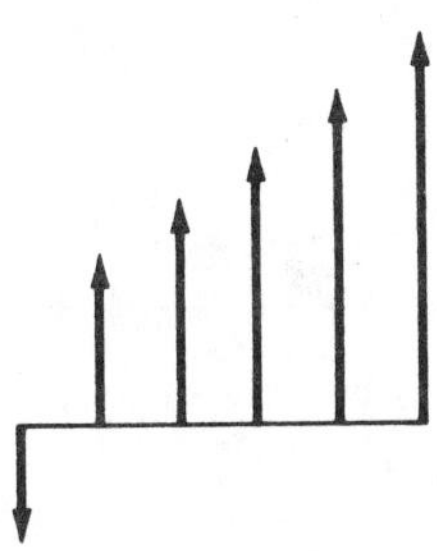

Cash flows of this form may be resolved into two components:

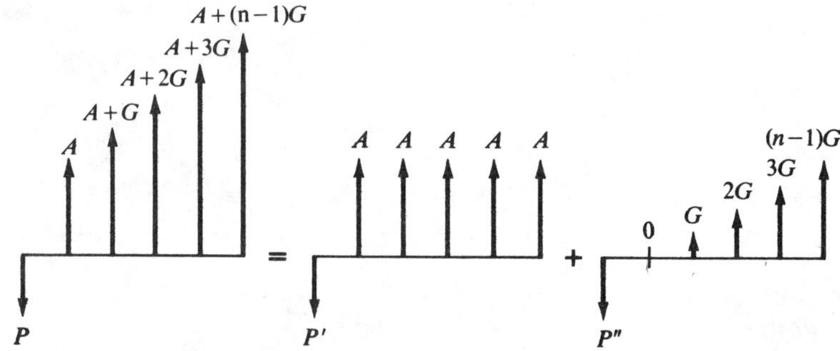

Note that by resolving the problem in this manner, it makes the first cash flow in the arithmetic gradient series equal to zero. We already have an equation for P', and we need to derive an equation for P''. In this way, we will be able to write

$$P = P' + P'' = A(P/A,i,n) + G(P/G,i,n)$$

Derivation of Arithmetic Gradient Factors

The arithmetic gradient is a series of increasing cash flows as follows:

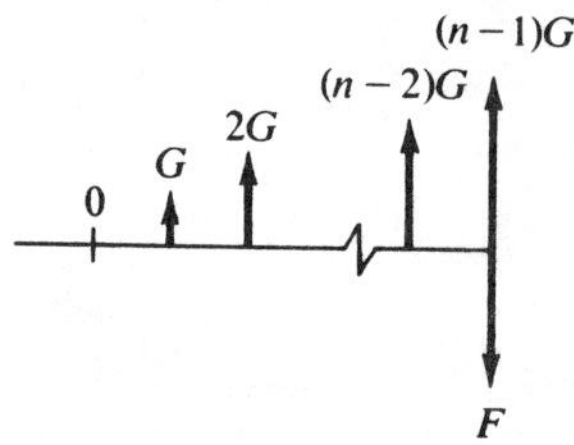

The arithmetic gradient series may be thought of as a series of individual cash flows:

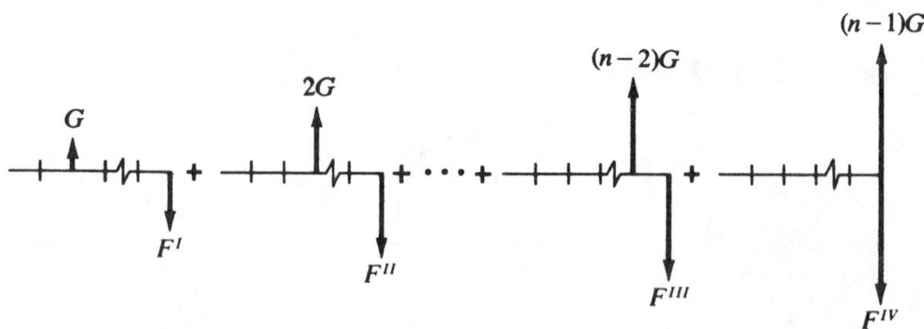

The value of F for the sum of the cash flows $= F^{\text{I}} + F^{\text{II}} + \cdots + F^{\text{III}} + F^{\text{IV}}$, or

$$F = G(1 + i)^{n-2} + 2G(1 + i)^{n-3} + \cdots + (n - 2)(G)(1 + i)^1 + (n - 1)G \qquad (4\text{-}11)$$

Multiply Eq. 4-11 by $(1 + i)$ and factor out G, or

$$(1 + i)F = G[(1 + i)^{n-1} + 2(1 + i)^{n-2} + \cdots + (n - 2)(1 + i)^2 + (n - 1)(1 + i)^1] \qquad (4\text{-}12)$$

Rewrite Eq. 4-11 to show other terms in the series,

$$F = G[(1 + i)^{n-2} + \cdots + (n - 3)(1 + i)^2 + (n - 2)(1 + i)^1 + n - 1] \qquad (4\text{-}13)$$

Subtracting Eq. 4-13 from Eq. 4-12, we obtain

$$F + iF - F = G[(1 + i)^{n-1} + (1 + i)^{n-2} + \cdots + (1 + i)^2 + (1 + i)^1 + 1] - nG \qquad (4\text{-}14)$$

In the derivation of Eq. 4-7, the terms within the brackets of Eq. 4-14 were shown to equal the series compound amount factor:

$$[(1 + i)^{n-1} + (1 + i)^{n-2} + \cdots + (1 + i)^2 + (1 + i)^1 + 1] = \frac{(1 + i)^n - 1}{i}$$

Thus, Equation 4-14 becomes

$$iF = G\left[\frac{(1 + i)^n - 1}{i}\right] - nG$$

Rearranging and solving for F,

$$F = \frac{G}{i}\left[\frac{(1 + i)^n - 1}{i} - n\right] \qquad (4\text{-}15)$$

Multiplying Eq. 4-15 by the single payment present worth factor,

$$P = \frac{G}{i}\left[\frac{(1 + i)^n - 1}{i} - n\right]\left[\frac{1}{(1 + i)^n}\right]$$

$$P = G\left[\frac{(1 + i)^n - in - 1}{i^2(1 + i)^n}\right]$$

$$\mathbf{(P/G,i,n)} = \left[\frac{(1 + i)^n - in - 1}{i^2(1 + i)^n}\right] \qquad (4\text{-}16)$$

Equation 4-16 is the ***arithmetic gradient present worth factor***. Multiplying Eq. 4-15 by the sinking fund factor,

$$A = \frac{G}{i}\left[\frac{(1+i)^n - 1}{i} - n\right]\left[\frac{i}{(1+i)^n - 1}\right]$$

$$= G\left[\frac{(1+i)^n - in - 1}{i(1+i)^n - i}\right]$$

$$(A/G,i,n) = \left[\frac{(1+i)^n - in - 1}{i(1+i)^n - i}\right] \qquad (4\text{-}17)$$

Equation 4-17 is the ***arithmetic gradient uniform series factor***.

EXAMPLE 4-13

A man purchased a new automobile. He wishes to set aside enough money in a bank account to pay the maintenance on the car for the first five years. It has been estimated that the maintenance cost of an automobile is as follows:

Year	*Maintenance cost*
1	$120
2	150
3	180
4	210
5	240

Assume the maintenance costs occur at the end of each year and that the bank pays 5% interest. How much should he deposit in the bank now?

Solution:

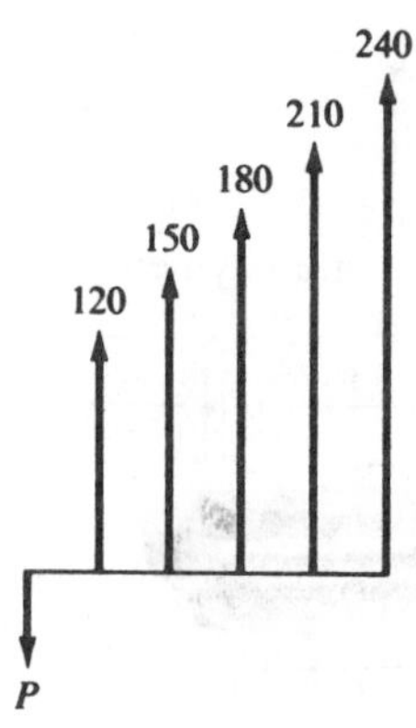

The cash flow may be broken into its two components:

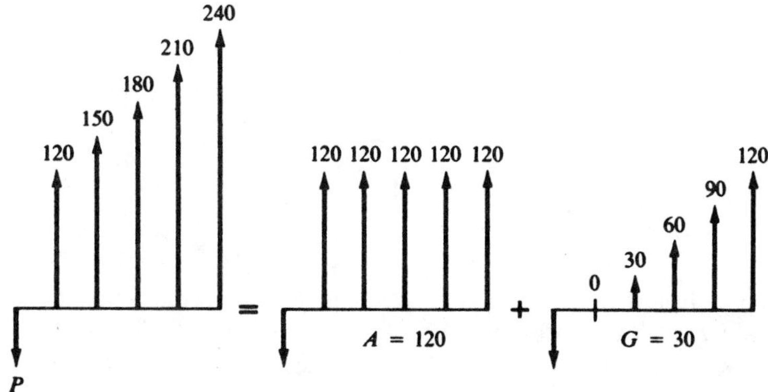

Both of the components represent cash flows for which compound interest factors have been derived. The first is uniform series present worth, and the second is arithmetic gradient series present worth.

$$P = A(P/A,5\%,5) + G(P/G,5\%,5)$$

Note that the value of n in the gradient factor is 5, not 4. In deriving the gradient factor, we had $(n - 1)$ values of G. Here we have 4 values of G. Thus, $(n - 1) = 4$, so $n = 5$.

$$\begin{aligned} P &= 120(P/A,5\%,5) + 30(P/G,5\%,5) \\ &= 120(4.329) + 30(8.237) = 519 + 247 \\ &= \$766 \end{aligned}$$

He should deposit $766 in the bank now. ◀

EXAMPLE 4-14

On a certain piece of machinery, it is estimated that the maintenance expense will be as follows:

Year	*Maintenance*
1	$100
2	200
3	300
4	400

What is the equivalent uniform annual maintenance cost for the machinery if 6% interest is used?

Solution:

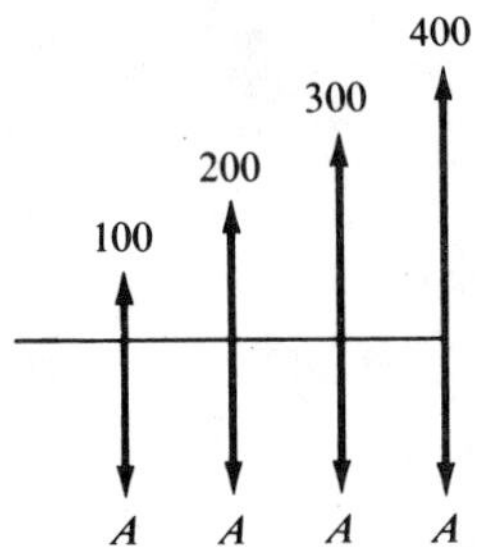

The first cash flow in the arithmetic gradient series is zero, hence the diagram above is *not* in proper form for the arithmetic gradient equation. The cash flow must be resolved into two components as is done in Example 4-13.

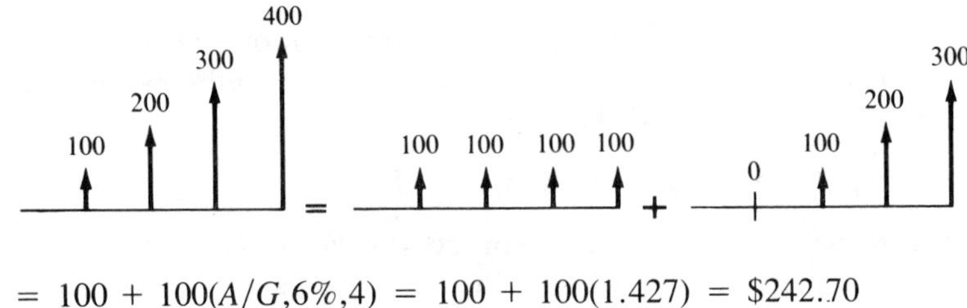

$$A = 100 + 100(A/G,6\%,4) = 100 + 100(1.427) = \$242.70$$

The equivalent uniform annual maintenance cost is $242.70. ◀

EXAMPLE 4-15

A textile mill in India installed a number of new looms. It is expected that initial maintenance and repairs will be high, but that they will then decline for several years. The projected cost is:

Year	*Maintenance and repair cost*
1	24,000 rupees
2	18,000
3	12,000
4	6,000

What is the projected equivalent annual maintenance and repair cost if interest is 10%?

Solution:

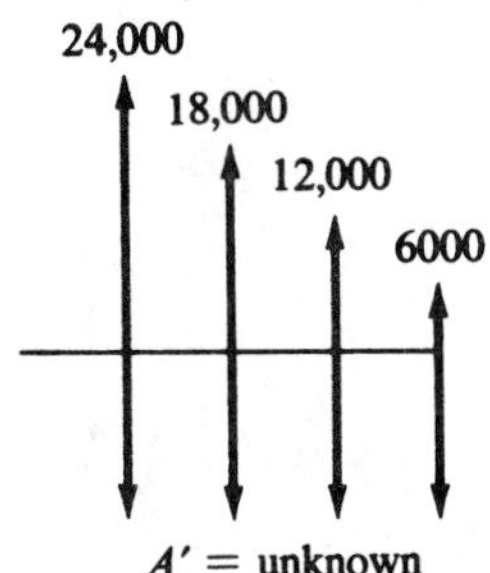

The projected cash flow is not in the form of the arithmetic gradient factors. Both factors were derived for an increasing gradient over time. The factors cannot be used directly for a declining gradient. Instead, we will subtract an increasing gradient from an assumed uniform series of payments.

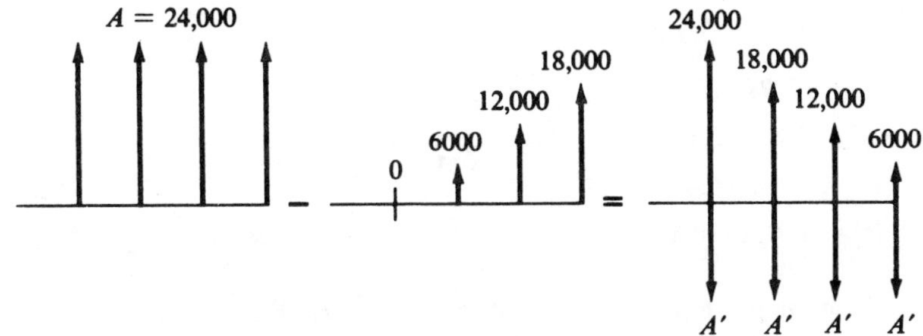

$$A' = 24{,}000 - 6000(A/G,10\%,4) = 24{,}000 - 6000(1.381)$$
$$= 15{,}714 \text{ rupees}$$

The projected equivalent uniform maintenance and repair cost is 15,714 rupees per year. ◀

EXAMPLE 4-16

Compute the value of P in the diagram below. Use a 10% interest rate.

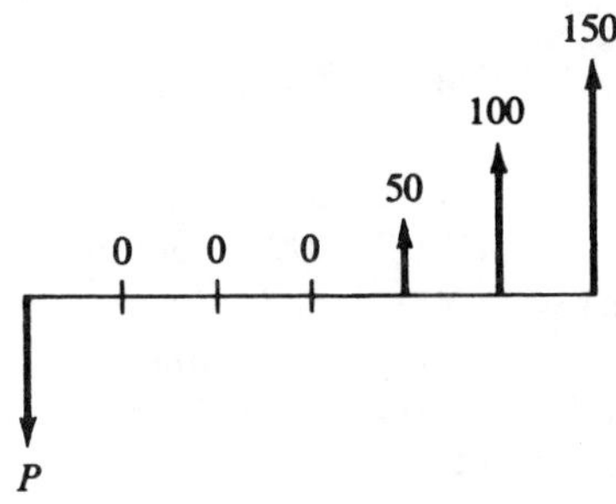

Solution: With the arithmetic gradient series present worth factor, we can compute a present sum J.

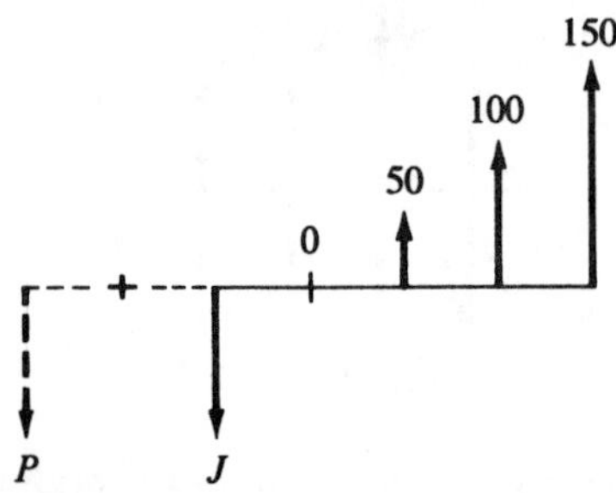

It is important that you closely examine the location of J. Based on the way the factor was derived, there will be one zero value in the gradient series to the right of J. (If this seems strange or incorrect, review the beginning of this Arithmetic Gradient section.)

$$J = G(P/G,i,n)$$

$$J = 50(P/G,10\%,4) \quad (\textit{Note:}\ 3\text{ would be incorrect.})$$

$$= 50(4.378) = 218.90$$

Then $P = J(P/F,10\%,2)$

To obtain the present worth of the future sum J, use the $(P/F,i,n)$ factor.

Combining,

$$P = 50(P/G,10\%,4)(P/F,10\%,2)$$

$$= 50(4.378)(0.8264) = \$180.90$$

The value of P is \$180.90. 

Geometric Gradient

In the previous section, we saw that the arithmetic gradient is applicable where the period-by-period change in a cash receipt or payment is a uniform amount. There are other situations where the period-by-period change is a ***uniform rate***, ***g***. An example of this would be where the maintenance costs for an automobile are \$100 the first year and increasing at a uniform rate, g, of 10% per year. For the first five years, the cash flow would be:

Year				*Cash flow*
1	100.00			= \$100.00
2	100.00 + 10%(100.00)	= $100(1 + 0.10)^1$	=	110.00
3	110.00 + 10%(110.00)	= $100(1 + 0.10)^2$	=	121.00
4	121.00 + 10%(121.00)	= $100(1 + 0.10)^3$	=	133.10
5	133.10 + 10%(133.10)	= $100(1 + 0.10)^4$	=	146.41

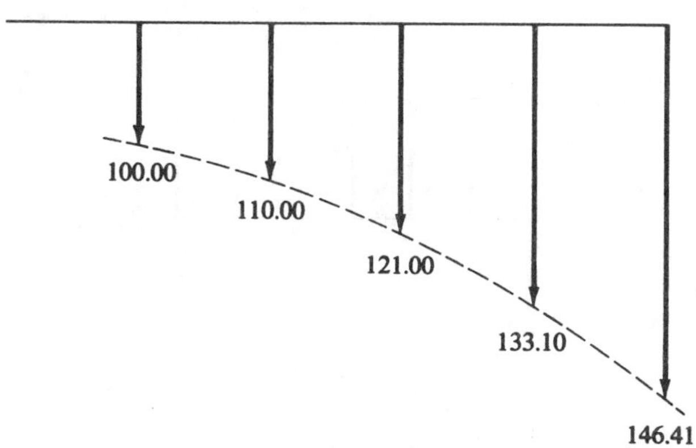

From the table, we can see that the maintenance cost in any year is

$$\$100(1 + g)^{n-1}$$

Stated in a more general form,

$$A_n = A_1(1 + g)^{n-1} \tag{4-18}$$

where

g = Uniform *rate* of cash flow increase/decrease from period to period, that is, the geometric gradient.

A_1 = Value of A at Year 1 (\$100 in the example).

A_n = Value of A at any Year n.

Since the present worth P_n of any cash flow A_n at interest rate i is

$$P_n = A_n(1 + i)^{-n} \tag{4-19}$$

we can substitute Eq. 4-18 into Eq. 4-19 to get

$$P_n = A_1(1 + g)^{n-1}(1 + i)^{-n}$$

This may be rewritten as

$$P_n = A_1(1 + i)^{-1}\left(\frac{1 + g}{1 + i}\right)^{n-1} \tag{4-20}$$

The present worth of the entire gradient series of cash flows may be obtained by expanding Eq. 4-20:

$$P = A_1(1 + i)^{-1} \sum_{x=1}^{n} \left(\frac{1 + g}{1 + i}\right)^{x-1} \tag{4-21}$$

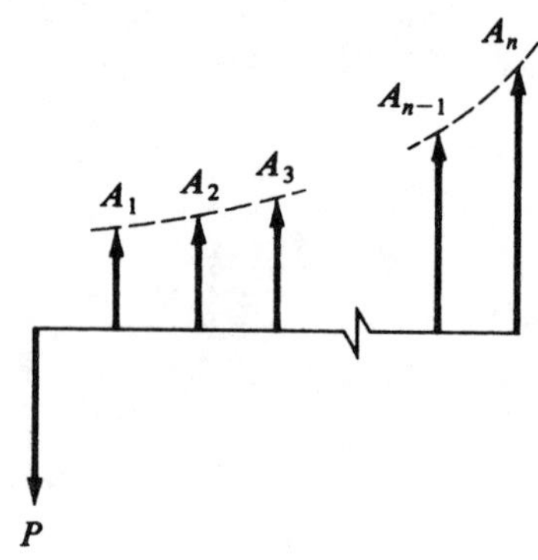

In the general case where $i \neq g$, Eq. 4-21 may be written out as

$$P = A_1(1 + i)^{-1} + A_1(1 + i)^{-1}\left(\frac{1 + g}{1 + i}\right) + A_1(1 + i)^{-1}\left(\frac{1 + g}{1 + i}\right)^2$$

$$+ \cdots + A_1(1 + i)^{-1}\left(\frac{1 + g}{1 + i}\right)^{n-1} \tag{4-22}$$

Let $a = A_1(1 + i)^{-1}$ and $b = \left(\frac{1 + g}{1 + i}\right)$. Equation 4-22 becomes

$$P = a + ab + ab^2 + \cdots + ab^{n-1} \tag{4-23}$$

Multiply Equation 4-23 by b:

$$bP = ab + ab^2 + ab^3 + \cdots + ab^{n-1} + ab^n \tag{4-24}$$

Subtract Eq. 4-24 from Eq. 4-23:

$$P - bP = a - ab^n$$

$$P(1 - b) = a(1 - b^n)$$

$$P = \frac{a(1 - b^n)}{1 - b}$$

Replacing the original values for a and b, we obtain:

$$P = A_1(1+i)^{-1}\left[\frac{1-\left(\frac{1+g}{1+i}\right)^n}{1-\left(\frac{1+g}{1+i}\right)}\right]$$

$$= A_1\left[\frac{1-\left(\frac{1+g}{1+i}\right)^n}{(1+i)-\left(\frac{1+g}{1+i}\right)(1+i)}\right]$$

$$= A_1\left[\frac{1-(1+g)^n(1+i)^{-n}}{1+i-1-g}\right]$$

$$P = A_1\left[\frac{1-(1+g)^n(1+i)^{-n}}{i-g}\right] \quad \text{where } i \neq g \qquad (4\text{-}25)$$

The expression in the brackets of Equation 4-25 is the ***geometric series present worth factor*** where $i \neq g$.

$$(P/A,g,i,n) = \left[\frac{1-(1+g)^n(1+i)^{-n}}{i-g}\right] \quad \text{where } i \neq g \qquad (4\text{-}26)$$

In the special case where $i = g$, Eq. 4-21 becomes

$$P = A_1 n(1+i)^{-1} \quad \text{where } i = g$$

$$(P/A,g,i,n) = [n(1+i)^{-1}] \quad \text{where } i = g \qquad (4\text{-}27)$$

EXAMPLE 4-17

The first year maintenance cost for a new automobile is estimated to be $100, and it increases at a uniform rate of 10% per year. What is the present worth of cost of the first five years of maintenance in this situation, using an 8% interest rate?

Step-by-Step Solution:

Year n			*Maintenance cost*		*(P/F,8%,n)*		*PW of maintenance*
1	100.00	=	100.00	×	0.9259	=	$ 92.59
2	100.00 + 10%(100.00)	=	110.00	×	0.8573	=	94.30
3	110.00 + 10%(110.00)	=	121.00	×	0.7938	=	96.05
4	121.00 + 10%(121.00)	=	133.10	×	0.7350	=	97.83
5	133.10 + 10%(133.10)	=	146.41	×	0.6806	=	99.65
							$480.42

Solution Using Geometric Series Present Worth Factor:

$$P = A_1\left[\frac{1 - (1 + g)^n(1 + i)^{-n}}{i - g}\right] \quad \text{where } i \neq g$$

$$= 100.00\left[\frac{1 - (1.10)^5(1.08)^{-5}}{-0.02}\right] = \$480.42$$

The present worth of cost of maintenance for the first five years is $480.42. 

Nominal And Effective Interest

EXAMPLE 4-18

Consider the situation of a person depositing $100 into a bank that pays 5% interest, compounded semi-annually. How much would be in the savings account at the end of one year?

Solution: Five percent interest, compounded semi-annually, means that the bank pays 2½% every six months. Thus, the initial amount P = $100 would be credited with 0.025(100) = $2.50 interest at the end of six months, or

$$P \rightarrow P + Pi = 100 + 100(0.025) = 100 + 2.50 = \$102.50$$

The $102.50 is left in the savings account; at the end of the second six-month period, the interest earned is 0.025(102.50) = $2.56, for a total in the account at the end of one year of 102.50 + 2.56 = $105.06, or

$$(P + Pi) \rightarrow (P + Pi) + i(P + Pi) = P(1 + i)^2 = 100(1 + 0.025)^2$$
$$= \$105.06$$ ◀

Nominal interest rate **per year, *r*, is the annual interest rate without considering the effect of any compounding.**

In the example, the bank pays 2½% interest every six months. The nominal interest rate per year, r, therefore, is 2 × 2½% = 5%.

Effective interest rate **per year, i_{eff}, is the annual interest rate taking into account the effect of any compounding during the year.**

In Example 4-18 we saw that $100 left in the savings account for one year increased to $105.06, so the interest paid was $5.06. The effective interest rate per year, i_{eff}, is $5.06/$100.00 = 0.0506 = 5.06%.

r = Nominal interest rate per interest period.

i = Effective interest rate per interest period.

m = Number of compounding subperiods per time period.

Using the method presented in Ex. 4-18, we can derive the equation for the effective interest rate. If a \$1 deposit were made to an account that compounded interest m times per year and paid a nominal interest rate per year, r, the *interest rate per compounding subperiod* would be r/m, and the total in the account at the end of one year would be

$$\$1\left(1+\frac{r}{m}\right)^m \quad \text{or simply} \quad \left(1+\frac{r}{m}\right)^m$$

If we deduct the \$1 principal sum, the expression would be

$$\left(1+\frac{r}{m}\right)^m - 1$$

Therefore,

$$\textbf{Effective interest rate per year, } i_{eff} = \left(1+\frac{r}{m}\right)^m - 1 \qquad (4\text{-}28)$$

where r = Nominal interest rate per year,
and m = Number of compounding subperiods per year.

Or, substituting the effective interest rate per compounding subperiod, $i = (r/m)$,

$$\textbf{Effective interest rate per year, } i_{eff} = (1+i)^m - 1 \qquad (4\text{-}29)$$

where i = Effective interest rate per compounding subperiod,
and m = Number of compounding subperiods per year.

Either Eq. 4-28 or 4-29 may be used to compute an effective interest rate per year.

One should note that i was described at the beginning of this chapter simply as the interest rate per interest period. We were describing the effective interest rate without making any fuss about it. A more precise definition, we now know, is that i is the *effective* interest rate per interest period. Although it seems more complicated, we are describing the same exact situation, but with more care.

The nominal interest rate r is often given for a one-year period (but it could be given for either a shorter or a longer time period). In the special case where the nominal interest rate is given per compounding subperiod, then the effective interest rate per compounding subperiod, i, equals the nominal interest rate per subperiod, r.

In the typical effective interest computation, there are multiple compounding subperiods ($m > 1$). The objective is to find the effective interest rate i_m for m compounding subperiods. The resulting effective interest rate is either the solution to the problem, or an intermediate solution, which allows us to use standard compound interest factors to proceed to solve the problem.

For ***continuous compounding*** (which is described in the next section),

$$\textbf{Effective interest rate per year, } i_{eff} = e^r - 1 \qquad (4\text{-}30)$$

EXAMPLE 4-19

If a savings bank pays 1½% interest every three months, what are the nominal and effective interest rates per year?

Solution:

Nominal interest rate per year, $r = 4 \times 1\frac{1}{2}\% = 6\%$

$$\text{Effective interest rate per year} = \left(1 + \frac{r}{m}\right)^m - 1$$

$$= \left(1 + \frac{0.06}{4}\right)^4 - 1 = 0.061$$

$$= 6.1\%$$

Alternately,

$$\text{Effective interest rate per year} = (1 + i)^m - 1$$

$$= (1 + 0.015)^4 - 1 = 0.061$$

$$= 6.1\%$$

◀

Table 4-3 tabulates the effective interest rate for a range of compounding frequencies and nominal interest rates. It should be noted that when a nominal interest rate is compounded annually, the nominal interest rate equals the effective interest rate. Also, it will be noted that increasing the frequency of compounding (for example, from monthly to continuously) has only a small impact on the effective interest rate.

Table 4-3 NOMINAL AND EFFECTIVE INTEREST

Nominal interest rate per year	*Effective interest rate per year, i_{eff}, when nominal rate is compounded*				
r	*Yearly*	*Semi-annually*	*Monthly*	*Daily*	*Continuously*
1%	1.0000%	1.0025%	1.0046%	1.0050%	1.0050%
2	2.0000	2.0100	2.0184	2.0201	2.0201
3	3.0000	3.0225	3.0416	3.0453	3.0455
4	4.0000	4.0400	4.0742	4.0809	4.0811
5	5.0000	5.0625	5.1162	5.1268	5.1271
6	6.0000	6.0900	6.1678	6.1831	6.1837
8	8.0000	8.1600	8.3000	8.3278	8.3287
10	10.0000	10.2500	10.4713	10.5156	10.5171
15	15.0000	15.5625	16.0755	16.1798	16.1834
25	25.0000	26.5625	28.0732	28.3916	28.4025

EXAMPLE 4-20

A loan shark lends money on the following terms:
"If I give you \$50 on Monday, you owe me \$60 on the following Monday."

a. What nominal interest rate per year (r) is the loan shark charging?
b. What effective interest rate per year is he charging?
c. If the loan shark started with \$50 and was able to keep it, as well as all the money he received, out in loans at all times, how much money would he have at the end of one year?

Solution to Ex. 4-20a:

$$F = P(F/P,i,n)$$
$$60 = 50(F/P,i,1)$$
$$(F/P,i,1) = 1.2$$
$$\text{Therefore, } i = 20\% \text{ per week}$$
$$\text{Nominal interest rate per year} = 52 \text{ weeks} \times 0.20 = 10.40 = 1040\%$$

Solution to Ex. 4-20b:

$$\text{Effective interest rate per year} = \left(1 + \frac{r}{m}\right)^m - 1$$
$$= \left(1 + \frac{10.40}{52}\right)^{52} - 1 = 13{,}105 - 1$$
$$= 13{,}104 = 1{,}310{,}400\%$$

Or,

$$\text{Effective interest rate per year} = (1 + i)^m - 1$$
$$= (1 + 0.20)^{52} - 1 = 13{,}104$$
$$= 1{,}310{,}400\%$$

Solution to Ex. 4-20c:

$$F = P(1 + i)^n = 50(1 + 0.20)^{52}$$
$$= \$655{,}200$$

With a nominal interest rate of 1040% per year and effective interest rate of 1,310,400% per year, if he started with \$50, the loan shark would have \$655,200 at the end of one year. ◀

When the various time periods in a problem match each other, we generally can solve the problem using simple calculations. In Ex. 4-8, for example, there is \$5000 in an account paying 8% interest, compounded annually. The desired five equal end-of-year withdrawals are simply computed as

$$A = P(A/P,8\%,5) = 5000(0.2505) = \$1252$$

Consider how this simple problem becomes more difficult if the compounding period is changed to no longer match the annual withdrawals.

EXAMPLE 4-21

On January 1st, a woman deposits $5000 in a credit union that pays 8% nominal annual interest, compounded quarterly. She wishes to withdraw all the money in five equal yearly sums, beginning December 31st of the first year. How much should she withdraw each year?

Solution: Since the 8% nominal annual interest rate r is compounded quarterly, we know that the effective interest rate per interest period, i, is 2%; and there are a total of $4 \times 5 = 20$ interest periods in five years. For the equation $A = P(A/P,i,n)$ to be used, there must be as many periodic withdrawals as there are interest periods, n. In this example we have five withdrawals and twenty interest periods.

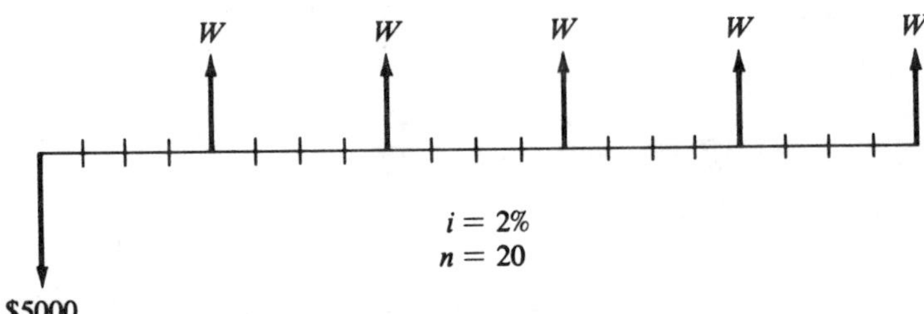

To solve the problem, we must adjust it so that it is in one of the standard forms for which we have compound interest factors. This means we must first either compute an equivalent A for each three-month interest period, or an effective i for each time period between withdrawals. Let's solve the problem both ways.

Ex. 4-21, Solution One:

Compute an equivalent *A* for each three-month time period. If we had been required to compute the amount that could be withdrawn quarterly, the diagram would be as follows:

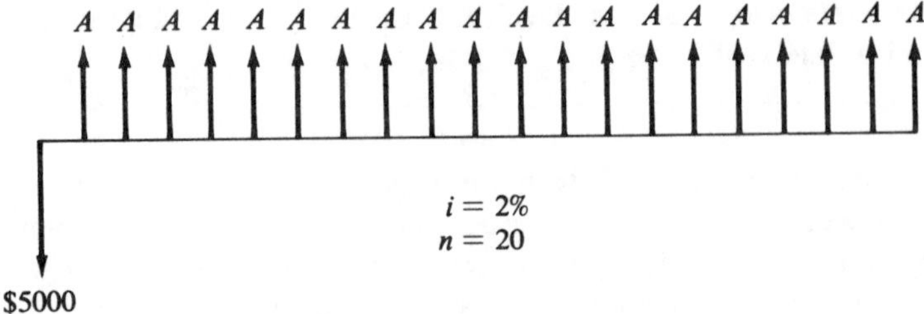

$$A = P(A/P,i,n) = 5000(A/P,2\%,20) = 5000(0.0612) = \$306$$

Now, since we know A, we can construct the diagram that relates it to our desired equivalent annual withdrawal, W:

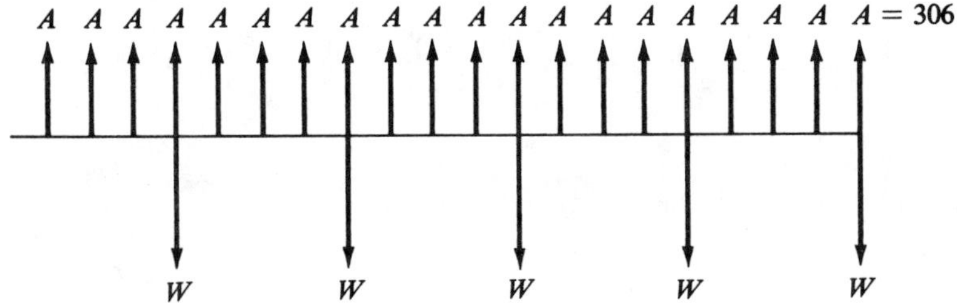

Looking at a one-year period,

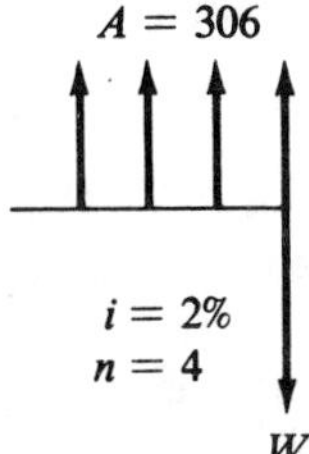

$$W = A(F/A,i,n) = 306(F/A,2\%,4) = 306(4.122)$$
$$= \$1260$$

Ex. 4-21, Solution Two:

Compute an effective i for the time period between withdrawals. Between withdrawals, W, there are four interest periods, hence $m = 4$ compounding subperiods per year. Since the nominal interest rate per year, r, is 8%, we can proceed to compute the effective interest rate per year.

$$\text{Effective interest rate } i_{eff} = \left(1 + \frac{r}{m}\right)^m - 1 = \left(1 + \frac{0.08}{4}\right)^4 - 1$$
$$= 0.0824 = 8.24\% \text{ per year}$$

Now the problem may be redrawn as:

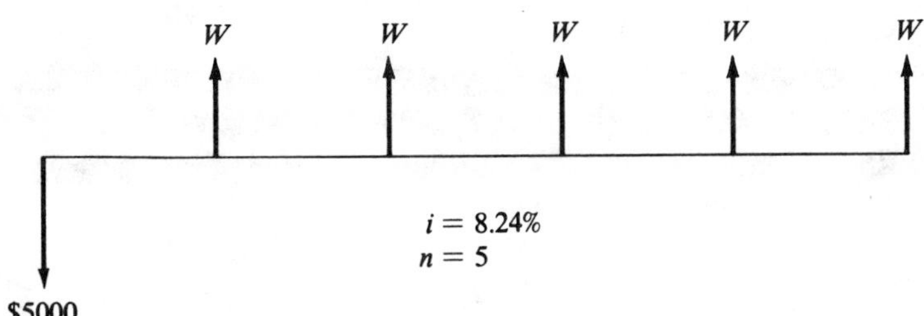

This diagram may be directly solved to determine the annual withdrawal W using the capital recovery factor:

$$W = P(A/P,i,n) = 5000(A/P,8.24\%,5)$$
$$= P\left[\frac{i(1+i)^n}{(1+i)^n - 1}\right] = 5000\left[\frac{0.0824(1+0.0824)^5}{(1+0.0824)^5 - 1}\right]$$
$$= 5000(0.2520) = \$1260$$

She should withdraw \$1260 per year. ◀

Continuous Compounding

Two variables we have introduced are:

r = Nominal interest rate per interest period
m = Number of compounding subperiods per time period

Since the interest period is normally one year, the definitions become:

r = Nominal interest rate per year
m = Number of compounding subperiods per year
$\frac{r}{m}$ = Interest rate per interest period
mn = Number of interest periods in n years

Single Payment Interest Factors–Continuous Compounding

The single payment compound amount formula (Equation 4-1)

$$F = P(1+i)^n$$

may be rewritten as

$$F = P\left(1 + \frac{r}{m}\right)^{mn}$$

If we increase m, the number of compounding subperiods per year, without limit, m becomes very large and approaches infinity, and r/m becomes very small and approaches zero.

This is the condition of ***continuous compounding***, that is, where the duration of the interest period decreases from some finite duration Δt to an infinitely small duration dt, and the number of interest periods per year becomes infinite. In this situation of continuous compounding:

$$F = P \lim_{m\to\infty}\left(1 + \frac{r}{m}\right)^{mn} \qquad (4\text{-}31)$$

An important limit in calculus is:

$$\lim_{x \to 0} (1 + x)^{1/x} = 2.71828 = e \qquad (4\text{-}32)$$

If we set $x = r/m$, then mn may be written as $(1/x)(rn)$. As m becomes infinite, x becomes 0. Equation 4-31 becomes

$$F = P[\lim_{x \to 0} (1 + x)^{1/x}]^{rn}$$

Equation 4-32 tells us the quantity within the brackets equals e, so Eq. 4-1

$$F = P(1 + i)^n \quad \text{becomes} \quad F = Pe^{rn} \qquad (4\text{-}33)$$

and $P = F(1 + i)^{-n}$ becomes $P = Fe^{-rn}$ (4-34)

We see that for continuous compounding,

$$(1 + i) = e^r$$

or, **Effective interest rate per year, $i_{eff} = e^r - 1$** (4-30)

Single Payment–Continuous Compounding

Compound Amount $F = P(e^{rn}) = P(F/P,r,n)$ (4-35)

Present Worth $P = F(e^{-rn}) = F(P/F,r,n)$ (4-36)

If your hand calculator does not have e^x, use the table of e^{rn} and e^{-rn} provided at the end of the Compound Interest Tables.

EXAMPLE 4-22

If you were to deposit \$2000 in a bank that pays 5% nominal interest, compounded continuously, how much would be in the account at the end of two years?

Solution: The single payment compound amount equation for continuous compounding is

$$F = Pe^{rn} \quad \text{where } r = \text{nominal interest rate} = 0.05$$
$$n = \text{number of years} = 2$$

$$F = 2000e^{(0.05 \times 2)} = 2000(1.1052) = \$2210.40$$

There would be \$2210.40 in the account at the end of two years. ◀

EXAMPLE 4-23

A bank offers to sell savings certificates that will pay the purchaser \$5000 at the end of ten years but will pay nothing to the purchaser in the meantime. If interest is computed at 6%, compounded continuously, at what price is the bank selling the certificates?

Solution:

$$P = Fe^{-rn} \quad \text{where } F = \$5000;\ r = 0.06;\ n = 10 \text{ years}$$

$$P = 5000e^{-(0.06 \times 10)} = 5000(0.5488) = \$2744$$

Therefore, the bank is selling the \$5000 certificates for \$2744. ◀

EXAMPLE 4-24

How long will it take for money to double at 10% nominal interest, compounded continuously?

$$F = Pe^{rn}$$

$$2 = 1e^{(0.10)n}$$

$$e^{0.10n} = 2$$

$$\text{or,}\quad 0.10n = \ln 2 = 0.693$$

$$n = 6.93 \text{ years}$$

It will take 6.93 years for money to double at 10% nominal interest. ◀

EXAMPLE 4-25

If the savings bank in Example 4-19 changed its interest policy to 6% interest, compounded continuously, what are the nominal and the effective interest rates?

Solution: The nominal interest rate remains at 6%.

$$\text{Effective interest rate} = e^{r} - 1$$

$$= e^{0.06} - 1 = 0.0618$$

$$= 6.18\%$$ ◀

Uniform Payment Series–Continuous Compounding At Nominal Rate *r* per Period

If we substitute the equation $i = e^{r} - 1$ into the equations for periodic compounding, we get:

Continuous Compounding Sinking Fund:

$$(A/F,r,n) = \frac{e^{r} - 1}{e^{rn} - 1} \tag{4-37}$$

Continuous Compounding Capital Recovery:

$$(A/P,r,n) = \frac{e^{rn}(e^{r} - 1)}{e^{rn} - 1} \tag{4-38}$$

Continuous Compounding Series Compound Amount:

$$(F/A,r,n) = \frac{e^{rn} - 1}{e^{r} - 1} \tag{4-39}$$

Continuous Compounding Series Present Worth:

$$(P/A,r,n) = \frac{e^{rn} - 1}{e^{rn}(e^r - 1)} \tag{4-40}$$

EXAMPLE 4-26

In Example 4-6, a man deposited \$500 per year into a credit union that paid 5% interest, compounded annually. At the end of five years, he had \$2763 in the credit union. How much would he have if they paid 5% nominal interest, compounded continuously?

Solution: $A = \$500 \qquad r = 0.05 \qquad n = 5 \text{ years}$

$$F = A(F/A,r,n) = A\left(\frac{e^{rn} - 1}{e^r - 1}\right) = 500\left(\frac{e^{0.05(5)} - 1}{e^{0.05} - 1}\right)$$

$$= \$2769.84$$

He would have \$2769.84. ◀

EXAMPLE 4-27

In Ex. 4-7, Jim Hayes wished to save a uniform amount each month so he would have \$1000 at the end of one year. Based on 6% nominal interest, compounded monthly, he had to deposit \$81.10 per month. How much would he have to deposit if his credit union paid 6% nominal interest, compounded continuously?

Solution: The deposits are made monthly; hence, there are twelve compounding subperiods in the one-year time period.

$$F = \$1000 \qquad r = \text{Nominal interest rate/interest period} = \frac{0.06}{12} = 0.005$$

$n = 12$ compounding subperiods in the one-year period of the problem

$$A = F(A/F,r,n) = F\left(\frac{e^r - 1}{e^{rn} - 1}\right) = 1000\left(\frac{e^{0.005} - 1}{e^{0.005(12)} - 1}\right)$$

$$= 1000\left(\frac{0.005013}{0.061837}\right) = \$81.07$$

He would have to deposit \$81.07 per month. Note that the difference between monthly and continuous compounding is just three cents per month. ◀

Continuous, Uniform Cash Flow (One Period) With Continuous Compounding At Nominal Interest Rate *r*

Equations for a continuous, uniform cash flow during one period only, with continuous compounding, can be derived as follows. Let the continuous, uniform cash flow $\overline{P}$ be distributed over m subperiods within one period ($n = 1$). Thus

$\bar{P}/m$ is the cash flow at the end of each subperiod. Since the nominal interest rate per period is r, the effective interest rate per subperiod is r/m. Substituting these values into the uniform series compound amount equation (Eq. 4-7) gives:

$$F = \frac{\bar{P}}{m}\left[\frac{[1 + (r/m)]^m - 1}{r/m}\right] \tag{4-41}$$

Setting $x = r/m$, we obtain:

$$F = \frac{\bar{P}}{m}\left[\frac{(1 + x)^{r/x} - 1}{r/m}\right] = \bar{P}\left[\frac{[(1 + x)^{1/x}]^r - 1}{r}\right] \tag{4-42}$$

As m increases, x approaches zero. Equation 4-32 says

$$\lim_{x \to 0} (1 + x)^{1/x} = e$$

hence Eq. 4-42 for one period becomes

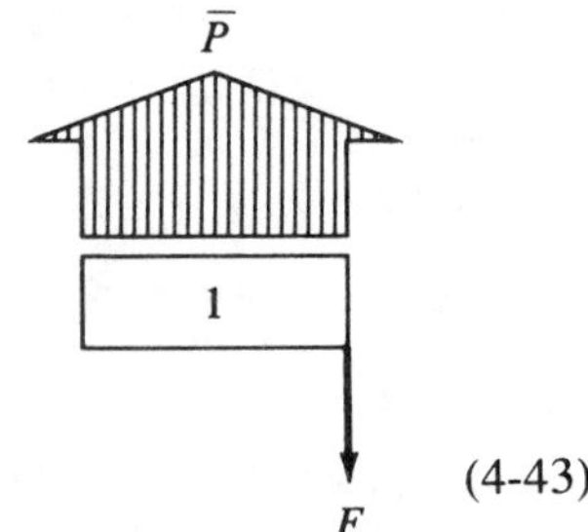

$$F = \bar{P}\left(\frac{e^r - 1}{r}\right) \tag{4-43}$$

Multiplying Eq. 4-43 by the single payment–continuous compounding factors:

For Any Future Time:

$$F = \bar{P}\left(\frac{e^r - 1}{r}\right)(e^{r(n-1)}) = \bar{P}\left[\frac{(e^r - 1)(e^{rn})}{re^r}\right]$$

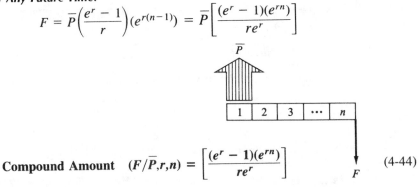

Compound Amount $\quad (F/\bar{P},r,n) = \left[\frac{(e^r - 1)(e^{rn})}{re^r}\right]$ (4-44)

For Any Present Time:

$$P = \bar{F}\left(\frac{e^r - 1}{r}\right)(e^{-rn}) = \bar{F}\left[\frac{e^r - 1}{re^{rn}}\right]$$

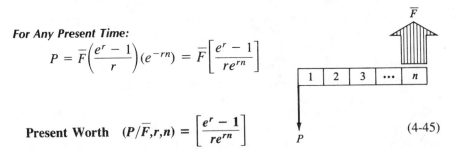

Present Worth $\quad (P/\bar{F},r,n) = \left[\frac{e^r - 1}{re^{rn}}\right]$ (4-45)

EXAMPLE 4-28

A self-service gasoline station has been equipped with an automatic teller machine (ATM). Customers may obtain gasoline simply by inserting their ATM bank card into the machine and filling their car with gasoline. When they have finished, the ATM unit automatically deducts the gasoline purchase from the customer's bank account and credits it to the gas station's bank account. The gas station receives $40,000 per month in this manner with the cash flowing uniformly throughout the month. If the bank pays 9% nominal interest, compounded continuously, how much will be in the gasoline station bank account at the end of the month?

Solution:

$\overline{P}$

1

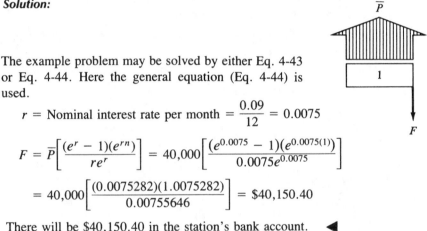

F

The example problem may be solved by either Eq. 4-43 or Eq. 4-44. Here the general equation (Eq. 4-44) is used.

$$r = \text{Nominal interest rate per month} = \frac{0.09}{12} = 0.0075$$

$$F = \overline{P}\left[\frac{(e^r - 1)(e^{rn})}{re^r}\right] = 40{,}000\left[\frac{(e^{0.0075} - 1)(e^{0.0075(1)})}{0.0075e^{0.0075}}\right]$$

$$= 40{,}000\left[\frac{(0.0075282)(1.0075282)}{0.00755646}\right] = \$40{,}150.40$$

There will be $40,150.40 in the station's bank account. ◀

SUMMARY

In this chapter, the concepts of the time value of money and equivalence are described in detail. Also, the various compound interest formulas are derived. It is essential that these concepts and the use of the interest formulas be carefully understood, as the remainder of this book is based upon them.

Time Value of Money. The continuing offer of banks to pay interest for the temporary use of other people's money is ample proof that there is a time value of money. Thus, we would always choose to receive $100 today rather than the promise of $100 to be paid at a future date.

Equivalence. What sum would a person be willing to accept a year hence instead of $100 today? If a 9% interest rate is considered to be appropriate, he would require $109 a year hence. If $100 today and $109 a year hence are considered equally desirable, we say the two sums of money are *equivalent.* But, if on further consideration, we decided that a 12% interest rate is applicable, then $109 a year hence would no longer be equivalent to $100 today. This illustrates that equivalence is dependent on the interest rate.

Compound Interest. The notation used is:

i = Effective interest rate per interest period*.

n = Number of interest periods.

P = A present sum of money.

F = A future sum of money. The future sum F is an amount, n interest periods from the present, that is equivalent to P with interest rate i.

A = An end-of-period cash receipt or disbursement in a uniform series continuing for n periods, the entire series equivalent to P or F at interest rate i.

G = Uniform period-by-period increase or decrease in cash receipts or disbursements; the arithmetic gradient.

g = Uniform *rate* of cash flow increase or decrease from period to period; the geometric gradient.

r = Nominal interest rate per interest period*.

m = Number of compounding subperiods per period*.

$\overline{P}, \overline{F}$ = Amount of money flowing continuously and uniformly during one given period.

Single Payment Formulas.

Compound amount:

$$F = P(1 + i)^n = P(F/P,i,n)$$

Present worth:

$$P = F(1 + i)^{-n} = F(P/F,i,n)$$

Uniform Series Formulas.

Compound amount:

$$F = A\left[\frac{(1 + i)^n - 1}{i}\right] = A(F/A,i,n)$$

Sinking fund:

$$A = F\left[\frac{i}{(1 + i)^n - 1}\right] = F(A/F,i,n)$$

*Normally the interest period is one year, but it could be something else.

Capital recovery:

$$A = P\left[\frac{i(1+i)^n}{(1+i)^n - 1}\right] = P(A/P,i,n)$$

Present worth:

$$P = A\left[\frac{(1+i)^n - 1}{i(1+i)^n}\right] = A(P/A,i,n)$$

Arithmetic Gradient Formulas.

Arithmetic gradient present worth:

$$P = G\left[\frac{(1+i)^n - in - 1}{i^2(1+i)^n}\right] = G(P/G,i,n)$$

Arithmetic gradient uniform series:

$$A = G\left[\frac{(1+i)^n - in - 1}{i(1+i)^n - i}\right] = G(A/G,i,n)$$

Geometric Gradient Formulas.

Geometric series present worth, where $i \neq g$:

$$P = A_1\left[\frac{1 - (1+g)^n(1+i)^{-n}}{i - g}\right] = A_1(P/A,g,i,n)$$

Geometric series present worth, where $i = g$:

$$P = A_1[n(1+i)^{-1}] = A_1(P/A,g,i,n)$$

Single Payment Formulas–Continuous Compounding at nominal rate r per period.

Compound amount:

$$F = P[e^{rn}] = P(F/P,r,n)$$

Present worth:

$$P = F[e^{-rn}] = F(P/F,r,n)$$

Uniform Payment Series–Continuous Compounding at nominal rate r per period.

Continuous compounding sinking fund:

$$A = F\left[\frac{e^r - 1}{e^{rn} - 1}\right] = F(A/F,r,n)$$

Continuous compounding capital recovery:

$$A = P\left[\frac{e^{rn}(e^r - 1)}{e^{rn} - 1}\right] = P(A/P,r,n)$$

Continuous compounding series compound amount:

$$F = A\left[\frac{e^{rn} - 1}{e^r - 1}\right] = A(F/A,r,n)$$

Continuous compounding series present worth:

$$P = A\left[\frac{e^{rn} - 1}{e^{rn}(e^r - 1)}\right] = A(P/A,r,n)$$

Continuous, Uniform Cash Flow (One Period) with Continuous Compounding at Nominal Interest Rate r.

Compound amount:

$$F = \overline{P}\left[\frac{(e^r - 1)(e^{rn})}{re^r}\right] = \overline{P}(F/\overline{P},r,n)$$

Present worth:

$$P = \overline{F}\left[\frac{e^r - 1}{re^{rn}}\right] = \overline{F}(P/\overline{F},r,n)$$

Nominal interest rate per year, r: The annual interest rate without considering the effect of any compounding.

Effective interest rate per year, i_{eff}: The annual interest rate taking into account the effect of any compounding during the year.

Effective interest rate per year (periodic compounding):

$$= \left(1 + \frac{r}{m}\right)^m - 1 \quad \text{or,}$$

$$= (1 + i)^m - 1$$

Effective interest rate per year (continuous compounding):

$$= e^r - 1$$

Problems

4-1 Solve Diagrams (a)–(d) below for the unknowns *Q, R, S,* and *T,* assuming a 10% interest rate.

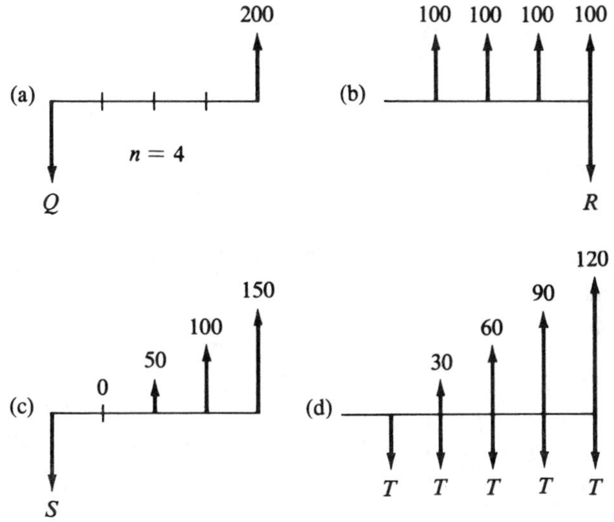

(*Answers:* Q = \$136.60; R = \$464.10; S = \$218.90; T = \$54.30)

4-2 A man borrowed \$500 from a bank on October 15th. He must repay the loan in 16 equal monthly payments, due on the 15th of each month, beginning November 15th. If interest is computed at 1% per month, how much must he pay each month?
(*Answer:* \$33.95)

4-3 A local finance company will loan \$10,000 to a homeowner. It is to be repaid in 24 monthly payments of \$499.00 each. The first payment is due thirty days after the \$10,000 is received. What interest rate per month are they charging? (*Answer:* 1½%)

4-4 For Diagrams (a)–(d) below, compute the unknown values: *B, i, V, x,* respectively.

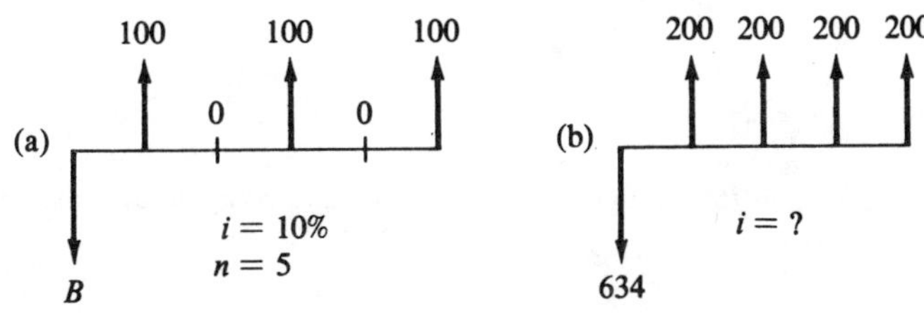

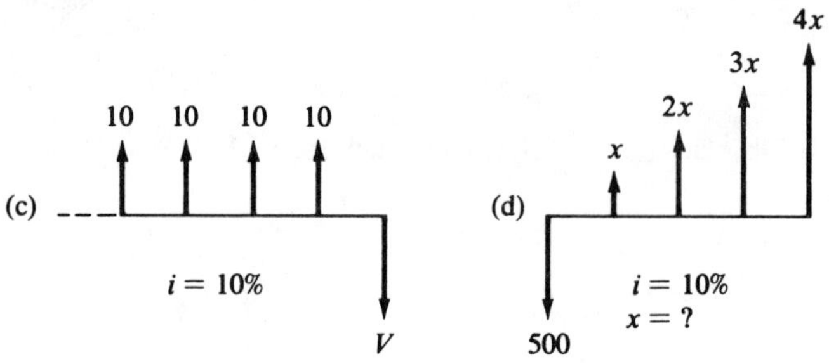

(*Answers:* B = \$228.13; i = 10%; V = \$51.05; x = \$66.24)

4-5 For Diagrams (a)–(d) below, compute the unknown values: C, i, F, A, respectively.

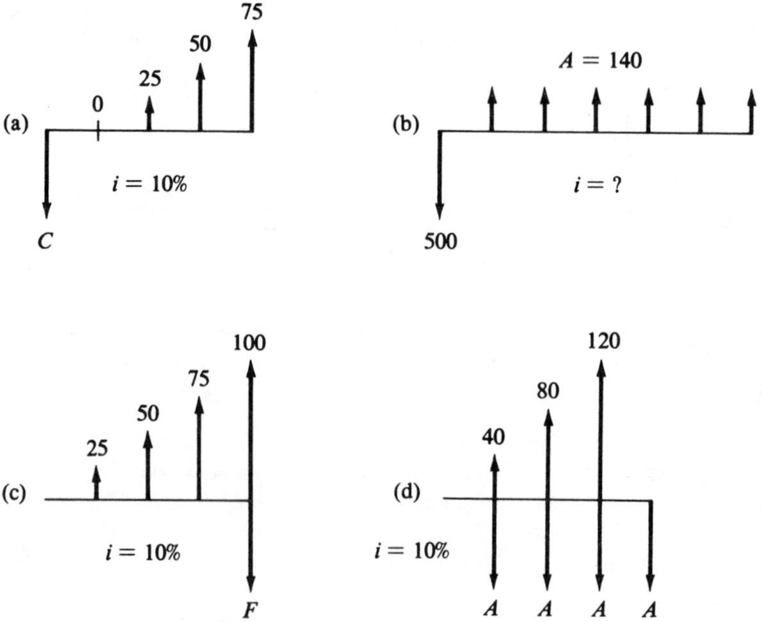

(*Answers:* C = \$109.45; i = 17.24%; F = \$276.37; A = \$60.78)

4-6 For Diagrams (a) through (d) on the next page, compute the unknown values: W, X, Y, Z, respectively.

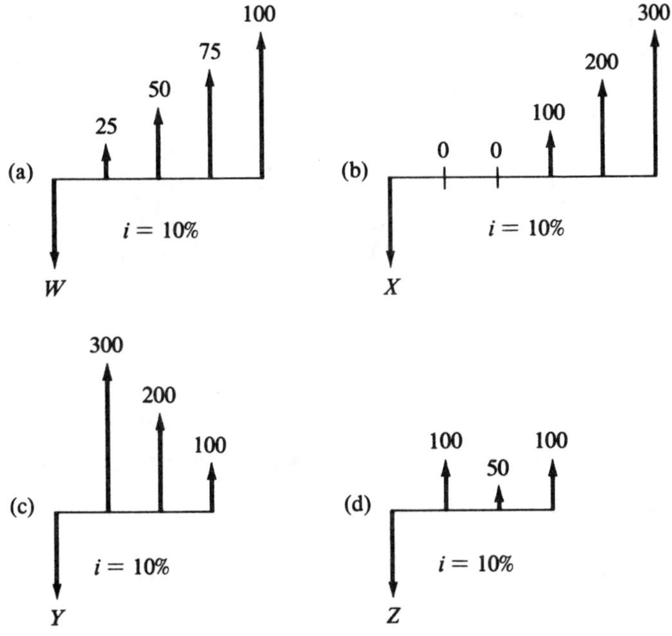

4-7 Using linear interpolation, determine the value of $(P/A,6\frac{1}{2}\%,10)$ from the Compound Interest Tables. Compute this same value using the equation. Why do the values differ?

4-8 Four plans have been presented for the repayment of \$5000 in five years with interest at 8%. Still another way to repay the \$5000 would be to make four annual end-of-year payments of \$1000 each, followed by a final payment at the end of the fifth year. How much would the final payment be?

4-9 Compute the value of P in the diagram below:

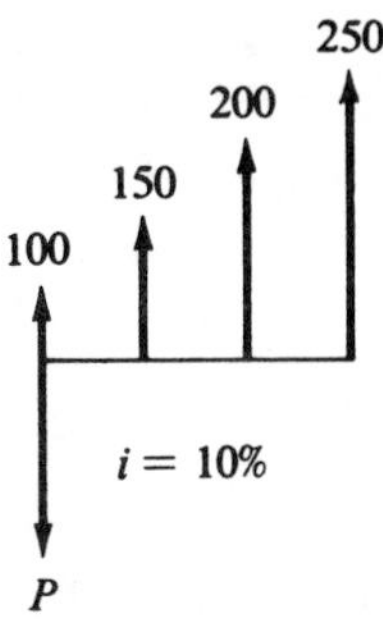

(*Answer:* \$589.50)

4-10 Compute the value of X in the diagram below using a 10% interest rate.

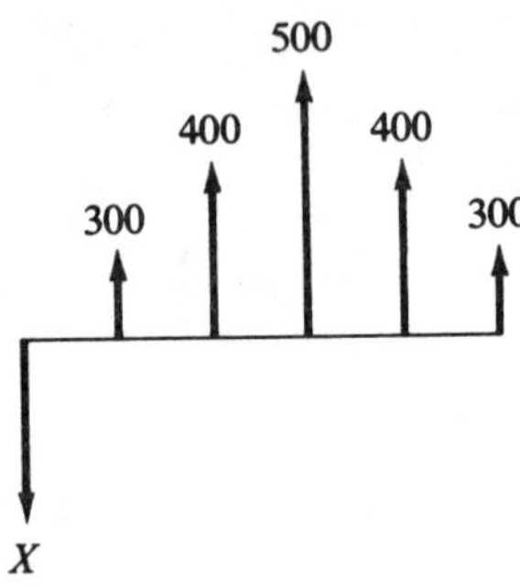

4-11 The following series of payments will repay a present sum of \$5000 at an 8% interest rate.

Year	*End-of-year payment*
1	\$1400
2	1320
3	1240
4	1160
5	1080

What present sum is equivalent to this series of payments at a 10% interest rate?

4-12 Compute the value of P in the diagram below using a 15% interest rate.

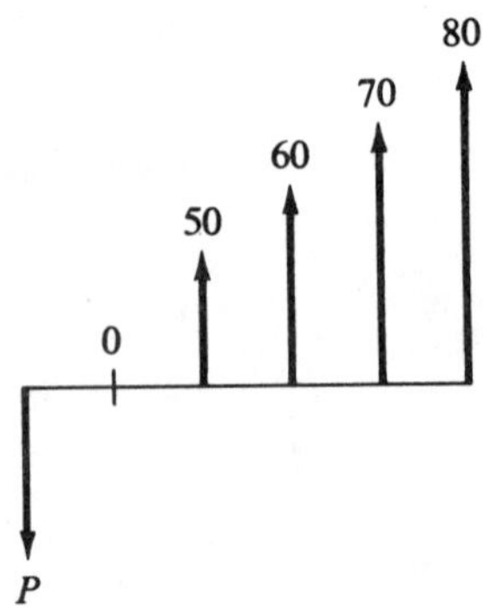

4-13 A man went to his bank and borrowed \$750. He agreed to repay the sum at the end of three years, together with the interest at 8% per year, compounded annually. How much will he owe the bank at the end of three years? (*Answer:* \$945)

4-14 What sum of money now is equivalent to \$8250 two years hence, if interest is 8% per annum, compounded semi-annually? (*Answer:* \$7052)

4-15 The local bank offers to pay 5% interest, compounded annually, on savings deposits. In a nearby town, the bank pays 5% interest, compounded quarterly. A man who has \$3000 to put in a savings account wonders if the increased interest paid in the nearby town justifies driving his car there to make the deposit. Assuming he will leave all money in the account for two years, how much additional interest would he obtain from the out-of-town bank over the local bank?

4-16 A sum of money invested at 4% interest, compounded semi-annually, will double in amount in approximately how many years? (*Answer:* 17½ years)

4-17 A local bank will lend a customer \$1000 on a two-year car loan as follows:

Money to pay for car	\$1000
Two years' interest at 7%: 2 × 0.07 × 1000	140
	\$1140

$$24 \text{ monthly payments} = \frac{1140}{24} = \$47.50$$

The first payment must be made in thirty days. What is the nominal annual interest rate the bank is receiving?

4-18 A local lending institution advertises the "51–50 Club." A person may borrow \$2000 and repay \$51 for the next fifty months, beginning thirty days after receiving the money. Compute the nominal annual interest rate for this loan. What is the effective interest rate?

4-19 A loan company has been advertising on television a plan where one may borrow \$1000 and make a payment of \$10.87 per month. This payment is for interest only and includes no payment on the principal. What is the nominal annual interest rate that they are charging?

4-20 What effective interest rate per annum corresponds to a nominal rate of 12% compounded monthly? (*Answer:* 12.7%)

4-21 Mr. Sansome withdrew \$1000 from a savings account and invested it in common stock. At the end of five years, he sold the stock and received a check for \$1307. If Mr. Sansome had left his \$1000 in the savings account, he would have received an interest rate of 5%, compounded quarterly. Mr. Sansome would like to compute a comparable interest rate on his common stock investment. Based on quarterly compounding, what nominal annual interest rate did Mr. Sansome receive on his investment in stock? What effective annual interest rate did he receive?

4-22 A woman opened an account in a local store. In the charge account agreement, the store indicated it charges 1½% each month on the unpaid balance. What nominal annual interest rate is being charged? What is the effective interest rate?

4-23 A man buys a car for $3000 with no money down. He pays for the car in thirty equal monthly payments with interest at 12% per annum, compounded monthly. What is his monthly loan payment? (*Answer:* $116.10)

4-24 What amount will be required to purchase, on a man's 40th birthday, an annuity to provide him with thirty equal semi-annual payments of $1000 each, the first to be received on his 50th birthday, if nominal interest is 4% compounded semi-annually?

4-25 Upon the birth of his first child, Dick Jones decided to establish a savings account to partly pay for his son's education. He plans to deposit $20 per month in the account, beginning when the boy is 13 months old. The savings and loan association has a current interest policy of 6% per annum, compounded monthly, paid quarterly. Assuming no change in the interest rate, how much will be in the savings account when Dick's son becomes sixteen years old?

4-26 An engineer borrowed $3000 from the bank, payable in six equal end-of-year payments at 8%. The bank agreed to reduce the interest on the loan if interest rates declined in the United States before the loan was fully repaid. At the end of three years, at the time of the third payment, the bank agreed to reduce the interest rate from 8% to 7% on the remaining debt. What was the amount of the equal annual end-of-year payments for each of the first three years? What was the amount of the equal annual end-of-year payments for each of the last three years?

4-27 On a new car, it is estimated that the maintenance cost will be $40 the first year. Each subsequent year, it is expected to be $10 more than the previous one. How much would you need to set aside when you bought a new car to pay all future maintenance costs if you planned to keep it seven years? Assume interest is 5% per annum. (*Answer:* $393.76)

4-28 A man decides to deposit $50 in the bank today and make ten additional deposits every six months beginning six months from now, the first of which will be $50 and increasing $10 per deposit after that. A few minutes after he makes the last deposit, he decides to withdraw all the money deposited. If the bank pays 6% nominal interest compounded semi-annually, how much money will he receive?

4-29 A young engineer wishes to become a millionaire by the time he is sixty years old. He believes that by careful investment he can obtain a 15% rate of return. He plans to add a uniform sum of money to his investment program each year, beginning on his 20th birthday and continuing through his 59th birthday. How much money must the engineer set aside in this project each year?

4-30 The councilmembers of a small town have decided that the earth levee that protects the town from a nearby river should be rebuilt and strengthened. The town engineer estimates that the cost of the work at the end of the first year will be $85,000. He estimates that in subsequent years the annual repair costs will decline by $10,000, making the second-year cost $75,000; the third-year $65,000, and so forth. The councilmembers want to know what the equivalent present cost is for the first five years of repair work if interest is 4%. (*Answer:* $292,870)

4-31 The Apex Company sold a water softener to Marty Smith. The price of the unit was $350. Marty asked for a deferred payment plan, and a contract was written. Under the contract, the buyer could delay paying for the water softener provided that he purchased the coarse salt for re-charging the softener from Apex. At the end of two years, the buyer was to pay for the unit in a lump sum, including 6% interest, compounded quarterly. The contract provided that, if the customer ceased buying salt from Apex at any time prior to two years, the full payment due at the end of two years would automatically become due.

Six months later, Marty decided to buy salt elsewhere and stopped buying from Apex, who thereupon asked for the full payment that was to have been due 18 months hence. Marty was unhappy about this, so Apex offered as an alternative to accept the $350 with interest at 20% per annum compounded semi-annually for the six months that Marty had had the water softener. Which of these alternatives should Marty accept? Explain.

4-32 A $150 bicycle was purchased on December 1st with a $15 down payment. The balance is to be paid at the rate of $10 at the end of each month, with the first payment due on December 31st. The last payment may be some amount less than $10. If interest on the unpaid balance is computed at 1½% per month, how many payments will there be, and what is the amount of the final payment?
(*Answers:* 16 payments; final payment: $1.99)

4-33 A company buys a machine for $12,000, which it agrees to pay for in five equal annual payments, beginning one year after the date of purchase, at an interest rate of 4% per annum. Immediately after the second payment, the terms of the agreement are changed to allow the balance due to be paid off in a single payment the next year. What is the final single payment? (*Answer:* $7778)

4-34 An engineering student bought a car at a local used car lot. Including tax and insurance, the total price was $3000. He is to pay for the car in twelve equal monthly payments, beginning with the first payment immediately (in other words, the first payment was the down payment). Nominal interest on the loan is 12%, compounded monthly. After he makes six payments (the down payment plus five additional payments), he decides to sell the car. A buyer agrees to pay a cash amount to pay off the loan in full at the time the next payment is due and also to pay the engineering student $1000. If there are no penalty charges for this early payment of the loan, how much will the car cost the new buyer?

4-35 A bank recently announced an "instant cash" plan for holders of its bank credit cards. A cardholder may receive cash from the bank up to a pre-set limit (about $500). There is a special charge of 4% made at the time the "instant cash" is sent the cardholder. The debt may be repaid in monthly installments. Each month the bank charges 1½% on the unpaid balance. The monthly payment, including interest, may be as little as $10. Thus, for $150 of "instant cash," an initial charge of $6 is made and added to the balance due. Assume the cardholder makes a monthly payment of $10 (this includes both principal and interest). How many months are required to repay the debt? If your answer includes a fraction of a month, round up to the next month.

4-36 The treasurer of a firm noted that many invoices were received by his firm with the following terms of payment:

"2%–10 days, net 30 days"

Thus, if he were to pay the bill within ten days of its date, he could deduct 2%. On the other hand, if he did not promptly pay the bill, the full amount would be due thirty days from the date of the invoice. Assuming a 20-day compounding period, the 2% deduction for prompt payment is equivalent to what effective interest rate per year?

4-37 In 1555, King Henry borrowed money from his bankers on the condition that he pay 5% of the loan at each fair (there were four fairs per year) until he had made forty payments. At that time the loan would be considered repaid. What effective annual interest did King Henry pay?

4-38 A man wants to help provide a college education for his young daughter. He can afford to invest \$600/yr for the next four years, beginning on the girl's fourth birthday. He wishes to give his daughter \$4000 on her 18th, 19th, 20th, and 21st birthdays, for a total of \$16,000. Assuming 5% interest, what uniform annual investment will he have to make on the girl's 8th through 17th birthdays? (*Answer:* \$792.73)

4-39 A man has \$5000 on deposit in a bank that pays 5% interest compounded annually. He wonders how much more advantageous it would be to transfer his funds to another bank whose dividend policy is 5% interest, compounded continuously. Compute how much he would have in his savings account at the end of three years under each of these situations.

4-40 A friend was left \$50,000 by his uncle. He has decided to put it into a savings account for the next year or so. He finds there are varying interest rates at savings institutions: 4⅜% compounded annually, 4¼% compounded quarterly, and 4⅛% compounded continuously. He wishes to select the savings institution that will give him the highest return on his money. What interest rate should he select?

4-41 One of the local banks indicates that it computes the interest it pays on savings accounts by the continuous compounding method. Suppose you deposited \$100 in the bank and they pay 4% per annum, compounded continuously. After five years, how much money will there be in the account?

4-42 A company expects to install smog control equipment on the exhaust of a gasoline engine. The local smog control district has agreed to pay to the firm a lump sum of money to provide for the first cost of the equipment and maintenance during its ten-year useful life. At the end of ten years the equipment, which initially cost \$10,000, is valueless. The company and smog control district have agreed that the following are reasonable estimates of the end-of-year maintenance costs:

Year		Year	
1	\$500	6	\$200
2	100	7	225
3	125	8	250
4	150	9	275
5	175	10	300

Assuming interest at 6% per year, how much should the smog control district pay to the company now to provide for the first cost of the equipment and its maintenance for ten years? (*Answer:* $11,693)

4-43 One of the largest automobile dealers in the city advertises a three-year-old car for sale as follows:

> Cash price $3575, or a down payment of $375 with 45 monthly payments of $93.41.

Susan DeVaux bought the car and made a down payment of $800. The dealer charged her the same interest rate used in his advertised offer. How much will Susan pay each month for 45 months? What effective interest rate is being charged?
(*Answers:* $81.03; 16.1%)

4-44 At the Central Furniture Company, customers who purchase on credit pay an effective annual interest rate of 16.1%, based on monthly compounding. What is the nominal annual interest rate that they pay?

4-45 Mary Lavor plans to save money at her bank for use in December. She will deposit $30 a month, beginning on March 1st and continuing through November 1st. She will withdraw all the money on December 1st. If the bank pays ½% interest each month, how much money will she receive on December 1st?

4-46 A man makes an investment every three months at a nominal annual interest rate of 28%, compounded quarterly. His first investment was $100, followed by investments *increasing* $20 each three months. Thus, the second investment was $120, the third investment $140, and so on. If he continues to make this series of investments for a total of twenty years, what will be the value of the investments at the end of that time?

4-47 A debt of $5000 can be repaid, with interest at 8%, by the following payments.

Year	*Payment*
1	$ 500
2	1000
3	1500
4	2000
5	*X*

The payment at the end of the fifth year is shown as *X*. How much is *X*?

4-48 Consider the cash flow:

Year	*Cash flow*
0	−$100
1	+50
2	+60
3	+70
4	+80
5	+140

Which one of the following is correct for this cash flow?

a. $100 = 50 + 10(A/G,i,5) + 50(P/F,i,5)$

b. $\dfrac{50(P/A,i,5) + 10(P/G,i,5) + 50(P/F,i,5)}{100} = 1$

c. $100(A/P,i,5) = 50 + 10(A/G,i,5)$

d. None of the equations are correct.

4-49 If $i = 12\%$, what is the value of B in the diagram below?

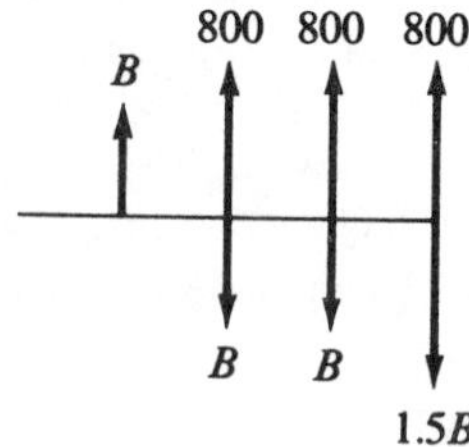

4-50 For a 10% interest rate, compute the value of n for the figure below.

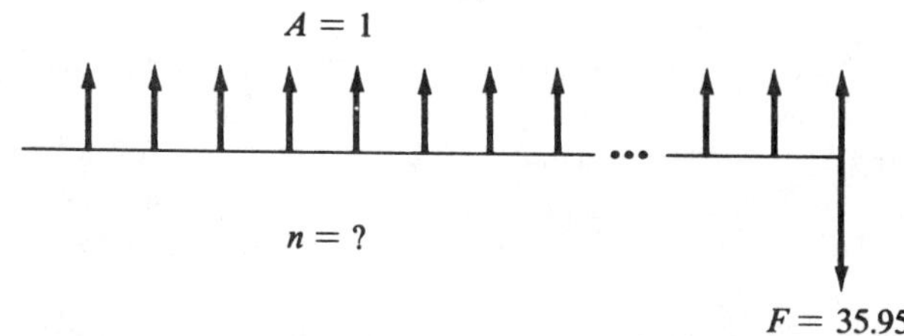

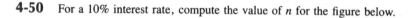

4-51 For the figure below, what is the value of n, based on a 3½% interest rate?

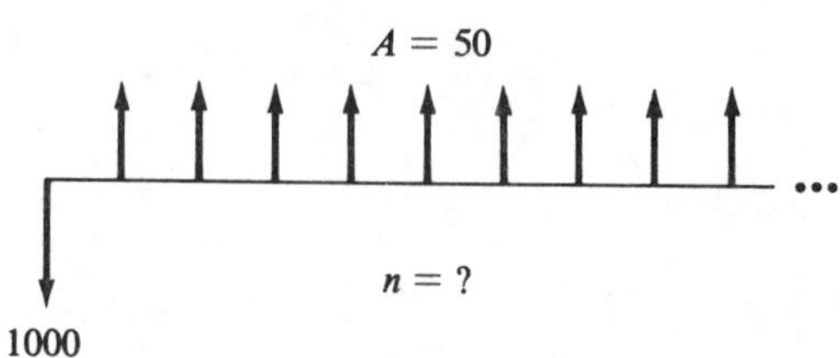

4-52 How many months will it take to pay off a \$525 debt, with monthly payments of \$15 at the end of each month if the interest rate is 18%, compounded monthly? (*Answer:* 50 months)

4-53 For the diagram below and a 10% interest rate, compute the value of J.

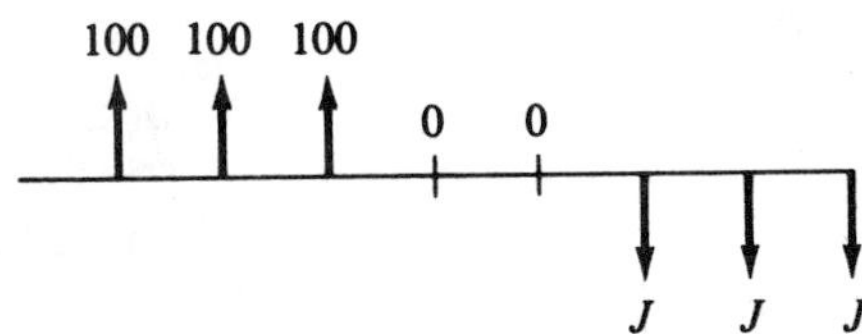

4-54 For the diagram below and a 10% interest rate, compute the value of C.

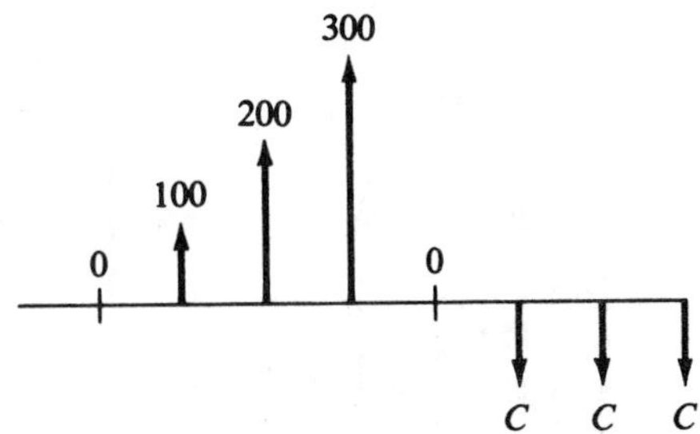

4-55 On January 1st, Frank Jenson bought a used car for $4200 and agreed to pay for it as follows: ⅓ down payment; the balance to be paid in 36 equal monthly payments; the first payment due February 1st; an annual interest rate of 9%, compounded monthly.

a. What is the amount of Frank's monthly payment?

b. During the summer, Frank made enough money that he decided to pay off the entire balance due on the car as of October 1st. How much did Frank owe on October 1st?

4-56 If $i = 12\%$, compute G in the diagram below:

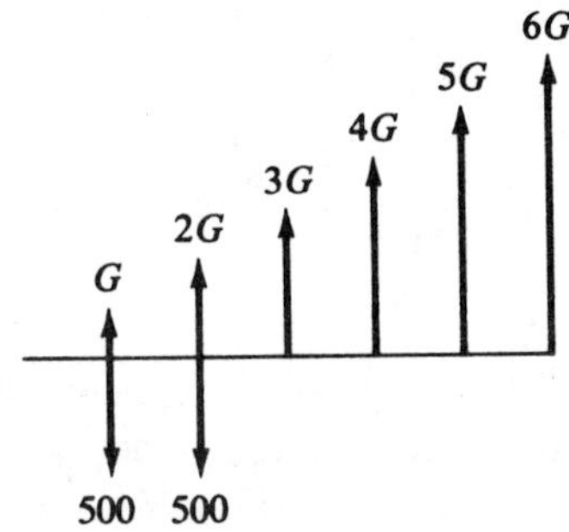

4-57 Compute the value of D in the diagram below:

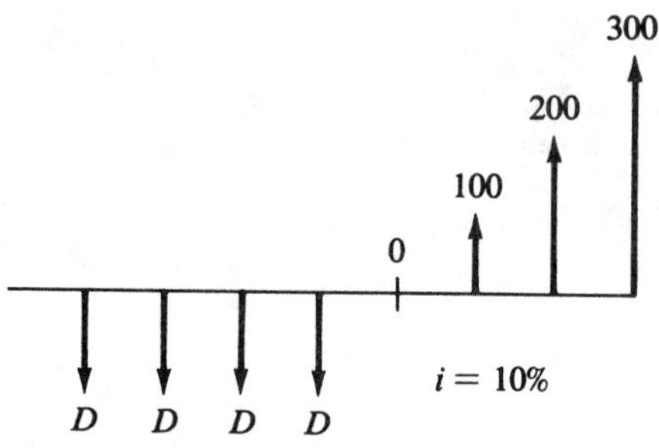

4-58 Compute E for the figure below:

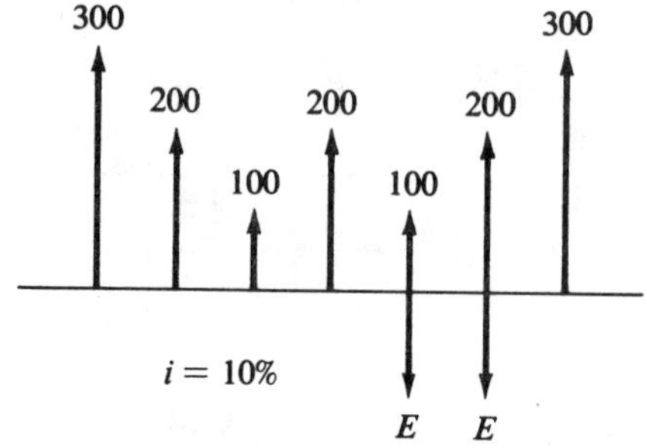

4-59 Compute B in the figure below, using a 10% interest rate.

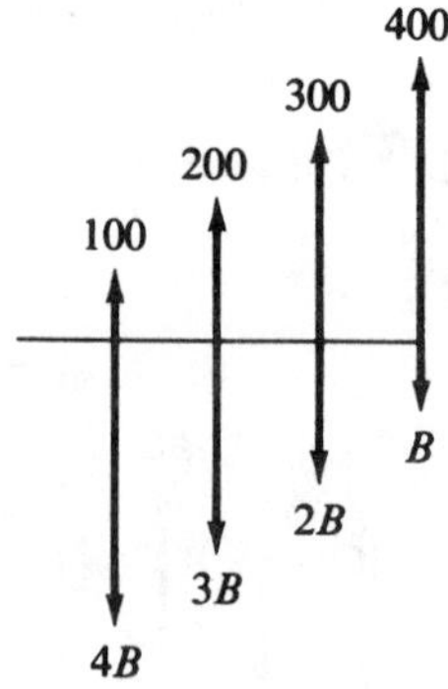

4-60 On January 1st, Laura Brown borrowed \$1000 from the Friendly Finance Company. The loan is to be repaid by four equal payments which are due at the end of March,

June, September, and December. If the finance company charges 18% interest, compounded quarterly, what is the amount of each payment? What is the effective annual interest rate? (*Answers:* $278.70; 19.3%)

4-61 If you want a 12% rate of return, continuously compounded, on a project that will yield $6000 at the end of 2½ years, how much must you be willing to invest now? (*Answer:* $4444.80)

4-62 What monthly interest rate is equivalent to an effective annual interest rate of 18%?

4-63 A department store charges 1¾% interest per month, compounded continuously, on its customer's charge accounts. What is the nominal annual interest rate? What is the effective interest rate? (*Answers:* 21%; 23.4%)

4-64 A bank is offering to sell six-month certificates of deposit for $9500. At the end of six months, the bank will pay $10,000 to the certificate owner. Based on a six-month interest period, compute the nominal annual interest rate and the effective annual interest rate.

4-65 Two savings banks are located across the street from each other. The West Bank put a sign in the window saying, "We pay 6.50%, compounded daily." The East Bank decided that they would do better, so they put a sign in their window saying, "We pay 6.50%, compounded continuously."

Jean Silva has $10,000 which she will put in the bank for one year. How much additional interest will Jean receive by placing her money in the East Bank rather than the West Bank?

4-66 A bank advertises it pays 7% annual interest, compounded daily, on savings accounts, provided the money is left in the account for four years. What effective annual interest rate do they pay?

4-67 To repay a $1000 loan, a man paid $91.70 at the end of each month for twelve months. Compute the nominal interest rate he paid.

4-68 Linda Dunlop will deposit $1500 in a bank savings account that pays 10% interest per year, compounded daily. How much will Linda have in her account at the end of two and one-half years?

4-69 Sally Struthers wants to have $10,000 in a savings account at the end of six months. The bank pays 8%, compounded continuously. How much should Sally deposit now? (*Answer:* $9608)

4-70 A student bought a $75 used guitar and agreed to pay for it with a single $85 payment at the end of six months. Assuming semi-annual (every six months) compounding, what is the nominal annual interest rate? What is the effective interest rate?

4-71 A firm charges its credit customers 1¾% interest per month. What is the effective interest rate?

4-72 One thousand dollars is invested for seven months at an interest rate of one percent per month. What is the nominal interest rate? What is the effective interest rate?
(*Answers:* 12%; 12.7%)

4-73 The ***Rule of 78's*** is a commonly used method of computing the amount of interest, when the balance of a loan is repaid in advance.

If one adds the numbers representing twelve months,

$$1 + 2 + 3 + 4 + 5 + \cdots + 11 + 12 = 78$$

If a twelve-month loan is repaid at the end of one month, for example, the interest the borrower would be charged is $^{12}\!/_{78}$ of the year's interest. If the loan is repaid at the end of two months, the total interest charged would be $(12 + 11)/78$, or $^{23}\!/_{78}$ of the year's interest. After eleven months the interest charge would therefore be $^{77}\!/_{78}$ of the total year's interest.

Helen Reddy borrowed $10,000 on January 1st at 9% annual interest, compounded monthly. The loan was to be repaid in twelve equal end-of-period payments. Helen made the first two payments and then decided to repay the balance of the loan when she pays the third payment. Thus she will pay the third payment plus an additional sum.

You are to calculate the amount of this additional sum

a. Based on the Rule of 78's;

b. Based on exact economic analysis methods.

4-74 Consider the following cash flow:

Year	*Cash flow*
0	−$ *P*
1	+1000
2	+850
3	+700
4	+550
5	+400
6	+400
7	+400
8	+400

Alice was asked to compute the value of *P* for the cash flow at 8% interest. She wrote the following three equations:

a. $P = 1000(P/A,8\%,8) - 150(P/G,8\%,8) + 150(P/G,8\%,4)(P/F,8\%,4)$

b. $P = 400(P/A,8\%,8) + 600(P/A,8\%,5) - 150(P/G,8\%,4)$

c. $P = 150(P/G,8\%,4) + 850(P/A,8\%,4) + 400(P/A,8\%,4)(P/F,8\%,4)$

Which of the equations is correct?

4-75 For the following cash flow, compute the interest rate at which the $100 cost is equivalent to the subsequent benefits.

Year	*Cash flow*
0	−\$100
1	+25
2	+45
3	+45
4	+30

4-76 Ann Landers deposits \$100 at the end of each month into her bank savings account. The bank pays 6% nominal interest, compounded and paid quarterly. No interest is paid on money not in the account for the full three-month period. How much will be in Ann's account at the end of three years? (*Answer:* \$3912.30)

4-77 A college professor just won \$85,000 in the state lottery. After paying income taxes, about half the money will be left. She and her husband plan to spend her sabbatical year on leave from the university on an around-the-world trip, but she must continue to teach three more years first. She estimates the trip will cost \$40,000 and they will spend the money as a continuous flow of funds during their year of travel. She will put enough of her lottery winnings in a bank account now to pay for the trip. The bank pays 7% nominal interest, compounded continuously. She asks you to compute how much she should set aside in the account for their trip.

4-78 Mark Johnson saves a fixed percentage of his salary at the end of each year. This year he saved \$1500. For the next five years, he expects his salary to increase at an 8% annual rate, and he plans to increase his savings at the same 8% annual rate. He invests his money in the stock market. Thus there will be six end-of-year investments (\$1500 plus five more). Solve the problem using the geometric gradient factor.

a. How much will his investments be worth at the end of six years if they increase in the stock market at a 10% annual rate?

b. How much will Mark have at the end of six years if his stock market investments only increase at an 8% annual rate?

4-79 The *Bawl Street Journal* costs \$206, payable now, for a two-year subscription. The newspaper is published 252 days per year (five days per week, except holidays). If a 10% nominal annual interest rate, compounded quarterly, is used:

a. What is the effective annual interest rate in this problem?

b. Compute the equivalent interest rate per 1/252 of a year.

c. What is a subscriber's cost per copy of the newspaper, taking interest into account?

4-80 Michael Jackson deposited \$500,000 into a bank for six months. At the end of that time, he withdrew the money and received \$520,000. If the bank paid interest based on continuous compounding:

a. What was the effective annual interest rate?

b. What was the nominal annual interest rate?

4-81 The I've Been Moved Corporation receives a constant flow of funds from its worldwide operations. This money (in the form of checks) is continuously deposited in many banks with the goal of earning as much interest as possible for "IBM." One billion dollars is deposited each month, and the money earns an average of ½% interest per month, compounded continuously. Assume all the money remains in the accounts until the end of the month.

a. How much interest does IBM earn each month?

b. How much interest would IBM earn each month if it held the checks and made deposits to its bank accounts just four times a month?

4-82 A married couple is opening an Individual Retirement Account (IRA) at a bank. Their goal is to accumulate $1,000,000 in the account by the time they retire from work in forty years. The bank manager estimates they may expect to receive 8% nominal annual interest, compounded quarterly, throughout the forty years. The couple believe their income will increase at a 7% annual rate during their working careers. They wish to start with as low a deposit as possible to their IRA now and increase it at a 7% rate each year. Assuming end-of-year deposits, how much should they deposit the first year?

4-83 The Macintosh Co. has an employee savings plan in which an employee may invest up to 5% of his or her annual salary. The money is invested in company common stock with the company guaranteeing the annual return will never be less than 8%. Jill was hired at an annual salary of $52,000. She immediately joined the savings plan investing the full 5% of her salary each year. If Jill's salary increases at an 8% uniform rate, and she continues to invest 5% of it each year, what amount of money is she guaranteed to have at the end of 20 years?

4-84 The football coach at a midwest university was given a five-year employment contract which paid $225,000 the first year, and increased at an 8% uniform rate in each subsequent year. At the end of the first year's football season, the alumni demanded that he be fired. The alumni agreed to buy the coach's remaining four years on the contract by paying him the equivalent present sum, computed using a 12% interest rate. How much will the coach receive?

4-85 A group of ten public-spirited citizens has agreed that they will support the local school hot lunch program. Each year one of the group is to pay the $15,000 years' cost that occurs continuously and uniformly during the year. Each member of the group is to underwrite the cost for one year. Slips of paper numbered Year 1 through Year 10 are put in a hat. As one of the group you draw the slip marked Year 6. Assuming an 8% nominal interest rate per year, how much do you need to set aside now to meet your obligation in Year 6?

CHAPTER 5

Present Worth Analysis

In the previous chapter two important tasks were accomplished. First, the concept of equivalence was presented. We are powerless to compare series of cash flows unless we can resolve them into some equivalent arrangement. Second, equivalence, with alteration of cash flows from one series to an equivalent sum or series of cash flows, created the need for compound interest factors. A whole series of compound interest factors were derived—some for periodic compounding and some for continuous compounding. This background sets the stage for the chapters that follow.

Economic Criteria

We have shown how to manipulate cash flows in a variety of ways, and in so doing we can now solve many kinds of compound interest problems. But engineering economic analysis is more than simply solving interest problems. The decision process (see Figure 2-2) requires that the outcomes of feasible alternatives be arranged so that they may be judged for ***economic efficiency*** in terms of the selection criterion. Depending on the situation, the economic criterion will be one of the following:*

Situation	*Criterion*
For fixed input	Maximize output
For fixed output	Minimize input
Neither input nor output fixed	Maximize (output − input)

We will now examine ways to resolve engineering problems, so that criteria for economic efficiency can be applied.

*This short table summarizes the discussion on selection of criteria, early in Chapter 3.

Equivalence provides the logic by which we may adjust the cash flow for a given alternative into some equivalent sum or series. To apply the selection criterion to the outcomes of the feasible alternatives, we must first resolve them into comparable units. The question is, how should they be compared? In this chapter we'll learn how analysis can resolve alternatives into *equivalent present consequences*, referred to simply as ***present worth analysis***. Chapter 6 will show how given alternatives are converted into an *equivalent uniform annual cash flow*, and Chapter 7 solves for the interest rate at which favorable consequences—that is, *benefits*—are equivalent to unfavorable consequences—or *costs*.

As a general rule, any economic analysis problem may be solved by the methods presented in this and in the two following chapters. This is true because *present worth, annual cash flow,* and *rate of return* are exact methods that will always yield the same solution in selecting the best alternative from among a set of mutually exclusive alternatives.* Some problems, however, may be more easily solved by one method than another. For this reason, we now focus on the kinds of problems that are most readily solved by present worth analysis.

Applying Present Worth Techniques

One of the easiest ways to compare mutually exclusive alternatives is to resolve their consequences to the present time. The three criteria for economic efficiency are restated in terms of present worth analysis in Table 5-1.

Present worth analysis is most frequently used to determine the present value of future money receipts and disbursements. It would help us, for example, to determine a present worth of income-producing property, like an oil well or an apartment house. If the future income and costs are known, then using a suitable interest rate, the present worth of the property may be calculated. This should provide a good estimate of the price at which the property could be bought or sold. Another application might be determining the valuation of stocks or bonds based on the anticipated future benefits from owning them.

In present worth analysis, careful consideration must be given to the time period covered by the analysis. Usually the task to be accomplished has a time period associated with it. In that case, the consequences of each alternative must be considered for this period of time which is usually called the ***analysis period***, or sometimes the ***planning horizon***.

**Mutually exclusive* is where selecting one alternative precludes selecting any other alternative. An example of mutually exclusive alternatives would be deciding between constructing a gas station or a drive-in restaurant on a particular piece of vacant land.

Table 5-1 PRESENT WORTH ANALYSIS

	Situation	*Criterion*
Fixed input	**Amount of money or other input resources are fixed**	**Maximize present worth of benefits or other outputs**
Fixed output	**There is a fixed task, benefit, or other output to be accomplished**	**Minimize present worth of costs or other inputs**
Neither input nor output is fixed	**Neither amount of money, or other inputs, nor amount of benefits, or other outputs, is fixed**	**Maximize (present worth of benefits *minus* present worth of costs), that is, maximize net present worth**

There are three different analysis-period situations that are encountered in economic analysis problems:

1. The useful life of each alternative equals the analysis period.
2. The alternatives have useful lives different from the analysis period.
3. There is an infinite analysis period, $n = \infty$.

1. Useful Lives Equal the Analysis Period

Since different lives and an infinite analysis period present some complications, we will begin with four examples where the useful life of each alternative equals the analysis period.

EXAMPLE 5-1

A firm is considering which of two mechanical devices to install to reduce costs in a particular situation. Both devices cost \$1000 and have useful lives of five years and no salvage value. Device *A* can be expected to result in \$300 savings annually. Device *B* will provide cost savings of \$400 the first year but will decline \$50 annually, making the second year savings \$350, the third year savings \$300, and so forth. With interest at 7%, which device should the firm purchase?

Solution: The analysis period can conveniently be selected as the useful life of the devices, or five years. Since both devices cost \$1000, we have a situation where, in choosing either *A* or *B*, there is a fixed input (cost) of \$1000. The appropriate decision criterion is to choose the alternative that maximizes the present worth of benefits.

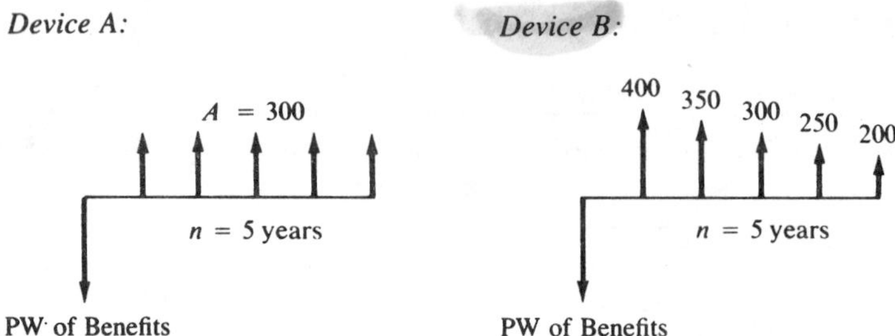

$$\text{PW of benefits } A = 300(P/A,7\%,5) = 300(4.100) = \$1230$$
$$\text{PW of benefits } B = 400(P/A,7\%,5) - 50(P/G,7\%,5)$$
$$= 400(4.100) - 50(7.647) = \$1257.65$$

Device *B* has the larger present worth of benefits and is, therefore, the preferred alternative. It is worth noting that, if we ignore the time value of money, both alternatives provide \$1500 worth of benefits over the five-year period. Device *B* provides greater benefits in the first two years and smaller benefits in the last two years. This more rapid flow of benefits from *B*, although the total magnitude equals that of *A*, results in a greater present worth of benefits. ◀

EXAMPLE 5-2

Wayne County will build an aqueduct to bring water in from the upper part of the state. It can be built at a reduced size now for \$300 million and be enlarged 25 years hence for an additional \$350 million. An alternative is to construct the full-sized aqueduct now for \$400 million.

Both alternatives would provide the needed capacity for the 50-year analysis period. Maintenance costs are small and may be ignored. At 6% interest, which alternative should be selected?

Solution: This problem illustrates stage construction. The aqueduct may be built in a single stage, or in a smaller first stage followed many years later by a second stage to provide the additional capacity when needed.

For the two-stage construction:

$$\text{PW of cost} = \$300 \text{ million} + 350 \text{ million}(P/F,6\%,25)$$
$$= \$300 \text{ million} + 81.6 \text{ million} = \$381.6 \text{ million}$$

For the single-stage construction:

$$\text{PW of cost} = \$400 \text{ million}$$

The two-stage construction has a smaller present worth of cost and is the preferred construction plan. ◀

EXAMPLE 5-3

A purchasing agent is considering the purchase of some new equipment for the mailroom. Two different manufacturers have provided quotations. An analysis of the quotations indicates the following:

Manufacturer	*Cost*	*Useful life, in years*	*End-of-useful-life salvage value*
Speedy	\$1500	5	\$200
Allied	1600	5	325

The equipment of both manufacturers is expected to perform at the desired level of (fixed) output. For a five-year analysis period, which manufacturer's equipment should be selected? Assume 7% interest and equal maintenance costs.

Solution: For fixed output, the criterion is to minimize the present worth of cost.

Speedy:

$$\text{PW of cost} = 1500 - 200(P/F,7\%,5) = 1500 - 200(0.7130)$$
$$= 1500 - 143 = \$1357$$

Allied:

$$\text{PW of cost} = 1600 - 325(P/F,7\%,5) = 1600 - 325(0.7130)$$
$$= 1600 - 232 = \$1368$$

Since it is only the *differences between alternatives* that are relevant, maintenance costs may be left out of the economic analysis. Although the PW of cost for each of the alternatives is nearly identical, we would, nevertheless, choose the one with minimum present worth of cost unless there were other tangible or intangible differences that would change the decision. Buy the Speedy equipment. ◀

EXAMPLE 5-4

A firm is trying to decide which of two alternate weighing scales it should install to check a package filling operation in the plant. The scales would allow better control of the filling operation and result in less overfilling. If both scales have lives equal to the six-year analysis period, which one should be selected? Assume an 8% interest rate.

Alternatives	*Cost*	*Uniform annual benefit*	*End-of-useful-life salvage value*
Atlas scales	\$2000	\$450	\$100
Tom Thumb scales	3000	600	700

Solution:

Atlas scales:

$$\begin{aligned}\text{PW of benefits} - \text{PW of cost} &= 450(P/A,8\%,6) + 100(P/F,8\%,6) - 2000\\ &= 450(4.623) + 100(0.6302) - 2000\\ &= 2080 + 63 - 2000 = \$143\end{aligned}$$

Tom Thumb scales:

$$\begin{aligned}\text{PW of benefits} - \text{PW of cost} &= 600(P/A,8\%,6) + 700(P/F,8\%,6) - 3000\\ &= 600(4.623) + 700(0.6302) - 3000\\ &= 2774 + 441 - 3000 = \$215\end{aligned}$$

The salvage value of the scales, it should be noted, is simply treated as another benefit of the alternative. Since the criterion is to maximize the present worth of benefits *minus* the present worth of cost, the preferred alternative is the Tom Thumb scales. ◀

Net Present Worth

In Example 5-4, we compared two alternatives and selected the one where present worth of benefits *minus* present worth of cost was a maximum. The criterion is called the ***net present worth criterion*** and written simply as **NPW**:

$$\begin{aligned}\text{Net present worth} &= \text{Present worth of benefits} - \text{Present worth of cost}\\ \text{NPW} &= \text{PW of benefits} - \text{PW of cost}\end{aligned}$$

2. Useful Lives Different from the Analysis Period

In present worth analysis, there always must be an identified analysis period. It follows, then, that each alternative must be considered for the entire period. In the previous Examples, the useful life of each alternative was equal to the analysis period. Often we can arrange it this way, but there will be many more situations where the alternatives have useful lives different from the analysis period. This section examines the problem and describes how to overcome this difficulty.

In Ex. 5-3, suppose that the Allied equipment was expected to have a ten-year useful life, or twice that of the Speedy equipment. Assuming the Allied salvage value would still be $325 ten years hence, which equipment should now be purchased? We will recompute the present worth of cost of the Allied equipment.

Allied:

$$\text{PW of cost} = 1600 - 325(P/F,7\%,10) = 1600 - 325(0.5083)$$
$$= 1600 - 165 = \$1435$$

The present worth of cost has increased. This is due, of course, to the more distant recovery of the salvage value. More importantly, we now find ourselves attempting to compare Speedy equipment, with its five-year life, against the Allied equipment with a ten-year life. This variation in the useful life of the equipment means we no longer have a situation of *fixed output*. Speedy equipment in the mailroom for five years is certainly not the same as ten years of service with Allied equipment. For present worth calculations, it is important that we select an analysis period and judge the consequences of each of the alternatives during the selected analysis period.

The analysis period for an economy study should be determined from the situation. In some industries with rapidly changing technologies, a rather short analysis period or planning horizon might be in order. Industries with more stable technologies (like steel making) might use a longer period (say, ten–twenty years), while government agencies frequently use analysis periods extending to fifty years or more.

Not only is the firm and its economic environment important in selecting an analysis period, but also the specific situation being analyzed is important. If the Allied equipment (Ex. 5-3) has a useful life of ten years, and the Speedy equipment will last five years, one method is to select an analysis period which is the **least common multiple** of their useful lives. Thus we would compare the ten-year life of Allied equipment against an initial purchase of Speedy equipment *plus* its replacement with new Speedy equipment in five years. The result is to judge the alternatives on the basis of a ten-year requirement in the mailroom. On this basis the economic analysis is as follows:

Speedy: Assuming the replacement Speedy equipment will also cost $1500 five years hence,

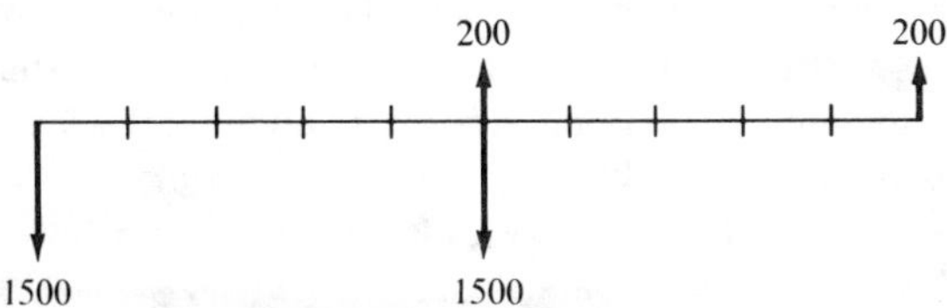

$$\text{PW of cost} = 1500 + (1500 - 200)(P/F,7\%,5) - 200(P/F,7\%,10)$$
$$= 1500 + 1300(0.7130) - 200(0.5083)$$
$$= 1500 + 927 - 102 = \$2325$$

Allied:

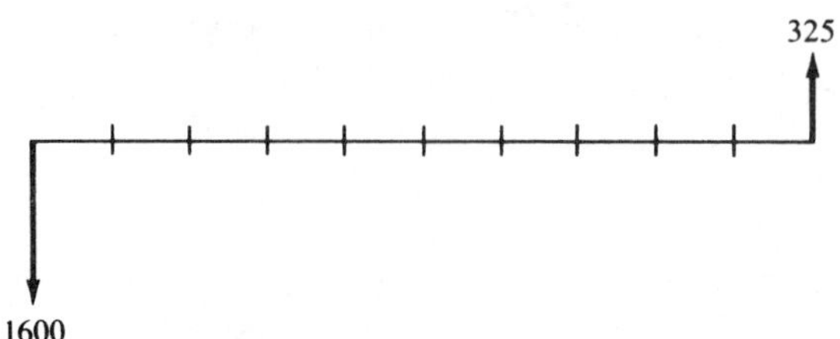

PW of cost = 1600 − 325(P/F,7%,10) = 1600 − 325(0.5083) = $1435

For the fixed output of ten years of service in the mailroom, the Allied equipment, with its smaller present worth of cost, is preferred. ◀

We have seen that setting the analysis period equal to the least common multiple of the lives of the two alternatives seems reasonable in the revised Ex. 5-3. What would one do, however, if in another situation the alternatives had useful lives of 7 and 13 years, respectively? Here the least common multiple of lives is 91 years. An analysis period of 91 years hardly seems realistic. Instead, a suitable analysis period should be based on how long the equipment is likely to be needed. This may require that terminal values be estimated for the alternatives at some point prior to the end of their useful lives.

Figure 5-1 graphically represents this concept. As Fig. 5-1 indicates, it is not necessary for the analysis period to equal the useful life of an alternative or some multiple of the useful life. To properly reflect the situation at the end of the analysis period, an estimate is required of the market value of the equipment at that time. The calculations might be easier if everything came out even, but it is not essential.

3. Infinite Analysis Period–Capitalized Cost

Another difficulty in present worth analysis arises when we encounter an infinite analysis period ($n = \infty$). In governmental analyses, at times there are circumstances where a service or condition is to be maintained for an infinite period. The need for roads, dams, pipelines, or whatever is sometimes considered permanent. In these situations a present worth of cost analysis would have an infinite analysis period. We call this particular analysis ***capitalized cost***.

Capitalized cost is the present sum of money that would need to be set aside now, at some interest rate, to yield the funds required to provide the service (or whatever) indefinitely. To accomplish this, the money set aside for future expenditures must not decline. The interest received on the money set aside can be spent, but not the principal. When one stops to think about an infinite analysis period (as opposed to something relatively short, like one hundred years), we

Alternative 1:

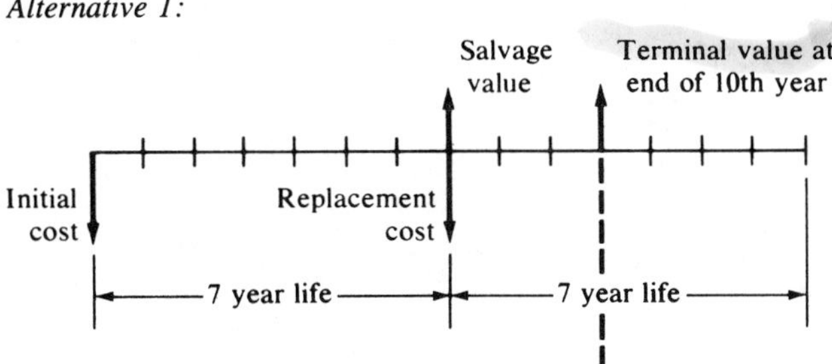

Alternative 2:

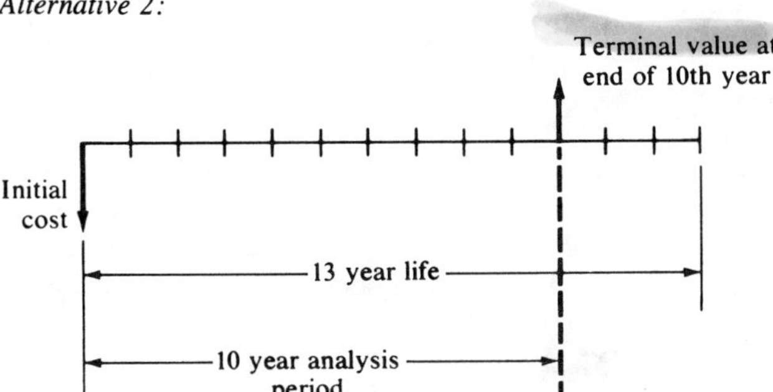

Figure 5-1 Superimposing an analysis period on 7- and 13-year alternatives.

see that an undiminished principal sum is essential, otherwise one will of necessity run out of money prior to infinity.

In Chapter 4 we saw:

$$\text{Principal sum} + \text{Interest for the period} = \text{Amount at end of period}$$

$$P \quad + \quad iP \quad = P + iP$$

If we spend iP, then in the next interest period the principal sum P will again increase to $(P + iP)$. Thus, we can again spend iP.

This concept may be illustrated by a numerical example: suppose you deposited $200 in a bank that paid 4% interest annually. How much money could be withdrawn each year without reducing the balance in the account below the initial $200? At the end of the first year, the $200 would earn 4%($200) = $8 interest. If this interest were withdrawn, the $200 would remain in the account. At the end of the second year, the $200 balance would again earn 4%($200) = $8. This $8 could also be withdrawn and the account would still have $200. This procedure could be continued indefinitely and the bank account would always contain $200.

The year-by-year situation would be depicted like this:

Year 1: \$200 initial $P \rightarrow 200 + 8 = 208$
Withdrawal $iP = -8$

Year 2: \$200 $\rightarrow 200 + 8 = 208$
Withdrawal $iP = -8$
\$200
and so on.

Thus, for any initial present sum *P*, there can be an end-of-period withdrawal of *A* equal to *iP* each period, and these withdrawals may continue forever without diminishing the initial sum *P*. This gives us the basic relationship:

For $n = \infty$, $A = Pi$

This relationship is the key to capitalized cost calculations. We previously defined capitalized cost as the present sum of money that would need to be set aside at some interest rate to yield the funds to provide the desired task or service forever. Capitalized cost is therefore the *P* in the equation $A = iP$. It follows that:

$$\textbf{Capitalized cost } P = \frac{A}{i}$$

If we can resolve the desired task or service into an equivalent *A*, the capitalized cost may be computed. The following examples illustrate such computations.

EXAMPLE 5-5

How much should one set aside to pay \$50 per year for maintenance on a gravesite if interest is assumed to be 4%? For perpetual maintenance, the principal sum must remain undiminished after making the annual disbursement.

Solution:

$$\text{Capitalized cost } P = \frac{\text{Annual disbursement } A}{\text{Interest rate } i}$$

$$P = \frac{50}{0.04} = \$1250$$

One should set aside \$1250. ◀

EXAMPLE 5-6

A city plans a pipeline to transport water from a distant watershed area to the city. The pipeline will cost \$8 million and have an expected life of seventy years.

The city anticipates it will need to keep the water line in service indefinitely. Compute the capitalized cost assuming 7% interest.

Solution: We have the capitalized cost equation:

$$P = \frac{A}{i}$$

that is simple to apply when there are end-of-period disbursements A. Here we have renewals of the pipeline every seventy years. To compute the capitalized cost, it is necessary to first compute an end-of-period disbursement A that is equivalent to \$8 million every seventy years.

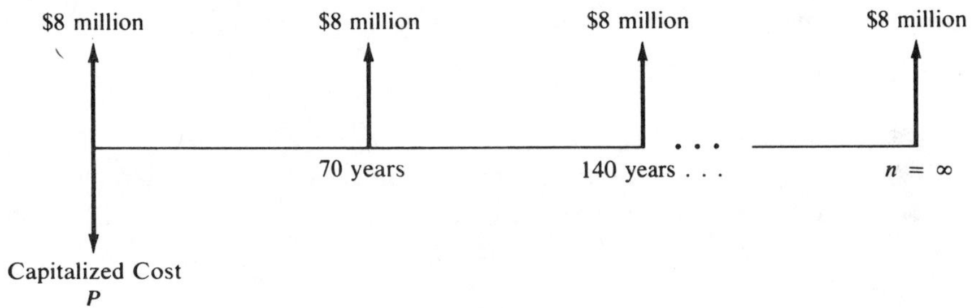

The \$8 million disbursement at the end of seventy years may be resolved into an equivalent A.

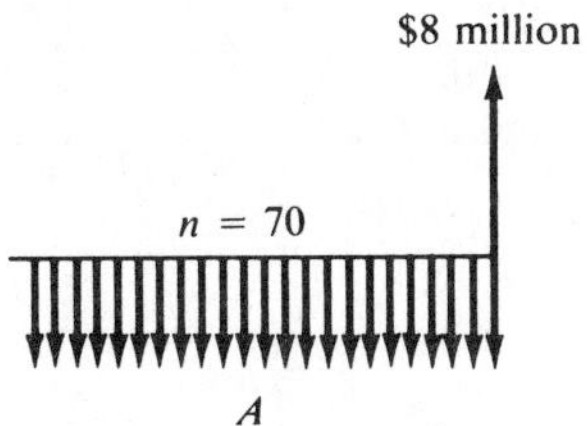

$$\begin{aligned} A = F(A/F,i,n) &= \$8 \text{ million}(A/F,7\%,70) \\ &= \$8 \text{ million}(0.00062) = \$4960 \end{aligned}$$

Each seventy-year period is identical to this one and the infinite series is shown in Fig. 5-2.

$$\begin{aligned} \text{Capitalized cost } P &= \$8 \text{ million} + \frac{A}{i} = \$8 \text{ million} + \frac{4960}{0.07} \\ &= \$8{,}071{,}000 \end{aligned}$$

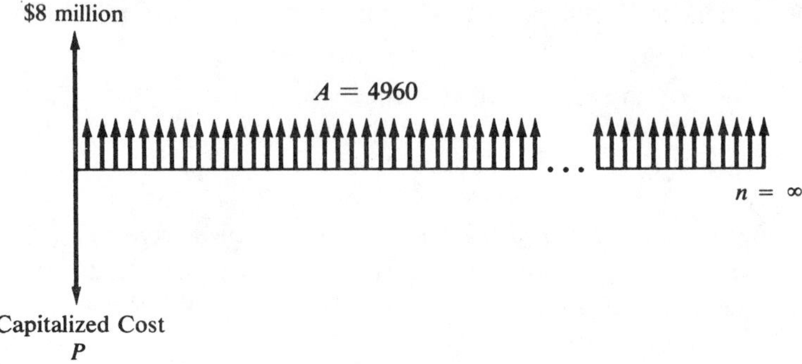

Figure 5-2 Infinite series computed using the sinking fund factor.

Alternate Solution: Instead of solving for an equivalent end-of-period payment A based on a *future* \$8 million disbursement, we could find A, given a *present* \$8 million disbursement.

$$\begin{aligned} A &= P(A/P,i,n) = \$8 \text{ million}(A/P,7\%,70) \\ &= \$8 \text{ million}(0.0706) = \$565{,}000 \end{aligned}$$

On this basis, the infinite series is shown in Fig. 5-3. Carefully note the difference between this and Fig. 5-2. Now:

$$\text{Capitalized cost } P = \frac{A}{i} = \frac{565{,}000}{0.07} = \$8{,}071{,}000 \quad ◀$$

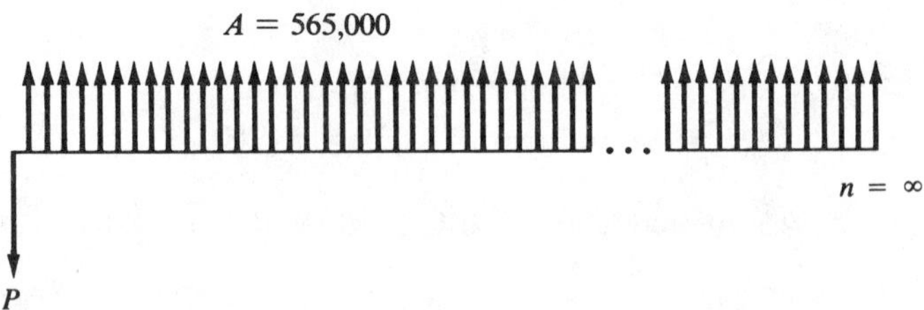

Figure 5-3 Infinite series computed using the capital recovery factor.

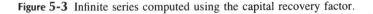

Multiple Alternatives

So far the discussion has been based on examples with only two alternatives. But multiple-alternative problems may be solved by exactly the same methods employed for problems with two alternatives. (The only reason for avoiding multiple alternatives was to simplify the examples.) Examples 5-7 and 5-8 have multiple alternatives.

EXAMPLE 5-7

A contractor has been awarded the contract to construct a six-miles-long tunnel in the mountains. During the five-year construction period, the contractor will need water from a nearby stream. He will construct a pipeline to convey the water to the main construction yard. An analysis of costs for various pipe sizes is as follows:

	Pipe size			
	2″	3″	4″	6″
Installed cost of pipeline and pump	\$22,000	\$23,000	\$25,000	\$30,000
Cost per hour for pumping	\$1.20	\$0.65	\$0.50	\$0.40

The pipe and pump will have a salvage value at the end of five years equal to the cost to remove them. The pump will operate 2000 hours per year. The lowest interest rate at which the contractor is willing to invest money is 7%. (The minimum required interest rate for invested money is called the ***minimum attractive rate of return***, or MARR.) Select the alternative with the least present worth of cost.

Solution: We can compute the present worth of cost for each alternative. For each pipe size, the Present worth of cost is *equal* to the Installed cost of the pipeline and pump *plus* the Present worth of five years of pumping costs.

	Pipe size			
	2″	3″	4″	6″
Installed cost of pipeline and pump	\$22,000	\$23,000	\$25,000	\$30,000
1.20 × 2000 hr × $(P/A,7\%,5)$	9,840			
0.65 × 2000 hr × 4.100		5,330		
0.50 × 2000 hr × 4.100			4,100	
0.40 × 2000 hr × 4.100				3,280
Present worth of cost	\$31,840	\$28,330	\$29,100	\$33,280

Select the 3″ pipe size. ◀

EXAMPLE 5-8

An investor paid $8000 to a consulting firm to analyze what he might do with a small parcel of land on the edge of town that can be bought for $30,000. In their report, the consultants suggested four alternatives:

Alternatives	*Total investment including land**	*Uniform net annual benefit*	*Terminal value at end of 20 yr*
A: Do nothing	$ 0	$ 0	$ 0
B: Vegetable market	50,000	5,100	30,000
C: Gas station	95,000	10,500	30,000
D: Small motel	350,000	36,000	150,000

*Includes the land and structures but does not include the $8000 fee to the consulting firm.

Assuming 10% is the minimum attractive rate of return, what should the investor do?

Solution: Alternative *A* represents the "do nothing" alternative. Generally, one of the feasible alternatives in any situation is to remain in the present status and do nothing. In this problem, the investor could decide that the most attractive alternative is not to purchase the property and develop it. This is clearly a do-nothing decision.

We note, however, that even if he does nothing, the total venture would not be a very satisfactory one. This is due to the fact that the investor spent $8000 for professional advice on the possible uses of the property. But because the $8000 is a past cost, it is a ***sunk cost***. Sunk cost is the name given to past costs. The only relevant costs in an economic analysis are *present* and *future* costs; past events and past costs are gone and cannot be allowed to affect future planning. (The only place where past costs may be relevant is in computing depreciation charges and income taxes.) It should not deter the investor from making the best decision now, regardless of the costs that brought him to this situation and point of time.

This problem is one of neither fixed input nor fixed output, so our criterion will be to maximize the Present worth of benefits *minus* the Present worth of cost, or, simply stated, maximize net present worth.

Alternative A, Do nothing:

$$\text{NPW} = 0$$

Alternative B, Vegetable market:

$$\begin{aligned}\text{NPW} &= -50{,}000 + 5100(P/A,10\%,20) + 30{,}000(P/F,10\%,20)\\ &= -50{,}000 + 5100(8.514) + 30{,}000(0.1486)\end{aligned}$$

$$= -50{,}000 + 43{,}420 + 4460$$
$$= -2120$$

Alternative C, Gas station:

$$\text{NPW} = -95{,}000 + 10{,}500(P/A,10\%,20) + 30{,}000(P/F,10\%,20)$$
$$= -95{,}000 + 89{,}400 + 4460$$
$$= -1140$$

Alternative D, Small motel:

$$\text{NPW} = -350{,}000 + 36{,}000(P/A,10\%,20) + 150{,}000(P/F,10\%,20)$$
$$= -350{,}000 + 306{,}500 + 22{,}290$$
$$= -21{,}210$$

The criterion is to maximize net present worth. In this situation, one alternative has NPW equal to zero, and three alternatives have negative values for NPW. We will select the best of the four alternatives, namely, the do-nothing Alt. *A* with NPW equal to zero. ◀

EXAMPLE 5-9

A piece of land may be purchased for $610,000 to be strip-mined for the underlying coal. Annual net income will be $200,000 per year for ten years. At the end of the ten years, the surface of the land will be restored as required by a federal law on strip mining. The cost of reclamation will be $1,500,000 more than the resale value of the land after it is restored. Using a 10% interest rate, determine whether the project is desirable.

Solution: The investment opportunity may be described by the following cash flow:

Year	*Cash flow, in thousands*
0	−$610
1–10	+200 (per year)
10	−1500

$$\text{NPW} = -610 + 200(P/A,10\%,10) - 1500(P/F,10\%,10)$$
$$= -610 + 200(6.145) - 1500(0.3855)$$
$$= -610 + 1229 - 578$$
$$= +41$$

Since NPW is positive, the project is desirable. ◀

EXAMPLE 5-10

Two pieces of construction equipment are being analyzed:

Year	*Alternative A*	*Alternative B*
0	−$2000	−$1500
1	+1000	+700
2	+850	+300
3	+700	+300
4	+550	+300
5	+400	+300
6	+400	+400
7	+400	+500
8	+400	+600

Based on an 8% interest rate, which alternative should be selected?

Solution:

Alternative A:

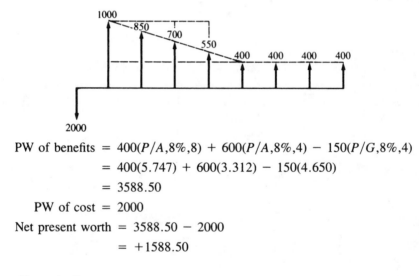

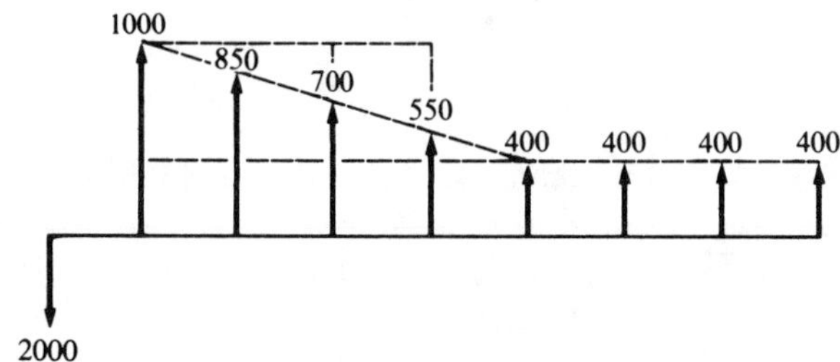

$$\begin{aligned}\text{PW of benefits} &= 400(P/A,8\%,8) + 600(P/A,8\%,4) - 150(P/G,8\%,4)\\ &= 400(5.747) + 600(3.312) - 150(4.650)\\ &= 3588.50\end{aligned}$$

$$\text{PW of cost} = 2000$$

$$\begin{aligned}\text{Net present worth} &= 3588.50 - 2000\\ &= +1588.50\end{aligned}$$

Alternative B:

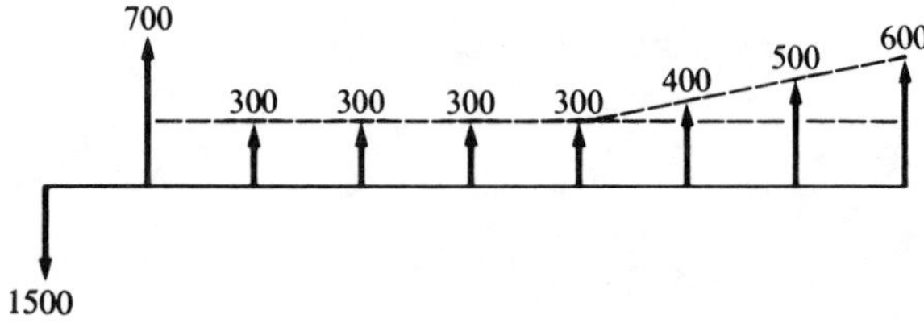

$$\text{PW of benefits} = 300\,(P/A,8\%,8) + (700 - 300)(P/F,8\%,1) + 100(P/G,8\%,4)(P/F,8\%,4)$$
$$= 300(5.747) + 400(0.9259) + 100(4.650)(0.7350)$$
$$= 2436.24$$
$$\text{PW of cost} = 1500$$
$$\text{Net present worth} = 2436.24 - 1500 = +936.24$$

To maximize NPW, choose Alt. *A*. ◀

Assumptions In Solving Economic Analysis Problems

One of the difficulties of problem solving is that most problems tend to be very complicated. It becomes apparent that *some* simplifying assumptions are needed to make such problems manageable. The trick, of course, is to solve the simplified problem and still be satisfied that the solution is applicable to the *real* problem! In the paragraphs that follow, we will consider six different items and explain the customary assumptions that are made.

End-of-Year Convention

As we indicated in Chapter 4, economic analysis textbooks follow the end-of-year convention. This makes "*A*" a series of end-of-period receipts or disbursements. (We generally assume in problems that all series of receipts or disbursements occur at the *end* of the interest period. This is, of course, a very self-serving assumption, for it allows us to use values from our Compound Interest Tables without any adjustments.)

A cash flow diagram of *P*, *A*, and *F* for the end-of-period convention is as follows:

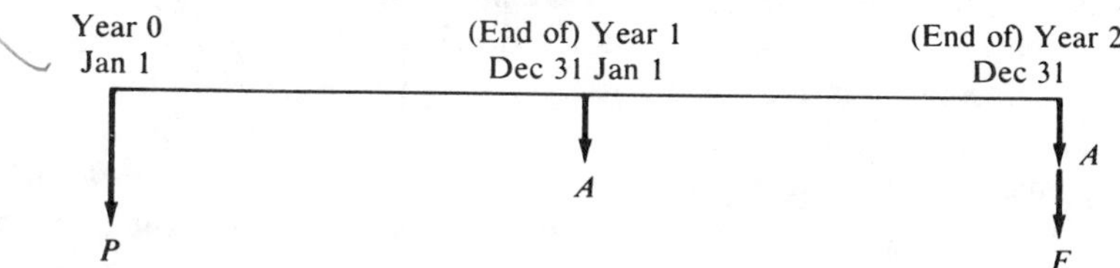

If one were to adopt a middle-of-period convention, the diagram would be:

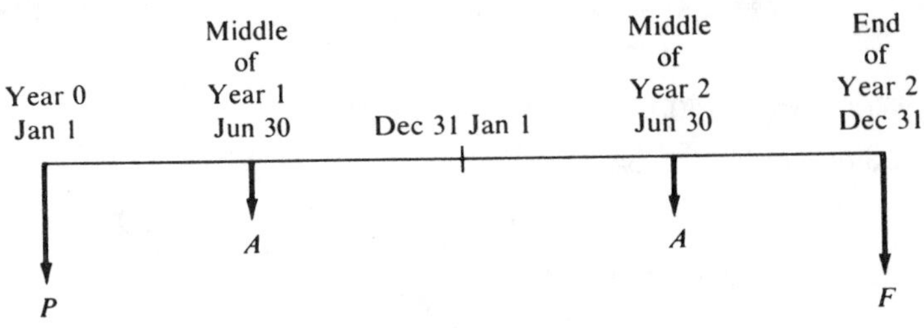

As the diagrams illustrate, only A shifts; P remains at the beginning-of-period and F at the end-of-period, regardless of the convention. The Compound Interest Tables in the Appendix are based on the end-of-period convention.

Viewpoint of Economic Analysis Studies

When we make economic analysis calculations, we must proceed from a point of reference. Generally, we will want to take the point of view of a total firm when doing industrial economic analyses. Example 3-1 vividly illustrated the problem: a firm's shipping department decided it could save money having its printing work done outside rather than by the in-house printing department. An analysis from the viewpoint of the shipping department supported this, as it could get for \$688.50 the same printing it was paying \$793.50 for in-house. Further analysis showed, however, that its printing department costs would decline *less* than using the commercial printer would save. From the viewpoint of the firm, the net result would be an increase in total cost.

From Ex. 3-1 we see it *is* important that the ***viewpoint of the study*** be carefully considered. Selecting a narrow viewpoint, like the shipping department, may result in a suboptimal decision from the viewpoint of the firm. For this reason, the viewpoint of the total firm is the normal point of reference in industrial economic analyses.

Sunk Costs

We know that it is the ***differences between alternatives*** that are relevant in economic analysis. Events that have occurred in the past really have no bearing on what we should do in the future. When the judge says, "\$200 fine or three days in jail," the events that led to these unhappy alternatives really are unimportant. What *is* important are the current and future differences between the alternatives. Past costs, like past events, have no bearing on deciding between alternatives unless the past costs somehow affect the present or future costs. In general, past costs do not affect the present or the future, so we refer to them as *sunk costs* and disregard them.

Borrowed Money Viewpoint

In most economic analyses, the proposed alternatives inevitably require money to be spent, and so it is natural to ask the source of that money. The source will vary from situation to situation. In fact, there are *two* aspects of money to determine: one is the ***financing***—the obtaining of money—problem; the other is the ***investment***—the spending of money—problem. Experience has shown that these two concerns should be distinguished. When separated, the problems of obtaining money and of spending it are both logical and straightforward. Failure to separate them sometimes produces confusing results and poor decision making.

The conventional assumption in economic analysis is that the money required to finance alternatives/solutions in problem solving is considered to be ***borrowed at interest rate i.***

Effect of Inflation and Deflation

For the present we will assume that ***prices are stable***. This means that a machine that costs $5000 today can be expected to cost the same amount several years hence. Inflation and deflation is a serious problem in many situations, but we will disregard it for now.

Income Taxes

This aspect of economic analyses, like inflation–deflation, must be considered if a realistic analysis is to be done. We will defer our introduction of income taxes into economic analyses until later.

SUMMARY

Present worth analysis is suitable for almost any economic analysis problem. But it is particularly desirable when we wish to know the present worth of future costs and benefits. And we frequently want to know the value today of such things as income-producing assets, stocks, and bonds.

For present worth analysis, the proper economic criteria are:

Fixed input	Maximize the PW of benefits
Fixed output	Minimize the PW of costs
Neither input nor output is fixed	Maximize (PW of benefits − PW of costs) or, more simply stated: Maximize NPW

To make valid comparisons, we need to analyze each alternative in a problem over the same *analysis period* or *planning horizon*. If the alternatives do not have equal lives, some technique must be used to achieve a common analysis period. One method is to select an analysis period equal to the least common multiple of the alternative lives. Another method is to select an analysis period and then compute end-of-analysis-period salvage values for the alternatives.

Capitalized cost is the present worth of cost for an infinite analysis period ($n = \infty$). When $n = \infty$, the fundamental relationship is $A = iP$. Some form of this equation is used whenever there is a problem with an infinite analysis period.

There are a number of assumptions that are routinely made in solving economic analysis problems. They include the following:

1. Present sums P are beginning of period and all series receipts or disbursements A and future sums F occur at the end of the interest period. The compound interest tables were derived on this basis.
2. In industrial economic analyses, the appropriate point of reference from which to compute the consequences of alternatives is the total firm. Taking a narrower view of the consequences can result in suboptimal solutions.
3. Only the differences between the alternatives are relevant. Past costs are *sunk costs* and generally do not affect present or future costs. For this reason they are ignored.
4. The investment problem should be isolated from the financing problem. We generally assume that all required money is borrowed at interest rate i.
5. For now, stable prices are assumed. The problem of inflation–deflation is deferred to Chapter 13. Similarly, income taxes are deferred to Chapter 11.

Problems

5-1 Compute P for the following diagram.

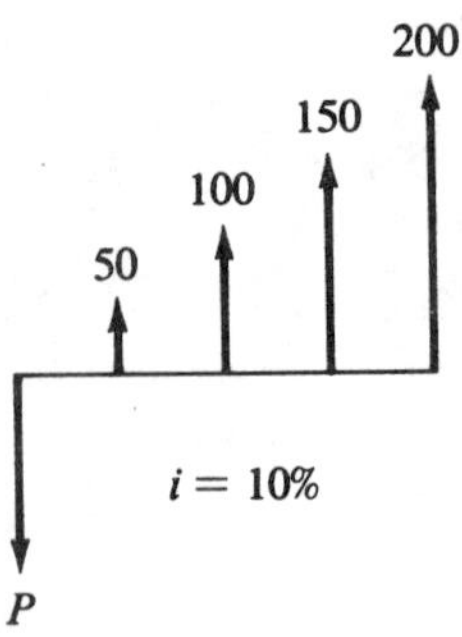

5-2 Compute the value of P that is equivalent to the four cash flows in the diagram on the next page.

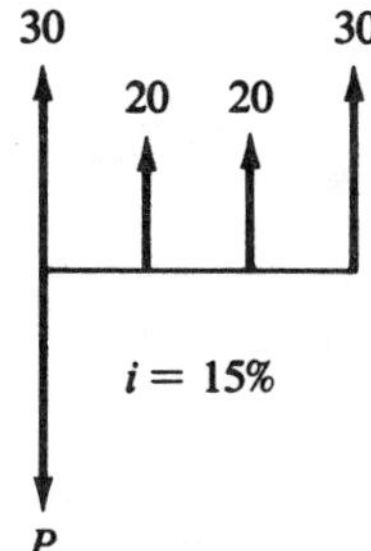

5-3 What is the value of P for the situation shown below?

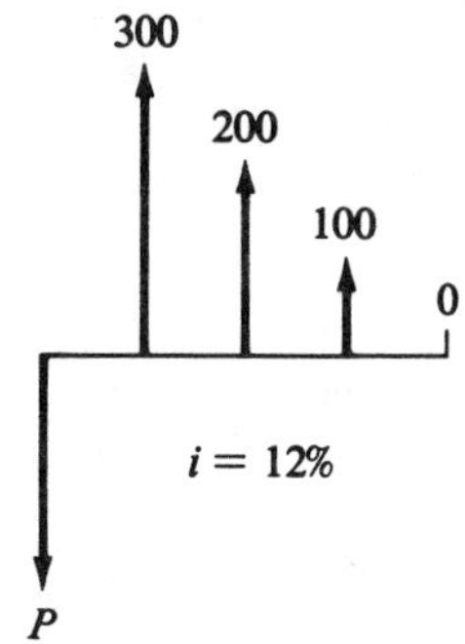

(*Answer:* P = \$498.50)

5-4 Compute the value of Q in the figure below.

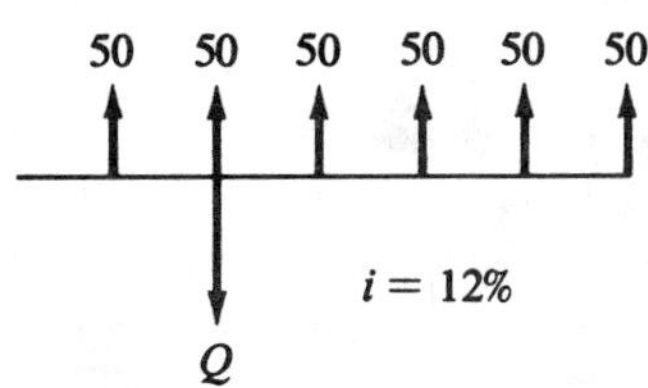

5-5 For the diagram, compute P.

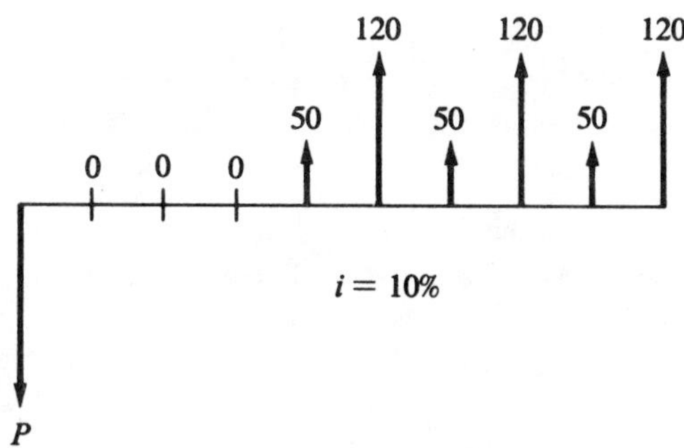

5-6 Compute P for the following diagram.

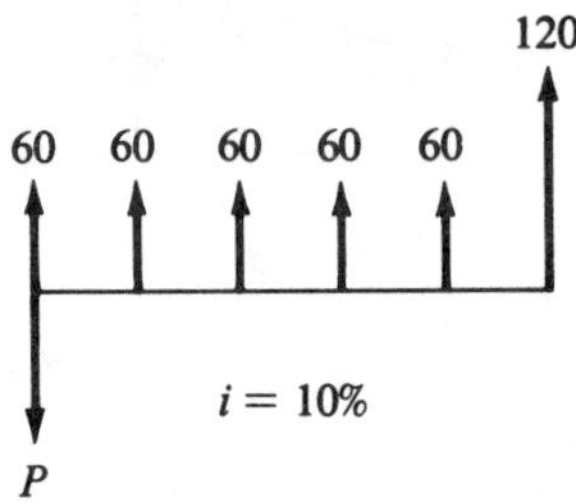

(*Answer:* P = \$324.71)

5-7 The annual income from a rented house is \$12,000. The annual expenses are \$3000. If the house can be sold for \$145,000 at the end of ten years, how much could you afford to pay for it now, if you considered 18% to be a suitable interest rate? (*Answer:* \$68,155)

5-8 Consider the following cash flow. At a 6% interest rate, what is the value of P, at the end of Year 1, that is equivalent to the benefits at the end of Years 2 through 7?

Year	*Cash flow*
1	$-P$
2	+100
3	+200
4	+300
5	+400
6	+500
7	+600

5-9 A rather wealthy man decided he would like to arrange for his descendants to be well educated. He would like each child to have \$60,000 for his or her education. He plans to set up a perpetual trust fund so that six children will receive this assistance each generation. He estimates that there will be four generations per century, spaced 25 years apart. He expects the trust to be able to obtain a 4% rate of return, and the first recipients to receive the money ten years hence. How much money should he now set aside in the trust? (*Answer:* \$389,150)

5-10 How much would the owner of a building be justified in paying for a sprinkler system that will save \$750 a year in insurance premiums if the system has to be replaced every twenty years and has a salvage value equal to 10% of its initial cost? Assume money is worth 7%. (*Answer:* \$8156)

5-11 A man had to have the muffler replaced on his two-year-old car. The repairman offered two alternatives. For \$50 he would install a muffler guaranteed for two years. But for \$65 he would install a muffler guaranteed "for as long as you own the car." Assuming the present owner expects to keep the car for about three more years, which muffler would you advise him to have installed if you thought 20% were a suitable interest rate and the less expensive muffler would only last two years?

5-12 A consulting engineer has been engaged to advise a town how best to proceed with the construction of a 200,000 km³ water supply reservoir. Since only 120,000 km³ of storage will be required for the next 25 years, an alternative to building the full capacity now is to build the reservoir in two stages. Initially, the reservoir could be built with 120,000 km³ of capacity and then, 25 years hence, the additional 80,000 km³ of capacity could be added by increasing the height of the reservoir. Estimated costs are as follows:

	Construction cost	*Annual maintenance cost*
Build in two stages.		
First stage: 120,000 km³ reservoir	$14,200,000	$75,000
Second stage: Add 80,000 km³ of capacity, additional construction and maintenance costs	12,600,000	25,000
Build full capacity now.		
200,000 km³ reservoir	$22,400,000	$100,000

If interest is computed at 4%, which construction plan is preferred?

5-13 An engineer has received two bids for an elevator to be installed in a new building. The bids, plus his evaluation of the elevators, are as follows:

	Bids	*Engineer's estimates*		
Alternatives	*Installed cost*	*Service life, in years*	*Annual operating cost, including repairs*	*Salvage value at end of service life*
Westinghome	$45,000	10	$2700/yr	$3000
Itis	54,000	15	2850/yr	4500

The engineer will make a present worth analysis using a 10% interest rate. Prepare the analysis and determine which bid should be accepted.

5-14 A railroad branch line is to be constructed to a missile site. It is expected the railroad line will be used for 15 years, after which the missile site will be removed and the land turned back to agricultural use. The railroad track and ties will be removed at that time.

In building the railroad line, either treated or untreated wood ties may be used. Treated ties have an installed cost of $6 and a ten-year life; untreated ties are $4.50 with a six-year life. If at the end of fifteen years the ties then in place have a remaining useful life of four years or more, they will be used by the railroad elsewhere and have an estimated salvage value of $3 each. Anytime ties are removed that are at the end of their service life, or are too close to the end of their service life to be used elsewhere, they are sold for $0.50 each.

Determine the most economical plan for the initial railroad ties and their replacement for the fifteen-year period. Make a present worth analysis assuming 8% interest.

5-15 A weekly business magazine offers a one-year subscription for $58 and a three-year subscription for $116. If you thought you would read the magazine for at least the next three years, and consider 20% as a minimum rate of return, which way would you purchase the magazine, with three one-year subscriptions or a single three-year subscription? (*Answer:* Choose the three-year subscription.)

5-16 A manufacturer is considering purchasing equipment which will have the following financial effects:

Year	*Disbursements*	*Receipts*
0	$4400	$ 0
1	660	880
2	660	1980
3	440	2420
4	220	1760

If money is worth 6%, should he invest in the equipment?

5-17 Jerry Stans, a young industrial engineer, prepared an economic analysis for some equipment to replace one production worker. The analysis showed that the present worth of benefits (of employing one less production worker) just equaled the present worth of the equipment costs, based on a ten-year useful life for the equipment. It was decided not to purchase the equipment.

A short time later, the production workers won a new three-year union contract that granted them an immediate 40¢-per-hour wage increase, plus an additional 25¢-per-hour wage increase in each of the two subsequent years. Assume that in each and every future year, a 25¢-per-hour wage increase will be granted.

Jerry Stans has been asked to revise his earlier economic analysis. The present worth of benefits of replacing one production employee will now increase. Assuming an interest rate of 8%, the justifiable cost of the automation equipment (with a ten-year useful life) will increase by how much? Assume the plant operates a single eight-hour shift, 250 days per year.

5-18 The management of an electronics manufacturing firm believes it is desirable to install some automation equipment in their production facility. They believe the equipment would have a ten-year life with no salvage value at the end of ten years. The plant engineering department has surveyed the plant and suggested there are eight mutually exclusive alternatives available.

Plan	*Initial cost, in thousands*	*Net annual benefit, in thousands*
1	$265	$51
2	220	39
3	180	26
4	100	15
5	305	57
6	130	23
7	245	47
8	165	33

If the firm expects a 10% rate of return, which alternative, if any, should they adopt? (*Answer:* Plan 1)

5-19 The president of the E. L. Echo Corporation thought it would be appropriate for his firm to "endow a chair" in the Industrial Engineering Department of the local university; that is, he was considering making a gift to the university of sufficient money to pay the salary of one professor forever. One professor in the department would be designated the E. L. Echo Professor of Industrial Engineering, and his salary would come from the fund established by the Echo Corporation. If the professor will receive $67,000 per year, and the interest received on the endowment fund is expected to remain at 8%, what lump sum of money will the Echo Corporation need to provide to establish the endowment fund? (*Answer:* $837,500)

5-20 A man who likes cherry blossoms very much would like to have an urn full of them put on his grave once each year forever after he dies. In his will, he intends to leave a certain sum of money in the trust of a local bank to pay the florist's annual bill. How much money should be left for this purpose? Make whatever assumptions you feel are justified by the facts presented. State your assumptions, and compute a solution.

5-21 A local symphony association offers memberships as follows:

Continuing membership, per year	$ 15
Patron lifetime membership	375

The patron membership has been based on the symphony association's belief that it can obtain a 4% rate of return on its investment. If you believed 4% to be an appropriate rate of return, would you be willing to purchase the patron membership? Explain why or why not.

5-22 A battery manufacturing plant has been ordered to cease discharging acidic waste liquids containing mercury into the city sewer system. As a result, the firm must now adjust the pH and remove the mercury from its waste liquids. Three firms have provided quotations on the necessary equipment. An analysis of the quotations provided the following table of costs.

Bidder	*Installed cost*	*Annual operating cost*	*Annual income from mercury recovery*	*Salvage value*
Foxhill Instrument	$ 35,000	$8000	$2000	$20,000
Quicksilver	40,000	7000	2200	0
Almaden	100,000	2000	3500	0

If the installation can be expected to last twenty years and money is worth 7%, which equipment should be purchased? (*Answer:* Almaden)

5-23 A firm is considering three mutually exclusive alternatives as part of a production improvement program. The alternatives are:

	A	*B*	*C*
Installed cost	$10,000	$15,000	$20,000
Uniform annual benefit	1,625	1,530	1,890
Useful life, in years	10	20	20

For each alternative, the salvage value at the end of its useful life is zero. At the end of ten years, *A* could be replaced with another *A* with identical cost and benefits. The minimum attractive rate of return is 6%. Which alternative should be selected?

5-24 A steam boiler is needed as part of the design of a new plant. The boiler can be fired either by natural gas, fuel oil, or coal. A decision must be made on which fuel to use. An analysis of the costs shows that the installed cost, with all controls, would be least for natural gas at $30,000; for fuel oil it would be $55,000; and for coal it would be $180,000. If natural gas is used rather than fuel oil, the annual fuel cost will increase by $7500. If coal is used rather than fuel oil, the annual fuel cost will be $15,000 per year less. Assuming 8% interest, a twenty-year analysis period, and no salvage value, which is the most economical installation?

5-25 An investor has carefully studied a number of companies and their common stock. From his analysis, he has decided that the stocks of six firms are the best of the many he has examined. They represent about the same amount of risk and so he would like to determine the one in which to invest. He plans to keep the stock for four years and requires a 10% minimum attractive rate of return.

Common stock	*Price per share*	*Annual end-of-year dividend per share*	*Estimated price at end of 4 years*
Western House	$23¾	$1.25	$32
Fine Foods	45	4.50	45
Mobile Motors	30⅝	0	42
Trojan Products	12	0	20
U.S. Tire	33⅜	2.00	40
Wine Products	52½	3.00	60

Which stock, if any, should the investor consider purchasing? (*Answer:* Trojan Products)

5-26 A home builder must construct a sewage treatment plant and deposit sufficient money in a perpetual trust fund to pay the $5000 per year operating cost and to replace the treatment plant every forty years. The plant will cost $150,000, and future replacement plants will also cost $150,000 each. If the trust fund earns 8% interest, what is the builder's capitalized cost to construct the plant and future replacements, and to pay the operating costs?

5-27 Using an eight-year analysis period and a 10% interest rate, determine which alternative should be selected:

	A	*B*
First cost	$5300	$10,700
Uniform annual benefit	1800	2,100
Useful life, in years	4	8

5-28 The local botanical society wants to ensure that the gardens in a local park are properly cared for. They just recently spent $100,000 to plant the gardens. They would like to set up a perpetual fund to provide $100,000 for future replantings of the gardens every ten years. If interest is 5%, how much money would be needed to forever pay the cost of replanting?

5-29 An elderly lady decided to distribute most of her considerable wealth to charity and to retain for herself only enough money to provide for her living. She feels that $1000 a month will amply provide for her needs. She will establish a trust fund at a bank which pays 6% interest, compounded monthly. At the end of each month she will withdraw \$1000. She has arranged that, upon her death, the balance in the account is to be paid to her niece, Susan. If she opens the trust fund and deposits enough money to pay her \$1000 a month forever, how much will Susan receive when her Aunt dies?

5-30 Solve the diagram below for P using a geometric gradient factor.

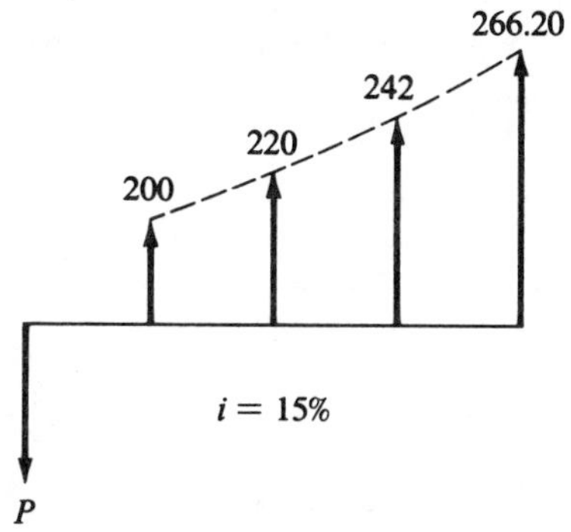

5-31 If $i = 10\%$, what is the value of P?

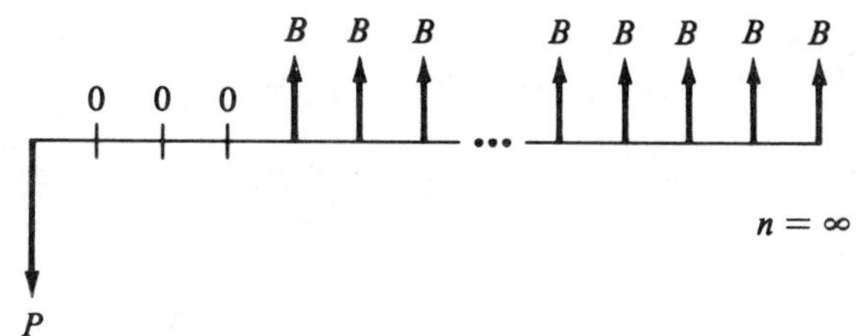

5-32 A stonecutter was carving the headstone for a well-known engineering economist.

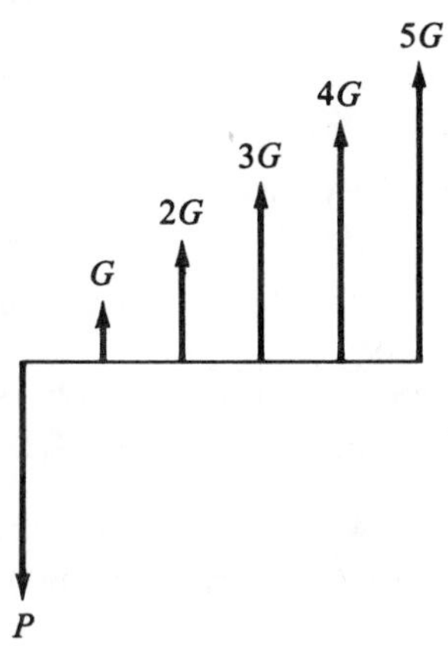

He carved the figure above and then started the equation as follows:

$$P = G(P/G,i,6) \qquad (1)$$

He realized he had made a mistake. The equation should have been

$$P = G(P/G,i,5) + G(P/A,i,5)$$

The stonecutter does not want to discard the stone and start over. He asks you to help him with his problem. The right side of Eq. 1 can be multiplied by one compound interest factor and then the equation will be correct for the carved figure. Equation 1 will be of the form:

$$P = G(P/G,i,6)(\quad ,i,\ )$$

Write the complete equation.

5-33 In a present worth analysis, one alternative has a Net Present Worth of +420, based on a six-year analysis period that equals the useful life of the alternative. A 10% interest rate was used in the computations.

The alternative is to be replaced at the end of the six years by an identical piece of equipment with the same cost, benefits, and useful life. Based on a 10% interest rate, compute the Net Present Worth of the equipment for the twelve-year analysis period.

(*Answer:* NPW = +657.09)

5-34 A project has a Net Present Worth (NPW) of −140 as of Jan. 1, 2000. If a 10% interest rate is used, what is the project NPW as of Dec. 31, 1997?

5-35 Consider the following four alternatives. Three are "do something" and one is "do nothing."

	Alternative			
	A	*B*	*C*	*D*
Cost	$0	$50	$30	$40
Net annual benefit	0	12	4.5	6
Useful life, in years		5	10	10

At the end of the five-year useful life of *B*, a replacement is not made. If a ten-year analysis period and a 10% interest rate are selected, which is the preferred alternative?

5-36 Six mutually exclusive alternatives are being examined. For an 8% interest rate, which alternative should be selected? Each alternative has a six-year useful life.

	Alternatives					
	A	*B*	*C*	*D*	*E*	*F*
Initial cost	$20.00	$35.00	$55.00	$60.00	$80.00	$100.00
Uniform annual benefit	6.00	9.25	13.38	13.78	24.32	24.32

5-37 A building contractor obtained bids for some asphalt paving, based on a specification. Three paving subcontractors quoted the following prices and terms of payment:

	Price	Payment schedule
1. *Quick Paving Co.*	$85,000	50% payable immediately; 25% payable in six months; 25% payable at the end of one year.
2. *Tartan Paving Co.*	$82,000	Payable immediately.
3. *Faultless Paving Co.*	$84,000	25% payable immediately; 75% payable in six months.

The building contractor uses a 12% nominal interest rate, compounded monthly, in this type of bid analysis. Which paving subcontractor should be awarded the paving job?

5-38 A cost analysis is to be made to determine what, if anything, should be done in a situation where there are three "do something" and one "do nothing" alternatives. Estimates of the cost and benefits are as follows:

Alternatives	*Cost*	*Uniform annual benefit*	*End-of-useful-life salvage value*	*Useful life, in years*
1	$500	$135	$ 0	5
2	600	100	250	5
3	700	100	180	10
4	0	0	0	0

Use a ten-year analysis period for the four mutually exclusive alternatives. At the end of five years, Alternatives 1 and 2 may be replaced with identical alternatives (with the same cost, benefits, salvage value, and useful life).

a. If an 8% interest rate is used, which alternative should be selected?

b. If a 12% interest rate is used, which alternative should be selected?

5-39 Consider five mutually exclusive alternatives:

	Alternatives				
	A	*B*	*C*	*D*	*E*
Initial cost	$600	$600	$600	$600	$600
Uniform annual benefits for first five years	100	100	100	150	150
Uniform annual benefits for last five years	50	100	110	0	50

The interest rate is 10%. If all the alternatives have a ten-year useful life, and no salvage value, which alternative should be selected?

5-40 On February 1, the Miro Company needs to purchase some office equipment. The company is presently short of cash and expects to be short for several months. The company treasurer has indicated that he could pay for the equipment as follows:

Date	*Payment*
April 1	$150
June 1	300
Aug. 1	450
Oct. 1	600
Dec. 1	750

A local office supply firm has been contacted, and they will agree to sell the equipment to the firm now and to be paid according to the treasurer's payment schedule. If interest will be charged at 3% every two months, with compounding once every two months, how much office equipment can the Miro Company buy now? (*Answer:* $2020)

5-41 Using 5% nominal interest, compounded continuously, solve for P.

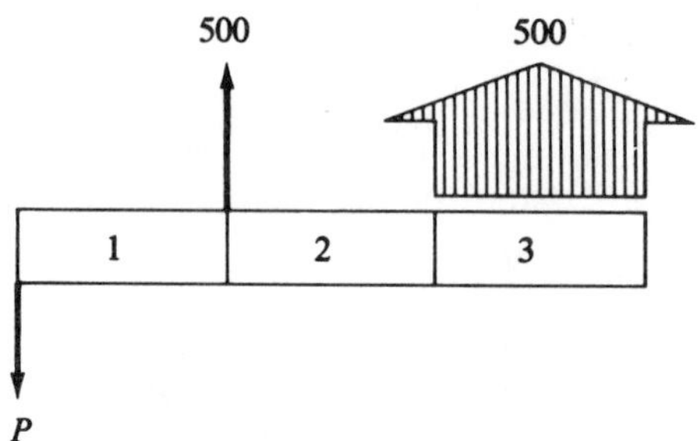

5-42 By installing some elaborate inspection equipment on its assembly line, the Robot Corp. can avoid hiring an extra worker. The worker would have earned $26,000 a year in wages, and Robot would have paid an additional $7500 a year in employee benefits. The inspection equipment has a six-year useful life and no salvage value. Use a nominal 18% interest rate in your calculations. How much can Robot afford to pay for the equipment if the wages and worker benefits are paid:

a. at the end of each year?

b. monthly?

c. continuously?

d. Explain why the answers in ***b*** and **c** are larger than in ***a***.

5-43 Using capitalized cost, determine which type of road surface is preferred on a particular section of highway. Use a 12% interest rate.

	A	*B*
Initial cost	$500,000	$700,000
Annual maintenance	35,000	25,000
Periodic resurfacing	350,000 every 10 years	450,000 every 15 years

CHAPTER 6

Annual Cash Flow Analysis

This chapter is devoted to annual cash flow analysis—the second of the three major analysis techniques. As we've said, alternatives must be resolved into such a form that they may be compared. This means we must use the equivalence concept to convert from a cash flow representing the alternative into some equivalent sum or equivalent cash flow.

With present worth analysis, we resolved an alternative into an equivalent cash sum. This might have been an equivalent present worth of cost, an equivalent present worth of benefit, or an equivalent net present worth. But instead of computing equivalent present sums, we could compare alternatives based on their equivalent annual cash flows. Depending on the particular situation, we may wish to compute the equivalent uniform annual cost (EUAC), the equivalent uniform annual benefit (EUAB), or their difference (EUAB − EUAC).

To prepare for a discussion of annual cash flow analysis, we will review some annual cash flow calculations, then examine annual cash flow criteria. Following this, we will proceed with annual cash flow analysis.

Annual Cash Flow Calculations

Resolving a Present Cost to an Annual Cost

Equivalence techniques were used in prior chapters to convert money, at one point in time, to some equivalent sum or series. In annual cash flow analysis, the goal is to convert money to an equivalent uniform annual cost or benefit. The simplest case is to convert a present sum P to a series of equivalent uniform end-of-period cash flows. This is illustrated in Example 6-1.

EXAMPLE 6-1

A woman bought $1000 worth of furniture for her home. If she expects it to last ten years, what will be her equivalent uniform annual cost if interest is 7%?

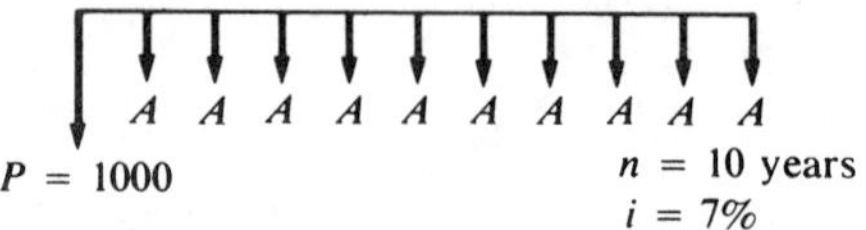

Solution:

$$\text{Equivalent uniform annual cost} = P(A/P, i, n)$$
$$= 1000(A/P, 7\%, 10)$$
$$= \$142.40$$

Her equivalent uniform annual cost is $142.40. ◀

Treatment of Salvage Value

In a situation where there is a salvage value, or future value at the end of the useful life of an asset, the result is to decrease the equivalent uniform annual cost.

EXAMPLE 6-2

The woman in Ex. 6-1 now believes she can resell the furniture at the end of ten years for $200. Under these circumstances, what is her equivalent uniform annual cost?

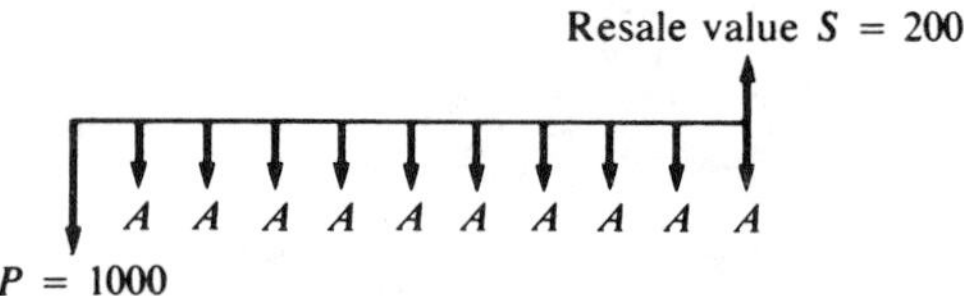

Solution: For this situation, the problem may be solved by each of three different calculations, as follows:

Ex. 6-2, Solution One:

$$\text{EUAC} = P(A/P,i,n) - S(A/F,i,n) \qquad (6\text{-}1)$$
$$= 100(A/P,7\%,10) - 200(A/F,7\%,10)$$
$$= 1000(0.1424) - 200(0.0724)$$
$$= 142.40 - 14.48 = \$127.92$$

This method reflects the Annual cost of the cash disbursement *minus* the Annual benefit of the future resale value.

Ex. 6-2, Solution Two: Equation 6-1 describes a relationship that may be modified by an identity presented in Chapter 4:

$$(A/P,i,n) = (A/F,i,n) + i \tag{6-2}$$

Substituting this into Eq. 6-1 gives:

$$\begin{aligned} \text{EUAC} &= P(A/F,i,n) + Pi - S(A/F,i,n) \\ &= (P - S)(A/F,i,n) + Pi \qquad (6\text{-}3) \\ &= (1000 - 200)(A/F,7\%,10) + 1000(0.07) \\ &= 800(0.0724) + 70 = 57.92 + 70 \\ &= \$127.92 \end{aligned}$$

This method computes the equivalent annual cost due to the unrecovered \$800 when the furniture is sold, and adds annual interest on the \$1000 investment.

Ex. 6-2, Solution Three: If the value for $(A/F,i,n)$ from Eq. 6-2 is substituted into Eq. 6-1, we obtain:

$$\begin{aligned} \text{EUAC} &= P(A/P,i,n) - S(A/P,i,n) + Si \\ &= (P - S)(A/P,i,n) + Si \qquad (6\text{-}4) \\ &= (1000 - 200)(A/P,7\%,10) + 200(0.07) \\ &= 800(0.1424) + 14 = 113.92 + 14 = \$127.92 \end{aligned}$$

This method computes the annual cost of the \$800 decline in value during the ten years, plus interest on the \$200 tied up in the furniture as the salvage value. ◀

Example 6-2 illustrates that when there is an initial disbursement P followed by a salvage value S, the annual cost may be computed in three different ways:

1. **EUAC = $P(A/P,i,n) - S(A/F,i,n)$** (6-1)
2. **EUAC = $(P - S)(A/F,i,n) + Pi$** (6-3)
3. **EUAC = $(P - S)(A/P,i,n) + Si$** (6-4)

Each of the three calculations gives the same results. In practice, the first and third methods are most commonly used.

EXAMPLE 6-3

Bill owned a car for five years. One day he wondered what his uniform annual cost for maintenance and repairs had been. He assembled the following data:

Year	Maintenance and repair cost for year
1	$ 45
2	90
3	180
4	135
5	225

Compute the equivalent uniform annual cost (EUAC) assuming 7% interest and end-of-year disbursements.

Solution: The EUAC may be computed for this irregular series of payments in two steps:

1. Compute the present worth of cost for the five years using single payment present worth factors.
2. With the PW of cost known, compute EUAC using the capital recovery factor.

$$\begin{aligned}\text{PW of cost} &= 45(P/F,7\%,1) + 90(P/F,7\%,2) + 180(P/F,7\%,3) \\ &\quad + 135(P/F,7\%,4) + 225(P/F,7\%,5) \\ &= 45(0.9346) + 90(0.8734) + 180(0.8163) \\ &\quad + 135(0.7629) + 225(0.7130) \\ &= \$531 \\ \text{EUAC} &= 531(A/P,7\%,5) = 531(0.2439) = \$130 \end{aligned}$$ ◀

EXAMPLE 6-4

Bill reexamined his calculations and found that he had reversed the Year 3 and 4 maintenance and repair costs in his table. The correct table is:

Year	Maintenance and repair cost for year
1	$ 45
2	90
3	135
4	180
5	225

Recompute the EUAC.

Solution: This time the schedule of disbursements is an arithmetic gradient series plus a uniform annual cost, as follows:

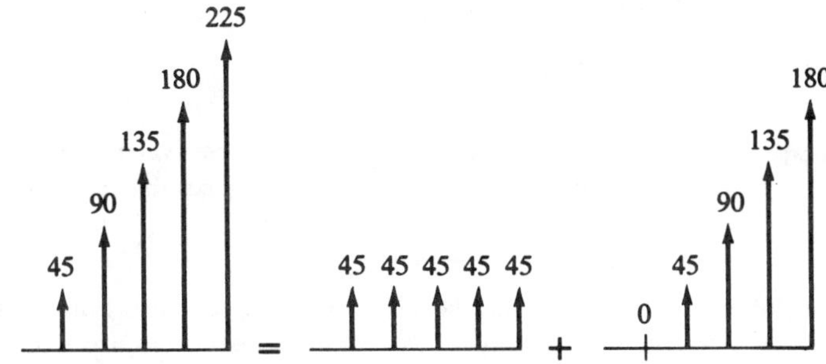

$$\begin{aligned}\text{EUAC} &= 45 + 45(A/G,7\%,5)\\ &= 45 + 45(1.865)\\ &= \$129\end{aligned}$$

Since the timing of the Ex. 6-3 and 6-4 expenditures is different, we would not expect to obtain the same EUAC. ◀

The examples have shown four essential points concerning cash flow calculations:

1. **There is a direct relationship between the present worth of cost and the equivalent uniform annual cost. It is**

   $$\textbf{EUAC} = \textbf{(PW of cost)}(A/P,i,n)$$

2. **In a problem, an expenditure of money increases the EUAC, while a receipt of money (like selling something for its salvage value) decreases EUAC.**
3. **When there are irregular cash disbursements over the analysis period, a convenient method of solution is to first determine the PW of cost; then, using the equation in Item 1 above, the EUAC may be calculated.**
4. **Where there is an increasing uniform gradient, EUAC may be rapidly computed using the arithmetic gradient uniform series factor, $(A/G,i,n)$.**

Annual Cash Flow Analysis

The criteria for economic efficiency are presented in Table 6-1. One notices immediately that the table is quite similar to Table 5-1. In the case of fixed input, for example, the present worth criterion is *maximize PW of benefits*, and the annual cost criterion is *maximize equivalent uniform annual benefits*. It is appar-

Table 6-1 ANNUAL CASH FLOW ANALYSIS

	Situation	*Criterion*
Fixed input	**Amount of money or other input resources is fixed**	**Maximize equivalent uniform benefits (maximize EUAB)**
Fixed output	**There is a fixed task, benefit, or other output to be accomplished**	**Minimize equivalent uniform annual cost (minimize EUAC)**
Neither input nor output is fixed	**Neither amount of money, or other inputs, nor amount of benefits, or other outputs, is fixed**	**Maximize (EUAB − EUAC)**

ent that, if you are maximizing the present worth of benefits, simultaneously you must be maximizing the equivalent uniform annual benefits. This is illustrated in Example 6-5.

EXAMPLE 6-5

A firm is considering which of two devices to install to reduce costs in a particular situation. Both devices cost \$1000 and have useful lives of five years with no salvage value. Device *A* can be expected to result in \$300 savings annually. Device *B* will provide cost savings of \$400 the first year but will decline \$50 annually, making the second year savings \$350, the third year savings \$300, and so forth. With interest at 7%, which device should the firm purchase?

Solution:

Device A:

$$\text{EUAB} = \$300$$

Device B:

$$\text{EUAB} = 400 - 50(A/G,7\%,5) = 400 - 50(1.865)$$
$$= \$306.75$$

To maximize EUAB, select device *B*. ◀

Example 6-5 was previously presented as Ex. 5-1 where we found:

$$\text{PW of benefits } A = 300(P/A,7\%,5) = 300(4.100) = \$1230$$

This is converted to EUAB by multiplying by the capital recovery factor:

$$\text{EUAB}_A = 1230(A/P,7\%,5) = 1230(0.2439) = \$300$$

$$\text{PW of benefits } B = 400(P/A,7\%,5) - 50(P/G,7\%,5)$$
$$= 400(4.100) - 50(7.647) = \$1257.65$$

and, hence,

$$\text{EUAB}_B = 1257.65(A/P,7\%,5) = 1257.65(0.2439)$$
$$= \$306.75$$

We see, therefore, that it is easy to convert the present worth analysis results into the annual cash flow analysis results. We could go from annual cash flow to present worth just as easily using the series present worth factor.

EXAMPLE 6-6

Three alternatives are being considered for improving an operation on the assembly line. The cost of the equipment varies as do their annual benefits compared to the present situation. Each of the alternatives has a ten-year life and a scrap value equal to 10% of its original cost.

	Plan A	*Plan B*	*Plan C*
Installed cost of equipment	$15,000	$25,000	$33,000
Material and labor savings per year	14,000	9,000	14,000
Annual operating expenses	8,000	6,000	6,000
End-of-useful life scrap value	1,500	2,500	3,300

If interest is 8%, which plan, if any, should be adopted?

Solution: Since neither installed cost nor output benefits are fixed, the economic criterion is to maximize (EUAB − EUAC).

	Plan A	*Plan B*	*Plan C*	*Do nothing*
Equivalent uniform annual benefit (EUAB):				
Material and labor per year	$14,000	$9,000	$14,000	$0
Scrap value $(A/F,8\%,10)$	104	172	228	0
EUAB =	$14,104	$9,172	$14,228	$0
Equivalent uniform annual cost (EUAC):				
Installed cost $(A/P,8\%,10)$	$ 2,235	$3,725	$ 4,917	$0
Annual operating expenses	8,000	6,000	6,000	0
EUAC =	$10,235	$9,725	$10,917	$0
(EUAB − EUAC) =	$ 3,869	−$553	$ 3,311	$0

Based on our criterion of maximizing (EUAB − EUAC), Plan *A* is the best of the four alternatives. We note, however, that since the do-nothing alternative has (EUAB − EUAC) = 0, it is a more desirable alternative than Plan *B*. ◀

Analysis Period

In the last chapter, we saw that the analysis period was an important consideration in computing present worth comparisons. It was essential that a common analysis period be used for each alternative. In annual cash flow comparisons, we again have the analysis period question. Example 6-7 will help in examining the problem.

EXAMPLE 6-7

Two pumps are being considered for purchase. If interest is 7%, which pump should be bought?

	Pump A	*Pump B*
Initial cost	$7000	$5000
End-of-useful-life salvage value	1500	1000
Useful life, in years	12	6

Solution: The annual cost for twelve years of Pump *A* can be found using Eq. 6-4:

$$\begin{aligned} \text{EUAC} &= (P - S)(A/P,i,n) + Si \\ &= (7000 - 1500)(A/P,7\%,12) + 1500(0.07) \\ &= 5500(0.1259) + 105 = \$797 \end{aligned}$$

Now compute the annual cost for six years of Pump *B*:

$$\begin{aligned} \text{EUAC} &= (5000 - 1000)(A/P,7\%,6) + 1000(0.07) \\ &= 4000(0.2098) + 70 = \$909 \end{aligned}$$

For a common analysis period of twelve years, we need to replace Pump *B* at the end of its six-year useful life. If we assume that another Pump *B′* can be obtained that has the same $5000 initial cost, $1000 salvage value and six-year life, the cash flow will be as follows:

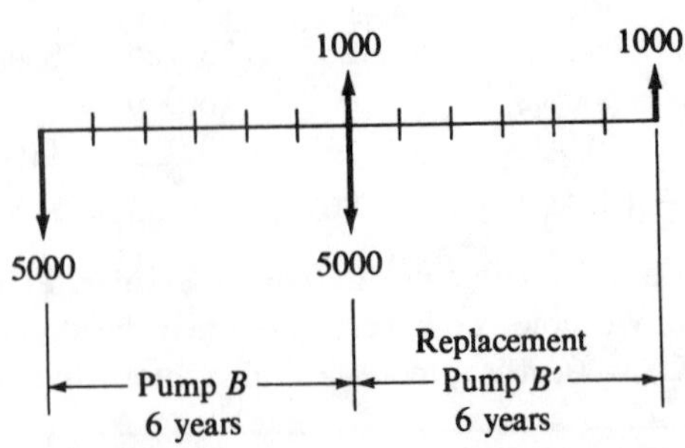

For the twelve-year analysis period, the annual cost for Pump B:

$$\begin{aligned}\text{EUAC} &= [5000 - 1000(P/F,7\%,6) + 5000(P/F,7\%,6) \\ &\quad - 1000(P/F,7\%,12)] \\ &\quad \times (A/P,7\%,12) \\ &= [5000 - 1000(0.6663) + 5000(0.6663) - 1000(0.4440)] \\ &\quad \times (0.1259) \\ &= (5000 - 666 + 3331 - 444)(0.1259) \\ &= (7211)(0.1259) = \$909\end{aligned}$$

The annual cost of B for the six-year analysis period is the same as the annual cost for the twelve-year analysis period. This is not a surprising conclusion when one recognizes that the annual cost of the first six-year period is repeated in the second six-year period. By assuming that the shorter-life equipment is replaced by equipment with identical economic consequences, we have avoided the analysis period problem. Select Pump A. ◀

Analysis Period Equal to Alternative Lives

When the analysis period for an economy study coincides with the useful life for each alternative, we have an ideal situation which causes no difficulties. The economy study is based on this analysis period.

Analysis Period a Common Multiple of Alternative Lives

When the analysis period is a common multiple of the alternative lives (for example, in Ex. 6-7, the analysis period was twelve years with six- and twelve-year alternative lives), a "replacement with an identical item with the same costs, performance, and so forth" is frequently assumed. This means that when an alternative has reached the end of its useful life, it is assumed to be replaced with an identical item. As shown in Ex. 6-7, the result is that the EUAC for Pump B with a six-year useful life is equal to the EUAC for the entire analysis period based on Pump B *plus* Replacement Pump B'.

Under these circumstances of identical replacement, it is appropriate to compare the annual cash flows computed for alternatives based on their own service lives. In Ex. 6-7, the annual cost for Pump A, based on its 12-year service life, was compared with the annual cost for Pump B, based on its six-year service life.

Analysis Period Is a Continuing Requirement

Many times the economic analysis is to determine how to provide for a more or less continuing requirement. One might need to pump water from a well as a continuing requirement. There is no distinct analysis period. In this situation, the analysis period is assumed to be long but undefined.

If, for example, we had a continuing requirement to pump water and alternative Pumps *A* and *B* had useful lives of seven and eleven years, respectively, what should we do? The customary assumption is that Pump *A*'s annual cash flow (based on a seven-year life) may be compared to Pump *B*'s annual cash flow (based on an eleven-year life). This is done without much concern that the least common multiple of the seven- and eleven-year lives is 77 years. This comparison of "different-life" alternatives assumes identical replacement (with identical costs, performance, and so forth) when an alternative reaches the end of its useful life. Example 6-8 illustrates the situation.

EXAMPLE 6-8

Pump *B* in Ex. 6-7 is now believed to have a nine-year useful life. Assuming the same initial cost and salvage value, compare it with Pump *A* using the same 7% interest rate.

Solution: If we assume that the need for *A* or *B* will exist for some continuing period, the comparison of annual costs for the unequal lives is an acceptable technique. For twelve years of Pump *A*:

$$\text{EUAC} = (7000 - 1500)(A/P,7\%,12) + 1500(0.07) = \$797$$

For nine years of Pump *B*:

$$\text{EUAC} = (5000 - 1000)(A/P,7\%,9) + 1000(0.07) = \$684$$

For minimum EUAC, select Pump *B*. ◀

Infinite Analysis Period

At times we have an alternative with a limited (finite) useful life in an infinite analysis period situation. The equivalent uniform annual cost may be computed for the limited life. The assumption of identical replacement (replacements have identical costs, performance, and so forth) is often appropriate. Based on this assumption, the same EUAC occurs for each replacement of the limited-life alternative. The EUAC for the infinite analysis period is therefore equal to the EUAC computed for the limited life. With identical replacement,

$$\text{EUAC}_{\substack{\text{for infinite}\\ \text{analysis period}}} = \text{EUAC}_{\substack{\text{for limited}\\ \text{life } n}}$$

A somewhat different situation occurs when there is an alternative with an infinite life in a problem with an infinite analysis period:

$$\text{EUAC}_{\substack{\text{for infinite}\\ \text{analysis period}}} = P(A/P,i,\infty) + \text{any other costs}$$

When $n = \infty$, we have $A = Pi$ and, hence, $(A/P,i,\infty)$ equals i.

$$\text{EUAC}_{\substack{\text{for infinite}\\\text{analysis period}}} = Pi + \text{any other costs}$$

EXAMPLE 6-9

In the construction of the aqueduct to expand the water supply of a city, there are two alternatives for a particular portion of the aqueduct. Either a tunnel can be constructed through a mountain, or a pipeline can be laid to go around the mountain. If there is a permanent need for the aqueduct, should the tunnel or the pipeline be selected for this particular portion of the aqueduct? Assume a 6% interest rate.

Solution:

	Tunnel through mountain	*Pipeline around mountain*
Initial cost	\$5.5 million	\$5 million
Maintenance	0	0
Useful life	Permanent	50 years
Salvage value	0	0

Tunnel: For the tunnel, with its permanent life, we want $(A/P,6\%,\infty)$. For an infinite life, the capital recovery is simply interest on the invested capital. So $(A/P,6\%,\infty) = i$,

$$\begin{aligned}\text{EUAC} &= Pi = \$5.5 \text{ million}(0.06)\\ &= \$330{,}000\end{aligned}$$

Pipeline:

$$\begin{aligned}\text{EUAC} &= \$5 \text{ million}(A/P,6\%,50)\\ &= \$5 \text{ million}(0.0634) = \$317{,}000\end{aligned}$$

For fixed output, minimize EUAC. Select the pipeline.

The difference in annual cost between a long life and an infinite life is small unless an unusually low interest rate is used. Here the tunnel is assumed to be permanent. At 6%, the annual cost = \$5.5 million × 0.06 = \$330,000. What would be the annual cost if an 85-year life is assumed for the tunnel?

Solution:

$$\begin{aligned}\text{EUAC} &= \$5.5 \text{ million}(A/P,6\%,85)\\ &= \$5.5 \text{ million}(0.0604) = \$332{,}000\end{aligned}$$

The difference in time between 85 years and infinity is great indeed, yet the difference in annual costs is slight. ◀

Some Other Analysis Period

The analysis period in a particular problem may be something other than one of the four we have so far described. It may be equal to the life of either the shorter-life alternative, the longer-life alternative, or something entirely different. One must carefully examine the consequences of each alternative throughout the analysis period and, in addition, see what differences there might be in salvage values, and so forth, at the end of the analysis period.

SUMMARY

Annual cash flow analysis is the second of the three major methods of resolving alternatives into comparable values. When an alternative has an initial cost P and salvage value S, there are three ways of computing the equivalent uniform annual cost:

- EUAC $= P(A/P,i,n) - S(A/F,i,n)$
- EUAC $= (P - S)(A/F,i,n) + Pi$
- EUAC $= (P - S)(A/P,i,n) + Si$

All three equations give the same answer.

The relationship between the present worth of cost and the equivalent uniform annual cost is:

- EUAC = (PW of cost)$(A/P,i,n)$

The three annual cash flow criteria are:

For fixed input	Maximize EUAB
For fixed output	Minimize EUAC
Neither input nor output fixed	Maximize (EUAB − EUAC)

In present worth analysis there must be a common analysis period. Annual cash flow analysis, however, allows some flexibility provided the necessary assumptions are suitable in the situation being studied. The analysis period may be different from the lives of the alternatives, and a valid cash flow analysis made, provided the following two criteria are met:

1. When an alternative has reached the end of its useful life, it is assumed to be replaced by an identical replacement (with the same costs, performance, and so forth).
2. The analysis period is a common multiple of the useful lives of the alternatives, or there is a continuing or perpetual requirement for the selected alternative.

If both these conditions do not apply, then it is necessary to make a detailed study of the consequences of the various alternatives over the entire analysis period with particular attention to the difference between the alternatives at the end of the analysis period.

There is very little numerical difference between a long-life alternative and a perpetual alternative. As the value of n increases, the capital recovery factor approaches i. At the limit, $(A/P,i,\infty) = i$.

Problems

6-1 On April 1st, \$100 is loaned to a man. The loan is to be repaid in three equal semi-annual (every six months) payments. If the annual interest rate is 7%, how much is each payment? (*Answer:* \$35.69)

6-2 Compute the value of C for the following diagram, based on a 10% interest rate.

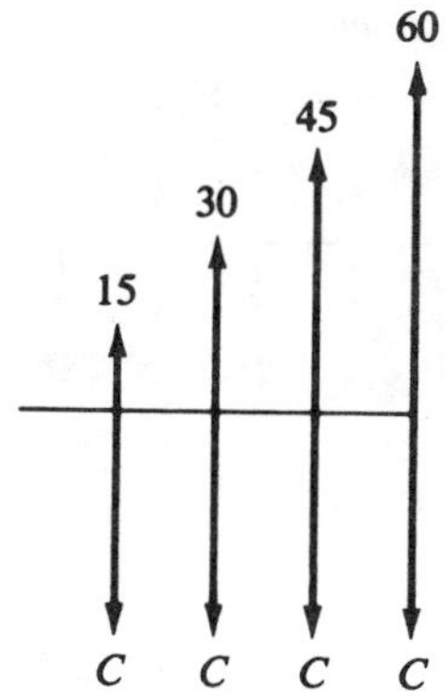

(*Answer:* C = \$35.72)

6-3 Compute the value of B for the following diagram:

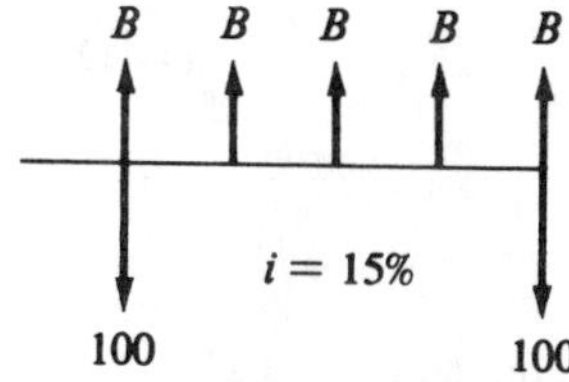

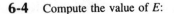

6-4 Compute the value of E:

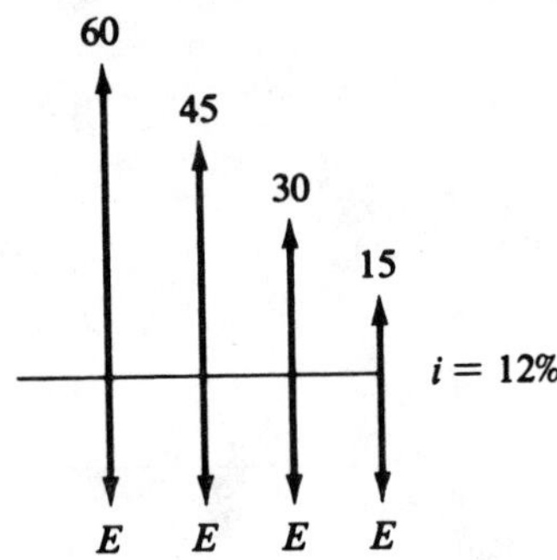

6-5 If $i = 6\%$, compute the value of D that is equivalent to the two disbursements shown.

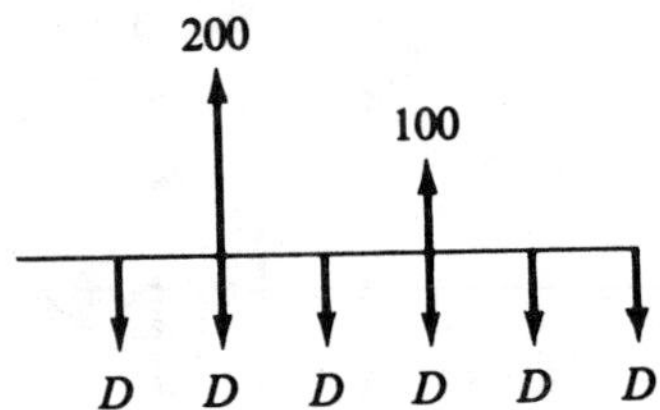

(*Answer:* D = \$52.31)

6-6 For the diagram, compute the value of D:

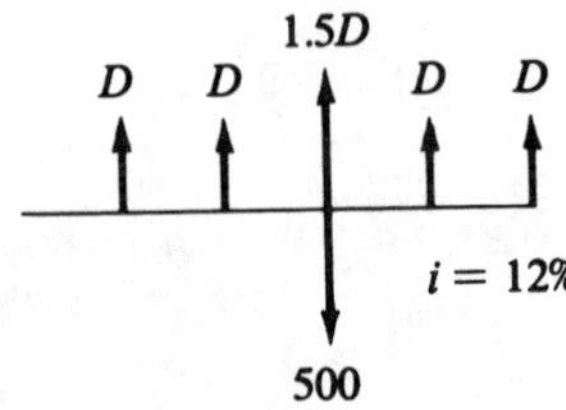

6-7 What is C in the figure on the next page?

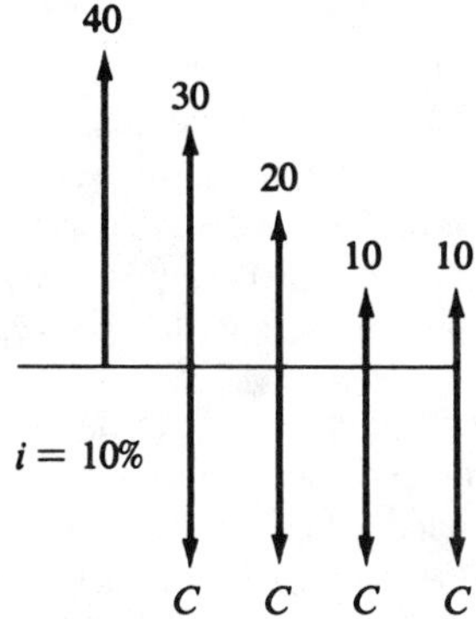

6-8 If interest is 10%, what is A?

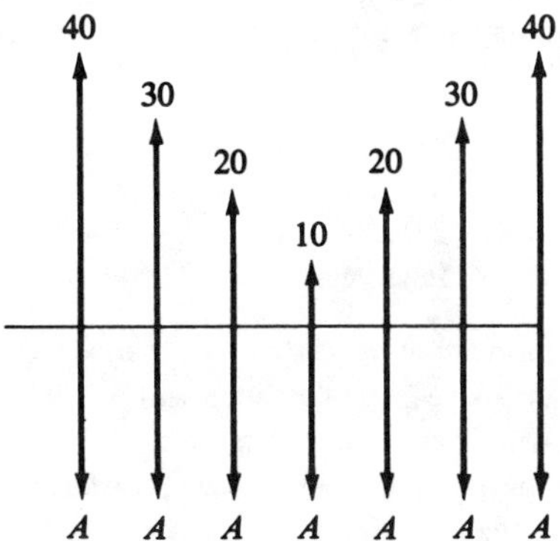

6-9 A certain industrial firm desires an economic analysis to determine which of two different machines should be purchased. Each machine is capable of performing the same task in a given amount of time. Assume the minimum attractive return is 8%. The following data are to be used in this analysis:

	Machine X	*Machine Y*
First cost	$5000	$8000
Estimated life, in years	5	12
Salvage value	0	$2000
Annual maintenance cost	0	150

Which machine would you choose? Base your answer on annual cost.
(*Answers:* X = $1252; Y = $1106)

6-10 An electronics firm invested $60,000 in a precision inspection device. It cost $4000 to operate and maintain in the first year, and $3000 in each of the subsequent years. At the end of four years, the firm changed their inspection procedure, eliminating the need for the device. The purchasing agent was very fortunate in being able to sell the inspection device for the $60,000 that had originally been paid for it. The plant manager asks you to compute the equivalent uniform annual cost of the device during the four years it was used. Assume interest at 10% per year. (*Answer:* $9287)

6-11 A firm is about to begin pilot plant operation on a process it has developed. One item of optional equipment that could be obtained is a heat exchanger unit. The company finds that one can be obtained now for $30,000, and that this unit can be used in other company operations. It is estimated that the heat exchanger unit will be worth $35,000 at the end of eight years. This seemingly high salvage value is due primarily to the fact that the $30,000 purchase price is really a rare bargain. If the firm believes 15% is an appropriate rate of return, what annual benefit is needed to justify the purchase of the heat exchanger unit? (*Answer:* $4135)

6-12 The maintenance foreman of a plant in reviewing his records found that a large press had the following maintenance cost record for the last five years:

5 years ago:	$ 600
4 years ago:	700
3 years ago:	800
2 years ago:	900
Last year:	1000

After consulting with a lubrication specialist, he changed the preventive maintenance schedule. He believes that this year maintenance will be $900 and will decrease $100 a year in each of the following four years. If his estimate of the future is correct, what will be the equivalent uniform annual maintenance cost for the ten-year period? Assume interest at 8%. (*Answer:* $756)

6-13 A firm purchased some equipment at a very favorable price of $30,000. The equipment resulted in an annual net saving of $1000 per year during the eight years it was used. At the end of eight years, the equipment was sold for $40,000. Assuming interest at 8%, did the equipment purchase prove to be desirable?

6-14 A manufacturer is considering replacing a production machine tool. The new machine would cost $3700, have a life of four years, have no salvage value, and save the firm $500 per year in direct labor costs and $200 per year indirect labor costs. The existing machine tool was purchased four years ago at a cost of $4000. It will last four more years and have no salvage value at the end of that time. It could be sold now for $1000 cash. Assume money is worth 8%, and that the difference in taxes, insurance, and so forth, for the two alternatives is negligible. Determine whether or not the new machine should be purchased.

6-15 Two possible routes for a power line are under study. Data on the routes are as follows:

	Around the lake	*Under the lake*
Length	15 km	5 km
First cost	\$5000/km	\$25,000/km
Maintenance	\$200/km/yr	\$400/km/yr
Useful life, in years	15	15
Salvage value	\$3000/km	\$5000/km
Yearly power loss	\$500/km	\$500/km
Annual property taxes	2% of first cost	2% of first cost

If 7% interest is used, should the power line be routed around the lake or under the lake? (*Answer:* Around the lake.)

6-16 Steve Lowe must pay his property taxes in two equal installments on December 1 and April 1. The two payments are for taxes for the fiscal year that begins on July 1 and ends the following June 30. Steve purchased a home on September 1. He estimates the annual property taxes will be \$850 per year. Assuming the annual property taxes remain at \$850 per year for the next several years, Steve plans to open a savings account and to make uniform monthly deposits the first of each month. The account is to be used to pay the taxes when they are due.

To begin the account, Steve deposits a lump sum equivalent to the monthly-payments-that-will-not-have-been-made for the first year's taxes. The savings account pays 9% interest, compounded monthly and payable quarterly (March 31, June 30, September 30, and December 31). How much money should Steve put into the account when he opens it on September 1? What uniform monthly deposit should he make from that time on? (A careful *exact* solution is expected.)

(*Answers:* Initial deposit \$350.28; Monthly deposit \$69.02)

6-17 An oil refinery finds that it is now necessary to process its waste liquids in a costly treating process before discharging them into a nearby stream. The engineering department estimates that the waste liquid processing will cost \$30,000 at the end of the first year. By making process and plant alterations, it is estimated that the waste treatment cost will decline \$3000 each year. As an alternate, a specialized firm, Hydro-Clean, has offered a contract to process the waste liquids for the ten years for a fixed price of \$15,000 per year, payable at the end of each year. Either way, there should be no need for waste treatment after ten years. If the refinery manager considers 8% a suitable interest rate, should he accept the Hydro-Clean offer or not?

6-18 Bill Anderson buys an automobile every two years as follows: initially he pays a downpayment of \$6000 on a \$15,000 car. The balance is paid in 24 equal monthly payments with annual interest at 12%. When he has made the last payment on the loan, he trades in the two-year old car for \$6000 on a new \$15,000 car, and the cycle begins over again.

Doug Jones decided on a different purchase plan. He thought he would be better off if he paid \$15,000 cash for a new car. Then he would make a monthly deposit in a savings account so that, at the end of two years, he would have \$9000 in the account. The \$9000 plus the \$6000 trade-in value of the car will allow Doug to replace his two-year-old car by paying \$9000 for a new one. The bank pays 6% interest, compounded quarterly.

a. What is Bill Anderson's monthly payment to pay off the loan on the car?

b. After he purchased the new car for cash, how much per month should Doug Jones deposit in his savings account to have sufficient money for the next car two years hence?

c. Why is Doug's monthly savings account deposit smaller than Bill's payment?

6-19 Claude James, a salesman, needs a new car for use in his business. He expects to be promoted to a supervisory job at the end of three years, and so his concern now is to have a car for the three years he expects to be "on the road." The company will reimburse their salesmen each month at the rate of 25¢ per mile driven. Claude has decided to drive a low-priced automobile. He finds, however, that there are three different ways of obtaining the automobile:

a. Purchase for cash; the price is $13,000.

b. Lease the car; the monthly charge is $350 on a 36-month lease, payable at the end of each month; at the end of the three-year period, the car is returned to the leasing company.

c. Lease the car with an option to purchase it at the end of the lease; pay $360 a month for 36 months; at the end of that time, Claude could purchase the car, is he chooses, for $3500.

Claude believes he should use a 12% interest rate in determining which alternative to select. If the car could be sold for $4000 at the end of three years, which method should he use to obtain it?

6-20 A college student has been looking for a new tire for his car and has located the following alternatives:

Tire warranty	*Price per tire*
12 mo.	$39.95
24 mo.	59.95
36 mo.	69.95
48 mo.	90.00

If the student feels that the warranty period is a good estimate of the tire life and that a 10% interest rate is appropriate, which tire should he buy?

6-21 A suburban taxi company is considering buying taxis with diesel engines instead of gasoline engines. The cars average 50,000 kilometers a year, with a useful life of three years for the taxi with the gas engine, and four years for the diesel taxi. Other comparative information is as follows:

	Diesel	*Gasoline*
Vehicle cost	$13,000	$12,000
Fuel cost per liter	48¢	51¢
Mileage, in km/liter	35	28
Annual repairs	300	200
Annual insurance premium	500	500
End-of-useful-life resale value	2,000	3,000

Determine the more economical choice if interest rate is 6%.

6-22 When he started work on his 22nd birthday, D. B. Cooper decided to invest money each month with the objective of becoming a millionaire by the time he reaches his 65th birthday. If he expects his investments to yield 18% per annum, compounded monthly, how much should he invest each month? (*Answer:* $6.92 a month.)

6-23 Linda O'Shay deposited $30,000 in a savings account as a perpetual trust. She believes the account will earn 7% annual interest during the first ten years and 5% interest thereafter. The trust is to provide a uniform end-of-year scholarship at the University. What *uniform* amount could be used for the student scholarship each year, beginning at the end of the first year and continuing forever?

6-24 A motorcycle is for sale for $2600. The motorcycle dealer is willing to sell it on the following terms:

- No downpayment; pay $44 at the end of each of the first four months; pay $84 at the end of each month after that, until the motorcycle is paid in full.

Based on these terms and a 12% annual interest rate compounded monthly, how many $84 payments will be required?

6-25 A machine costs $20,000 and has a five-year useful life. At the end of the five years, it can be sold for $4000. If annual interest is 8%, compounded semi-annually, what is the equivalent uniform annual cost of the machine? (An *exact* solution is expected.)

6-26 The average age of engineering students when they graduate is a little over 23 years. This means the working career of most engineers is almost exactly 500 months.

How much would an engineer need to save each month to become a millionaire by the end of his working career? Assume a 15% interest rate, compounded monthly.

6-27 As shown in the cash flow diagram, there is an annual disbursement of money that varies from year to year from $100 to $300 in a fixed pattern that repeats forever. If interest is 10%, compute the value of A, also continuing forever, that is equivalent to the fluctuating disbursements.

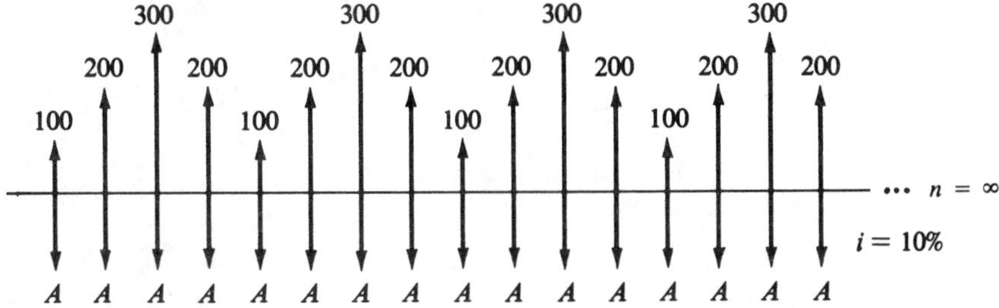

6-28 Alice Cooper has arranged to buy some home recording equipment. She estimates that it will have a five-year useful life and no salvage value. The dealer, who is a friend, has offered Alice two alternative ways to pay for the equipment:

a. Pay $2000 immediately and $500 at the end of one year.

b. Pay nothing until the end of four years when a single payment of $3000 must be made.

If Alice believes 12% is a suitable interest rate, which method of payment should she select? (*Answer:* Select ***b***)

6-29 A company must decide whether to buy Machine *A* or Machine *B*:

	Machine A	*Machine B*
Initial cost	$10,000	$20,000
Useful life, in years	4	10
End-of-useful-life salvage value	$10,000	$10,000
Annual maintenance	1,000	0

At a 10% interest rate, which machine should be installed? (*Answer:* Machine *A*)

6-30 The Johnson Company pays $200 a month to a trucker to haul wastepaper and cardboard to the city dump. The material could be recycled if the company would buy a $6000 hydraulic press bailer and spend $3000 a year for labor to operate the bailer. The bailer has an estimated useful life of thirty years and no salvage value. Strapping material would cost $200 per year for the estimated 500 bales a year that would be produced. A wastepaper company will pick up the bales at the plant and pay the Johnson Co. $2.30 per bale for them.

a. If interest is 8%, is it economical to install and operate the bailer?

b. Would you recommend that the bailer be installed?

6-31 Consider the following:

	Alternative	
	A	*B*
Cost	$50	$180
Uniform annual benefit	15	60
Useful life, in years	10	5

The analysis period is ten years, but there will be no replacement for Alternative *B* at the end of five years. Based on a 15% interest rate, determine which alternative should be selected.

6-32 Consider the following two mutually exclusive alternatives:

	A	*B*
Cost	$100	$150
Uniform annual benefit	16	24
Useful life, in years	∞	20

Alternative *B* may be replaced with an identical item every twenty years at the same $150 cost and will have the same $24 uniform annual benefit. Using a 10% interest rate, determine which alternative should be selected.

6-33 Some equipment will be installed in a warehouse that a firm has leased for seven years. There are two alternatives:

	A	B
Cost	\$100	\$150
Uniform annual benefit	55	61
Useful life, in years	3	4

At any time after the equipment is installed, it has no salvage value. Assume that Alternatives A and B will be replaced at the end of their useful lives by identical equipment with the same costs and benefits. For a seven-year analysis period and a 10% interest rate, determine which alternative should be selected.

6-34 When he purchased his home, Al Silva borrowed \$80,000 at 10% interest to be repaid in 25 equal annual end-of-year payments. Ten years later, after making ten payments, Al found he could refinance the balance due on his loan at 9% interest for the remaining 15 years.

To refinance the loan, Al must pay the original lender the balance due on the loan, plus a penalty charge of 2% of the balance due; to the new lender he also must pay a \$1000 service charge to obtain the loan. The new loan would be made equal to the balance due on the old loan, plus the 2% penalty charge, and the \$1000 service charge. Should Al refinance the loan, assuming that he will keep the house for the next 15 years?

6-35 Consider the following three mutually exclusive alternatives:

	A	B	C
Cost	\$100	\$150.00	\$200.00
Uniform annual benefit	10	17.62	55.48
Useful life, in years	∞	20	5

Assuming that Alternatives B and C are replaced with identical replacements at the end of their useful lives, and an 8% interest rate, which alternative should be selected?
(*Answer:* Select C)

6-36 When Sandra began working, she resolved to save \$1000 a year from her income. After working a couple of months, she realized that it is easier to set a goal than to follow it. Her overall goal is to have saved a "reasonable sum of money" after ten years of work; she decides she can accomplish the same goal in a less painful way by changing her savings pattern: instead of saving \$1000 a year in a bank account that pays 6% annual interest, she will save an annual sum based on the geometric gradient. She believes her salary will increase 7% each year. Thus, she can save a fixed percentage of her salary each year, and still achieve her goal of saving \$1000/year, by making a smaller deposit at the end of the first year and increasing the amount of the deposit each year by the same 7% rate her salary increases.

- **a.** What amount does she want to have in the bank at the end of ten years?
- **b.** Following her revised savings plan, how much should she deposit in her savings account at the end of the first year?

6-37 An engineer has a fluctuating future budget for the maintenance of a particular machine. During each of the first five years, \$1000 per year will be budgeted. During the second five years, the annual budget will be \$1500 per year. In addition, \$3500 will be budgeted for an overhaul of the machine at the end of the fourth year, and another $3500 for an overhaul at the end of the eighth year.

The engineer asks you to compute what uniform annual expenditure would be equivalent to these fluctuating amounts, assuming interest at 6% per year.

6-38 An engineer wishes to have five million dollars by the time he retires in 40 years. Assuming 15% nominal interest, compounded continuously, what annual sum must he set aside? (*Answer:* $2011)

6-39 Two mutually exclusive alternatives are being considered.

Year	*A*	*B*
0	−$3000	−$5000
1	+845	+1400
2	+845	+1400
3	+845	+1400
4	+845	+1400
5	+845	+1400

One of the alternatives must be selected. Using a 15% nominal interest rate, compounded continuously, determine which one. Solve by annual cash flow analysis.

6-40 A company must decide whether to provide their salesmen with company-owned automobiles, or to pay the salesmen a mileage allowance and have them drive their own automobiles. New automobiles would cost about $18,000 each and could be resold four years later for about $7000 each. Annual operating costs would be $600 per year plus 12¢ per mile.

If the salesmen drive their own automobiles, the company probably would pay them 30¢ per mile. Calculate the number of miles each salesman would have to drive each year for it to be economically practical for the company to provide the automobiles. Assume a 10% annual interest rate.

6-41 (20-35) A pump is required for ten years at a remote location. The pump can be driven by an electric motor if a powerline is extended to the site. Otherwise, a gasoline engine will be used. Based on the following data and a 10% interest rate, how should the pump be powered?

	Gasoline	*Electric*
First cost	$2400	$6000
Annual operating cost	1200	750
Annual maintenance	300	50
Salvage value	300	600
Life in years	5 yrs	10 yrs

CHAPTER 7

Rate Of Return Analysis

The third of the three major analysis methods is rate of return. In this chapter we will examine three aspects of rate of return. First, we describe the meaning of "rate of return"; then, the calculation of rate of return is illustrated; finally, rate of return analysis problems will be presented. In an Appendix to this chapter, we describe difficulties sometimes encountered in attempting to compute an interest rate for certain kinds of cash flows.

Rate Of Return

In Chapter 4 we examined four plans to repay $5000 in five years with interest at 8% (Table 4-1). In each of the four plans the amount loaned ($5000) and the loan duration (five years) was the same. Yet the total interest paid to the lender varied from $1200 to $2347, depending on the loan repayment plan. We saw, however, that the lender received 8% interest each year on the amount of money actually owed. And, at the end of five years, the principal and interest payments exactly repaid the $5000 debt with interest at 8%. We say the lender received an "8% rate of return."

> ***Rate of return* is defined as the interest rate paid on the unpaid balance of a *loan* such that the payment schedule makes the unpaid loan balance equal to zero when the final payment is made.**

Instead of lending money, we might invest $5000 in a machine tool with a five-year useful life and an equivalent uniform annual benefit of $1252. An appropriate question is, "What rate of return would we receive on this investment?" The cash flow would be as follows:

Year	*Cash flow*
0	−$5000
1	+1252
2	+1252
3	+1252
4	+1252
5	+1252

We recognize the cash flow as Plan 3 of Table 4-1. We know that five payments of $1252 are equivalent to a present sum of $5000 when interest is 8%. Therefore, the rate of return on this investment is 8%. Stated in terms of an investment, we may define rate of return as follows:

***Rate of return* is the interest rate earned on the unrecovered *investment* such that the payment schedule makes the unrecovered investment equal to zero at the end of the life of the investment.**

It must be understood that the 8% rate of return does not mean an annual return of 8% on the $5000 investment, or $400 in each of the five years. Instead, each $1252 payment represents an 8% return on the Unrecovered investment *plus* the Partial return of the investment. This may be tabulated as follows:

Year	*Cash flow*	*Unrecovered investment at beginning of year*	*8% return on unrecovered investment*	*Investment repayment at end of year*	*Unrecovered investment at end of year*
0	−$5000				
1	+1252	$5000	$ 400	$ 852	$4148
2	+1252	4148	331	921	3227
3	+1252	3227	258	994	2233
4	+1252	2233	178	1074	1159
5	+1252	1159	93	1159	0
			$1260	$5000	

This cash flow represents a situation where the $5000 investment has benefits that produce an 8% rate of return. But, in the five-year period, the total return is only $1260, far less than $400 per year for five years. The reason, we can see, is because rate of return is defined as the interest rate earned on the unrecovered investment.

Although the two definitions of rate of return are stated differently, one in terms of a loan and the other in terms of an investment, there is only one fundamental concept being described. It is that the ***rate of return is the interest rate at which the benefits are equivalent to the costs.***

Calculating Rate Of Return

To calculate a rate of return on an investment, we must convert the various consequences of the investment into a cash flow. Then we will solve the cash flow for the unknown value of i, which is the rate of return. Five forms of the cash flow equation are:

$$\textbf{PW of benefits} - \textbf{PW of costs} = 0 \qquad (7\text{-}1)$$

$$\frac{\textbf{PW of benefits}}{\textbf{PW of costs}} = 1 \qquad (7\text{-}2)$$

$$\textbf{Net Present Worth} = 0 \qquad (7\text{-}3)$$

$$\textbf{EUAB} - \textbf{EUAC} = 0 \qquad (7\text{-}4)$$

$$\textbf{PW of costs} = \textbf{PW of benefits} \qquad (7\text{-}5)$$

The five equations represent the same concept in different forms. They can relate costs and benefits with rate of return i as the only unknown. The calculation of rate of return is illustrated by the following Examples.

EXAMPLE 7-1

An $8200 investment returned $2000 per year over a five-year useful life. What was the rate of return on the investment?

Solution: Using Equation 7-2,

$$\frac{\text{PW of benefits}}{\text{PW of costs}} = 1 \qquad \frac{2000(P/A,i,5)}{8200} = 1$$

Rewriting the equation, we see that

$$(P/A,i,5) = \frac{8200}{2000} = 4.1$$

Then look at the Compound Interest Tables for the value of i where $(P/A,i,5) = 4.1$; if no tabulated value of i gives this value, we will then find values on either side of the desired value (4.1) and interpolate to find the rate of return i.

From interest tables we find:

i	$(P/A,i,5)$
6%	4.212
7%	4.100
8%	3.993

In this example, no interpolation is needed as the rate of return, for this investment is exactly 7%. ◀

EXAMPLE 7-2

An investment resulted in the following cash flow. Compute the rate of return.

Year	*Cash flow*
0	−\$700
1	+100
2	+175
3	+250
4	+325

Solution:

$$\text{EUAB} - \text{EUAC} = 0$$
$$100 + 75(A/G,i,4) - 700(A/P,i,4) = 0$$

In this situation, we have two different interest factors in the equation. We will not be able to solve it as easily as Ex. 7-1. Since there is no convenient direct method of solution, we will solve the equation by trial and error. Try $i = 5\%$ first:

$$\text{EUAB} - \text{EUAC} = 0$$
$$100 + 75(A/G,5\%,4) - 700(A/P,5\%,4) = 0$$
$$100 + 75(1.439) - 700(0.2820) = 0$$
$$\text{At } i = 5\%,\ \text{EUAB} - \text{EUAC} = 208 - 197 = +11$$

The EUAC is too low. If the interest rate is increased, EUAC will increase. Try $i = 8\%$:

$$\text{EUAB} - \text{EUAC} = 0$$
$$100 + 75(A/G,8\%,4) - 700(A/P,8\%,4) = 0$$
$$100 + 75(1.404) - 700(0.3019) = 0$$
$$\text{At } i = 8\%,\ \text{EUAB} - \text{EUAC} = 205 - 211 = -6$$

This time the EUAC is too large. We see that the true rate of return is between 5% and 8%. Try $i = 7\%$:

$$\text{EUAB} - \text{EUAC} = 0$$
$$100 + 75(A/G,7\%,4) - 700(A/P,7\%,4) = 0$$
$$100 + 75(1.416) - 700(0.2952) = 0$$
$$\text{At } i = 7\%,\ \text{EUAB} - \text{EUAC} = 206 - 206 = 0$$

The rate of return is 7%. ◀

EXAMPLE 7-3

Given the cash flow below, calculate the rate of return on the investment.

Year	*Cash flow*
0	−$100
1	+20
2	+30
3	+20
4	+40
5	+40

Solution: Using NPW = 0, try i = 10%:

$$\begin{aligned}\text{NPW} &= -100 + 20(P/F,10\%,1) + 30(P/F,10\%,2) + 20(P/F,10\%,3)\\&\quad + 40(P/F,10\%,4) + 40(P/F,10\%,5)\\&= -100 + 20(0.9091) + 30(0.8264) + 20(0.7513) + 40(0.6830)\\&\quad + 40(0.6209)\\&= -100 + 18.18 + 24.79 + 15.03 + 27.32 + 24.84\\&= -100 + 110.16\\&= +10.16\end{aligned}$$

The trial interest rate i is too low. Select a second trial, i = 15%:

$$\begin{aligned}\text{NPW} &= -100 + 20(0.8696) + 30(0.7561) + 20(0.6575) + 40(0.5718)\\&\quad + 40(0.4972)\\&= -100 + 17.39 + 22.68 + 13.15 + 22.87 + 19.89\\&= -100 + 95.98\\&= -4.02\end{aligned}$$

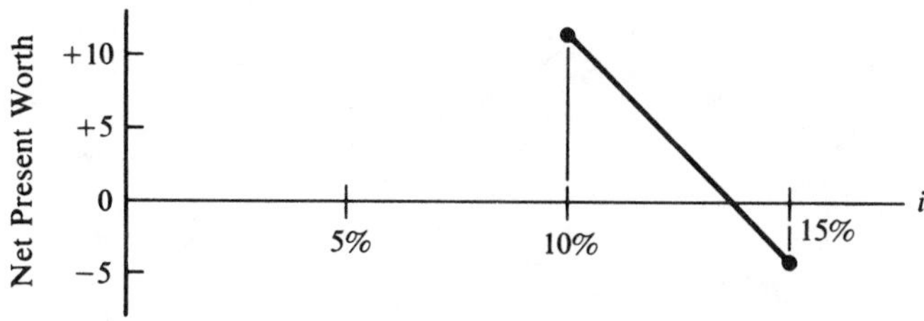

Figure 7-1 Plot of NPW *vs.* interest rate i.

These two points are plotted in Figure 7-1. By linear interpolation we compute the rate of return as follows:

$$i = 10\% + (15\% - 10\%)\left(\frac{10.16}{10.16 + 4.02}\right) = 13\tfrac{1}{2}\%$$

We can prove that the rate of return is very close to 13½% by showing that the unrecovered investment is very close to zero at the end of the life of the investment.

Year	*Cash flow*	*Unrecovered investment at beginning of year*	*13½% return on unrecovered investment*	*Investment repayment at end of year*	*Unrecovered investment at end of year*
0	−$100				
1	+20	$100.0	$13.5	$ 6.5	$93.5
2	+30	93.5	12.6	17.4	76.1
3	+20	76.1	10.3	9.7	66.4
4	+40	66.4	8.9	31.1	35.3
5	+40	35.3	4.8	35.2	0.1*

*This small unrecovered investment indicates that the rate of return is slightly less than 13½%.

If in Figure 7-1 NPW had been computed for a broader range of values of *i*, Figure 7-2 would have been obtained. From this figure it is apparent that the error resulting from linear interpolation increases as the interpolation width increases. ◀

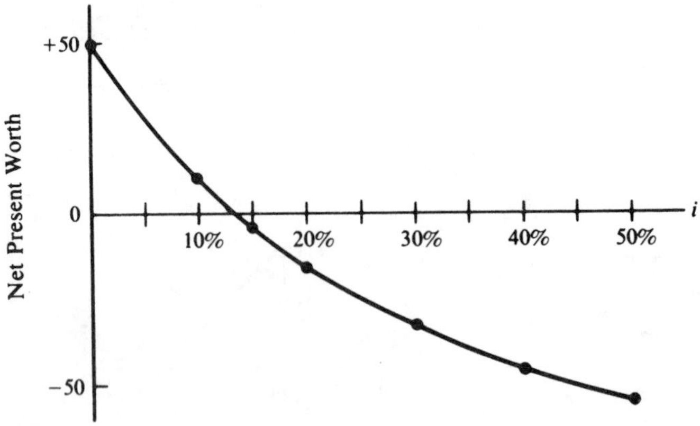

Figure 7-2 Replot of NPW *vs.* interest rate *i* over a larger range of values.

Plot of NPW *vs.* Interest Rate *i*

Figure 7-2—the plot of NPW *vs.* interest rate *i*—is an important source of information. A cash flow representing an investment followed by benefits from the investment would have an NPW *vs.* *i* plot (we will call it an **NPW *plot*** for convenience) in the form of Fig. 7-3.

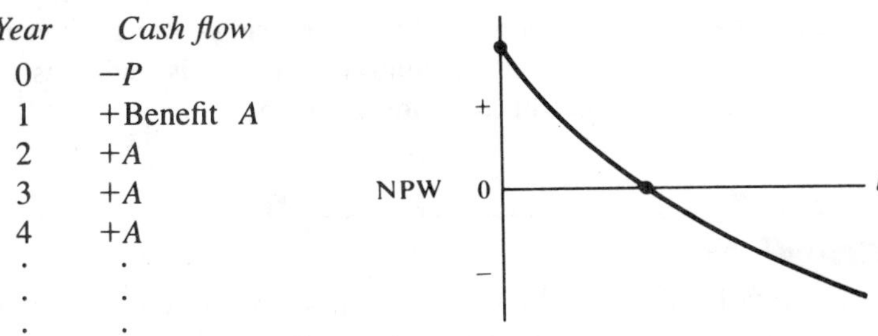

Year	*Cash flow*
0	$-P$
1	+Benefit A
2	$+A$
3	$+A$
4	$+A$
.	.
.	.
.	.

Figure 7-3 Typical NPW plot for an investment.

If, on the other hand, borrowed money was involved, the NPW plot would appear as in Fig. 7-4. This form of cash flow typically results when one is a borrower of money. In such a case, the usual pattern is a receipt of borrowed money early in the time period with a later repayment of an equal sum, plus payment of interest on the borrowed money. In all cases where interest is charged, the NPW at 0% will be negative.

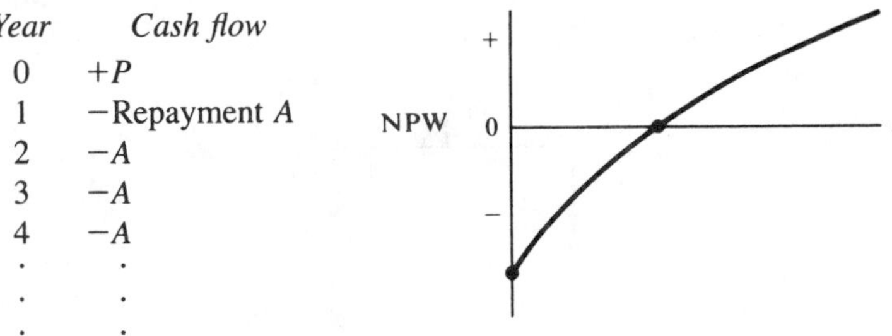

Year	*Cash flow*
0	$+P$
1	−Repayment A
2	$-A$
3	$-A$
4	$-A$
.	.
.	.
.	.

Figure 7-4 Typical NPW plot for borrowed money.

How do we determine the interest rate paid by the borrower in this situation? Typically we would write an equation, such as PW of income = PW of disbursements, and solve for the unknown *i*. Is the resulting *i* positive or negative from the borrower's point of view? If the lender said he was receiving, say,

+11% on the debt, it seems reasonable to state that the borrower is faced with −11% interest. Yet this is not the way interest is discussed; rather, interest is referred to in absolute terms without associating a positive or negative sign to it. A banker says he pays 5% interest on savings accounts and charges 11% on personal loans.

Thus, we implicitly recognize interest as a charge for the use of someone else's money and a receipt for letting others use our money. In determining the interest rate in a particular situation, we solve for a single unsigned value of it. We then view this value of i in the customary way, that is, either as a charge for borrowing money, or a receipt for lending money.

EXAMPLE 7-4

A new corporate bond was initially sold by a stockbroker to an investor for \$1000. The issuing corporation promised to pay the bondholder \$40 interest on the \$1000 face value of the bond every six months, and to repay the \$1000 at the end of ten years. After one year the bond was sold by the original buyer for \$950.

a. What rate of return did the original buyer receive on his investment?

b. What rate of return can the new buyer (paying \$950) expect to receive if he keeps the bond for its remaining nine-year life?

Solution to Ex. 7-4a:

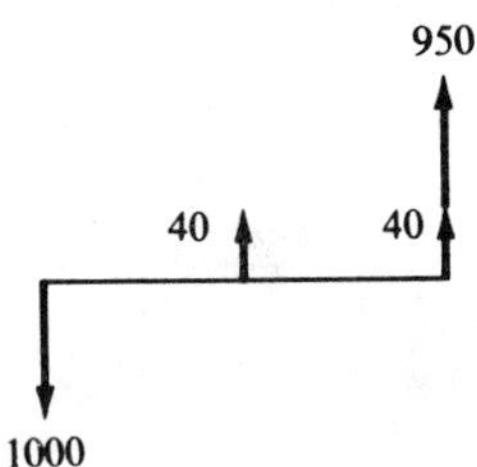

Since \$40 is received each six months, we will solve the problem using a six-month interest period.

$$\text{Let PW of cost} = \text{PW of benefits}$$

$$1000 = 40(P/A,i,2) + 950(P/F,i,2)$$

Try $i = 1\frac{1}{2}\%$:

$$1000 \stackrel{?}{=} 40(1.956) + 950(0.9707) = 78.24 + 922.17$$

$$\stackrel{?}{=} 1000.41$$

The interest rate per six months is very close to $1\frac{1}{2}\%$. This means the nominal (annual) interest rate is $2 \times 1.5\% = 3\%$. The effective (annual) interest rate $= (1 + 0.015)^2 - 1 = 3.02\%$.

Solution to Ex. 7-4b:

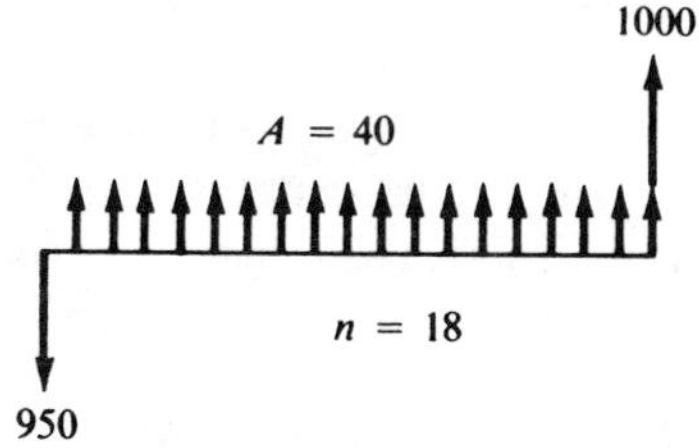

We have the same $40 semi-annual interest payments. For six-month interest periods:

$$950 = 40(P/A,i,18) + 1000(P/F,i,18)$$

Try $i = 5\%$:

$$950 \stackrel{?}{=} 40(11.690) + 1000(0.4155) = 467.60 + 415.50$$
$$\stackrel{?}{=} 883.10$$

The PW of benefits is too low. Try a lower interest rate, say, $i = 4\%$:

$$950 \stackrel{?}{=} 40(12.659) + 1000(0.4936) = 506.36 + 493.60$$
$$\stackrel{?}{=} 999.96$$

The value of i is between 4% and 5%. By interpolation,

$$i = 4\% + (1\%)\left(\frac{999.96 - 950.00}{999.96 - 883.10}\right) = 4.43\%$$

The nominal interest rate is $2 \times 4.43\% = 8.86\%$. The effective interest rate is $(1 + 0.0443)^2 - 1 = 9.05\%$. ◀

Rate Of Return Analysis

Rate of return analysis is probably the most frequently used exact analysis technique in industry. Although problems in computing rate of return sometimes occur, its major advantage outweighs the occasional difficulty. The major advantage is that we can compute a single figure of merit that is readily understood.

Consider these statements:

- The net present worth on the project is \$32,000.
- The equivalent uniform annual net benefit is \$2800.
- The project will produce a 23% rate of return.

While none of these statements tells the complete story, the third one gives a measure of desirability of the project in terms that are widely understood. It is this acceptance by engineers and businessmen alike of rate of return that has promoted its more frequent use than present worth or annual cash flow methods.

There is another advantage to rate of return analysis. In both present worth and annual cash flow calculations, one must select an interest rate for use in the calculations—and this may be a difficult and controversial item. In rate of return analysis, no interest rate is introduced into the calculations (except as described in Chapter 7A). Instead, we compute a rate of return (more accurately called *internal rate of return*) from the cash flow. To decide how to proceed, the calculated rate of return is compared with a preselected ***minimum attractive rate of return***, or simply MARR. This is the same value of i used for present worth and annual cash flow analysis.

When there are two alternatives, rate of return analysis is performed by computing the ***incremental rate of return***—ΔROR—on the difference between the alternatives. Since we want to look at increments of investment, the cash flow for the difference between the alternatives is computed by taking the higher initial-cost alternative *minus* the lower initial-cost alternative. If the ΔROR is ≥ the MARR, choose the higher-cost alternative. If the ΔROR is < the MARR, choose the lower-cost alternative.

Two-alternative situation	*Decision*
ΔROR ≥ MARR	Choose the higher-cost alternative
ΔROR < MARR	Choose the lower-cost alternative

Rate of return analysis is illustrated by Examples 7-5 through 7-8.

EXAMPLE 7-5

If an electromagnet is installed on the input conveyor of a coal processing plant, it will pick up scrap metal in the coal. The removal of this metal will save an estimated \$1200 per year in machinery damage being caused by metal. The electromagnetic equipment has an estimated useful life of five years and no salvage value. Two suppliers have been contacted: Leaseco will provide the equipment in return for three beginning-of-year annual payments of \$1000 each; Saleco will provide the equipment for \$2783. If the MARR is 10%, which supplier should be selected?

Solution: Since both suppliers will provide equipment with the same useful life and benefits, this is a fixed-output situation. In rate of return analysis, the method of solution is to examine the differences between the alternatives. By taking (Saleco − Leaseco) we obtain an increment of investment.

Year	*Leaseco*	*Saleco*	*Difference between alternatives Saleco − Leaseco*
0	−$1000	−$2783	−$1783
1	−1000 +1200	+1200	+1000
2	−1000 +1200	+1200	+1000
3	+1200	+1200	0
4	+1200	+1200	0
5	+1200	+1200	0

Compute the NPW at various interest rates on the increment of investment represented by the difference between the alternatives.

Year n	*Cash flow Saleco − Leaseco*	*PW* at 0%*	*PW* at 8%*	*PW* at 20%*	*PW* at ∞%*
0	−$1783	−$1783	−$1783	−$1783	−$1783
1	+1000	+1000	+926	+833	0
2	+1000	+1000	+857	+694	0
3	0	0	0	0	0
4	0	0	0	0	0
5	0	0	0	0	0
	NPW =	+217	0	−256	−1783

*Each year the cash flow is multiplied by $(P/F,i,n)$.

At 0%: $(P/F,0\%,n) = 1$ for all values of n

At ∞%: $(P/F,\infty\%,0) = 1$

$(P/F,\infty\%,n) = 0$ for all other values of n

These data are plotted in Figure 7-5. From the figure we see that NPW = 0 at $i = 8\%$.

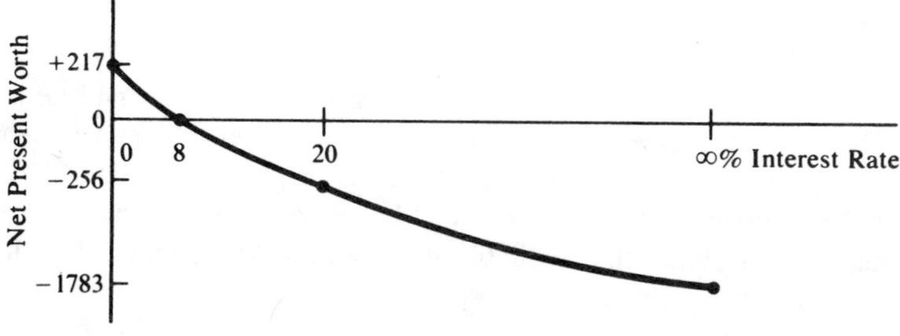

Figure 7-5 NPW plot for Example 7-5.

Thus, the incremental rate of return—ΔROR—of selecting Saleco rather than Leaseco is 8%. This is less than the 10% MARR. Select Leaseco. ◀

EXAMPLE 7-6

You are given the choice of selecting one of two mutually exclusive alternatives. The alternatives are as follows:

Year	*Alternative 1*	*Alternative 2*
0	−\$10	−\$20
1	+15	+28

Any money not invested here may be invested elsewhere at the MARR of 6%. If you can only choose one alternative one time, which one would you select?

Solution: We will select the lesser-cost Alt. 1, unless we find the additional cost of Alt. 2 produces sufficient additional benefits that we would prefer it. If we consider Alt. 2 in relation to Alt. 1, then

$$\begin{bmatrix}\text{Higher-cost}\\ \text{Alt. 2}\end{bmatrix} = \begin{bmatrix}\text{Lower-cost}\\ \text{Alt. 1}\end{bmatrix} + \text{Differences between Alt. 1 and 2}$$

or

$$\text{Differences between Alt. 1 and 2} = \begin{bmatrix}\text{Higher-cost}\\ \text{Alt. 2}\end{bmatrix} - \begin{bmatrix}\text{Lower-cost}\\ \text{Alt. 1}\end{bmatrix}$$

The choice between the two alternatives reduces to an examination of the differences between them. We can compute the rate of return on the differences between the alternatives. Writing the alternatives again,

Year	*Alt. 1*	*Alt. 2*	*Alt. 2 − Alt. 1*
0	−\$10	−\$20	−\$20 − (−\$10) = −\$10
1	+15	+28	+28 − (+15) = +13

$$\text{PW of cost} = \text{PW of benefit}$$
$$10 = 13(P/F,i,1)$$
$$(P/F,i,1) = \frac{10}{13} = 0.7692$$

One can see that if \$10 increases to \$13 in one year, the interest rate must be 30%. The Interest Tables confirm this conclusion. The 30% rate of return on the difference between the alternatives is far higher than the 6% MARR. The additional \$10 investment to obtain Alt. 2 is superior to investing the \$10 elsewhere at 6%. To obtain this desirable increment of investment, with its 30% rate of return, Alt. 2 is selected. ◀

To understand more about Example 7-6, compute the rate of return for each alternative.

Alternative 1:

$$\text{PW of cost} = \text{PW of benefit}$$
$$\$10 = \$15(P/F,i,1)$$
$$(P/F,i,1) = \frac{10}{15} = 0.6667$$

From the Interest Tables: rate of return = 50%.

Alternative 2:

$$\text{PW of cost} = \text{PW of benefit}$$
$$\$20 = \$28(P/F,i,1)$$
$$(P/F,i,1) = \frac{20}{28} = 0.7143$$

From the Interest Tables: rate of return = 40%.

One is tempted to select Alt. 1, based on these rate of return computations. We have already seen, however, that this is not the correct solution. Solve the problem again, this time using present worth analysis.

Present Worth Analysis:

Alternative 1:

$$\text{NPW} = -10 + 15(P/F,6\%,1) = -10 + 15(0.9434) = +\$4.15$$

Alternative 2:

$$\text{NPW} = -20 + 28(P/F,6\%,1) = -20 + 28(0.9434) = +\$6.42$$

Alternative 1 has a 50% rate of return and an NPW (at the 6% MARR) of +$4.15. Alternative 2 has a 40% rate of return on a larger investment, with the result that its NPW (at the 6% MARR) is +$6.42. Our economic criterion is to maximize the return, rather than the rate of return. To maximize NPW, select Alt. 2. This agrees with the rate of return analysis on the differences between the alternatives.

EXAMPLE 7-7

Solve Ex. 7-6 again, but this time compute the interest rate on the increment (Alt. 1 − Alt. 2). How do you interpret the results?

Solution: This time the problem is being viewed as:

Alt. 1 = Alt. 2 + [Alt.1 − Alt. 2]

Year	*Alt. 1*	*Alt. 2*	*[Alt. 1 − Alt. 2]*
0	−$10	−$20	−$10 − (−$20) = +$10
1	+15	+28	+15 − (+28) = −13

We can write one equation in one unknown:

$$\text{NPW} = \text{PW of benefit} - \text{PW of cost} = 0$$

$$+10 - 13(P/F,i,1) = 0$$

$$(P/F,i,1) = \frac{10}{13} = 0.7692$$

Once again the interest rate is found to be 30%. The critical question is, what does the 30% represent? Looking at the increment again:

Year	*Alt. 1 − Alt. 2*
0	+$10
1	−13

The cash flow does *not* represent an investment; instead, it represents a loan. It is as if we borrowed $10 in Year 0 (+$10 represents a receipt of money) and repaid it in Year 1 (−$13 represents a disbursement). The 30% interest rate means this is the amount *we would pay* for the use of the $10 borrowed in Year 0 and repaid in Year 1.

Is this a desirable borrowing? Since the MARR on investments is 10%, it is reasonable to assume our maximum interest rate on borrowing would also be 10%. Here the interest rate is 30%, which means the borrowing is undesirable. Since Alt. 1 = Alt. 2 + (Alt. 1 − Alt. 2), and we do not like the (Alt. 1 − Alt. 2) increment, we should reject Alternative 1 as it contains the undesirable increment. This means we should select Alternative 2—the same conclusion reached in Ex. 7-6. ◀

This example illustrated that one can analyze either ***increments of investment*** or ***increments of borrowing***. When looking at increments of investment, we accept the increment when the incremental rate of return equals or exceeds the minimum attractive rate of return (ΔROR $\geq$ MARR). When looking at increments of borrowing, we accept the increment when the incremental interest rate is less than or equal to the minimum attractive rate of return ($\Delta i \leq$ MARR). One way to avoid much of the possible confusion is to organize the solution to any problem so that one is examining increments of investment. This is illustrated in the next example.

EXAMPLE 7-8

A firm is considering which of two devices to install to reduce costs in a particular situation. Both devices cost \$1000, have useful lives of five years and no salvage value. Device *A* can be expected to result in \$300 savings annually. Device *B* will provide cost savings of \$400 the first year but will decline \$50 annually, making the second-year savings \$350, the third-year savings \$300, and so forth. For a 7% MARR, which device should the firm purchase?

Solution: This problem has been solved by present worth analysis (Ex. 5-1) and annual cost analysis (Ex. 6-5). This time we will use rate of return analysis. The example has fixed input (\$1000) and differing outputs (savings).

In determining whether to use an $(A - B)$ or $(B - A)$ difference between the alternatives, we seek an increment of investment. By looking at both $(A - B)$ and $(B - A)$, we find that $(A - B)$ is the one that represents an increment of investment.

Year	*Device A*	*Device B*	*Difference between alternatives* *Device A − Device B*
0	−\$1000	−\$1000	\$0
1	+300	+400	−100
2	+300	+350	−50
3	+300	+300	0
4	+300	+250	+50
5	+300	+200	+100

For the difference between the alternatives, write a single equation with i as the only unknown.

$$\text{EUAC} = \text{EUAB}$$

$$[100(P/F,i,1) + 50(P/F,i,2)](A/P,i,5) = [50(F/P,i,1) + 100](A/F,i,5)$$

The equation is cumbersome, but need not be solved. Instead, we observe that the sum of the costs (−100 and −50) equals the sum of the benefits (+50 and +100). This indicates that 0% is the ΔROR on the $A - B$ increment of *investment*. This is less than the 7% MARR; therefore, the increment is undesirable. Reject Device *A* and choose Device *B*.

As described in Ex. 7-7, if the increment examined is $(B - A)$, the interest rate would again be 0%, indicating a desirable *borrowing* situation. We would choose Device *B*. ◀

Analysis Period

In discussing present worth analysis and annual cash flow analysis, an important consideration is the analysis period. This is also true in rate of return analysis. The method of solution for two alternatives is to examine the differences between the alternatives. The examination must necessarily cover the selected analysis period. An assumption that an alternative can be replaced with one of identical costs and performance appears dubious at best. For now, we can only suggest that the assumptions made should reflect one's perception of the future as accurately as possible.

Example 7-9 is a problem where the analysis period is a common multiple of the alternative service lives, and where identical replacement is assumed. It will illustrate an analysis of the differences between the alternatives over the analysis period.

EXAMPLE 7-9

Two machines are being considered for purchase. If the MARR (here, the minimum required interest rate) is 10%, which machine should be bought?

	Machine X	*Machine Y*
Initial cost	$200	$700
Uniform annual benefit	95	120
End-of-useful-life salvage value	50	150
Useful life, in years	6	12

Solution: The solution is based on a twelve-year analysis period and a Replacement Machine *X* that is identical to the present Machine *X*. The cash flow for the differences between the alternatives is as follows:

Year	*Machine X*	*Machine Y*	*Mach. Y − Mach. X*
0	−$200	−$700	−$500
1	+95	+120	+25
2	+95	+120	+25
3	+95	+120	+25
4	+95	+120	+25
5	+95	+120	+25
6	+95 +50 −200	+120	+25 +150
7	+95	+120	+25
8	+95	+120	+25
9	+95	+120	+25
10	+95	+120	+25
11	+95	+120	+25
12	+95 +50	+120 +150	+25 +100

$$\text{PW of cost} = \text{PW of benefits}$$
$$500 = 25(P/A,i,12) + 150(P/F,i,6) + 100(P/F,i,12)$$

The sum of the benefits over the twelve years is $550 which is only a little greater than the $500 additional cost. This indicates that the rate of return is quite low. Try $i = 1\%$:

$$500 \stackrel{?}{=} 25(11.255) + 150(0.942) + 100(0.887)$$
$$\stackrel{?}{=} 281 + 141 + 89 = 511$$

The interest rate is too low. Try $i = 1\frac{1}{2}\%$:

$$500 \stackrel{?}{=} 25(10.908) + 150(0.914) + 100(0.836)$$
$$\stackrel{?}{=} 273 + 137 + 84 = 494$$

The rate of return on the $Y - X$ increment is about 1.3%, far below the 10% minimum attractive rate of return. The additional investment to obtain Y yields an unsatisfactory rate of return, therefore X is the preferred alternative. ◀

SUMMARY

Rate of return may be defined as the interest rate paid on the unpaid balance of a loan such that the loan is exactly repaid by the schedule of payments. On an investment, rate of return is the interest rate earned on the unrecovered investment such that the payment schedule makes the unrecovered investment equal to zero at the end of the life of the investment. Although the two definitions of rate of return are stated differently, there is only one fundamental concept being described. It is that the rate of return is the interest rate at which the benefits are equivalent to the costs.

There are a variety of ways of writing the cash flow equation in which the rate of return i may be the single unknown. Five of them are:

$$\text{PW of benefits} - \text{PW of costs} = 0$$
$$\frac{\text{PW of benefits}}{\text{PW of costs}} = 1$$
$$\text{NPW} = 0$$
$$\text{EUAB} - \text{EUAC} = 0$$
$$\text{PW of costs} = \text{PW of benefits}$$

Rate of Return Analysis: Rate of return analysis is the most frequently used method in industry, as the resulting rate of return is readily understood. Also, the difficulties in selecting a suitable interest rate to use in present worth and annual cash flow analysis are avoided.

Criteria:

Two alternatives.

Compute the incremental rate of return—ΔROR—on the increment of *investment* between the alternatives. Then,

- If ΔROR ≥ MARR, choose the higher-cost alternative; or,
- If ΔROR < MARR, choose the lower-cost alternative.

When an increment of *borrowing* is examined,

- If $\Delta i \leq$ MARR, the increment is acceptable; or,
- If $\Delta i >$ MARR, the increment is not acceptable.

Three or more alternatives.

Incremental analysis is needed, which we move to in Chapter 8.

Rate of return is further described in Chapter 7A and still further in Chapter 18. These two chapters concentrate on the difficulties that occur with some cash flows that yield more than one rate of return.

Problems

7-1 A woman went to the Beneficial Loan Company and borrowed $3000. She must pay $119.67 at the end of each month for the next thirty months.

a. Calculate the nominal annual interest rate she is paying to within ±0.15%.

b. What effective annual interest rate is she paying?

7-2 For the diagram shown, compute the interest rate to within ½%.

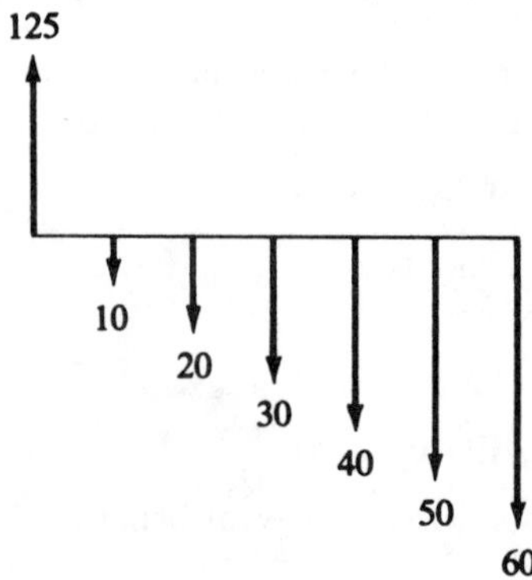

7-3 Helen is buying a $12,375 automobile with a $3000 down payment, followed by 36 monthly payments of $325 each. The down payment is paid immediately, and the monthly payments are due at the end of each month. What nominal annual interest rate is Helen paying? What effective interest rate? (*Answers:* 15%; 16.08%)

7-4 Consider the following cash flow:

Year	*Cash flow*
0	−\$500
1	+200
2	+150
3	+100
4	+50

Compute the rate of return represented by the cash flow.

7-5 For the diagram below, compute the interest rate at which the costs are equivalent to the benefits.

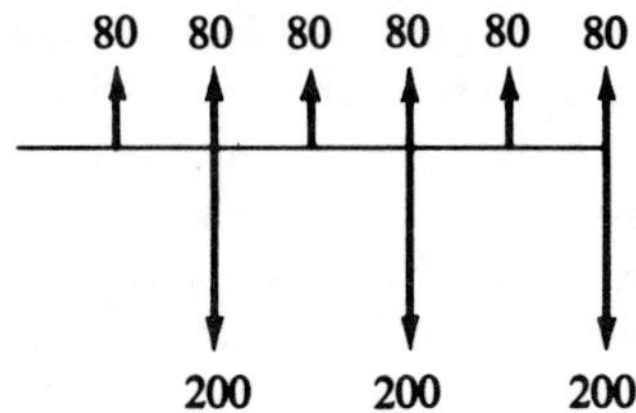

(*Answer:* 50%)

7-6 For the diagram below, compute the rate of return.

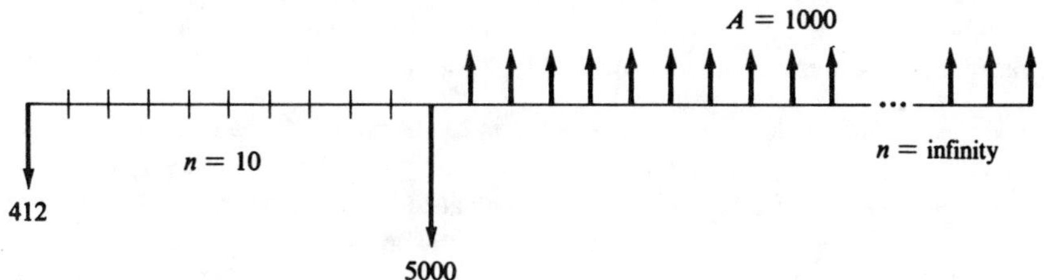

7-7 Consider the following cash flow:

Year	*Cash flow*
0	−\$1000
1	0
2	+300
3	+300
4	+300
5	+300

Compute the rate of return on the \$1000 investment to within 0.1%. (*Answer:* 5.4%)

7-8 Peter Minuit bought an island from the Manhattoes Indians in 1626 for \$24 worth of glass beads and trinkets. The 1991 estimate of the value of land on this island is \$12 billion. What rate of return would the Indians have received if they had retained title to the island rather than selling it for \$24?

7-9 A man buys a corporate bond from a bond brokerage house for $925. The bond has a face value of $1000 and pays 4% of its face value each year. If the bond will be paid off at the end of ten years, what rate of return will the man receive? (*Answer:* 4.97%)

7-10 A well-known industrial firm has issued $1000 bonds that carry a 4% nominal annual interest rate paid semi-annually. The bonds mature twenty years from now, at which time the industrial firm will redeem them for $1000 plus the terminal semi-annual interest payment. From the financial pages of your newspaper you learn that the bonds may be purchased for $715 each ($710 for the bond plus a $5 sales commission). What nominal annual rate of return would you receive if you purchased the bond now and held it to maturity twenty years hence? (*Answer:* 6.6%)

7-11 One aspect of obtaining a college education is the prospect of improved future earnings, compared to non-college graduates. Sharon Shay estimates that a college education has a $28,000 equivalent cost at graduation. She believes the benefits of her education will occur throughout forty years of employment. She thinks she will have a $3000-per-year higher income during the first ten years out of college, compared to a non-college graduate. During the subsequent ten years, she projects an annual income that is $6000-per-year higher. During the last twenty years of employment, she estimates an annual salary that is $12,000 above the level of the non-college graduate. Assuming her estimates are correct, what rate of return will she receive as a result of her investment in a college education?

7-12 An investor purchased a one-acre lot on the outskirts of a city for $9000 cash. Each year he paid $80 of property taxes. At the end of four years, he sold the lot. After deducting his selling expenses, the investor received $15,000. What rate of return did he receive on his investment? (*Answer:* 12.92%)

7-13 A popular reader's digest offers a lifetime subscription to the magazine for $200. Such a subscription may be given as a gift to an infant at birth (the parents can read it in those early years), or taken out by an individual for himself. Normally, the magazine costs $12.90 per year. Knowledgeable people say it probably will continue indefinitely at this $12.90 rate. What rate of return would be obtained if a life subscription were purchased for an infant, rather than paying $12.90 per year beginning immediately? You may make any reasonable assumptions, but the compound interest factors must be *correctly* used.

7-14 On April 2, 1988, an engineer buys a $1000 bond of an American airline for $875. The bond pays 6% on its principal amount of $1000, half in each of its April 1 and October 1 semi-annual payments; it will repay the $1000 principal sum on October 1, 2001. What nominal rate of return will the engineer receive from the bond if he holds it to its maturity (on October 1, 2001)? (*Answer:* 7.5%)

7-15 The cash price of a machine tool is $3500. The dealer is willing to accept a $1200 down payment and 24 end-of-month monthly payments of $110 each. At what effective interest rate are these terms equivalent? (*Answer:* 14.4%)

7-16 A local bank makes automobile loans. It charges 4% per year in the following manner: if $3600 is borrowed to be repaid over a three-year period, the bank interest charge is \$3600 × 0.04 × 3 years = \$432. The bank deducts the \$432 of interest from the \$3600 loan and gives the customer \$3168 in cash. The customer must repay the loan by paying 1/36 of \$3600, or \$100, at the end of each month for 36 months. What nominal annual interest rate is the bank actually charging for this loan?

7-17 Upon graduation every engineer must decide whether or not to go on to graduate school. Estimate the costs of going full time to the university to obtain a Master of Science degree. Then estimate the resulting costs and benefits. Combine the various consequences into a cash flow table and compute the rate of return.

7-18 In his uncle's will, Frank is to choose one of two alternatives:

Alternative 1: \$2000 cash.

Alternative 2: \$150 cash now plus \$100 per month for twenty months beginning the first day of next month.

a. At what rate of return are the two alternatives equivalent?

b. If Frank thinks the rate of return in ***a*** is too low, which alternative should he select?

7-19 A man buys a table saw at a local store for \$175. He may either pay cash for it, or pay \$35 now and \$12.64 a month for twelve months beginning thirty days hence. If the man chooses the time payment plan, what is the nominal annual interest rate he will be charged? (*Answer:* 15%)

7-20 In January, 1983, an investor purchased a convertible debenture bond issued by the XLA Corporation. The bond cost \$1000 and paid \$60 per year interest in annual payments on December 31. Under the convertible feature of the bond, it could be converted into twenty shares of common stock by tendering the bond, together with \$400 cash. The day after the investor received the December 31, 1985, interest payment, he submitted the bond together with \$400 to the XLA Corporation. In return, he received the twenty shares of common stock. The common stock paid no dividends. On December 31, 1987, the investor sold the stock for \$1740, terminating his five-year investment in XLA Corporation. What rate of return did he receive?

7-21 A man owns a corner lot. He must decide which of several alternatives to select in trying to obtain a desirable return on his investment. After much study and calculation, he decides that the two best alternatives are:

	Build gas station	*Build soft ice cream stand*
First cost	\$80,000	\$120,000
Annual property taxes	3,000	5,000
Annual income	11,000	16,000
Life of building, in years	20	20
Salvage value	0	0

If the owner wants a minimum attractive rate of return on his investment of 6%, which of the two alternatives would you recommend to him?

7-22 Two alternatives are as follows:

Year	*A*	*B*
0	−$2000	−$2800
1	+800	+1100
2	+800	+1100
3	+800	+1100

If 5% is considered the minimum attractive rate of return, which alternative should be selected?

7-23 The Southern Guru Copper Company operates a large mine in a South American country. A legislator in the National Assembly said in a speech that most of the capital for the mining operation was provided by loans from the World Bank; in fact, Southern Guru has only $500,000 of its own money actually invested in the property. The cash flow for the mine is:

Year	*Cash flow*
0	$0.5 million investment
1	3.5 million profit
2	0.9 million profit
3	3.9 million profit
4	8.6 million profit
5	4.3 million profit
6	3.1 million profit
7	6.1 million profit

The legislator divided the $30.4 million total profit by the $0.5 million investment. This produced, he said, a 6080% rate of return on the investment. Southern Guru claims their actual rate of return is much lower. They ask you to compute their rate of return.

7-24 Two alternatives are being considered:

	A	*B*
First cost	$9200	$5000
Uniform annual benefit	1850	1750
Useful life, in years	8	4

If the minimum attractive rate of return is 7%, which alternative should be selected?

7-25 Jean has decided it is time to purchase a new battery for her car. Her choices are:

	Zappo	*Kicko*
First cost	$56	$90
Guarantee period, in months	12	24

Jean believes the batteries can be expected to last only for the guarantee period. She does not want to invest extra money in a battery unless she can expect a 50% rate of return. If she plans to keep her present car another two years, which battery should she buy?

7-26 For the diagram below, compute the rate of return on the $3810 investment.

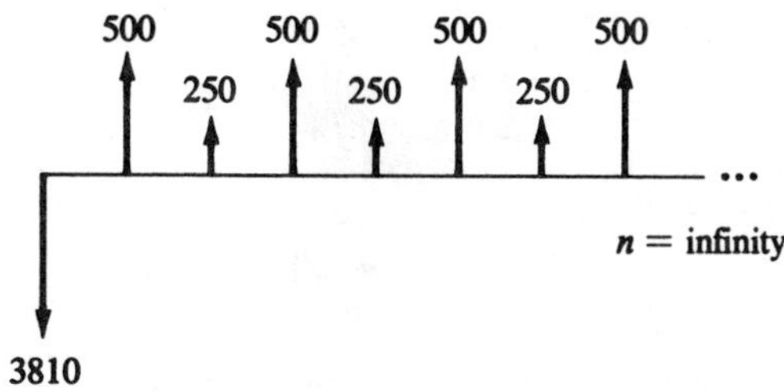

7-27 Consider the following cash flow:

Year	*Cash flow*
0	−$400
1	0
2	+200
3	+150
4	+100
5	+50

Write one equation, with i as the only unknown, for the cash flow. In the equation you are not to use more than two single payment compound interest factors. (You may use as many other factors as you wish.) Then solve your equation for i.

7-28 Compute the rate of return for the following cash flow to within ½%.

Year	*Cash flow*
0	−$100
1–10	+27

(*Answer:* 23.9%)

7-29 For the following diagram, compute the rate of return.

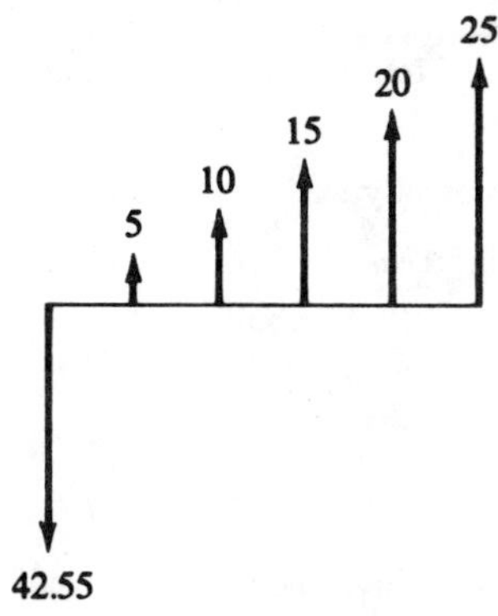

7-30 Solve the cash flow below for the rate of return to within 0.5%.

Year	*Cash flow*
0	−$500
1	−100
2	+300
3	+300
4	+400
5	+500

7-31 For the cash flow below, compute the rate of return.

Year	*Cash flow*
1–5	−$223
6–10	+1000

7-32 Compute the rate of return for the following cash flow to within 0.5%.

Year	*Cash flow*
0	−$640
1	0
2	100
3	200
4	300
5	300

(*Answer:* 9.3%)

7-33 Consider two mutually exclusive alternatives:

Year	*X*	*Y*
0	−$100	−$50.0
1	35	16.5
2	35	16.5
3	35	16.5
4	35	16.5

If the minimum attractive rate of return is 10%, which alternative should be selected?

7-34 Consider these two mutually exclusive alternatives:

Year	*A*	*B*
0	−$50	−$53
1	17	17
2	17	17
3	17	17
4	17	17

At a MARR of 10%, which alternative should be selected? (*Answer:* A)

7-35 Two mutually exclusive alternatives are being considered. Both have a ten-year useful life. If the MARR is 8%, which alternative is preferred?

	A	*B*
Initial cost	$100.00	$50.00
Uniform annual benefit	19.93	11.93

7-36 Two alternatives are being considered:

	A	*B*
Initial cost	$9200	$5000
Uniform annual benefit	1850	1750
Useful life, in years	8	4

Base your computations on a MARR of 7% and an eight-year analysis period. If identical replacement is assumed, which alternative should be selected?

7-37 Two investment opportunities are as follows:

	A	*B*
First cost	$150	$100
Uniform annual benefit	25	22.25
End-of-useful-life salvage value	20	0
Useful life, in years	15	10

At the end of ten years, Alt. *B* is not replaced. Thus, the comparison is 15 years of *A* *vs*. 10 years of *B*. If the MARR is 10%, which alternative should be selected?

7-38 An insurance company is offering to sell an annuity for $20,000 cash. In return they will guarantee to pay the purchaser 20 annual end-of-year payments, with the first payment amounting to $1100. Subsequent payments will increase at a uniform 10% rate each year (second payment is $1210; third payment is $1331, and so on). What rate of return will the purchaser receive if he buys the annuity?

7-39 Consider two mutually exclusive alternatives:

Year	*X*	*Y*
0	−$5000	−$5000
1	−3000	+2000
2	+4000	+2000
3	+4000	+2000
4	+4000	+2000

If the MARR is 8%, which alternative should be selected?

7-40 A bank proudly announces that they have changed their interest computation method to continuous compounding. Now $2000 left in the bank for nine years will double to $4000.

a. What nominal interest rate, compounded continuously, are they paying?

b. What effective interest rate are they paying?

7-41 Fifteen families live in Willow Canyon. Although several water wells have been drilled, none has produced water. The residents take turns driving a water truck to a fire hydrant in a nearby town. They fill the truck with water and then haul it to a storage tank in Willow Canyon. Last year it cost $3180 in truck fuel and maintenance costs. This year the residents are seriously considering spending $100,000 to install a pipeline from the nearby town to their storage tank. What rate of return would the Willow Canyon residents receive on their new water supply pipeline if the pipeline is considered to last:

a. forever?

b. 100 years?

c. 50 years?

d. Would you recommend that the pipeline be installed? Explain.

7-42 Compute the rate of return for the following cash flow.

Year	*Cash flow*
0	−$ 500
1–3	0
4	+4500

7-43 Jan purchased 100 shares of Peach Computer stock for $18 per share, plus a $45 brokerage commission. Every six months she received a 50 cents per share dividend from Peach. At the end of two years, just after receiving the fourth dividend, she sold the stock for $23 per share, and paid a $58 brokerage commission from the proceeds. What annual rate of return did she receive on her investment?

7-44 The Diagonal Stamp Co., which sells used postage stamps to collectors, advertises that their average price has increased from $1 to $5 in the last five years. Thus, they say, investors would have received a 100% rate of return each year if they had purchased stamps from Diagonal.

a. To check their calculations, compute the annual rate of return.

b. Why is your computed rate of return less than 100%?

7-45 (20-37) An investor purchased 100 shares of Omega common stock for $9000. He held the stock for nine years. For the first four years he received annual end-of-year dividends of $800. For the next four years he received annual dividends of $400. He received no dividend for the ninth year. At the end of the ninth year he sold his stock for $6000. What rate of return did he receive on his investment?

7-46 (20-38) You spend $1000 and in return receive two payments of $1094.60—one at the end of three years and the other at the end of six years. Calculate the resulting rate of return.

CHAPTER 7A

Difficulties Solving For An Interest Rate

Occasionally we encounter a cash flow that cannot be solved for a single positive interest rate. In this appendix to Chapter 7, we examine ways to resolve this difficulty. Example 7A-1 illustrates the situation.

EXAMPLE 7A-1

The Going Aircraft Company has an opportunity to supply a large airplane to Interair, a foreign airline. Interair will pay \$19 million when the contract is signed and \$10 million one year later. Going estimates its second- and third-year net cash flows at \$50 million each when the airplane is being produced. Interair will take delivery of the airplane during Year 4, and agrees to pay \$20 million at the end of that year and the \$60 million balance at the end of Year 5. Compute the rate of return on this project.

Solution: Computation of NPW at various interest rates, using single payment present worth factors,* is presented:

Year	*Cash flow*	0%	10%	20%	40%	50%
0	+\$19	+\$19	+\$19	+\$19	+\$19	+\$19
1	+10	+10	+9.1	+8.3	+7.1	+6.7
2	−50	−50	−41.3	−34.7	−25.5	−22.2
3	−50	−50	−37.6	−28.9	−18.2	−14.8
4	+20	+20	+13.7	+9.6	+5.2	+4.0
5	+60	+60	+37.3	+24.1	+11.2	+7.9
	NPW =	+\$9	+\$0.2	−\$2.6	−\$1.2	+\$0.6

*For example, for Year 2 and $i = 10\%$: PW $= -50(P/F, 10\%, 2) = -50(0.826) = -41.3$.

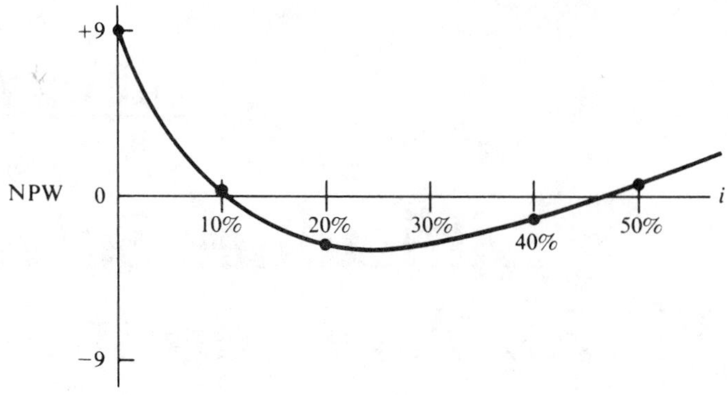

Figure 7A-1 NPW plot.

The NPW plot for this cash flow is represented in Fig. 7A-1. We see this cash flow produces *two* points at which NPW = 0. Thus, there are *two* positive rates of return, one at about 10.1% and the other at about 47%. ◀

Example 7A-1 produced unexpected and undesirable results. We want to know when we may expect this kind of result, what it means, and how we may resolve the difficulty. In the next section, we find that the solution of an economic analysis problem is really the solution of a mathematical equation.

Converting a Cash Flow to a Mathematical Equation

Given a simple cash flow,

Year	*Cash flow*
0	$-P$
1	$+A_1$
2	$+A_2$
·	·
·	·
·	·
n	$+A_n$

Setting NPW = PW of benefits *minus* PW of cost = 0, and using single payment present worth factors, the cash flow may be rewritten as

$$+A_1(1 + i)^{-1} + A_2(1 + i)^{-2} + \cdots + A_n(1 + i)^{-n} - P = 0 \tag{7A-1}$$

If we let $X = (1 + i)^{-1}$, then Eq. 7A-1 may be written

$$+A_1X + A_2X^2 + \cdots + A_nX^n - P = 0 \tag{7A-2}$$

Rearranging terms,

$$A_nX^n + \cdots + A_2X^2 + A_1X - P = 0 \tag{7A-3}$$

Equation 7A-3 is an nth order polynomial to which ***Descartes' Rule*** may be applied. The rule is:

> **If a polynomial with real coefficients has m sign changes, then the number of positive roots will be $m - 2k$, where k is a positive integer or zero ($k = 0, 1, 2, 3, \ldots$).**

A sign change is where successive nonzero terms, written according to descending powers of X, have different signs (that is, change from + to − or *vice versa*). Descartes' Rule means that the number of positive roots of the polynomial cannot exceed m, the number of sign changes; the number of positive roots must either be equal to m or less by an even integer.

Descartes' Rule for polynomials gives the following:

Number of sign changes, m	*Number of positive values of X*
0	0
1	1
2	2 or 0
3	3 or 1
4	4, 2, or 0

But Equation 7A-3's polynomial is not the equation that represents the economic analysis problem. By substituting $(1 + i)^{-1}$ in place of X, we return to Eq. 7A-1. We see from the relationship $X = (1 + i)^{-1}$ that a positive value of X does not ensure a positive value of i. In fact, whenever X is greater than 1, i is negative. Thus, in a particular situation when m equals two sign changes, there would either be two or zero positive values of X. But this means there could be two, one, or zero positive values of i for the corresponding economic analysis equation with two sign changes.

From this discussion, we form a ***cash flow rule of signs***.

Cash Flow Rule Of Signs

> **There may be as many positive values of i as there are sign changes in the cash flow.**

Sign changes are computed the same way as for Descartes' Rule. A sign change is where successive nonzero values in the cash flow have different signs (that is, change from + to −, or *vice versa*). A zero cash flow is ignored. The five cash flows in Table 7A-1 illustrate the counting of the number of sign changes.

Table 7A-1 SIGN CHANGES IN CASH FLOWS

	Cash flow				
Year	*A*	*B*	*C*	*D*	*E*
0	+\$100	−\$100	−\$100	+\$50	+\$50
1	+10	+10	0	+40	−50
2	+50	+50	+50	−100	+50
3	+20	+20	0	+10	−10
4	+40	+40	+80	+10	−30
Sign changes:	0	1	1	2	3

The cash flow rule of signs indicates the following possibilities:

Number of sign changes, m	*Number of positive values of i*
0	0
1	1 or 0
2	2, 1, or 0
3	3, 2, 1, or 0

Thus there are three possibilities that need examination: zero sign changes, one sign change, and two or more sign changes.

Zero Sign Changes

There are two situations that produce zero sign changes in a cash flow. Either all terms have a positive sign, representing receipts, or all terms have a negative sign, reflecting a series of disbursements.

The first case would be like walking into a store and finding oneself their millionth customer and, hence, the recipient of money and gifts. This is the utopian something-for-nothing situation. There is no value of i that can be computed, for there are no disbursements to offset the receipts.

The second case represents the less-happy situation of disbursements without any compensating receipts. It would be like the periodic purchase of lottery tickets—but never winning anything. There is no value of i that would reflect this economic situation.

One Sign Change

Unless there are rather unusual circumstances, one sign change is the normal cash flow pattern. Given one sign change, it is very likely that we have a situation where there will be a single positive value of i.

There is no positive value of i whenever *in an investment situation* the subsequent benefits do not equal the magnitude of the investment. An example would be:

Year	*Cash flow*
0	−$50
1	+20
2	+20

Also, there is no positive value of i whenever *in a borrowing situation* the subsequent repayments do not equal the magnitude of the borrowed money.

Year	*Cash flow*
0	+$50
1	−20
2	−20

These circumstances can be readily identified and, hence, present no confusion or particular problem.

Two or More Sign Changes

When there are two or more sign changes in the cash flow, we know that there are several possibilities concerning the number of positive values of i. Probably the greatest danger in this situation is to fail to recognize the multiple possibilities and to solve for a value of i. Then one might assume—incorrectly or correctly—that the value of i that is obtained is the only positive value of i. In this multiple sign change situation, one approach is to prepare an NPW plot like Fig. 7A-1. This may be a tedious procedure, yet it graphically portrays the exact situation.

What the Difficulties Mean

If there is a single positive value of i, we have no problem. On the other hand, a situation with no positive value of i or multiple positive values represents a situation that may be attractive, unattractive, or confusing. Where there are multiple values of i, none of them should be considered a suitable measure of the rate of return or attractiveness of the cash flow.

We classify a cash flow as an investment situation if we put money into a project and benefits of the project come back to us. The rate of return tells us something about the attractiveness of the project. But what was the situation in Example 7A-1?

Year	*Cash flow*
0	+$19
1	+10
2	−50
3	−50
4	+20
5	+60

At the beginning, money is generated and flows out of the project. In Years 2 and 3, money is spent on production, followed by receipts in the final two years. We know that 10.1% is a value of i at which NPW = 0. What does this mean? If the initial outflows from the project are invested in some *external investment* at 10.1% and then their compound amount returned to the project in Year 2, the project, or internal investment, will have i = 10.1%. Similarly, if the initial outflows from the project can be put into an external investment at 47%, then the project or internal investment will also show a 47% interest rate. This sounds somewhat unbelievable, so let's demonstrate both situations.

For i = 10.1%:

Year	*Cash flow*
0	+$19 invest for 2 years in an *external investment* at 10.1%
	$F = +19(F/P,10.1\%,2)$
	$F = +19(1.21) = +23$ →
1	+10 invest for 1 year in an *external investment* at 10.1%
	$F = +10(F/P,10.1\%,1)$
	$F = +10(1.101) = +11$ →
2	+11 return to internal investment ←
	+23 return to internal investment ←
	−50
3	−50
4	+20
5	+60

When the external investment phase is completed, the cash flow for the internal investment is as follows:

Year	*Cash flow*
0	$0
1	0
2	−16
3	−50
4	+20
5	+60

We see that for the internal investment there is one sign change, so we can solve for the internal value of i directly. It will be 10.1% if, when we use that value, NPW = 0.

$$\begin{aligned}\text{NPW} &= -16(P/F,10.1\%,2) - 50(P/F,10.1\%,3) + 20(P/F,10.1\%,4) \\ &\quad + 60(P/F,10.1\%,5) \\ &= -16(0.825) - 50(0.749) + 20(0.681) + 60(0.618) \\ &= -13.2 - 37.5 + 13.6 + 37.1 \\ &= 0\end{aligned}$$

Similarly, for i = 47%:

Year	*Cash flow*
0	+\$19 invest for 2 years in an *external investment* at 47%
	$F = +19(F/P,47\%,2) = +19(2.161) = +41.1$ → (returned at Year 2)
1	+10 invest for 1 year in an *external investment* at 47%
	$F = +10(F/P,47\%,1) = +10(1.47) = +14.7$ → (returned at Year 2)
2	+14.7 return to project ←
	+41.1 return to project ←
	−50
3	−50
4	+20
5	+60

When the cash outflows to the external investment are returned, the transformed cash flow is tabulated below:

Year	*Cash flow*
0	\$0
1	0
2	+5.8
3	−50
4	+20
5	+60

There is the need for external investment of the 5.8 at Year 2 that will be needed in the project at Year 3.

$$F = +5.8(F/P,47\%,1) = 5.8(1.47) = +8.5$$

When the 8.5 is returned to the project in Year 3, the resulting net required investment in Year 3 is $-50 + 8.5 = -41.5$. Now the internal investment cash flow is,

Year	*Cash flow*
0	\$0
1	0
2	0
3	−41.5
4	+20
5	+60

Solving for the rate of return on the internal investment,

$$\text{PW of costs} = \text{PW of benefits}$$

Thus, at Year 3,

$$41.5 = 20(P/F,i,1) + 60(P/F,i,2)$$

Try $i = 47\%$:

$$\begin{aligned}41.5 &= 20(0.680) + 60(0.464)\\ &= 13.6 + 27.9\\ &= 41.5\end{aligned}$$

From the computations, we have seen that the two positive interest rates (10.1% and 47%) require that the internal investment and the external investment both earn the same interest rate. Thus, if one were prepared to agree that the appropriate interest rate for external investments is 47%, then we would have to agree that the resulting interest rate on the internal investment is 47%. But if a suitable interest rate on external investments (say, putting the money in a savings account) were only 6%, then what is the rate of return on the internal investment? None of the calculations we have made so far tell us. But, in the next section, we find that this *is* a practical approach for solving the multiple interest rate problem.

External Interest Rate

From the discussion of the meaning of multiple interest rates, we see an important general situation.

> **Solving a cash flow for an unknown interest rate means that money in any required external investment is assumed to earn the same interest rate as money invested in the internal investment.**

This occurs regardless of whether there is only one or more than one positive interest rate.

There can be no particular reason why we would assume that external investments earn the same rate of return as internal investments. The required external investment may be of short duration, like a year or two. And, as will be discussed later, it is very likely that the rate of return available on a capital investment in the business is two or three times the rate of return available on short-duration external deposits, or other external investments of money. Thus, two interest rates are reasonable, one on the internal investment and one on the temporary external investment.

By separating the interest rates, we have also provided the means for resolving any difficulties that arise from multiple positive rates of return. What we desire is to determine the rate of return on the internal investment assuming a realistic value for the rate of return available on the external investment. With the external interest rate we can compute the effect of any required external investments. The results can be introduced back into the cash flow in the same manner as was done in our detailed examination of the external investment assumed in Ex. 7A-1. Example 7A-1 will now be presented again, this time with a preselected external interest rate of 6%.

EXAMPLE 7A-2

Take the Ex. 7A-1 cash flow and assume that any money held outside of the project earns 6% interest (that is, the external interest rate = 6%).

Year	*Cash flow*
0	+$19
1	+10
2	−50
3	−50
4	+20
5	+60

At both Year 0 and Year 1, there is a flow of money resulting from the advance payments before the aircraft is manufactured. The money will be needed later to help pay the production costs. If the external interest rate is 6%, the +19 (million dollars) will be invested externally for two years and the +10 for one year. Their compound amount at the end of Year 2 will be:

$$\begin{aligned}\text{Compound amount at end of Year 2} &= +19(F/P,6\%,2) + 10(F/P,6\%,1)\\ &= +19(1.124) + 10(1.06)\\ &= +21.4 + 10.6\\ &= +32\end{aligned}$$

When this amount is returned to the project, the net cash flow for Year 2 becomes $-50 + 32 = -18$. The resulting cash flow for the project is,

Year	Cash flow	*Computation of NPW at various interest rates using single payment present worth factors* 0%	8%	10%
0	\$0	\$0	\$0	\$0
1	0	0	0	0
2	−18	−18	−15.4	−14.9
3	−50	−50	−39.7	−37.6
4	+20	+20	+14.7	+13.7
5	+60	+60	+40.8	+37.3
	NPW =	+\$12	+\$0.4	−\$1.5

The cash flow has one sign change indicating there is either zero or one positive interest rate. We have located a point where NPW = 0 at

$$i = 8\% + 2\%\left(\frac{0.4}{1.5 + 0.4}\right) = 8\% + 2\%(0.21) = 8.4\%$$

Thus, we have identified the single positive root for the cash flow. Assuming an external interest rate of 6%, the rate of return on the Interair plane contract is 8.4%. ◀

In Ex. 7A-2 we accomplished two tasks:

1. A realistic interest rate was used to find equivalent sums when money must be invested externally. This external interest rate should reflect the rate on external investment opportunities and, therefore, be independent of the rate of return on any particular internal investment.
2. Through the use of an external interest rate, the number of sign changes in the cash flow was reduced to one, ensuring that there will not be multiple positive rates of return.

Resolving Multiple Rate of Return Problems

Example 7A-1 contains a cash flow that produces two positive rates of return. Yet, on closer examination, we saw a cash flow for external investment in the initial part of the problem. The outflow of cash was invested at the external interest rate until it was needed in the project. It was returned to the project at the end of Year 2. In so doing, the number of sign changes in the cash flow was reduced from two to one. A single positive rate of return was then computed. In this way, the multiple rate of return difficulty was resolved.

The general method for handling cash flows with multiple sign changes is: alter the cash flow through the use of an external interest rate to reduce the number of sign changes to one, thereby ensuring no more than one positive rate of return. We must point out, however, that the changes made with the external interest rate affect the project, or internal rate of return. To keep the sensitivity of the internal rate of return to the external interest rate as small as possible, the cash flow adjustments should be kept to a minimum.

A Further Look at the Computation of Rate of Return

The cash flow rule of signs tells us we can have cash flows with multiple sign changes but only one positive rate of return. Using the approach in this subchapter, we may alter a cash flow when no alteration is needed (or be making too great an alteration when one is needed). There is thus more to be said about solving a cash flow for a rate of return, but for now, we continue our broader examination of engineering economic analysis. We will take a further look at the computation of rate of return in Chapter 18.

SUMMARY

In some situations, we find that solution of a cash flow equation results in more than one positive rate of return. This is possible by the cash flow rule of signs. A sign change is where successive nonzero values in the cash flow have different signs (that is, change from + to −, or *vice versa*).

Zero sign changes indicates there is no rate of return, as the cash flow is either all disbursements or all receipts.

One sign change is the usual situation and a single positive rate of return generally results. There will, however, be no rate of return whenever loan repayments are less than the loan or an investment fails to return benefits equal to the investment.

Multiple sign changes may result in multiple positive rates of return. The difficulty is not that it will happen, but that the analyst may not recognize that the cash flow has multiple sign changes and may have multiple positive rates of return. When they occur, none of the multiple rates of return are a suitable measure of the economic desirability of the project represented by the cash flow.

Multiple positive rates of return indicate a project that at some time has money invested outside the project. Since investments outside the project may earn interest at a different rate from the internal project rate of return, an external rate should be selected. This approach leaves the rate of return on the money actually invested in the project as the single unknown. The number of sign changes are thereby reduced to one, eliminating the possibility of multiple positive rates of return. This topic is also discussed further in Chapter 18.

Problems

7A-1 The owner of a walnut orchard wished to enjoy some of his wealth and yet not sell the orchard for ten years. He negotiated an agreement with the Omega Insurance Company as follows. Omega would pay the owner $4000 per year in twenty equal annual payments beginning immediately. At the end of the tenth year, the owner is obligated to sell the orchard and with the proceeds pay Omega $75,000 at that time. Omega will, of course, continue the $4000 annual payments to the retired orchardist for nine more years. What interest rate was used in devising the agreement between Omega and the orchardist?

7A-2 A group of businessmen formed a partnership to buy and race an Indianapolis-type racing car. They agreed to pay an individual $50,000 for the car and associated equipment. The payment was to be in a lump sum at the end of the year. In what must have been "beginner's luck," the group won a major race the first week and $80,000. The rest of the first year, however, was not so good: at the end of the first year, the group had to pay out $35,000 for expenses plus the $50,000 for the car and equipment. The second year was a poor one: the group had to pay $70,000 at the end of the second year just to clear up the racing debts. During the third and fourth years, racing income just equalled costs. When the group was approached by a prospective buyer for the car, they readily accepted $80,000 cash, which was paid at the end of the fourth year. What rate of return did the businessmen obtain from their racing venture?

7A-3 A student organization, at the beginning of the Fall quarter, purchased and operated a soft-drink vending machine as a means of helping finance its activities. The vending machine cost $75 and was installed at a gasoline station near the university. The student organization pays $75 every three months to the station owner for the right to keep their vending machine at the station. During the year the student organization owned the machine, they received the following quarterly income from it, before making the $75 quarterly payment to the station owner:

	Income
Fall quarter	$150
Winter quarter	25
Spring quarter	125
Summer quarter	150

At the end of one year, the student group resold the machine for $50. Determine the quarterly cash flow. Then answer ***a***, ***b***, and **c** below.

a. Assume the cash flow has a single positive rate of return. Proceed to compute the nominal annual rate of return the organization received on their investment.

b. Using a nominal external interest rate of 12% (3% per quarter-year) transform the cash flow to one sign change. Then compute the nominal rate of return.

c. Why do the answers for ***a*** and ***b*** differ? Which one is the "correct" answer?

7A-4 Given the following cash flow:

Year	*Cash flow*
0	−$500
1	+2000
2	−1200
3	−300

Determine the rate of return on the internal investment. If necessary, assume external investments earn 6% interest. (*Answer:* 20.2%)

7A-5 Given the following cash flow:

Year	*Cash flow*
0	−$500
1	+200
2	−500
3	+1200

Determine the rate of return on the internal investment. If necessary, assume external investments earn 6% interest. (*Answer:* 19.6%)

7A-6 Given the following cash flow:

Year	*Cash flow*
0	−$100
1	+360
2	−570
3	+360

Determine the rate of return on the internal investment. If necessary, assume external investments earn 6% interest.

7A-7 Consider the following cash flow:

Year	*Cash flow*
0	−$110
1	−500
2	+300
3	−100
4	+400
5	+500

Assume external investment of money is at a 10% interest rate. Compute the rate of return on the internal investment.

7A-8 Consider the following cash flow:

Year	*Cash flow*
0	−$50.0
1	+20.0
2	−40.0
3	+36.8
4	+36.8
5	+36.8

Assume any external investment of money is at a 10% interest rate. Compute the rate of return on the internal investment. (*Answer:* 15%)

7A-9 A firm invested $15,000 in a project that appeared to have excellent potential. Unfortunately, a lengthy labor dispute in Year 3 resulted in costs that exceeded benefits by $8000. The cash flow for the project is as follows:

Year	*Cash flow*
0	−$15,000
1	+10,000
2	+6,000
3	−8,000
4	+4,000
5	+4,000
6	+4,000

Compute the rate of return for the project, assuming a 12% interest rate on external investments.

7A-10 The textbook tells us that for the following cash flow there is no positive interest rate.

Year	*Cash flow*
0	−$50
1	+20
2	+20

There is, however, a negative interest rate. Compute its value.
(*Answer:* i = −13.7% or −146%)

7A-11 For the following cash flow compute the internal rate of return, assuming 15% interest on external investments. Do not transform the cash flow any more than is essential.

Year	*Cash flow*
0	$0
1	0
2	−20
3	0
4	−10
5	+20
6	−10
7	+100

7A-12 Given the following cash flow:

Year	*Cash flow*
0	−\$800
1	+500
2	+500
3	−300
4	+400
5	+275

If external investments earn 10%, what is the rate of return on the internal investment? (*Answer:* 25%)

7A-13 Consider the following cash flow.

Year	*Cash flow*
0	−\$100
1	+240
2	−143

a. Solve the cash flow for all positive values of i.

b. Assuming any external investment of money will earn 12% per year, compute the rate of return on the internal investment.

c. If the minimum attractive rate of return is 12%, should the project be undertaken?

7A-14 Refer to the strip-mining project in Example 5-9. Compute the rate of return for the project, assuming if necessary a 10% interest rate on external investments.

7A-15 Consider the following cash flow.

Year	*Cash flow*
0	−\$500
1	+800
2	+170
3	−550

Compute the rate of return on the internal investment. Assume any external investment of money is at 10%.

7A-16 (20-42) Consider the cash flow:

Year	*Cash flow*
0	−\$100
1	+360
2	−428
3	+168

a. Solve the cash flow for all positive rates of return. Make a plot of NPW vs. i.

b. Determine the rate of return on the internal investment. If necessary, assume any external investments earn 10% interest.

7A-17 (20-43) Consider the cash flow:

Year	*Cash flow*
0	−$1200
1	+358
2	+358
3	+358
4	+358
5	+358
6	−394

Assume any external investment of money is at a 10% interest rate. Compute the rate of return on the internal investment.

7A-18 (20-44) Consider the cash flow:

Year	*Cash flow*
0	−$3570
1–3	+1000
4	−3170
5–8	+1500

Determine the rate of return on the internal investment. If necessary, any external investments earn 8% interest.

7A-19 (20-45) Bill purchased a vacation lot he saw advertised on television for an $800 downpayment and monthly payments of $55. When he visited the lot he had purchased, he found it was not something he wanted to own. After 40 months he was finally able to sell the lot. The new purchaser assumed the balance of the loan on the lot and paid Bill $2500. What rate of return did Bill receive on his investment?

7A-20 (20-46) Consider the cash flow:

Year	*Cash flow*
0	−$ 850
1	+600
2–9	+200
10	−1800

a. Compute a rate of return for the cash flow.

b. Compute the Net Present Worth (NPW) of the cash flow at the firm's 10% minimum attractive rate of return.

c. Plot a graph of NPW vs. i for the cash flow.

d. Compute the rate of return on the internal investment. Assume any external investment of money is at 10%.

CHAPTER 8

Incremental Analysis

We now see how to solve problems by each of three major methods, with one exception: for three or more alternatives, no rate of return solution was given. The reason is that under these circumstances, incremental analysis is required and it has not been discussed. This chapter will show how to solve that problem.

Incremental analysis can be defined as the examination of the differences between alternatives. By emphasizing alternatives, we are really deciding whether or not differential costs are justified by differential benefits.

In retrospect, we see that the simplest form of incremental analysis was presented in Chapter 7. We did incremental analysis by the rate of return evaluation of the differences between two alternatives. We recognized that the two alternatives could be related as follows:

$$\begin{matrix}\text{Higher-cost} \\ \text{alternative}\end{matrix} = \begin{matrix}\text{Lower-cost} \\ \text{alternative}\end{matrix} + \begin{matrix}\text{Differences} \\ \text{between them}\end{matrix}$$

We will see that incremental analysis can be examined either graphically or numerically. We will first look at graphical representations of problems, proceed with numerical solutions of rate of return problems, and see that a graphical representation may be useful in examining problems whether using incremental analysis or not.

Graphical Solutions

In the last chapter, we examined problems with two alternatives. Our method of solution represented a form of incremental analysis. A graphical review of that situation will help to introduce incremental analysis.

EXAMPLE 8-1

This is a review of Ex. 7-6. There were two mutually exclusive alternatives:

Year	*Alt. 1*	*Alt. 2*
0	−$10	−$20
1	+15	+28

If 6% interest is assumed, which alternative should be selected?

Solution: For this problem, we will plot the two alternatives on a PW of benefits *vs.* PW of cost graph.

Alternative 1:

$$\text{PW of cost} = \$10$$

$$\text{PW of benefit} = \$15(P/F,6\%,1) = 15(0.9434) = \$14.15$$

Alternative 2:

$$\text{PW of cost} = \$20$$

$$\text{PW of benefit} = \$28(P/F,6\%,1) = 28(0.9434) = \$26.42$$

The alternatives are plotted in Figure 8-1, which looks very simple yet tells us a great deal about the situation.

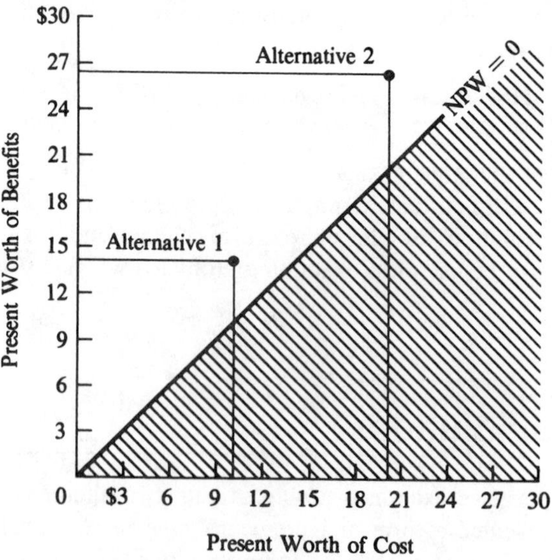

Figure 8-1 PW of benefits *vs.* PW of cost graph.

On a graph of PW of benefits *vs*. PW of cost (for convenience, we will call it a ***benefit–cost graph***), there will be a line where NPW = 0. Where the scales used on the two axes are identical, as in this case, the resulting line will be at a 45° angle. If unequal scales are used, the line will be at some other angle.

For the chosen interest rate (6% in this example), this NPW = 0 line divides the graph into an area of desirable alternatives and undesirable alternatives. To the left (or above) the line is desirable, while to the right (or below) the line is undesirable. We see that to the left of the line, PW of benefits exceeds the PW of cost or, we could say, NPW is positive. To the right of the line, in the shaded area, PW of benefits is less than PW of cost; thus, NPW is negative.

In this example, both alternatives are to the left of the NPW = 0 line. Therefore, both alternatives will have a rate of return greater than the 6% interest rate used in constructing the graph. In fact, other rate of return lines could also be computed and plotted on the graph *for this special case of a one-year analysis period*. We must emphasize at the outset that the additional rate of return lines shown in Fig. 8-2 can be plotted only for this special situation. For analysis periods greater than one year, the NPW = 0 line is the only line that can be accurately drawn. The graphical results in Fig. 8-2 agree with the calculations made in Ex. 7-6, where the rates of return for the two alternatives were 50% and 40%, respectively.

Figure 8-2 shows that the slope of a line on the graph represents a particular rate of return for this special case of a one-year analysis period. Between the

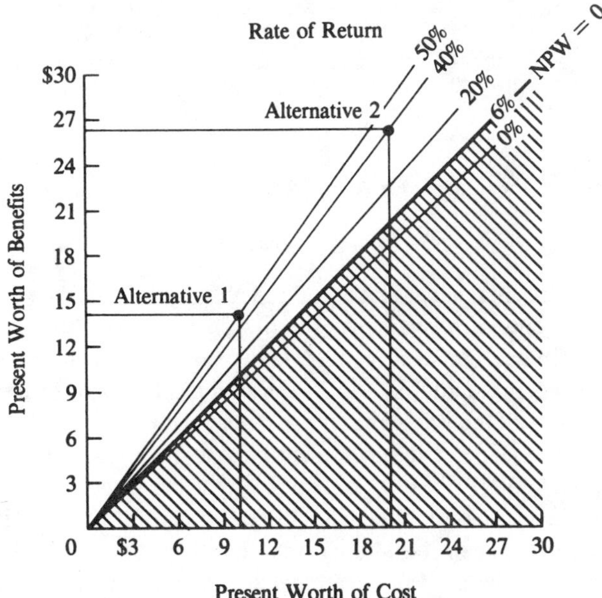

Figure 8-2 Benefit–cost graph for Example 8-1 with a one-year analysis period.

origin and Alt. 1, the slope represents a 50% rate of return; while from the origin to Alt. 2, the slope represents a 40% rate of return. Since

$$\begin{matrix}\text{Higher-cost} \\ \text{Alt. 2}\end{matrix} = \begin{matrix}\text{Lower-cost} \\ \text{Alt. 1}\end{matrix} + \begin{matrix}\text{Differences} \\ \text{between them}\end{matrix}$$

the differences between the alternatives can be represented by a line shown in Fig. 8-3.

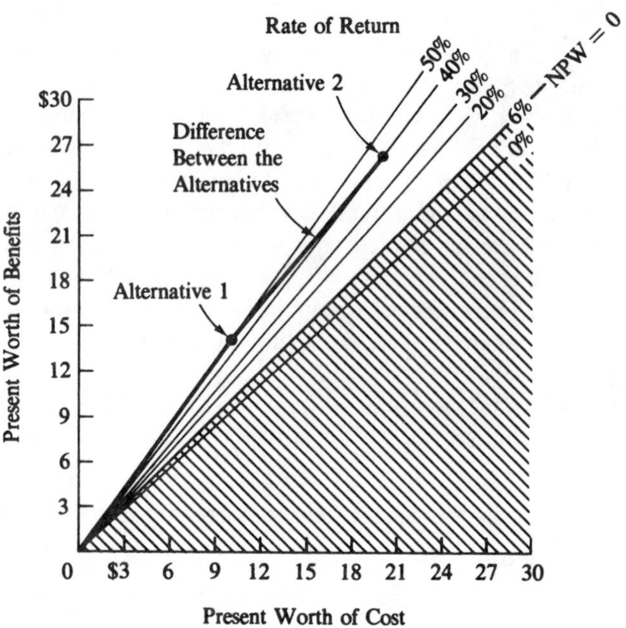

Figure 8-3 Benefit–cost graph for Alternatives 1 and 2 with a one-year analysis period.

Viewed in this manner, we clearly see that Alt. 2 may be considered two separate increments of investment. The first increment is Alt. 1 and the second one is the difference between the alternatives. Thus we will select Alt. 2 if the difference between the alternatives is a desirable increment of investment.

Since the slope of the line represents a rate of return, we see that the increment is desirable if the slope of the increment is greater than the slope of the 6% line that corresponds to NPW = 0. We can see that the slope is greater; hence, the increment of investment is attractive. In fact, a careful examination shows that the "Difference between the alternatives" line has the same slope as the 30% rate of return line. We can say, therefore, that the incremental rate of return from selecting Alt. 2 rather than Alt. 1 is 30%. This is the same as was computed in Ex. 7-6. We conclude that Alt. 2 is the preferred alternative. ◀

EXAMPLE 8-2

Solve Ex. 7-9 by means of a benefit–cost graph. Two machines are being considered for purchase. If the minimum attractive rate of return (MARR) is 10%, which machine should be bought?

	Machine X	*Machine Y*
Initial cost	\$200	\$700
Uniform annual benefit	95	120
End-of-useful-life salvage value	50	150
Useful life, in years	6	12

Solution: Using a 12-year analysis period,

Machine X:

$$\text{PW of cost*} = 200 + (200 - 50)(P/F,10\%,6) - 50(P/F,10\%,12)$$
$$= 200 + 150(0.5645) - 50(0.3186) = 269$$
$$\text{PW of benefit} = 95(P/A,10\%,12) = 95(6.814) = 647$$

Machine Y:

$$\text{PW of cost*} = 700 - 150(P/F,10\%,12) = 700 - 150(0.3186) = 652$$
$$\text{PW of benefit} = 120(P/A,10\%,12) = 120(6.814) = 818$$

*Salvage value is considered a reduction in cost rather than a benefit.

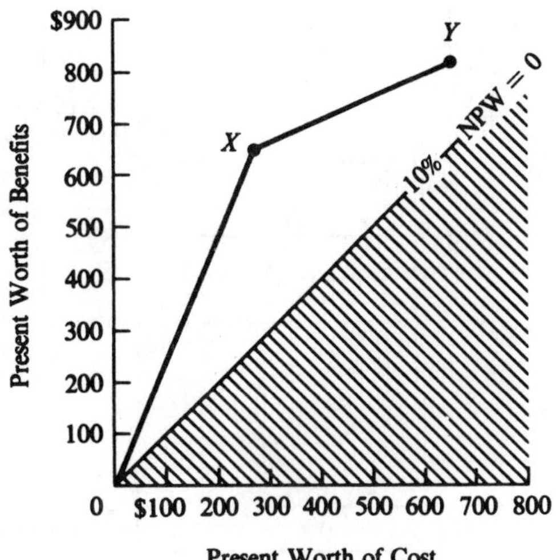

Figure 8-4 Benefit–cost graph.

The two alternatives are plotted in Fig. 8-4; we see that the increment Y–X has a slope much less than the 10% rate of return line. The rate of return on the increment of investment is less than 10%; hence, the increment is undesirable. This means that Machine X should be selected rather than Machine Y. ◀

The two example problems show us some aspects of incremental analysis. We now will examine some multiple-alternative problems. We can solve multiple-alternative problems by present worth and annual cash flow analysis without any difficulties. Rate of return analysis requires that, for two alternatives, the differences between them must be examined to see whether or not they are desirable. Now, if we can choose between two alternatives, then by a successive examination we can choose from multiple alternatives. Figure 8-5 illustrates the method:

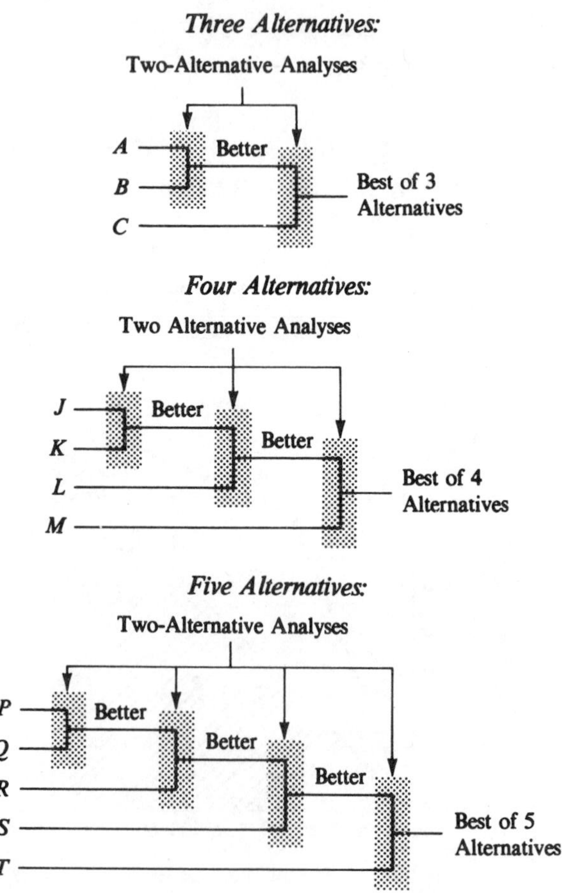

Figure 8-5 Solving multiple-alternative problems by successive two-alternative analyses.

EXAMPLE 8-3

Consider the three mutually exclusive alternatives below.

	A	*B*	*C*
Initial cost	\$2000	\$4000	\$5000
Uniform annual benefit	410	639	700

Each alternative has a twenty-year life and no salvage value. If the MARR is 6%, which alternative should be selected?

Solution: At 6%, (PW of benefits) = (Uniform annual benefit) × (Series present worth factor).

$$\text{PW of benefits} = (\text{Uniform annual benefit})(P/A,6\%,20)$$

$$\text{PW of benefits for Alt. } A = \$410(11.470) = \$4703$$

$$\text{Alt. } B = \$639(11.470) = \$7329$$

$$\text{Alt. } C = \$700(11.470) = \$8029$$

Figure 8-6 is a plot of the situation; we see that the slope of the line from the origin to *A* is greater than the 6% line (NPW = 0). Thus the rate of return for *A* is greater than 6%. For the increment of additional cost of *B* over *A*, the slope of Line *B–A* is greater than the 6% line. This indicates that the rate of return on the increment of investment also exceeds 6%. But the slope of Increment *C–B* indicates its rate of return is less than 6%, hence, undesirable. We conclude that the *A* investment is satisfactory as well as the *B–A* increment; therefore, *B* is satisfactory. The *C–B* increment is unsatisfactory; so *C* is undesirable compared to *B*. Our decision is to select Alternative *B*. ◀

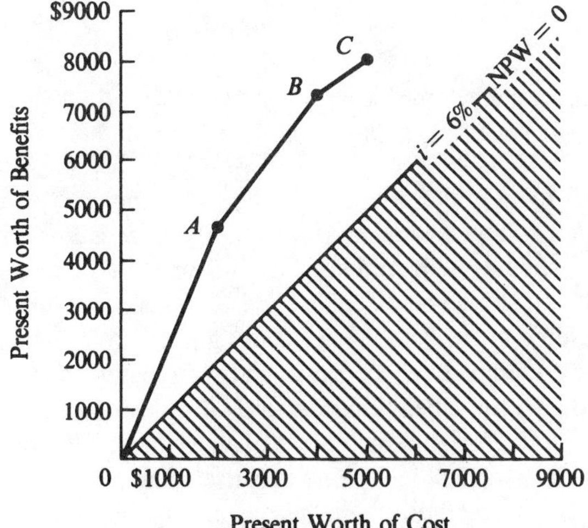

Figure 8-6 Benefit–cost graph for Example 8-3.

EXAMPLE 8-4

Further study of the three alternatives of Ex. 8-3 reveals that Alt. *A*'s uniform annual benefit was overstated. It is now projected to be 122 rather than 410. Replot the benefit–cost graph for this changed situation.

Solution:

$$\text{Alt. } A'\text{:} \quad \text{PW of benefits} = 122(P/A,6\%,20)$$
$$= 122(11.470) = 1399$$

Figure 8-7 shows the revised plot of the three alternatives. The graph shows that the revised Alt. *A'* is no longer desirable. We see that it has a rate of return less than 6%.

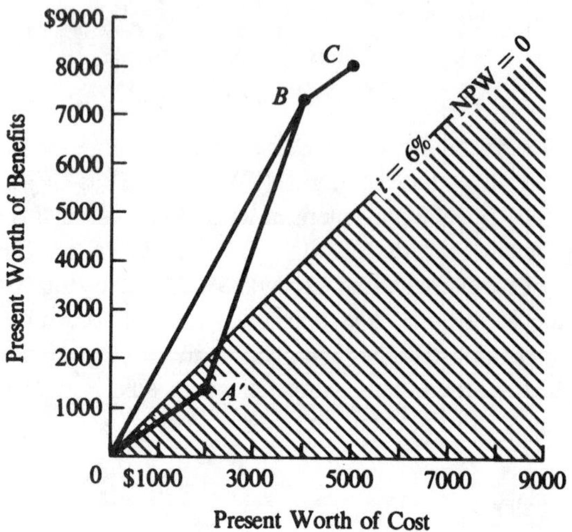

Figure 8-7 Benefit–cost graph for Example 8-4.

Now we wish to examine Alt. *B*. Should we compare it to the do-nothing alternative (which is represented by the origin), or as a *B–A'* increment over *A'*? Graphically, should we examine Line 0–*B* or *B–A'*? Since *A'* is an undesirable alternative, it should be discarded and not considered further. Thus ignoring *A'*, we should compare *B* to the do-nothing alternative, which is Line 0–*B*. Alternative *B* is preferred over the do-nothing alternative since it has a rate of return greater than 6%. Then Increment *C–B* is examined and, as we saw previously, it is an undesirable increment of investment. The decision to select *B* has not changed, which should be no surprise: if an inferior *A* has become an even less attractive *A'*, we still would select the superior alternative, *B*. ◀

The graphical solution of the four example problems has helped us visualize the mechanics of incremental analysis. While problems could be solved this way, in practice they are solved mathematically rather than graphically. We will now proceed to solve problems mathematically by incremental rate of return analysis.

Incremental Rate Of Return Analysis

To illustrate incremental rate of return analysis, we will solve three of the examples again by mathematical, rather than graphical, methods.

EXAMPLE 8-5

Solve Ex. 8-1 mathematically. With two mutually exclusive alternatives and a 6% MARR, which alternative should be selected?

Year	*Alt. 1*	*Alt. 2*
0	−\$10	−\$20
1	+15	+28

Solution: Examine the differences between the alternatives.

Year	*Alt. 2 − Alt. 1*
0	−20 − (−10) = −10
1	+28 − (+15) = +13

Incremental rate of return (ΔROR),

$$10 = 13(P/F,i,1)$$

$$(P/F,i,1) = \frac{10}{13} = 0.7692$$

$$\Delta\text{ROR} = 30\%$$

The ΔROR is greater than the MARR; hence, we will select the alternative that gives this increment.

$$\begin{bmatrix}\text{Higher-cost}\\ \text{Alt. 2}\end{bmatrix} = \begin{bmatrix}\text{Lower-cost}\\ \text{Alt. 1}\end{bmatrix} + \begin{bmatrix}\text{Increment}\\ \text{between them}\end{bmatrix}$$

Select Alternative 2. ◀

EXAMPLE 8-6

Recompute Example 8-3. MARR = 6%. Each alternative has a twenty-year life and no salvage value.

	A	*B*	*C*
Initial cost	\$2000	\$4000	\$5000
Uniform annual benefit	410	639	700

Solution: A practical first step is to compute the rate of return for each alternative.

Alternative A:

$$2000 = 410(P/A,i,20)$$

$$(P/A,i,20) = \frac{2000}{410} = 4.878 \qquad i = 20\%$$

Alternative B:

$$4000 = 639(P/A,i,20)$$

$$(P/A,i,20) = \frac{4000}{639} = 6.259 \qquad i = 15\%$$

Alternative C:

$$5000 = 700(P/A,i,20)$$

$$(P/A,i,20) = \frac{5000}{700} = 7.143$$

The rate of return is between 12% and 15%:

$$i = 12\% + \left(\frac{7.469 - 7.143}{7.469 - 6.259}\right)(3\%) = 12.8\%$$

At this point, we would reject any alternative that fails to meet the MARR criterion of 6%. All three alternatives exceed the MARR in this example.

Next, we arrange the alternatives in order of increasing initial cost. Then we can examine the increments between the alternatives.

	A	*B*	*C*
Initial cost	\$2000	\$4000	\$5000
Uniform annual benefit	410	639	700
Rate of return	20%	15%	12.8%

	Increment B–A	*Increment C–B*
Incremental cost	\$2000	\$1000
Incremental uniform annual benefit	229	61

Incremental rate of return:

$$2000 = 229(P/A,i,20)$$

$$(P/A,i,20) = \frac{2000}{229} \qquad \Delta\text{ROR } 9.6\%$$

$$1000 = 61(P/A,i,20)$$

$$(P/A,i,20) = \frac{1000}{61} \qquad \Delta\text{ROR } 2.0\%$$

The *B–A* increment is satisfactory; therefore, *B* is preferred over *A*. The *C–B* increment has an unsatisfactory 2% rate of return; therefore, *B* is preferred over *C*. Conclusion: select Alternative *B*. ◀

EXAMPLE 8-7

Solve Ex. 8.4 mathematically. Alternative *A* in the previous example was believed to have an overstated benefit. The new situation for *A* (we will again call it *A′*) is a uniform annual benefit of 122. Compute the rate of return for *A′*.

$$2000 = 122(P/A,i,20)$$

$$(P/A,i,20) = \frac{2000}{122} = 16.39 \qquad i = 2\%$$

This time Alt. *A′* has a rate of return less than the MARR of 6%. Alternative *A′* is rejected, and the problem now becomes selecting the better of *B* and *C*. In Ex. 8-6 we saw that the increment *C–B* had a ΔROR of 2% and it, too, was undesirable. Thus, we again select Alternative *B*. ◀

EXAMPLE 8-8

The following information is provided for five mutually exclusive alternatives that have twenty-year useful lives. If the minimum attractive rate of return is 6%, which alternative should be selected?

	A	*B*	*C*	*D*	*E*
Cost	$4000	$2000	$6000	$1000	$9000
Uniform annual benefit	639	410	761	117	785
PW of benefit*	7330	4700	8730	1340	9000
Rate of return	15%	20%	11%	10%	6%

*PW of benefit = (Uniform annual benefit)$(P/A,6\%,20)$ = 11.470(Uniform annual benefit). These values will be used later to plot a PW of cost *vs*. PW of benefit curve.

Solution: We see that the rate of return for each alternative equals or exceeds the MARR, therefore, no alternatives are rejected at this point. Next, we rearrange the alternatives to put them in order of increasing cost:

	D	*B*	*A*	*C*	*E*
Cost	$1000	$2000	$4000	$6000	$9000
Uniform annual benefit	117	410	639	761	785
Rate of return	10%	20%	15%	11%	6%

	Increment B–D	*Increment A–B*	*Increment C–A*
ΔCost	$1000	$2000	$2000
ΔAnnual benefit	293	229	122
ΔRate of return	29%	10%	2%

Beginning with the analysis of Increment *B–D*, we compute a ΔROR of 29%. Alternative *B* is thus preferred to Alt. *D* and *D* may be discarded at this point. The ΔROR for *A–B* is also satisfactory, so *A* is retained and *B* is now discarded. The *C–A* increment has a rate of return less than the MARR. Therefore, *C* is discarded and *A* continues to be retained.

At this point, we have examined four alternatives—*D, B, A, C*—and retained *A* after discarding the other three. Now we must decide whether *A* or *E* is the superior alternative. The increment we will examine is *E–A*. (*Note:* Increment *E–C* would have no particular meaning for we have already discarded *C*.)

	Increment E–A
ΔCost	$5000
ΔAnnual benefit	146

Over the twenty-year useful life, the total benefits (20 × 146 = 2920) are less than the cost. There is no rate of return on this increment (or one might say the ΔROR < 0%). This is an unsatisfactory increment, so *E* is discarded. Alternative *A* is the best of the five alternatives. ◀

The benefit–cost graph (Fig. 8-8) of this example problem illustrates an interesting situation. All five alternatives have rates of return equal to or greater than the MARR of 6%. Yet, on detailed examination, we see that Alternatives *C* and *E* contain increments of investment that are unsatisfactory. Even though *C* has an 11% rate of return, it is unsatisfactory when compared to Alt. *A*.

Also noteworthy is the fact that the project with the greatest rate of return—Alternative *B*—is *not* the best alternative, because the proper economic criterion in this situation is to accept all separable increments of investment that have a rate of return greater than the 6% MARR. We found a desirable *A–B* increment with a 10% ΔROR. A relationship between Alternatives *A* and *B*, and the computed rates of return, are:

$$\frac{\text{Higher-cost Alt. } A}{\text{15\% rate of return}} = \frac{\text{Alt. } B}{\text{20\% rate of return}} + \frac{\text{Differences between } A \text{ and } B}{\text{10\% rate of return}}$$

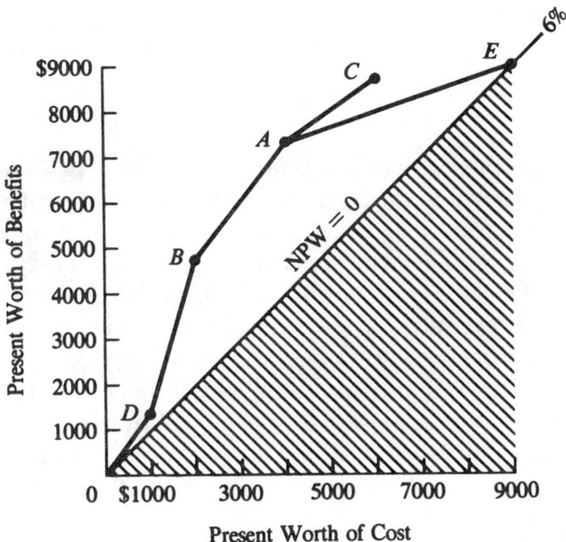

Figure 8-8 Benefit–cost graph for Example 8-8 data.

By selecting *A* we have, in effect, acquired a 20% rate of return on $2000 and a 10% rate of return on an additional $2000. Both of these are desirable. Taken together as *A*, the result is a 15% rate of return on a $4000 investment. This is economically preferable to a 20% rate of return on a $2000 investment, assuming we seek to invest money whenever we can identify an investment opportunity that meets our 6% MARR criterion. This implies that we have sufficient money to accept *all* investment opportunities that come to our attention where the MARR is exceeded. This abundant supply of money is considered appropriate in most industrial analyses, but it is not likely to be valid for individuals. The selection of an appropriate MARR is discussed in Chapter 15.

Elements In Incremental Rate Of Return Analysis

1. ***Be sure all the alternatives are identified.*** In textbook problems the alternatives will be well-defined, but industrial problems may be less clear. Before proceeding, one must have all the mutually exclusive alternatives tabulated, including the "do-nothing" or the "keep doing the same thing" alternative, if appropriate.

2. ***(Optional) Compute the rate of return for each alternative.*** If one or more alternatives has a rate of return equal to or greater than the minimum attractive

rate of return (ROR ≥ MARR), then any other alternatives with a ROR < MARR may be immediately rejected. This optional step requires calculations that may or may not eliminate alternatives. In an exam, this step probably should be skipped, but in less pressing situations, these are logical computations.

3. *Arrange the remaining alternatives in ascending order of investment.* The goal is to organize the alternatives so that the incremental analysis will be of separable increments of investment when we analyze the Higher-cost alternative *minus* the Lower-cost alternative. In textbook problems, this is usually easy—but there are lots of potential difficulties in following this simple rule.

The ordering of alternatives is not a critical element in incremental analysis; in fact, the differences between any two alternatives, like X and Y, can be examined either as an X–Y increment or a Y–X increment. If this is done in a random fashion, however, we will be looking sometimes at an increment of borrowing and sometimes at an increment of investment. It can get a little difficult to keep all of this straight; the basic goal is to restrict the incremental analysis, where possible, to increments of investment.

4. *Make a two-alternative analysis of the first two alternatives.* In the typical situation, we have:

$$\begin{bmatrix}\text{Higher-cost}\\ \text{Alt. } Y\end{bmatrix} = \begin{bmatrix}\text{Lower-cost}\\ \text{Alt. } X\end{bmatrix} + \begin{bmatrix}\text{Differences between}\\ \text{them } (Y - X)\end{bmatrix}$$

so the increment examined is $(Y - X)$, which represents an increment of investment.

Compute the ΔROR on the increment of *investment*. The criterion is:

- If ΔROR ≥ MARR, retain the higher-cost Alt. Y.
- If ΔROR < MARR, retain the lower-cost Alt. X.
- Reject the other alternative used in the analysis.

Sometimes the two alternatives being examined cannot be described as "higher cost" and "lower cost." In Example 7-8, we encountered two alternatives, A and B, with equal investments. There we selected the $(A - B)$ increment because it was an increment of investment.

In other situations one may encounter cash flows of the differences between alternatives that have multiple sign changes (like Cash flows D and E in Table 7A-1). The logical approach is to look at both possible differences between alternatives, for example, $(X - Y)$ and $(Y - X)$, and select the one for analysis where the investment component dominates.

In situations where an increment of *borrowing* is examined, the criterion is:

- If $\Delta i \leq$ MARR, the increment is acceptable.
- If $\Delta i >$ MARR, the increment is not acceptable.

5. Take the preferred alternative from Step 4, and the next alternative from the list created in Step 3. Proceed with another two-alternative comparison.

6. Continue until all alternatives have been examined and the best of the multiple alternatives has been identified.

Incremental Analysis Where There Are Unlimited Alternatives

There are situations where the possible alternatives are a more or less continuous function. For example, an analysis to determine the economical height of a dam represents a situation where the number of alternatives could be infinite. If the alternatives were limited, however, to heights in even feet, the number of alternatives would still be large and have many of the qualities of a continuous function, as in the following example.

EXAMPLE 8-9

A careful analysis has been made of the consequences of constructing a dam in the Blue Canyon. It would be feasible to construct a dam at this site with a height anywhere from 200 to 500 feet. Using a 4% MARR and a 75-year life, the various data have been used to construct Figure 8-9. Note particularly that the dam heights are plotted on the x-axis along with the associated PW of cost. What height of dam should be constructed?

Solution: Five points have been labelled on Fig. 8-9 to aid in the discussion. Dam heights below Point A have a PW of cost > PW of benefit; hence, the rate of return is less than MARR, and we would not build a dam of these heights. In the region of Point B, an increment of additional PW of cost—ΔC—produces a larger increment of PW of benefit—ΔB. These are, therefore, desirable increments of additional investment and, hence, dam height. At Point D and also at Point E, the reverse is true. An increment of additional investment—ΔC—produces a smaller increment of PW of benefit—ΔB; this is undesirable. We do not want these increments, so the dam should not be built to these heights.

At Point C, we are at the point where $\Delta B = \Delta C$. Lower dam heights have desirable increments of investment and higher dam heights have unfavorable increments of investment. The optimal dam height, therefore, is where $\Delta B = \Delta C$. On the figure, this corresponds to a height of approximately 250 feet. Another way of defining the point where $\Delta B = \Delta C$ is to describe it as the point where the slope of the curve equals the slope of the NPW = 0 line. ◀

The techniques for solving discrete alternatives or continuous function alternatives are really the same. We proceed by adding increments whenever ΔROR $\geq$ MARR and discarding increments when ΔROR < MARR.

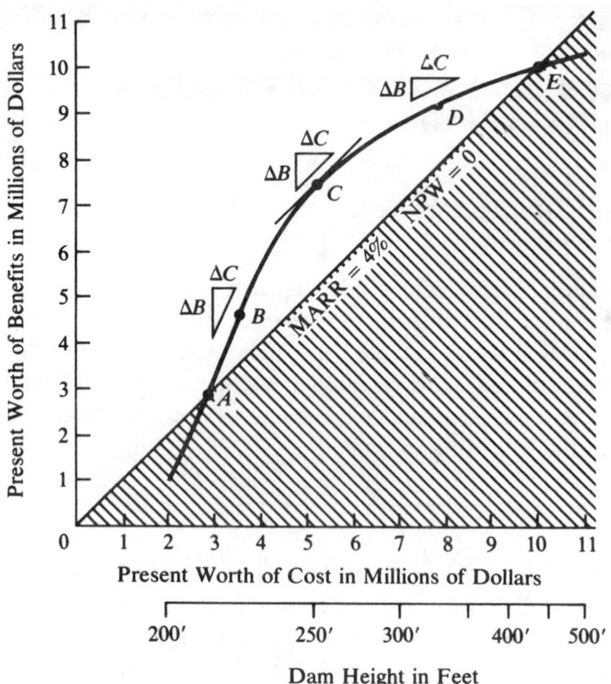

Figure 8-9 Benefit–cost graph for Example 8-9.

Present Worth Analysis With Benefit–Cost Graphs

Any of the example problems presented so far in this chapter could be solved by the present worth method. The benefit–cost graphs we introduced earlier can be used to graphically solve problems by present worth analysis.

In present worth analysis, where neither input nor output is fixed, the economic criterion is to maximize NPW. In Ex. 8-1, we had the case of two alternatives and the MARR equal to 6%:

Year	*Alt. 1*	*Alt. 2*
0	−$10	−$20
1	+15	+28

In Example 8-1 we computed,

PW of cost	$10	$20
PW of benefit	14.15	26.42

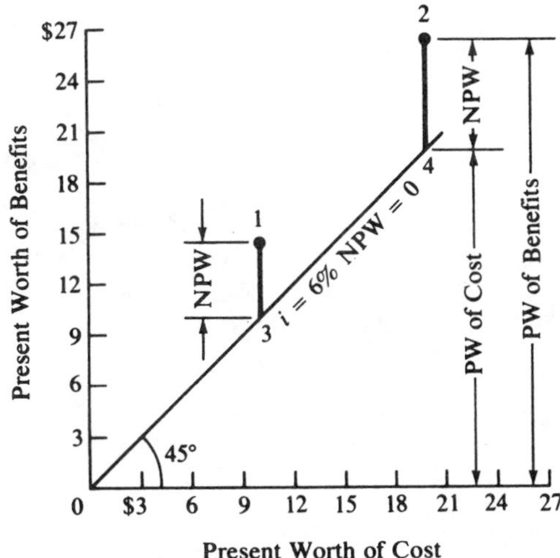

Figure 8-10 Benefit–cost graph for PW analysis.

These points are plotted in Figure 8-10. Looking at Fig. 8-10, we see that the NPW = 0 line is at 45° since identical scales were used on both axes. The point for Alt. 2 is plotted at the coordinates (PW of cost, PW of benefits). We drop a vertical line from Alt. 2 to the diagonal NPW = 0 line. The coordinates of any point on the graph are (PW of cost, PW of benefits), but along the NPW = 0 line (45° line), the x and y coordinates are equal. Thus, the coordinates of Point 4 are also (PW of cost, PW of cost). Since

NPW = PW of benefits − PW of cost

the vertical distance from Point 4 to Alt. 2 represents NPW. Similarly, the vertical distance between Point 3 and Alt. 1 presents NPW for Alt. 1. Since the criterion is to maximize NPW, we select Alt. 2 with larger NPW. The same technique is used in situations where there are multiple alternatives or continuous alternatives. In Ex.8-8 there were five alternatives; Fig. 8-11 shows a plot of these five alternatives. We can see that A has the greatest NPW and, therefore, is the preferred alternative.

Choosing An Analysis Method

At this point, we have examined in detail the three major economic analysis techniques: present worth analysis, annual cash flow analysis, and rate of return analysis. A practical question is, "Which method should be used for a particular

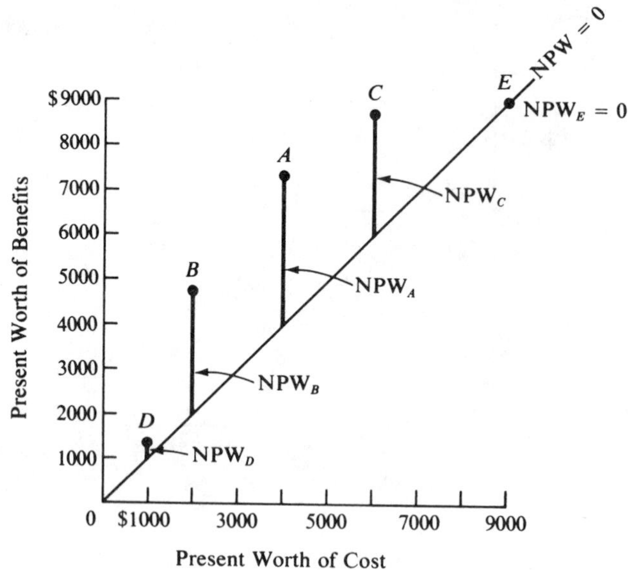

Figure 8-11 Present worth analysis of Example 8-8 using a benefit–cost plot.

problem?" While the obvious answer is to use the method requiring the least computations, there are a number of factors that may affect the decision:

1. Unless the MARR—minimum attractive rate of return (or minimum required interest rate for invested money)—is known, neither present worth analysis nor annual cash flow analysis can be done.
2. Present worth analysis and annual cash flow analysis often require far less computations than rate of return analysis.
3. In some situations, a rate of return analysis is easier to explain to people unfamiliar with economic analysis. At other times, an annual cash flow analysis may be easier to explain.
4. Business enterprises generally adopt one, or at most two, analysis techniques for broad categories of problems. If you work for a corporation, and the policy manual specifies rate of return analysis, you would appear to have no choice in the matter.

Since one may not always be able to choose the analysis technique computationally best-suited to the problem, this book illustrates how to use each of the three methods in all feasible situations. Ironically, the most difficult method—rate of return analysis—is the one most frequently used by engineers in industry!

SUMMARY

A graph of the PW of benefits *vs*. PW of cost (called a benefit–cost graph) can be an effective way to examine two alternatives by incremental analysis. And since multiple-alternative incremental analysis is done by the successive analysis of two alternatives, benefit–cost graphs can be used to solve multiple alternative and continuous-function alternatives as easily as two alternative problems.

The important steps in incremental rate of return analysis are:

1. Check to see that all the alternatives in the problem are identified.
2. (Optional) Compute the rate of return for each alternative. If one or more alternatives has ROR $\geq$ MARR, reject any alternatives with ROR < MARR.
3. Arrange the remaining alternatives in ascending order of investment.
4. Make a two-alternative analysis of the first two alternatives.
5. Take the preferred alternative from Step 4, and the next alternative from the list in Step 3. Proceed with another two-alternative comparison.
6. Continue until all alternatives have been examined and the best of the multiple alternatives has been identified.

Decision Criteria for Increments of *Investment*:

- If ΔROR $\geq$ MARR, retain the higher-cost alternative.
- If ΔROR < MARR, retain the lower-cost alternative.
- Reject the other alternative used in the analysis.

Decision Criteria for Increments of *Borrowing*:

- If $\Delta i \leq$ MARR, the increment is acceptable.
- If $\Delta i >$ MARR, the increment is not acceptable.

Benefit–cost graphs, being a plot of the PW of benefits *vs*. the PW of cost, can also be used in present worth analysis to graphically show the NPW for each alternative.

Problems

Unless otherwise noted, all Ch. 8 problems should be solved by rate of return analysis.

8-1 A firm is considering moving its manufacturing plant from Chicago to a new location. The Industrial Engineering Department was asked to identify the various alternatives together with the costs to relocate the plant, and the benefits. They examined six likely sites, together with the do-nothing alternative of keeping the plant at its present location. Their findings are summarized below:

Plant location	First cost	Uniform annual benefit
Denver	$300 thousand	$ 52 thousand
Dallas	550	137
San Antonio	450	117
Los Angeles	750	167
Cleveland	150	18
Atlanta	200	49
Chicago	0	0

The annual benefits are expected to be constant over the eight-year analysis period. If the firm uses 10% annual interest in its economic analysis, where should the manufacturing plant be located? (*Answer:* Dallas)

8-2 In a particular situation, four mutually exclusive alternatives are being considered. Each of the alternatives costs $1300 and has no end-of-useful-life salvage value.

Alternative	*Annual benefit*	*Useful life, in years*	*Calculated rate of return*
A	$100 at end of first year; *increasing* $30 per year thereafter	10	10.0%
B	$10 at end of first year; *increasing* $50 per year thereafter	10	8.8%
C	Annual end of year benefit = $260	10	15.0%
D	$450 at end of first year; *declining* $50 per year thereafter	10	18.1%

If the MARR is 8%, which alternative should be selected? (*Answer:* Alt. *C*)

8-3 A more detailed examination of the situation in Problem 8-2 reveals that there are two additional mutually exclusive alternatives to be considered. Both cost more than the $1300 for the four original alternatives.

Alternative	*Cost*	*Annual end-of-year benefit*	*Useful life, in years*	*Calculated rate of return*
E	$3000	$ 488	10	10.0%
F	5850	1000	10	11.2%

If the MARR remains at 8%, which one of the six alternatives should be selected? Neither Alt. *E* nor *F* has any end-of-useful-life salvage value. (*Answer:* Alt. *F*)

8-4 The owner of a downtown parking lot has employed a civil engineering consulting firm to advise him whether or not it is economically feasible to construct an office building on the site. Bill Samuels, a newly hired civil engineer, has been assigned to make the analysis. He has assembled the following data:

Alternative	*Total investment**	*Total net annual revenue from property*
Sell parking lot	$ 0	$ 0
Keep parking lot	200,000	22,000
Build 1-story building	400,000	60,000
Build 2-story building	555,000	72,000
Build 3-story building	750,000	100,000
Build 4-story building	875,000	105,000
Build 5-story building	1,000,000	120,000

*Includes the value of the land.

The analysis period is to be 15 years. For all alternatives, the property has an estimated resale (salvage) value at the end of 15 years equal to the present total investment. If the MARR is 10%, what recommendation should Bill make?

8-5 An oil company plans to purchase a piece of vacant land on the corner of two busy streets for $70,000. The company has four different types of businesses that it installs on properties of this type.

Plan	*Cost of improvements**	
A	$ 75,000	Conventional gas station with service facilities for lubrication, oil changes, etc.
B	230,000	Automatic carwash facility with gasoline pump island in front
C	30,000	Discount gas station (no service bays)
D	130,000	Gas station with low-cost quick-carwash facility

*Cost of improvements does not include the $70,000 cost of land.

In each case, the estimated useful life of the improvements is 15 years. The salvage value for each is estimated to be the $70,000 cost of the land. The net annual income, after paying all operating expenses, is projected as follows:

Plan	*Net annual income*
A	$23,300
B	44,300
C	10,000
D	27,500

If the oil company expects a 10% rate of return on its investments, which plan (if any) should be selected?

8-6 A firm is considering three mutually exclusive alternatives as part of a production improvement program. The alternatives are:

	A	*B*	*C*
Installed cost	$10,000	$15,000	$20,000
Uniform annual benefit	1,625	1,625	1,890
Useful life, in years	10	20	20

For each alternative, the salvage value at the end-of-useful-life is zero. At the end of ten years, Alt. *A* could be replaced by another *A* with identical cost and benefits. The MARR is 6%. If the analysis period is twenty years, which alternative should be selected?

8-7 Given the following four mutually exclusive alternatives:

	A	*B*	*C*	*D*
First cost	$75	$50	$50	$85
Uniform annual benefit	16	12	10	17
Useful life, in years	10	10	10	10
End-of-useful-life salvage value	0	0	0	0
Computed rate of return	16.8%	20.2%	15.1%	15.1%

If the MARR is 8%, which alternative should be selected? (*Answer: A*)

8-8 Consider the following three mutually exclusive alternatives:

	A	*B*	*C*
First cost	$200	$300	$600
Uniform annual benefit	59.7	77.1	165.2
Useful life, in years	5	5	5
End-of-useful-life salvage value	0	0	0
Computed rate of return	15%	9%	11.7%

For what range of values of MARR is Alt. *C* the preferred alternative? Put your answer in the following form: "Alt. *C* is preferred when ______% ≤ MARR ≤ ______%."

8-9 Consider four mutually exclusive alternatives that each have an 8-year useful life:

	A	*B*	*C*	*D*
First cost	$1000	$800	$600	$500
Uniform annual benefit	122	120	97	122
Salvage value	750	500	500	0

If the minimum attractive rate of return is 8%, which alternative should be selected?

8-10 Three mutually exclusive projects are being considered:

	A	*B*	*C*
First cost	$1000	$2000	$3000
Uniform annual benefit	150	150	0
Salvage value	1000	2700	5600
Useful life, in years	5	6	7

When each project reaches the end of its useful life, it would be sold for its salvage value and there would be no replacement. If 8% is the desired rate of return, which project should be selected?

8-11 Consider three mutually exclusive alternatives:

Year	*Buy X*	*Buy Y*	*Do nothing*
0	−$100.0	−$50.0	0
1	+31.5	+16.5	0
2	+31.5	+16.5	0
3	+31.5	+16.5	0
4	+31.5	+16.5	0

Which alternative should be selected:

a. if the minimum attractive rate of return equals 6%?

b. if MARR = 9%?

c. if MARR = 10%?

d. if MARR = 14%?

(*Answers:* **a.** *X*; **b.** *Y*; **c.** *Y*; **d.** Do nothing)

8-12 Consider the three alternatives:

Year	*A*	*B*	*Do nothing*
0	−$100	−$150	0
1	+30	+43	0
2	+30	+43	0
3	+30	+43	0
4	+30	+43	0
5	+30	+43	0

Which alternative should be selected:

a. if MARR = 6%?

b. if MARR = 8%?

c. if MARR = 10%?

8-13 A firm is considering two alternatives:

	A	*B*
Initial cost	$10,700	$5,500
Uniform annual benefits	2,100	1,800
Salvage value at end of useful life	0	0
Useful life, in years	8	4

At the end of four years, another *B* may be purchased with the same cost, benefits, and so forth. If the MARR is 10%, which alternative should be selected?

8-14 Consider the following alternatives:

	A	*B*	*C*
Initial cost	$300	$600	$200
Uniform annual benefits	41	98	35

Each alternative has a ten-year useful life and no salvage value. If the MARR is 8%, which alternative should be selected?

8-15 Given the following:

Year	*X*	*Y*
0	−$10	−$20
1	+15	+28

Over what range of values of MARR is *Y* the preferred alternative?

8-16 Consider four mutually exclusive alternatives:

	A	*B*	*C*	*D*
Initial cost	$770.00	$1406.30	$2563.30	0
Uniform annual benefit	420.00	420.00	420.00	0
Useful life, in years	2	4	8	0
Computed rate of return	6.0%	7.5%	6.4%	0

The analysis period is eight years. At the end of two years, four years, and six years, Alt. *A* will have an identical replacement. Alternative *B* will have a single identical replacement at the end of four years. Over what range of values of MARR is Alt. *B* the preferred alternative?

8-17 Consider the three alternatives:

	A	*B*	*C*
Initial cost	$1500	$1000	$2035
Annual benefit in each of first 5 years	250	250	650
Annual benefit in each of subsequent 5 years	450	250	145

Each alternative has a ten-year useful life and no salvage value. Based on a MARR of 15%, which alternative should be selected? Where appropriate, use an external interest rate of 10% to transform a cash flow to one sign change before proceeding with rate of return analysis.

8-18 A new 10,000 sq. meter warehouse next door to the Tyre Corporation is for sale for $450,000. The terms offered are a $100,000 down payment with the balance being paid in sixty equal monthly payments based on 15% interest. It is estimated that the warehouse would have a resale value of $600,000 at the end of five years.

Tyre has the needed cash available and could buy the warehouse, but does not need all the warehouse space at this time. The Johnson Company has offered to lease half the new warehouse for $2500 a month.

Tyre presently rents and utilizes 7000 sq. meters of warehouse space for $2700 a month. It has the option of reducing the rented space to 2000 sq. meters, in which case the monthly rent would be $1000 a month. Further, Tyre could cease renting warehouse space entirely. Tom Clay, the Tyre Corp. plant engineer, is considering three alternatives:

1. Buy the new warehouse and lease the Johnson Company half the space. In turn, the Tyre-rented space would be reduced to 2000 sq. meters.
2. Buy the new warehouse and cease renting any warehouse space.
3. Continue as is, with 7000 sq. meters of rented warehouse space.

Based on a 20% minimum attractive rate of return, which alternative should be selected?

8-19 Consider the alternatives below:

	A	*B*	*C*
Initial cost	$100.00	$150.00	$200.00
Uniform annual benefit	10.00	17.62	55.48
Useful life, in years	infinite	20	5

Use present worth analysis, an 8% interest rate, and an infinite analysis period. Which alternative should be selected in each of the two following situations?

1. Alternatives *B* and *C* are replaced at the end of their useful lives with identical replacements.
2. Alternatives *B* and *C* are replaced at the end of their useful lives with alternatives that provide an 8% rate of return.

8-20 A problem often discussed in the engineering economy literature is the "oil-well pump problem":* Pump 1 is a small pump; Pump 2 is a larger pump that costs more, will produce slightly more oil, and will produce it more rapidly. If the MARR is 20%, which pump should be selected? Assume any temporary external investment of money earns 10% per year.

Year	*Pump 1* (\$000*s*)	*Pump 2* (\$000*s*)
0	−$100	−$110
1	+70	+115
2	+70	+30

8-21 Three mutually exclusive alternatives are being studied. If the MARR is 12%, which alternative should be selected?

Year	*A*	*B*	*C*
0	−$20,000	−$20,000	−$20,000
1	+10,000	+10,000	+5,000
2	+5,000	+10,000	+5,000
3	+10,000	+10,000	+5,000
4	+6,000	0	+15,000

8-22 The South End bookstore has an annual profit of $170,000. The owner is considering opening a second bookstore on the north side of the campus. He can lease an existing building for five years with an option to continue the lease for a second five-year period. If he opens the second bookstore he expects the existing store will lose some business that will be gained by the new "The North End" bookstore. It will take $500,000 of store fixtures and inventory to open The North End. He believes that the two stores will have a combined profit of $260,000 a year after paying all the expenses of both stores.

*One of the more interesting exchanges of opinion about this problem is in Prof. Martin Wohl's "Common Misunderstandings About the Internal Rate of Return and Net Present Value Economic Analysis Methods," and the associated discussion by Professors Winfrey, Leavenworth, Steiner, and Bergmann, published in *Evaluating Transportation Proposals*, Transportation Research Record 731, Transportation Research Board, Washington, D.C., 1979.

The owner's economic analysis is based on a five year period. He will be able to recover this $500,000 investment at the end of five years by selling the store fixtures and inventory. The owner will not open The North End unless he can expect a 15% rate of return. What should he do? Show computations to justify your decision.

8-23 A paper mill is considering two types of pollution control equipment.

	Neutralization	*Precipitation*
Initial cost	$700,000	$500,000
Annual chemical cost	40,000	110,000
Salvage value	175,000	125,000
Useful life	5 years	5 years

The firm wants a 12% rate of return on any avoidable increments of investment. Which equipment should be purchased?

8-24 A stockbroker has proposed two investments in low-rated corporate bonds paying high interest rates and selling below their stated value (in other words, *junk bonds*). Both bonds are rated as equally risky. Which, if any, of the bonds should you buy if your MARR is 25%?

Bond	*Stated value*	*Annual interest payment*	*Current market price, including buying commission*	*Bond maturity**
Gen Dev	$1000	$ 94	$480	15 years
RJR	1000	140	630	15

8-25 (20-47) Three mutually exclusive alternatives are being considered.

	A	*B*	*C*
Initial investment	$50,000	$22,000	$15,000
Annual net income	5,093	2,077	1,643
Computed rate of return	8%	7%	9%

Each alternative has a 20-year useful life with no salvage value. If the minimum attractive rate of return is 7%, which alternative should be selected?

8-26 (20-48) A firm is considering five alternatives.

	1	*2*	*3*	*4*	*5*
Initial cost	$100.00	$130.00	$200.00	$330.00	Do nothing
Uniform annual net income	26.38	38.78	47.48	91.55	
Computed rate of return	10%	15%	6%	12%	

Each alternative has a five-year useful life. The firm's minimum attractive rate of return is 8%. Which alternative should be selected?

---

*At maturity the bondholder receives the last interest payment plus the bond stated value.

CHAPTER 9

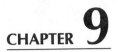

Other Analysis Techniques

Chapter 9 examines four topics. They are:

- Future Worth Analysis;
- Benefit–Cost Ratio Analysis;
- Payback Period;
- Sensitivity and Breakeven Analysis.

Future worth analysis is very much like present worth analysis, dealing with *then*—future worth—rather than with *now*—present worth—situations.

Previously, we have written economic analysis relationships based on either:

$$\text{PW of cost} = \text{PW of benefit} \quad \text{or} \quad \text{EUAC} = \text{EUAB}$$

Instead of writing it in this form, we could define these relationships as

$$\frac{\text{PW of benefit}}{\text{PW of cost}} = 1 \quad \text{or} \quad \frac{\text{EUAB}}{\text{EUAC}} = 1$$

When economic analysis is based on these ratios, the calculations are called benefit–cost ratio analysis.

Payback period is an approximate analysis technique, generally defined as the time required for cumulative benefits to equal cumulative costs. Sensitivity describes the relative magnitude of a particular variation in one or more elements of a problem that is sufficient to change a particular decision. Closely related is *breakeven analysis*, which determines the conditions where two alternatives are equivalent. Thus, breakeven analysis is a form of *sensitivity analysis*.

Future Worth Analysis

We have seen how economic analysis techniques resolved alternatives into comparable units. In present worth analysis, the comparison was made in terms of the present consequences of taking the feasible courses of action. In annual cash flow analysis, the comparison was in terms of equivalent uniform annual costs (or benefits). We saw that we could easily convert from present worth to annual cash flow, and *vice versa*. But the concept of resolving alternatives into comparable units is not restricted to a present or annual comparison. The comparison may be made at any point in time. There are many situations where we would like to know what the *future* situation will be, if we take some particular course of action *now*. This is called ***future worth analysis***.

EXAMPLE 9-1

Ron Jamison, a twenty-year-old college student, considers himself an average cigarette smoker, for he consumes about a carton a week. He wonders how much money he could accumulate by the time he reaches 65 if he quit smoking now and put his cigarette money into a savings account. Cigarettes cost $9 per carton. Ron expects that a savings account would earn 5% interest, compounded semi-annually. Compute Ron's future worth at age 65.

Solution:

Semi-annual saving = \$9/carton × 26 weeks = \$234

Future worth (FW) = $A(F/A,2\frac{1}{2}\%,90)$ = 234(329.2) = \$77,000 ◀

EXAMPLE 9-2

An East Coast firm has decided to establish a second plant in Kansas City. There is a factory for sale for $850,000 that, with extensive remodeling, could be used. As an alternative, the company can buy vacant land for $85,000 and have a new plant constructed on the property. Either way, it will be three years before the company will be able to get the plant into production. The timing and cost of the various components for the factory are given in the cash flow table below.

Year	*Construct new plant*		*Remodel available factory*	
0	Buy land	$ 85,000	Purchase factory	$850,000
1	Design and initial construction costs	200,000	Design and remodeling costs	250,000
2	Balance of construction costs	1,200,000	Additional remodeling costs	250,000
3	Setup production equipment	200,000	Setup production equipment	250,000

If interest is 8%, which alternative results in the lower equivalent cost when the firm begins production at the end of the third year?

Solution:

New plant:

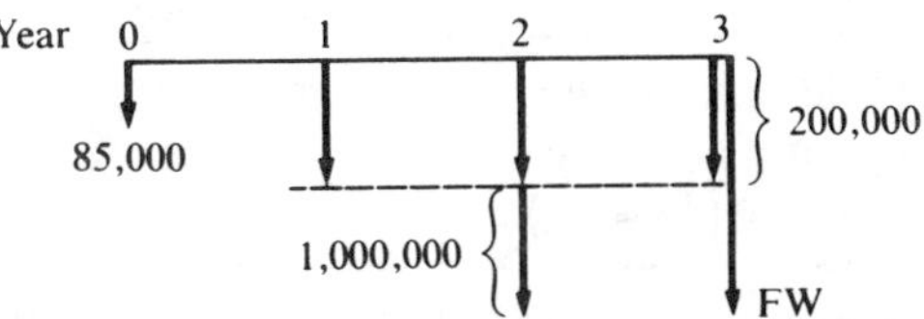

$$\text{Future worth (FW)} = 85{,}000(F/P,8\%,3) + 200{,}000(F/A,8\%,3) + 1{,}000{,}000(F/P,8\%,1) = \$1{,}836{,}000$$

Remodel available factory:

$$\text{Future worth (FW)} = 850{,}000(F/P,8\%,3) + 250{,}000(F/A,8\%,3) = \$1{,}882{,}000$$

The total cost of remodeling the available factory ($1,600,000) is smaller than the total cost of a new plant ($1,685,000). The timing of the expenditures, however, is less favorable than building the new plant. The new plant is projected to have the smaller future worth of cost, and thus is the preferred alternative. ◀

Benefit–Cost Ratio Analysis

At a given minimum attractive rate of return (MARR), we would consider an alternative acceptable, provided:

$$\text{PW of benefits} - \text{PW of costs} \geq 0 \quad \text{or} \quad \text{EUAB} - \text{EUAC} \geq 0$$

These could also be stated as a ratio of benefits to costs, or

$$\text{Benefit–cost ratio } \frac{\text{B}}{\text{C}} = \frac{\text{PW of benefits}}{\text{PW of costs}} = \frac{\text{EUAB}}{\text{EUAC}} \geq 1$$

Rather than solving problems using present worth or annual cash flow analysis, we can base the calculations on the benefit–cost ratio, B/C. The criteria are presented in Table 9-1. We will illustrate "B/C analysis" by solving the same example problems worked by other economic analysis methods.

Table 9-1 BENEFIT–COST RATIO ANALYSIS

	Situation	Criterion
Fixed input	**Amount of money or other input resources are fixed**	**Maximize B/C**
Fixed output	**Fixed task, benefit, or other output to be accomplished**	**Maximize B/C**
Neither input nor output fixed	**Neither amount of money or other inputs nor amount of benefits or other outputs are fixed**	***Two alternatives:*** **Compute incremental benefit–cost ratio (ΔB/ΔC) on the increment of investment between the alternatives. If ΔB/ΔC ≥ 1, choose higher-cost alternative; otherwise, choose lower-cost alternative.** ***Three or more alternatives:*** **Solve by benefit–cost ratio incremental analysis**

EXAMPLE 9-3

A firm is trying to decide which of two devices to install to reduce costs in a particular situation. Both devices cost $1000 and have useful lives of five years and no salvage value. Device *A* can be expected to result in $300 savings annually. Device *B* will provide cost savings of $400 the first year, but will decline $50 annually, making the second-year savings $350, the third-year savings $300, and so forth. With interest at 7%, which device should the firm purchase?

Solution: This problem was previously solved by present worth (Ex. 5-1), annual cash flow (Ex. 6-5), and rate of return (Ex. 7-7) analyses.

Device A:

$$\text{PW of cost} = \$1000$$

$$\text{PW of benefits} = 300(P/A,7\%,5) = 300(4.100) = \$1230$$

$$\frac{\text{B}}{\text{C}} = \frac{\text{PW of benefits}}{\text{PW of cost}} = \frac{1230}{1000} = 1.23$$

Device B:

$$\text{PW of cost} = \$1000$$

$$\text{PW of benefit} = 400(P/A,7\%,5) - 50(P/G,7\%,5)$$

$$= 400(4.100) - 50(7.647) = 1640 - 382 = 1258$$

$$\frac{\text{B}}{\text{C}} = \frac{\text{PW of benefits}}{\text{PW of cost}} = \frac{1258}{1000} = 1.26$$

To maximize the benefit–cost ratio, select Device *B*. ◀

EXAMPLE 9-4

Two machines are being considered for purchase. Assuming 10% interest, which machine should be bought?

	Machine X	*Machine Y*
Initial cost	\$200	\$700
Uniform annual benefit	95	120
End-of-useful-life salvage value	50	150
Useful life, in years	6	12

Solution: Assuming a twelve-year analysis period, the cash flow table is:

Year	*Machine X*	*Machine Y*
0	−\$200	−\$700
1–5	+95	+120
6	+95 −200 +50	+120
7–11	+95	+120
12	+95 +50	+120 +150

We will solve the problem using

$$\frac{\text{B}}{\text{C}} = \frac{\text{EUAB}}{\text{EUAC}}$$

and considering the salvage value of the machines to be reductions in cost, rather than increases in benefits.

Machine X:

$$\text{EUAC} = 200(A/P,10\%,6) - 50(A/F,10\%,6)$$
$$= 200(0.2296) - 50(0.1296) = 46 - 6 = \$40$$
$$\text{EUAB} = \$95$$

Note that this assumes the replacement for the last six years has identical costs. Under these circumstances, the EUAC for the first six years equals the EUAC for all twelve years.

Machine Y:

$$\text{EUAC} = 700(A/P,10\%,12) - 150(A/F,10\%,12)$$
$$= 700(0.1468) - 150(0.0468) = 103 - 7 = \$96$$
$$\text{EUAB} = \$120$$

Machine Y − Machine X:

$$\frac{\Delta\text{B}}{\Delta\text{C}} = \frac{120 - 95}{96 - 40} = \frac{25}{56} = 0.45$$

Since the incremental benefit–cost ratio is less than 1, it represents an undesirable increment of investment. We therefore choose the lower cost alternative—Machine *X*. If we had computed benefit–cost ratios for each machine, they would have been:

Machine X	*Machine Y*
$\frac{\text{B}}{\text{C}} = \frac{95}{40} = 2.38$	$\frac{\text{B}}{\text{C}} = \frac{120}{96} = 1.25$

The fact that B/C = 1.25 for Machine *Y* (the higher-cost alternative) must not be used as the basis for suggesting that the more expensive alternative should be selected. The incremental benefit–cost ratio, ΔB/ΔC, clearly shows that *Y* is a less desirable alternative than *X*. Also, we must not jump to the conclusion that the best alternative is always the one with the largest B/C ratio. This, too, may lead to incorrect decisions—as we shall see when we examine problems with three or more alternatives. ◀

EXAMPLE 9-5

Consider the five mutually exclusive alternatives from Ex. 8-8 plus an additional alternative, *F*. They have twenty-year useful lives and no salvage value. If the minimum attractive rate of return is 6%, which alternative should be selected?

	A	*B*	*C*	*D*	*E*	*F*
Cost	$4000	$2000	$6000	$1000	$9000	$10,000
PW of benefit	7330	4700	8730	1340	9000	9,500
$\frac{\text{B}}{\text{C}} = \frac{\text{PW of benefits}}{\text{PW of cost}}$	1.83	2.35	1.46	1.34	1.00	0.95

Solution: Incremental analysis is needed to solve the problem. The steps in the solution are the same as the ones presented for incremental rate of return, except here the criterion is ΔB/ΔC, rather than ΔROR.

1. Be sure all the alternatives are identified.
2. (Optional) Compute the B/C ratio for each alternative. Since there are alternatives whose B/C ≥ 1, we will discard any with a B/C < 1. Discard Alt. *F*.
3. Arrange the remaining alternatives in ascending order of investment.

	D	*B*	*A*	*C*	*E*
Cost (= PW of cost)	$1000	$2000	$4000	$6000	$9000
PW of benefits	1340	4700	7330	8730	9000
B/C	1.34	2.35	1.83	1.46	1.00

	Increment B–D	*Increment A–B*	*Increment C–A*
ΔCost	$1000	$2000	$2000
ΔBenefit	3360	2630	1400
ΔB/ΔC	3.36	1.32	0.70

4. Examine each separable increment of investment. If $\Delta B/\Delta C < 1$, the increment is not attractive. If $\Delta B/\Delta C \geq 1$, the increment of investment is desirable. The increments *B–D* and *A–B* are desirable. Thus, of the first three alternatives (*D*, *B*, and *A*), Alt. *A* is the preferred alternative. Increment *C–A* is not attractive as $\Delta B/\Delta C = 0.70$, which indicates that of the first four alternatives (*D*, *B*, *A*, and *C*), *A* continues as the best of the four. Now we want to decide between *A* and *E*, which we'll do by examining the increment of investment that represents the difference between these alternatives.

	Increment E–A
ΔCost	$5000
ΔBenefit	1670
$\Delta B/\Delta C$	0.33

The increment is undesirable. We choose Alt. *A* as the best of the six alternatives. One should note that the best alternative in this example does not have the highest B/C ratio. ◀

Benefit–cost ratio analysis may be graphically represented. Figure 9-1 is a graph of Ex. 9-5. We see that *F* has a B/C ratio < 1 and can be discarded. Alt. *D* is the starting point for examining the separable increments of investment. The

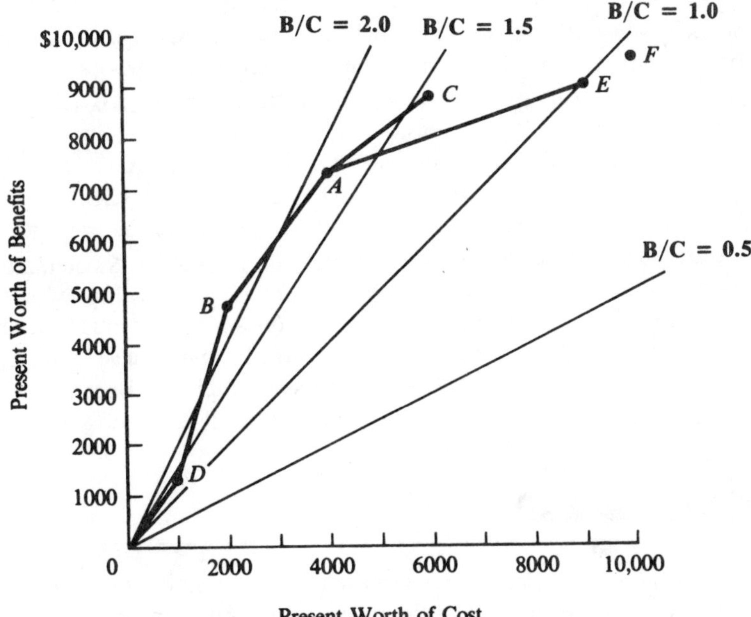

Figure 9-1 Benefit–cost ratio graph of Example 9-5.

slope of Line *B–D* indicates a $\Delta B/\Delta C$ ratio of > 1. This is also true for Line *A–B*. Increment *C–A* has a slope much flatter than $B/C = 1$, indicating an undesirable increment of investment. Alt. *C* is therefore discarded and *A* retained. Increment *E–A* is similarly unattractive. Alt. *A* is therefore the best of the six alternatives.

Note particularly two additional things about Fig. 9-1: first, even if alternatives with a B/C ratio < 1 had not been initially excluded, they would have been systematically eliminated in the incremental analysis. Since this is the case, it is not essential that the B/C ratio be computed for each alternative as an initial step in incremental analysis. Nevertheless, it seems like an orderly and logical way to approach a multiple-alternative problem. Second, Alt. *B* had the highest B/C ratio ($B/C = 2.35$), but it is not the best of the six alternatives. We saw this same situation in rate of return analysis of three or more alternatives. The reason is the same in both analysis situations. We seek to maximize the *total* profit, not the profit rate.

Continuous Alternatives

There are times when the feasible alternatives are a continuous function. The height of a dam in Chapter 8 was an example of this situation. It was possible to build the dam anywhere from 200 to 500 feet high.

In many situations, the projected capacity of an industrial plant can be varied continuously over some feasible range. In these cases, we seek to add increments of investment where $\Delta B/\Delta C \geq 1$ and avoid increments where $\Delta B/\Delta C < 1$. The optimal size of such a project is where $\Delta B/\Delta C = 1$. Figure 9-2a shows the line of feasible alternatives with their costs and benefits. This may represent a lot of calculations to locate points through which the line passes.

Figure 9-2b shows how the incremental benefit–cost ratio ($\Delta B/\Delta C$) changes as one moves along the line of feasible alternatives. In Fig. 9-2b, the ratio of Incremental net present worth *to* Incremental cost ($\Delta NPW/\Delta C$) is also plotted. As expected, we are adding increments of NPW as long as $\Delta B/\Delta C > 1$. Finally, in Fig. 9-2c, we see the plot of (total) NPW *vs.* the size of the project.

This three-part figure demonstrates that both present worth analysis and benefit–cost ratio analysis lead to the same optimal decision. We saw in Chapter 8 that rate of return and present worth analysis led to identical decisions. Any of the exact analysis methods—present worth, annual cash flow, rate of return, or benefit–cost ratio—will lead to the same decision. Benefit–cost ratio analysis is extensively used in economic analysis at all levels of government.

Payback Period

Payback period is the period of time required for the profit or other benefits from an investment to equal the cost of the investment. This is the general definition for payback period, but there are other definitions. Others consider depreciation

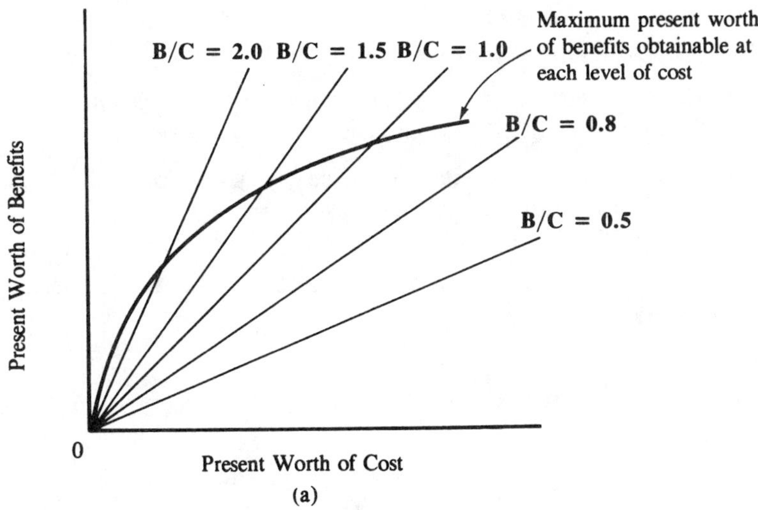

(a)

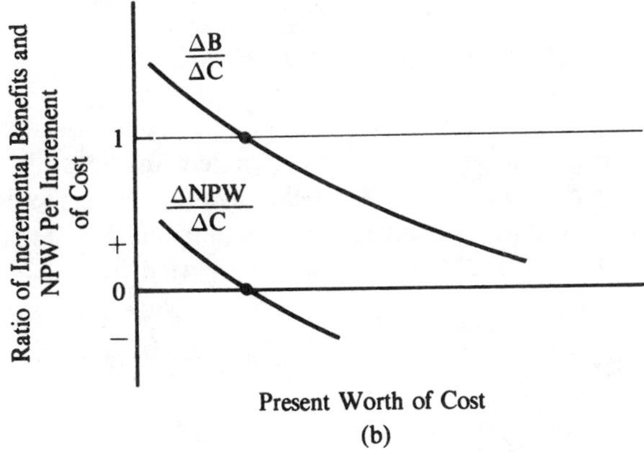

(b)

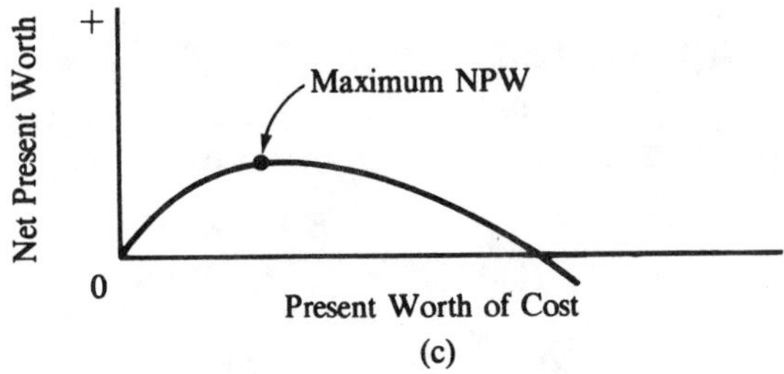

(c)

Figure 9-2a–c Selecting optimal size of project.

of the investment and income taxes; they, too, are simply called "payback period." For now, we will limit our discussion to the simplest form.

> ***Payback period* is the period of time required for the profit or other benefits of an investment to equal the cost of the investment.**

The criterion in all situations is to minimize the payback period. The computation of payback period is illustrated in Examples 9-6 and 9-7.

EXAMPLE 9-6

The cash flows for two alternatives are as follows:

Year	*A*	*B*
0	−$1000	−$2783
1	+200	+1200
2	+200	+1200
3	+1200	+1200
4	+1200	+1200
5	+1200	+1200

Based on payback period, which alternative should be selected?

Solution:

Alternative A: Payback period is the period of time required for the profit or other benefits of an investment to equal the cost of the investment. In the first two years, only $400 of the $1000 cost is recovered. The remaining $600 cost is recovered in the first half of Year 3. Thus the payback period for Alt. *A* is 2.5 years.

Alternative B: Since the annual benefits are uniform, the payback period is simply

$2783/$1200 per year = 2.3 years

To minimize the payback period, choose Alt. *B*. ◀

EXAMPLE 9-7

A firm is trying to decide which of two alternate weighing scales it should install to check a package-filling operation in the plant. If both scales have a six-year life, which one should be selected? Assume an 8% interest rate.

Alternative	*Cost*	*Uniform annual benefit*	*End-of-useful-life salvage value*
Atlas scales	$2000	$450	$100
Tom Thumb scales	3000	600	700

Solution:

Atlas scales:

$$\text{Payback period} = \frac{\text{Cost}}{\text{Uniform annual benefit}} = \frac{2000}{450}$$
$$= 4.4 \text{ years}$$

Tom Thumb scales:

$$\text{Payback period} = \frac{\text{Cost}}{\text{Uniform annual benefit}} = \frac{3000}{600}$$
$$= 5 \text{ years}$$

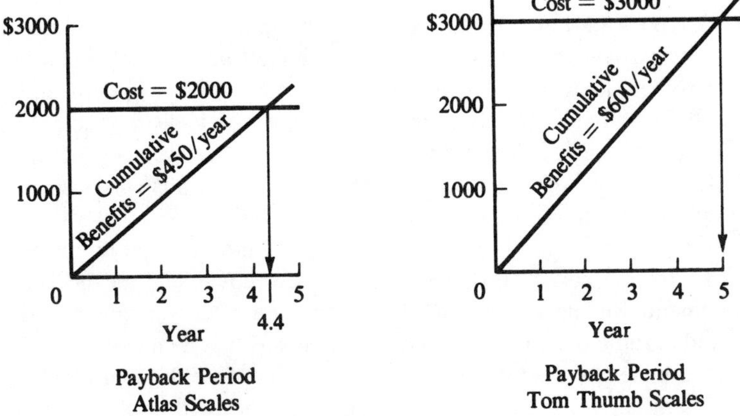

Figure 9-3 Payback period plots for Example 9-7.

Figure 9-3 illustrates the situation. To minimize payback period, select the Atlas scales. ◀

There are four important points to be understood about payback period calculations:

1. **This is an approximate, rather than an exact, economic analysis calculation.**
2. **All costs and all profits, or savings of the investment, prior to payback are included *without* considering differences in their timing.**
3. **All the economic consequences beyond the payback period are completely ignored.**

4. **Being an approximate calculation, payback period may or may not select the correct alternative. That is, the payback period calculations may select a different alternative from that found by exact economic analysis techniques.**

This last point—that payback period may select the *wrong* alternative—was illustrated by Ex. 9-7. Using payback period, the Atlas scales appear to be the more attractive alternative. Yet, when this same problem was solved by the present worth method (Ex. 5-4), the Tom Thumb scales were the chosen alternative. A review of the problem reveals the reason for the different conclusions. The $700 salvage value at the end of six years for the Tom Thumb scales is a significant benefit. The salvage value occurs after the payback period, so it was ignored in the payback calculation. It *was* considered in the present worth analysis, with the result that Tom Thumb scales were more desirable.

But if payback period calculations are approximate, and are even capable of selecting the wrong alternative, why is the method used at all? There are two primary answers: first, the calculations can be readily made by people unfamiliar with economic analysis. One does not need to know how to use gradient factors, or even to have a set of Compound Interest Tables. Second, payback period is a readily understood concept. Earlier we pointed out that this was also an advantage to rate of return.

Moreover, payback period *does* give us a useful measure, telling us how long it will take for the cost of the investment to be recovered from the benefits of the investment. Businesses and industrial firms are often very interested in this time period: a rapid return of invested capital means that it can be re-used sooner for other purposes by the firm. But one must not confuse the *speed* of the return of the investment, as measured by the payback period, with economic *efficiency*. They are two distinctly separate concepts! The former emphasizes the quickness with which invested monies return to a firm; the latter considers the overall profitability of the investment.

We can create another situation to illustrate how selecting between alternatives by the payback period criterion may result in an unwise decision.

EXAMPLE 9-8

A firm is purchasing production equipment for a new plant. Two alternative machines are being considered for a particular operation.

	Tempo machine	*Dura machine*
Installed cost	$30,000	$35,000
Net annual benefit after deducting all annual expenses	$12,000 the first year, *declining* $3000 per year thereafter	$1000 the first year, *increasing* $3000 per year thereafter
Useful life, in years	4	8

Neither machine has any salvage value. Compute the payback period for each of the alternatives.

Solution:

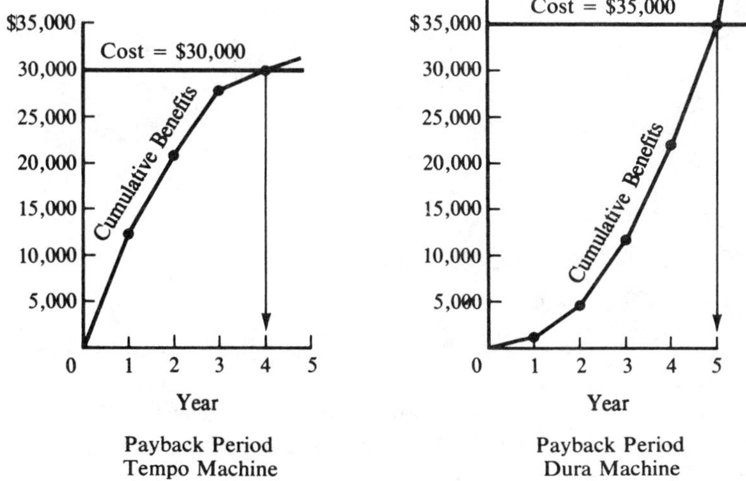

Figure 9-4 Payback period plots for Example 9-7.

The Tempo machine has a declining annual benefit, while the Dura has an increasing annual benefit. Figure 9-4 shows the Tempo has a four-year payback period and the Dura has a five-year payback period. To minimize the payback period, the Tempo is selected.

Now, as a check on the payback period analysis, compute the rate of return for each alternative. Assume the minimum attractive rate of return is 10%.

Solution based on rate of return: The cash flows for the two alternatives are as follows:

Year	*Tempo machine*	*Dura machine*
0	−$30,000	−$35,000
1	+12,000	+1,000
2	+9,000	+4,000
3	+6,000	+7,000
4	+3,000	+10,000
5	0	+13,000
6	0	+16,000
7	0	+19,000
8	0	+22,000
$\Sigma =$	0	+57,000

Tempo machine: Since the sum of the cash flows for the Tempo machine is zero, we see immediately that the \$30,000 investment just equals the subsequent benefits. The resulting rate of return is 0%.

Dura machine:

$$35{,}000 = 1000(P/A,i,8) + 3000(P/G,i,8)$$

Try $i = 20\%$:

$$35{,}000 \stackrel{?}{=} 1000(3.837) + 3000(9.883)$$
$$\stackrel{?}{=} 3837 + 29{,}649 = 33{,}486$$

The 20% interest rate is too high. Try $i = 15\%$:

$$35{,}000 \stackrel{?}{=} 1000(4.487) + 3000(12.481)$$
$$\stackrel{?}{=} 4487 + 37{,}443 = 41{,}930$$

This time, the interest rate is too low. Linear interpolation would show that the rate of return is approximately 19%.

Using an exact calculation—rate of return—it is clear that Tempo is not economically very attractive. Yet it was this alternative, and not the Dura machine, that was preferred based on the payback period calculations. On the other hand, the shorter payback period for Tempo does give a measure of the speed of the return of the investment not found in the Dura. The conclusion to be drawn is that ***liquidity*** and ***profitability*** may be two quite different criteria. ◀

From the discussion and the examples, we see that payback period can be helpful in providing a measure of the speed of the return of the investment. This might be quite important, for example, for a company that is short of working capital, or one where there are rapid changes in technology. This must not, however, be confused with a careful economic analysis. We have shown that a short payback period does not always mean that the investment is desirable. Thus, payback period should not be considered a suitable replacement for accurate economic analysis calculations.

Sensitivity And Breakeven Analysis

Since many data gathered in solving a problem represent *projections* of future consequences, there may be considerable uncertainty regarding the accuracy of that data. As the desired result of the analysis is decision making, an appropriate question is, "To what extent do variations in the data affect my decision?" When small variations in a particular estimate would change selection of the alternative, the decision is said to be ***sensitive to the estimate***. To better evaluate the impact of any particular estimate, we compute "what variation to a particular estimate would be necessary to change a particular decision." This is called ***sensitivity analysis***.

An analysis of the sensitivity of a problem's decision to its various parameters highlights the important and significant aspects of that problem. For example, one might be concerned that the estimates for annual maintenance and future salvage value in a particular problem may vary substantially. Sensitivity analysis might indicate that the decision is insensitive to the salvage-value estimate over the full range of possible values. But, at the same time, we might find that the decision is sensitive to changes in the annual-maintenance estimate. Under these circumstances, one should place greater emphasis on improving the annual-maintenance estimate and less on the salvage-value estimate.

As indicated at the beginning of this chapter, breakeven analysis is a form of sensitivity analysis. To illustrate the sensitivity of a decision between alternatives to particular estimates, breakeven analysis is often presented as a ***breakeven chart***.

Sensitivity and breakeven analysis frequently are useful in engineering problems called *stage construction*. Should a facility be constructed now to meet its future full-scale requirement, or should it be constructed in stages as the need for the increased capacity arises? Three examples of this situation are:

- Should we install a cable with 400 circuits now or a 200-circuit cable now and another 200-circuit cable later?
- A 10-cm water main is needed to serve a new area of homes. Should the 10-cm main be installed now, or should a 15-cm main be installed to later provide an adequate water supply to adjoining areas when other homes are built?
- An industrial firm needs a 10,000-m^2 warehouse now and estimates that it will need an additional 10,000 m^2 in four years. The firm could have a 10,000-m^2 warehouse built now and later enlarged, or have the 20,000-m^2 warehouse built right away.

Examples 9-9 and 9-10 illustrate sensitivity and breakeven analysis.

EXAMPLE 9-9

Consider the following situation where a project may be constructed to full capacity now or may be constructed in two stages.

Construction costs:

Two-stage construction	
Construct first stage now	\$100,000
Construct second stage *n* years from now	120,000
Full-capacity construction	
Construct full capacity now	140,000

Other factors:

1. All facilities will last until forty years from now regardless of when they are installed; at that time they will have zero salvage value.
2. The annual cost of operation and maintenance is the same for both two-stage construction and full-capacity construction.
3. Assume an 8% interest rate.

Plot a graph showing "age when second stage is constructed" *vs.* "costs for both alternatives." Mark the breakeven point. What is the sensitivity of the decision to second-stage construction sixteen or more years in the future?

Solution: Since we are dealing with a common analysis period, the calculations may be either annual cost or present worth. Present worth calculations appear simpler and are used here:

Construct full capacity now:

$$\text{PW of cost} = \$140{,}000$$

Two-stage construction:

First stage constructed now and the second stage to be constructed n years hence. Compute the PW of cost for several values of n (years).

$$\text{PW of cost} = 100{,}000 + 120{,}000(P/F,8\%,n)$$

$$n = 5 \quad \text{PW} = 100{,}000 + 120{,}000(0.6806) = \$181{,}700$$

$$n = 10 \quad \text{PW} = 100{,}000 + 120{,}000(0.4632) = 155{,}600$$

$$n = 20 \quad \text{PW} = 100{,}000 + 120{,}000(0.2145) = 125{,}700$$

$$n = 30 \quad \text{PW} = 100{,}000 + 120{,}000(0.0994) = 111{,}900$$

These data are plotted in the form of a breakeven chart in Fig. 9-5.

Figure 9-5 portrays the PW of cost for the two alternatives. The x-axis variable is the *time* when the second stage is constructed. We see that the PW of cost for two-stage construction naturally decreases as the time for the second stage is deferred. The one-stage construction (full capacity now) is unaffected by the x-axis variable and, hence, is a horizontal line on the graph.

The breakeven point on the graph is the point at which both alternatives have equivalent costs. We see that if, in two-stage construction, the second stage is deferred for 15 years, then the PW of cost of two-stage construction is equal to one-stage construction; Year 15 is the breakeven point. The graph also shows that if the second stage was needed prior to Year 15, then one-stage construction, with its smaller PW of cost, would be preferred. On the other hand, if the second stage would not be required until after fifteen years, two-stage construction is preferred.

The decision on how to construct the project is sensitive to the age at which the second stage is needed *only* if the range of estimates includes 15 years. For

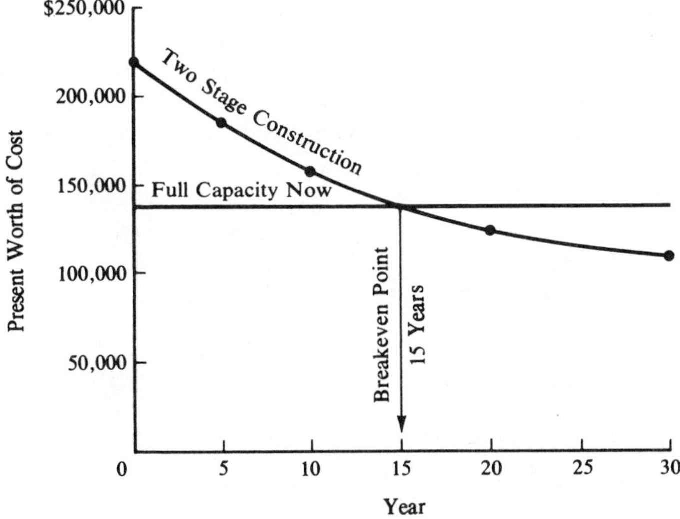

Figure 9-5 Breakeven chart for Example 9-9.

example, if one estimated that the second stage capacity would be needed somewhere between five and ten years hence, the decision is insensitive to that estimate. For any value within that range, the decision does not change. The more economical thing to do is to build the full capacity now. But, if the second-stage capacity were needed sometime between, say, 12 and 18 years, the decision would be sensitive to the estimate of when the full capacity would be needed. ◀

One question posed by Ex. 9-9 is *how* sensitive the decision is to the need for the second stage at 16 years or beyond. The graph shows that the decision is insensitive. In all cases for construction on or after 16 years, two-stage construction has a lower PW of cost.

EXAMPLE 9-10

Example 8-3 posed the following situation. Three mutually exclusive alternatives are given, each with a 20-year life and no salvage value. The minimum attractive rate of return is 6%.

	A	*B*	*C*
Initial cost	$2000	$4000	$5000
Uniform annual benefit	410	639	700

In Ex. 8-3 we found that Alt. *B* was the preferred alternative. Here we would like to know how sensitive the decision is to our estimate of the initial cost of *B*. If *B* is preferred at an initial cost of \$4000, it will continue to be preferred at any smaller initial cost. But *how much* higher than \$4000 can the initial cost be and still have *B* the preferred alternative? The computations may be done several different ways. With neither input nor output fixed, maximizing net present worth is a suitable criterion.

Alternative A:

$$\begin{aligned}\text{NPW} &= \text{PW of benefit} - \text{PW of cost}\\ &= 410(P/A,6\%,20) - 2000\\ &= 410(11.470) - 2000 = 2703\end{aligned}$$

Alternative B: Let x = Initial cost of *B*.

$$\begin{aligned}\text{NPW} &= 639(P/A,6\%,20) - x\\ &= 639(11.470) - x\\ &= 7329 - x\end{aligned}$$

Alternative C:

$$\begin{aligned}\text{NPW} &= 700(P/A,6\%,20) - 5000\\ &= 700(11.470) - 5000 = 3029\end{aligned}$$

For the three alternatives, we see that *B* will only maximize NPW as long as its NPW is greater than 3029.

$$\begin{aligned}3029 &= 7329 - x\\ x &= 7329 - 3029 = 4300\end{aligned}$$

Therefore, *B* is the preferred alternative if its initial cost does not exceed \$4300.

Figure 9-6 is a breakeven chart for the three alternatives. Here the criterion is to maximize NPW; as a result, the graph shows that *B* is preferred if its initial cost is less than \$4300. At an initial cost above \$4300, *C* is preferred. We have a breakeven point at \$4300. When *B* has an initial cost of \$4300, *B* and *C* are equally desirable. ◀

Sensitivity analysis and breakeven point calculations can be very useful in identifying how different estimates affect the calculations. It must be recognized that these calculations assume all parameters except one are held constant, and the sensitivity of the decision to that one variable is evaluated. Later we will look further at the impact of parameter estimates on decision making.

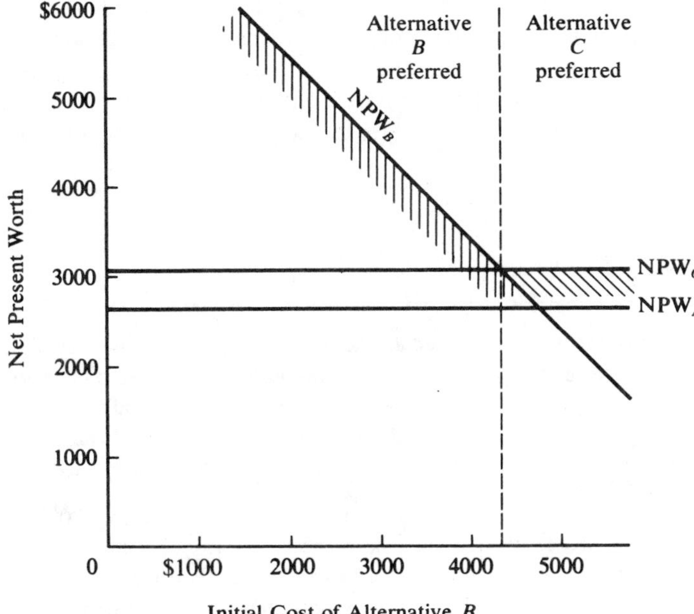

Figure 9-6 Breakeven chart.

SUMMARY

In this chapter, we have looked at four new analysis techniques.

Future Worth. When the point in time at which the comparison between alternatives will be made is in the future, the calculation is called future worth. This is very similar to present worth, which is based on the present, rather than a future point in time.

Benefit–Cost Ratio Analysis. This technique is based on the ratio of benefits to costs using either present worth or annual cash flow calculations. The method is graphically similar to present worth analysis. When neither input nor output is fixed, incremental benefit–cost ratios ($\Delta B/\Delta C$) are required. The method is similar in this respect to rate of return analysis. Benefit–cost ratio analysis is often used at the various levels of government.

Payback Period. Here we define payback as the period of time required for the profit or other benefits of an investment to equal the cost of the investment. Although simple to use and simple to understand, payback is a poor analysis technique for ranking alternatives. While it provides a measure of the speed of the return of the investment, it is not an accurate measure of the profitability of an investment.

Sensitivity and Breakeven Analysis. These techniques are used to see how sensitive a decision is to estimates for the various parameters. Breakeven analysis is done to locate conditions where the alternatives are equivalent. This is often presented in the form of breakeven charts. Sensitivity analysis is an examination of a range of values for some parameter to determine their effect on a particular decision.

Problems

9-1 A twenty-year-old student decided to set aside \$100 on his 21st birthday for investment. Each subsequent year through his 55th birthday, he plans to increase the sum for investment on a \$100 arithmetic gradient. He will not set aside additional money after his 55th birthday. If he can achieve a 12% rate of return on his investment, how much will he have accrued on his 65th birthday? (*Answer:* \$1,160,700)

9-2 You have an opportunity to purchase a piece of vacant land for \$30,000 cash. If you bought it, you would plan to hold the property for 15 years and then sell it at a profit. During this period, you would have to pay annual property taxes of \$600. You would have no income from the property. Assuming that you would want a 10% rate of return from the investment, at what net price would you have to sell it 15 years hence? (*Answer:* \$144,373)

9-3 An individual's salary is now \$32,000 per year and he anticipates retiring in thirty more years. If his salary is increased by \$600 each year and he deposits 10% of his yearly salary into a fund that earns 7% interest compounded annually, what will be the amount accumulated at the time of his retirement?

9-4 A business executive is offered a management job at Generous Electric Company. They offer to give him a five-year contract which calls for a salary of \$62,000 per year, plus 600 shares of their stock at the end of the five years. This executive is currently employed by Fearless Bus Company and they, too, have offered him a five-year contract. It calls for a salary of \$65,000, plus 100 shares of Fearless stock each year. The stock is currently worth \$60 per share and pays an annual dividend of \$2 per share. Assume end-of-year payments of salary and stock. Stock dividends begin one year after the stock is received. The executive believes that the value of the stock and the dividend will remain constant. If the executive considers 9% a suitable rate of return in this situation, what must the Generous Electric stock be worth per share to make the two offers equally attractive? (*Answer:* \$83.76)

9-5 Tom Jackson is preparing to buy a new car. He knows it represents a large expenditure of money, so he wants to do an analysis to see which of two cars is more economical. Alternative *A* is an American-built compact car. It has an initial cost of \$8900 and operating costs of 9¢ per km, excluding depreciation. Tom checked automobile resale statistics. From them he estimates the American automobile can be resold at the end of three years for \$1700. Alt. *B* is a foreign-built Fiasco. Its initial cost is \$8000. The operating cost, also excluding depreciation, is 8¢ per km. How low could the resale value of the Fiasco be to provide equally economical transportation? Assume Tom will drive 12,000 km per year and considers 8% as an appropriate interest rate.

(*Answer:* \$175)

9-6 A newspaper is considering purchasing locked vending machines to replace open newspaper racks for the sale of its newspapers in the downtown area. The newspaper vending machines cost \$45 each. It is expected that the annual revenue from selling the same quantity of newspapers will increase \$12 per vending machine. The useful life of the vending machine is unknown.

a. To determine the sensitivity of rate of return to useful life, prepare a graph for rate of return *vs.* useful life for lives up to eight years.

b. If the newspaper requires a 12% rate of return, what minimum useful life must it obtain from the vending machines?

c. What would be the rate of return if the vending machines were to last indefinitely?

9-7 Able Plastics, an injection molding firm, has negotiated a contract with a national chain of department stores. A plastic pencil box is to be produced for a two-year period. Able Plastics has never produced this item before and, therefore, requires all new dies. If the firm invests \$67,000 for special removal equipment to unload the completed pencil boxes from the molding machine, one machine operator can be eliminated. This would save the firm \$26,000 per year. The removal equipment has no salvage value and is not expected to be used after the two-year production contract is completed. The equipment, although useless, would be serviceable for about 15 years. You have been asked to do a payback period analysis on whether or not to purchase the special removal equipment. What is the payback period? Should Able Plastics buy the removal equipment?

9-8 A cannery is considering installing an automatic case-sealing machine to replace current hand methods. If they purchase the machine for \$3800 in June, at the beginning of the canning season, they will save \$400 per month for the four months each year that the plant is in operation. Maintenance costs of the case-sealing machine is expected to be negligible. The case-sealing machine is expected to be useful for five annual canning seasons and wili have no salvage value at the end of that time. What is the payback period? Calculate the nominal annual rate of return based on the estimates.

9-9 Consider three alternatives:

	A	*B*	*C*
First cost	\$50	\$150	\$110
Uniform annual benefit	28.8	39.6	39.6
Useful life, in years*	2	6	4
Computed rate of return	10%	15%	16.4%

*At the end of its useful life, an identical alternative (with the same cost, benefits, and useful life) may be installed.

All of the alternatives have no salvage value. If the MARR is 12%, which alternative should be selected?

a. Solve the problem by future worth analysis.

b. Solve the problem by benefit–cost ratio analysis.

c. Solve the problem by payback period.

d. If the answers in parts *a*, *b*, and *c* differ, explain why this is the case.

9-10 An investor is considering buying some land for $100,000 and constructing an office building on it. Three different buildings are being analyzed.

	Building height		
	2 stories	*5 stories*	*10 stories*
Cost of building (excluding cost of land)	$400,000	$800,000	$2,100,000
Resale value* of land and building at end of 20-year analysis period	200,000	300,000	400,000
Annual rental income after deducting all operating expenses	70,000	105,000	256,000

*Resale value to be considered a reduction in cost, rather than a benefit.

Using benefit–cost ratio analysis and an 8% MARR, determine which alternative, if any, should be selected.

9-11 Using benefit–cost ratio analysis, determine which one of the three mutually exclusive alternatives should be selected.

	A	*B*	*C*
First cost	$560	$340	$120
Uniform annual benefit	140	100	40
Salvage value	40	0	0

Each alternative has a six-year useful life. Assume a 10% MARR.

9-12 Compute *F* for the diagram below.

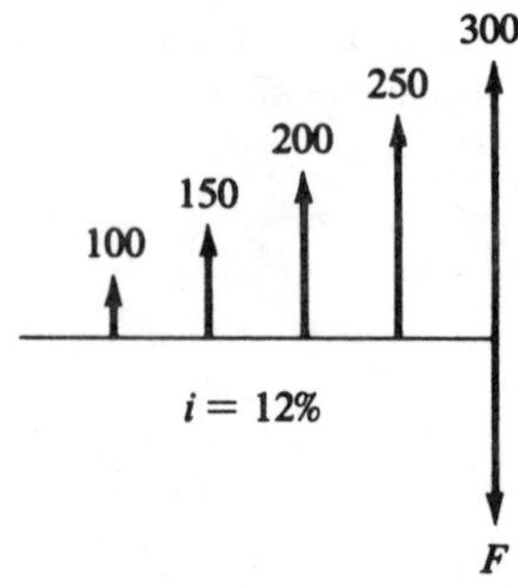

(*Answer:* $F = \$1199$)

9-13 Sally deposited $100 a month in her savings account for 24 months. For the next five years she made no deposits. How much was in the savings account at the end of the seven years, if the account earned 6% annual interest, compounded monthly?

(*Answer:* $3430.78)

9-14 For the diagram, compute *F*.

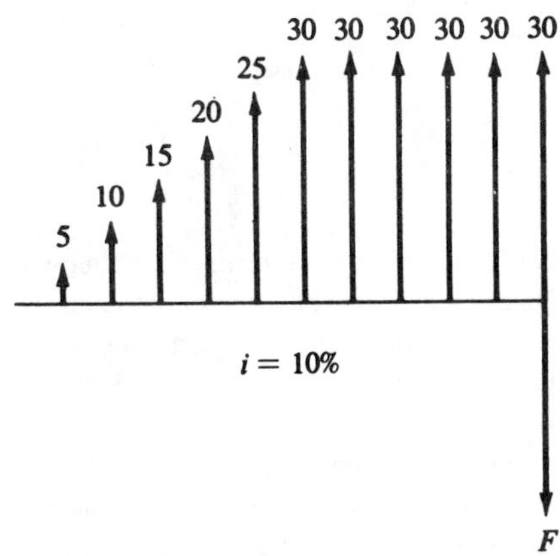

9-15 For a 12% interest rate, compute the value of *F* in the diagram.

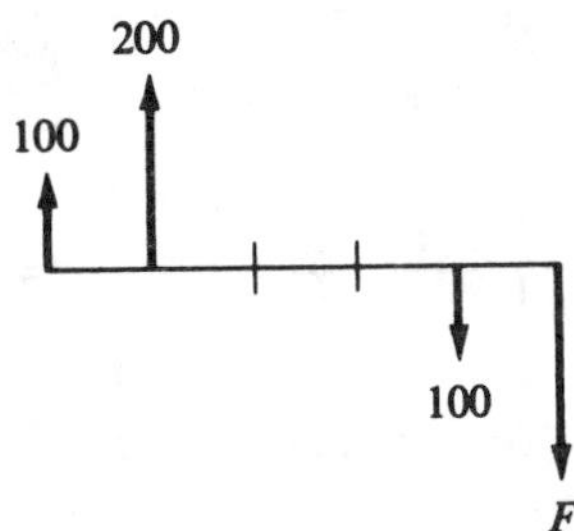

9-16 Stamp collecting has become an increasingly popular—and expensive—hobby. One favorite method is to save plate blocks (usually four stamps with the printing plate number in the margin) of each new stamp as it is issued by the post office. But with the rising postage rates and increased numbers of new stamps being issued, this collecting plan costs more each year.

Stamp collecting, however, may have been a good place to invest money over the last ten years, as the demand for stamps previously issued has caused resale prices to increase 18% each year. Suppose a collector purchased $100 worth of stamps ten years ago, and increased his purchases by $50 per year in each subsequent year. After ten years of stamp collecting, how much would you now estimate his collection could be sold for today?

9-17 For the diagram, compute F.

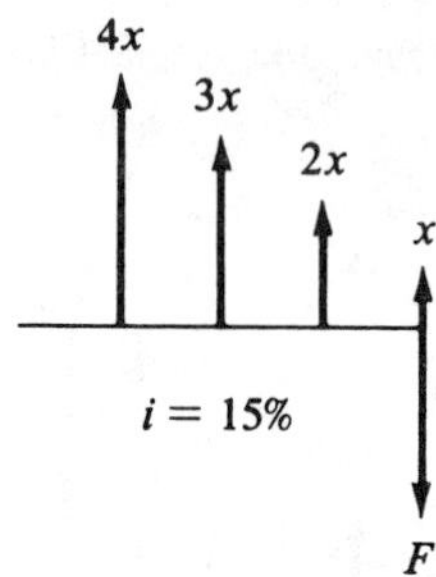

9-18 In the early 1980's, planners were examining alternate sites for a new airport to serve London. In their economic analysis, they computed the value of the structures that would need to be removed from various airport sites. At one airport site, the 12th Century Norman church of St. Michaels, in the village of Stewkley, would be demolished. The planners used the value of the fire insurance policy on the church—a few thousand pounds sterling—as the value of the structure.

An outraged antiquarian wrote to the London *Times* that an equally plausible computation would be to assume that the original cost of the church (estimated at 100 pounds sterling) be increased at the rate of 10% per year for 800 years. Based on his proposal, what would be the present value of St. Michaels? (*Note:* There was such public objection to tearing down the church, it was spared.)

9-19 Compute F for the figure.

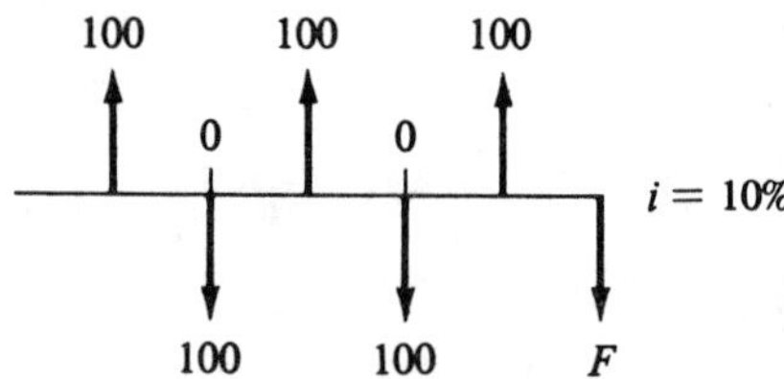

9-20 Bill made a budget and planned to deposit $150 a month in a savings account, beginning September 1st. He did this, but on the following January 1st, he reduced the monthly deposits to $100 a month. In all he made 18 deposits, four at $150 and 14 at $100. If the savings account paid 6% interest, compounded monthly, how much was in his savings account immediately after he made the last deposit? (*Answer:* $2094.42)

9-21 A company deposits $1000 in a bank at the beginning of each year for six years. The account earns 8% interest, compounded every six months. How much will be in the account at the end of six years? Make a careful, accurate computation.

9-22 Don Ball is a 55-year-old engineer. According to mortality tables, a male at 55 has an average life expectancy of 21 more years. In prior years, Don has accumulated

$48,500 including interest, toward his retirement. He is now adding $5000 per year to his retirement fund. The fund earns 12% interest. Don's goal is to retire when he can obtain an annual income from his retirement fund of $20,000 per year, assuming he lives to age 76. He will make no provision for a retirement income after age 76. What is the youngest age at which Don can retire, based on his criteria?

9-23 The three alternatives shown each have a five-year useful life. If the MARR is 10%, which alternative should be selected? Solve the problem by benefit–cost ratio analysis.

	A	*B*	*C*
Cost	$600.0	$500.0	$200.0
Uniform annual benefit	158.3	138.7	58.3

(*Answer: B*)

9-24 Consider four alternatives that each have an eight-year useful life:

	A	*B*	*C*	*D*
Cost	$100.0	$80.0	$60.0	$50.0
Uniform annual benefit	12.2	12.0	9.7	12.2
Salvage value	75.0	50.0	50.0	0

If the MARR is 8%, which alternative should be selected? Solve the problem by benefit–cost ratio analysis.

9-25 Three mutually exclusive alternatives are being considered:

	A	*B*	*C*
Initial cost	$500	$400	$300
Benefit at end of the first year	200	200	200
Uniform benefit at end of subsequent years	100	125	100
Useful life, in years	6	5	4

At the end of its useful life, an alternative is *not* replaced. If the MARR is 10%, which alternative should be selected:

a. based on the payback period?

b. based on benefit–cost ratio analysis?

9-26 Consider three alternatives that have a ten-year useful life. If the MARR is 10%, which alternative should be selected? Solve the problem by benefit–cost ratio analysis.

	A	*B*	*C*
Cost	$800	$300	$150
Uniform annual benefit	142	60	33.5

9-27 Using benefit–cost ratio analysis, a five-year useful life, and a 15% MARR, determine which of the following five alternatives should be selected.

	A	*B*	*C*	*D*	*E*
Cost	$100	$200	$300	$400	$500
Uniform annual benefit	37	69	83	126	150

9-28 The cash flows for three alternatives are as shown:

Year	*A*	*B*	*C*
0	−$500	−$600	−$900
1	−400	−300	0
2	200	350	200
3	250	300	200
4	300	250	200
5	350	200	200
6	400	150	200

a. Based on payback period, which alternative should be selected?

b. Using future worth analysis, and a 12% interest rate, determine which alternative should be selected.

9-29 A project has the following costs and benefits. What is the payback period?

Year	*Costs*	*Benefits*
0	$1400	
1	500	
2	300	$400
3–10		300 per year

9-30 Two alternatives are being considered:

	A	*B*
Initial cost	$500	$800
Uniform annual cost	200	150
Useful life, in years	8	8

Both alternatives provide an identical benefit.

a. Compute the payback period if Alt. *B* is purchased rather than Alt. *A*.

b. Using a MARR of 12% and benefit–cost ratio analysis, which alternative should be selected?

9-31 Consider three mutually exclusive alternatives. The MARR is 10%.

Year	*X*	*Y*	*Z*
0	−$100	−$50	−$50
1	25	16	21
2	25	16	21
3	25	16	21
4	25	16	21

a. For Alt. *X*, compute the benefit–cost ratio.

b. Based on the payback period, which alternative should be selected?

c. Determine the preferred alternative based on an exact economic analysis method.

9-32

Year	E	F	G	H
0	−$90	−$110	−$100	−$120
1	20	35	0	0
2	20	35	10	0
3	20	35	20	0
4	20	35	30	0
5	20	0	40	0
6	20	0	50	180

a. Using future worth analysis, which of the above four alternatives is preferred at 6% interest?

b. Using future worth analysis, which alternative is preferred at 15% interest?

c. Based on the payback period, which alternative is preferred?

d. At 7% interest, what is the benefit–cost ratio for Alt. *G*?

9-33 Consider four mutually exclusive alternatives:

	A	B	C	D
Cost	$75.0	$50.0	$15.0	$90.0
Uniform annual benefit	18.8	13.9	4.5	23.8

Each alternative has a five-year useful life and no salvage value. The MARR is 10%. Which alternative should be selected, based on:

a. future worth analysis?

b. benefit–cost ratio analysis?

c. the payback period?

9-34 Tom Sewel has gathered data on the relative costs of a solar water heater system and on a conventional electric water heater. The data are based on a mid-American city and assume that during cloudy days an electric heating element in the solar heating system will provide the necessary heat.

The installed cost of a conventional electric water tank and heater is $200. A family of four uses an average of 300 liters of hot water a day, which takes $230 of electricity per year. The tank is glass-lined and has a twenty-year guarantee. This is probably a reasonable estimate of its actual useful life.

The installed cost of two solar panels, small electric pump, and storage tank with auxiliary electric heating element is $1400. It will cost $60 a year for electricity to run the pump and heat water on cloudy days. The solar system will require $180 of maintenance work every four years. Neither the conventional electric water heater nor the solar water heater will have any salvage value at the end of their useful lives.

a. Using Tom's data, what is the payback period if the solar water heater system is installed, rather than the conventional electric water heater?

b. Chris Cook studied the same situation and decided that all the data are correct, except that he believes the solar system will *not* require the $180 of maintenance every four years. Chris believes future replacements of either the conventional electric water heater, or the solar water heater system can be made at the same costs and useful lives as the initial installation. Based on a 10% interest rate, what must be the useful life of the solar system to make it no more expensive than the electric water heater system?

9-35 Data for two alternatives are as follows:

	A	*B*
Cost	\$800	\$1000
Uniform annual benefit	230	230
Useful life, in years	5	*X*

If the MARR is 12%, compute the value of *X* that makes the two alternatives equally desirable.

9-36 What is the cost of Alt. *B* that will make it at the breakeven point with Alt. *A*, assuming a 12% interest rate?

	A	*B*
Cost	\$150	\$ *X*
Uniform annual benefit	40	65
Salvage value	100	200
Useful life, in years	6	6

9-37 Consider two alternatives:

	A	*B*
Cost	\$500	\$300
Uniform annual benefit	50	75
Useful life, in years	infinity	*X*

Assume that Alt. *B* is *not* replaced at the end of its useful life. If the MARR is 10%, what must be the useful life of *B* to make Alternatives *A* and *B* equally desirable?

9-38 A project will cost \$50,000. The benefits at the end of the first year are estimated to be \$10,000, increasing at a 10% uniform rate in subsequent years. Using an eight-year analysis period and a 10% interest rate, compute the benefit–cost ratio.

9-39 Jane Chang is making plans for a summer vacation. She will take \$1000 with her in the form of traveller's checks. From the newspaper, she finds that if she purchases the checks by May 31st, she will receive them without paying a service charge. That is, she will obtain \$1000 worth of traveller's checks for \$1000. But if she waits and buys the checks immediately before starting her summer trip, she must pay a 1% service charge. (It will cost her \$1010 for \$1000 of traveller's checks.)

Jane can obtain a 13% interest rate, compounded weekly, on her money. To help her with her planning, Jane decides to compute how many weeks after May 31st she can begin her trip and still justify buying the traveller's checks on May 31st. She asks you to make the computations for her. What is the answer?

9-40 A machine costs $5240 and produces benefits of $1000 at the end of each year for eight years. Assume continuous compounding and a nominal annual interest rate of 10%.

a. What is the payback period (in years)?

b. What is the breakeven point (in years)?

c. Since the answers in ***a*** and ***b*** are different, which one is "correct"?

9-41 Jean invests $100 in year one, and doubles the amount each year after that (so the investment is $100, 200, 400, 800, . . .). If she continues to do this for 10 years, and the investment pays 10% annual interest, how much will her investment be worth at the end of 10 years?

9-42 Fence posts for a particular job cost $10.50 each to install, including the labor cost. They will last 10 years. If the posts are treated with a wood preservative they can be expected to have a 15-year life. Assuming a 10% interest rate, how much could one afford to pay for the wood preservative treatment?

9-43 A motor with a 200-horsepower output is needed in the factory for intermittant use. A Graybar motor costs $7000, and has an electrical efficiency of 89%. A Blueball motor costs $6000 and has an 85% efficiency. Neither motor would have any salvage value as the cost to remove it would equal its scrap value. The annual maintenance cost for either motor is estimated at $300 per year. Electric power costs $0.072/kilowatt-hour (1 hp = 0.746 kW). If a 10% interest rate is used in the calculations, what is the minimum number of hours the higher initial cost Graybar motor must be used each year to justify its purchase?

9-44 Five mutually exclusive investment alternatives have been proposed. Based on benefit-cost ratio analysis, and a MARR of 15%, which alternative should be selected?

Year	*A*	*B*	*C*	*D*	*E*	*F*
0	−$200	−$125	−$100	−$125	−$150	−$225
1–5	+68	+40	+25	+42	+52	+68

9-45 (20-49) Plan *A* requires a $100,000 investment now. Plan *B* requires an $80,000 investment now and an additional $40,000 investment at a later time. At 8% interest, compute the breakeven point for the timing of the $40,000 investment.

9-46 (20-50) A piece of property is purchased for $10,000 and yields a $1000 yearly net profit. If the property is sold after five years, what is its minimum price to breakeven with interest at 10%?

9-47 (20-51) If you invested $2500 in a bank 24-month certificate of deposit paying 8.65%, compounded monthly, how much would you receive when the certificate of deposit matures in two years?

9-48 (20-52) Rental equipment is for sale for $110,000. A prospective buyer estimates he would keep the equipment for 12 years and spend $6000 a year on maintaining the equipment. Estimated annual net receipts from equipment rentals would be $14,400.

It is estimated the rental equipment could be sold for $80,000 at the end of 12 years. If the buyer wants a 7% rate of return on his investment, what is the maximum price he should pay for the equipment?

9-49 (20-53) A low carbon steel machine part, operating in a corrosive atmosphere, lasts six years and costs $350 installed. If the part is treated for corrosion resistance it will cost $500 installed. How long must the treated part last to be the preferred alternative, if 10% interest is used?

9-50 (20-54) A car dealer presently leases a small computer with software for $5000 per year. As an alternative he could purchase the computer for $7000 and lease the software for $3500 per year. Any time he would decide to switch to some other computer system he could cancel the software lease and sell the computer for $500. If he purchases the computer and leases the software,

a. what is the payback period?

b. and kept them for six years, what would be the benefit-cost ratio, based on a 10% interest rate?

9-51 (20-55) Given the following data for two machines:

	A	*B*
Original cost	$55,000	$75,000
Annual expenses		
Operation	9,500	7,200
Maintenance	5,000	3,000
Taxes and insurance	1,700	2,250

The machines have no net salvage value. At what useful life are the machines equivalent if

a. 10% interest is used in the computations?

b. 0% interest is used in the computations?

CHAPTER 10

Depreciation

We have so far dealt with a variety of economic analysis problems and many techniques for their solution. In the process we have carefully avoided an important element of most economic analyses—income taxes. Now that the essential concepts have been presented, we can move to more realistic—and, unfortunately, more complex—situations.

Our Government taxes individuals and businesses to support its processes—lawmaking, domestic and foreign economic policy-making, even the making and issuing of money itself. The omnipresence of taxes requires that we take them into account in our economic analyses, which means we must understand something about the *way* taxes are imposed. For capital equipment, depreciation is an important component in computing income taxes. Chapter 10 examines various aspects of depreciation, and Chapter 11 illustrates how depreciation is used in income tax computations.

Basic Aspects Of Depreciation

The word ***depreciation*** is defined as a "decrease in value." This is not an entirely satisfactory definition, for *value* has several meanings. In the context of economic analysis, value may refer either to *market value*—that is, the monetary value others place on property—or *value to the owner*. Thus, we now have two definitions of depreciation: a decrease in market value, or a decrease in value to the owner.

Deterioration and Obsolescence

A machine may depreciate (decline in value) because it is wearing out and no longer performing its function as well as when it was new. This situation is called ***deterioration***. Many kinds of machinery require increased maintenance as they age, reflecting a slow but continuing failure of individual parts. In other types of equipment, the quality of output may decline due to wear on components and resulting poorer mating of parts. Anyone who has worked to maintain the mechanical components of an automobile will testify to deterioration due to both the failure of individual parts (such as, fan belt, muffler, or battery) and the wear on components (such as, bearings, piston rings, alternator brushes).

Another aspect of depreciation is that caused by ***obsolescence***. A machine is described as obsolete when the function it performs can be done in some better manner. A machine may be in excellent working condition, yet may still be obsolete. In the 1970s, for example, there was a major shift in the construction of business calculators. Previously these devices were complex machines, with hundreds of gears and levers. But the advancement of integrated circuits resulted in a completely different approach to calculator design. Electronic circuitry was such an immense improvement on metal mechanisms that mechanical calculators became obsolete in a very short time. Thus, mechanical calculators, even though they had not physically deteriorated, declined in value—depreciated—rapidly.

If your automobile depreciated in the last year, that meant it had declined in market value. On the other hand, a manager who indicates a piece of machinery has depreciated may be describing a machine that has deteriorated because of use or because it has become obsolete compared to newer machinery. Both situations indicate the machine has declined in value to the owner.

The accounting profession defines depreciation in yet a third way. Although in everyday conversation we are likely to use depreciation to mean a decline in market value, accountants talk of depreciation as the allocation of the cost of an asset divided by its ***useful***, or ***depreciable***, ***life***. Thus, we now have *three distinct definitions of depreciation*:

1. Decline in market value of an asset.
2. Decline in value of an asset to its owner.
3. Systematic allocation of the Cost of an asset *divided by* its Depreciable life.

In determining taxable income—hence, income taxes—it is the accountant's definition of depreciation that is used in the computations. Chapter 11, therefore, will deal only with this third definition of depreciation.

Almost everything seems to decline in value as time proceeds. Machines become obsolete or wear out. Automobiles, if they are driven far enough, can literally fall apart! Buildings, too, show the heavy signs of age. There are, however, exceptions: Manhattan Island—like "land" generally—has not only *not* declined in value, its market value has continued to increase. Yet land is not considered to be an asset subject to depreciation. Regardless of whether the market value is going up, down, or remains unchanged, accountants consider land to be an asset that does not have a limited useful life. Land is considered, therefore, a nondepreciable asset, and its cost is not allocated over a useful life.

Depreciation Calculation Fundamentals

To understand the complexities of depreciation, the first step is to examine the fundamentals of depreciation calculations. Depreciation is the allocation of the cost of an asset over its depreciable life. While it is sometimes possible to project that an asset will have no salvage value at the end of its life, the various depreciation calculations below show how salvage value, if present, is treated.

Figure 10-1 illustrates the general depreciation problem of allocating the total depreciation charges over the life of the asset. The vertical axis is labelled *Money*, but when we plot the curve of cost *minus* depreciation charges made, the vertical axis is more appropriately called ***book value***, or

Book value **= Cost − Depreciation charges made**

Looked at another way, one can also define book value as *the remaining unallocated cost of an asset.*

The six principal depreciation methods are presented in turn in this chapter:

1. Straight Line depreciation.
2. Sum-Of-Years Digits depreciation.
3. Declining Balance depreciation.
4. Declining Balance with conversion to Straight Line depreciation.
5. Unit-Of-Production depreciation.
6. Accelerated Cost Recovery System depreciation.

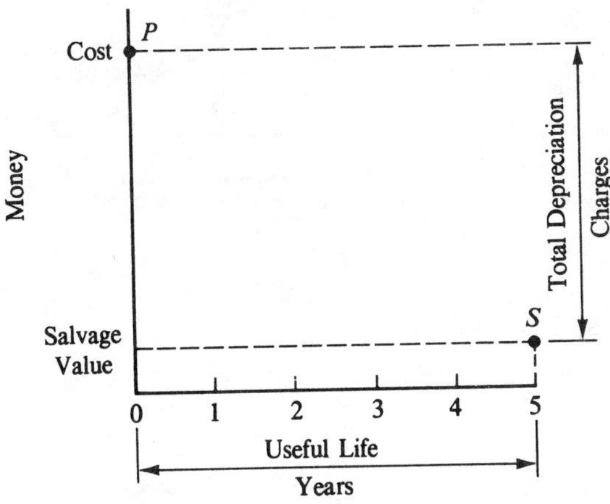

Figure 10-1 General depreciation.

Straight Line Depreciation

The simplest, and best known, of the various depreciation methods is ***straight line depreciation***. In this method a constant depreciation charge is made. To obtain the ***annual depreciation charge***, the total amount to be depreciated, $P - S$, is *divided by* the useful life, in years, N:

$$\textit{Annual depreciation charge} = \frac{1}{N}(P - S) \tag{10-1}$$

EXAMPLE 10-1

Consider the following:

Cost of the asset, P	$900
Useful life, in years, N	5
End-of-useful-life salvage value, S	70

Compute the straight line depreciation schedule for this situation.

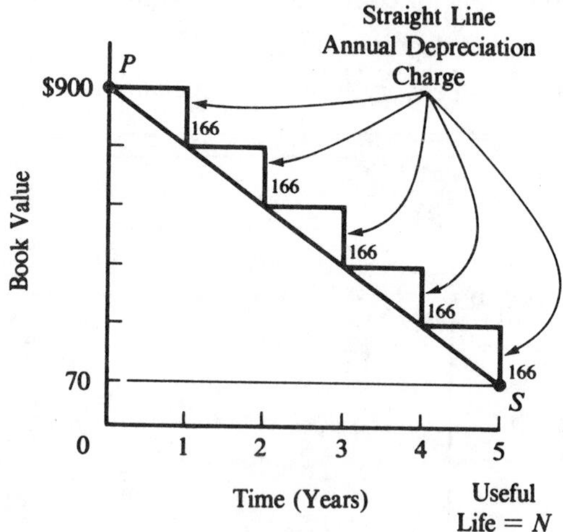

Figure 10-2 Straight line depreciation.

Solution:

$$\text{Annual depreciation charge} = \frac{1}{N}(P - S) = \frac{1}{5}(900 - 70)$$
$$= \$166$$

Year	*Book value before depreciation charge*	*Depreciation for year*	*Book value after depreciation charge*
1	Cost = $900	$166	$734
2	734	166	568
3	568	166	402
4	402	166	236
5	236	166	70 = Salvage Value
	Total Depreciation =	$830	

This situation is illustrated in Figure 10-2. ◀

There is an alternate way of computing the straight-line depreciation charge in any year:

$$\textbf{\textit{Straight line depreciation charge for any year}} = \frac{\textbf{Book value at beginning of year} - \textbf{Salvage value}}{\textbf{Remaining useful life at beginning of year}} \qquad (10\text{-}2)$$

The depreciation charge in the third year of Ex. 10-1 could be computed as:

$$\text{Straight line depreciation for third year} = \frac{568 - 70}{3} = \$166$$

Both methods give the same result. At any time, the book value of the asset would be the cost *minus* the depreciation to that point in time. For Ex. 10-1, the book value at the end of three years would be $900 − 3($166) = $402. In general,

$$\textbf{\textit{Book value, end of Jth year}} = P - \frac{J}{N}(P - S)$$

Sum-Of-Years Digits Depreciation

Another method for allocating the Cost of an asset *minus* Salvage value *over* its Useful life is called ***Sum-Of-Years Digits***—SOYD—***depreciation***. This method results in larger-than-straight-line depreciation charges during the early years of an asset and, necessarily, smaller charges as the asset nears the end of its estimated useful life. Each year, the depreciation charge is computed as the remaining useful life at the beginning of the year *divided by* the sum of the years digits for the total useful life, with this ratio *multiplied by* the total amount to be depreciated $(P - S)$, or:

$$\textbf{\textit{Sum-of-years digits depreciation charge for any year}} = \frac{\textbf{Remaining useful life at beginning of year}}{\textbf{Sum-of-years digits for total useful life}}(P - S) \qquad (10\text{-}3)$$

where

$$\text{SOYD for total useful life} = 1 + 2 + 3 + \cdots + (N - 1) + N$$

An equation for the sum-of-years digits—**SUM**—may be derived:

$$\text{SUM} = 1 \quad + 2 \quad + \cdots + (N - 1) + N \qquad (10\text{-}4)$$

Write the terms in Eq. 10-4 in reverse order,

$$\text{SUM} = N \quad + (N - 1) + \cdots + 2 \quad + 1 \qquad (10\text{-}5)$$

Add Equations 10-4 and 10-5,

$$2(\text{SUM}) = (N + 1) \ + (N + 1) + \cdots + (N + 1) + (N + 1)$$

For N terms on the right side,

$$2(\text{SUM}) = N(N + 1)$$

$$\textbf{\textit{Sum-of-years digits}} = \frac{N}{2}(N + 1) \qquad (10\text{-}6)$$

EXAMPLE 10-2

Compute the SOYD depreciation schedule for the situation in Ex. 10-1:

Cost of the asset, P	\$900
Useful life, in years, N	5
End-of-useful-life salvage value, S	70

Solution:

$$\text{Sum-of-years digits} = 5 + 4 + 3 + 2 + 1 = 15, \quad \text{or,}$$

$$\text{SUM} = \frac{N}{2}(N + 1) = \frac{5}{2}(5 + 1) = 15$$

$$\text{1st year SOYD depreciation} = \frac{5}{15}(900 - 70) = \$277$$

$$\text{2nd year} = \frac{4}{15}(900 - 70) = 221$$

$$\text{3rd year} = \frac{3}{15}(900 - 70) = 166$$

$$4\text{th year} = \frac{2}{15}(900 - 70) = 111$$

$$5\text{th year} = \frac{1}{15}(900 - 70) = \underline{55}$$

$$\$830$$

These data are plotted in Figure 10-3. ◀

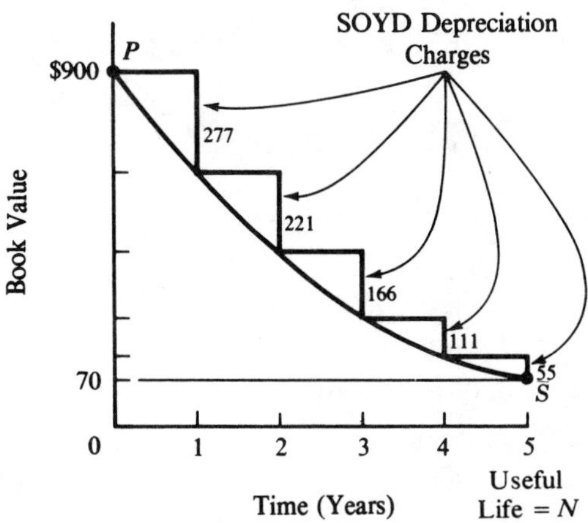

Figure 10-3 Sum-of-years digits depreciation.

Declining Balance Depreciation

A third major method is ***declining balance depreciation***. Here a *constant depreciation rate* is applied to the book value of the property. Two rates specified in the Tax Reform Act of 1986 are 150% and 200% of the straight-line rate. Since 200% is twice the straight-line rate, it is called ***double declining balance***, or DDB; the general equation is:

$$\text{Double declining balance depreciation in any year} = \frac{2}{N}(\text{Book value})$$

Since book value equals cost *minus* depreciation charges to date,

$$\textbf{DDB}\ \textbf{\textit{depreciation in any year}} = \frac{2}{N}(\textbf{Cost} - \textbf{Depreciation charges to date})$$

EXAMPLE 10-3

Compute the DDB depreciation schedule for the situations in Examples 10-1 and 10-2.

Cost of the asset, P	\$900
Useful life, in years, N	5
End-of-useful-life salvage value, S	70

Solution:

$$\text{DDB depreciation} = \frac{2}{N}(P - \text{Depreciation charges to date})$$

$$\text{1st year} = \frac{2}{5}(900 - 0) = \$360$$

$$\text{2nd year} = \frac{2}{5}(900 - 360) = 216$$

$$\text{3rd year} = \frac{2}{5}(900 - 576) = 130$$

$$\text{4th year} = \frac{2}{5}(900 - 706) = 78$$

$$\text{5th year} = \frac{2}{5}(900 - 784) = 46$$

$$\overline{\$830}$$

Figure 10-4 illustrates the situation. ◀

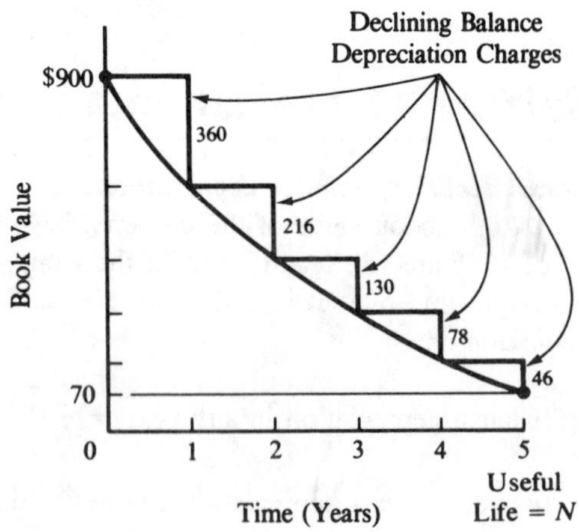

Figure 10-4 Declining balance depreciation.

Declining Balance Depreciation in Any Year

For double declining balance, the depreciation schedule is

$$\text{1st year DDB depreciation} = \frac{2}{N}(P) = \frac{2P}{N}$$

$$\text{2nd year DDB depreciation} = \frac{2}{N}\left(P - \frac{2P}{N}\right) = \frac{2P}{N}\left(1 - \frac{2}{N}\right)$$

$$\text{3rd year DDB depreciation} = \frac{2}{N}\left[P - \frac{2P}{N} - \frac{2P}{N}\left(1 - \frac{2}{N}\right)\right]$$

$$= \frac{2P}{N}\left[1 - 2\left(\frac{2}{N}\right) + \left(\frac{2}{N}\right)^2\right]$$

$$= \frac{2P}{N}\left(1 - \frac{2}{N}\right)^2$$

And, **in any year *n***,

$$\textbf{DDB } \textit{depreciation} = \frac{2P}{N}\left(1 - \frac{2}{N}\right)^{n-1} \qquad (10\text{-}7)$$

For 150% declining balance depreciation, the "2" appearing in the two factors of Eq. 10-7 would be replaced by "1.50".

Total Declining Balance Depreciation at End of *n* Years

For double declining balance depreciation,

Total DDB depreciation

$$= \frac{2P}{N}\left[1 + \left(1 - \frac{2}{N}\right) + \left(1 - \frac{2}{N}\right)^2 + \cdots + \left(1 - \frac{2}{N}\right)^{n-1}\right] \qquad (10\text{-}8)$$

Multiply by $\left(1 - \frac{2}{N}\right)$:

$$(\text{Total DDB depreciation})\left(1 - \frac{2}{N}\right) = \frac{2P}{N}\left[\left(1 - \frac{2}{N}\right)\right.$$

$$\left. + \left(1 - \frac{2}{N}\right)^2 + \cdots + \left(1 - \frac{2}{N}\right)^{n-1} + \left(1 - \frac{2}{N}\right)^n\right] \qquad (10\text{-}9)$$

Subtract Eq. 10-8 from Eq. 10-9:

$$-\frac{2}{N}(\text{Total DDB depreciation}) = \frac{2P}{N}\left[-1 + \left(1 - \frac{2}{N}\right)^n\right]$$

$$\textbf{\textit{Total}}\ \textbf{DDB}\ \textbf{\textit{depreciation}} = \mathbf{P}\left[\mathbf{1} - \left(\mathbf{1} - \frac{\mathbf{2}}{\mathbf{N}}\right)^{\mathbf{n}}\right] \quad (10\text{-}10)$$

For $1.50/N$ depreciation, substitute this value in Eq. 10-10 for the $2/N$ term.

Book Value of an Asset at End of *n* Years

The book value at the end of n years will be the cost of the asset P *minus* the total depreciation at the end of n years. For DDB depreciation,

$$\begin{aligned}\text{Book value} &= P - \text{Total DDB depreciation at end of } n \text{ years}\\ &= P - P\left[1 - \left(1 - \frac{2}{N}\right)^n\right] = P\left(1 - \frac{2}{N}\right)^n \quad (10\text{-}11)\end{aligned}$$

Effect of Salvage Value on Declining Balance Depreciation

In Example 10-3, the salvage value was estimated to be \$70. And we see that the calculations produced a depreciation schedule that resulted in a \$70 salvage value. This is hard to explain, for the calculations did not consider our salvage value estimate. The answer is that the example problem was devised to come out this way! Since the depreciation schedule is independent of the estimated salvage value, any of the three situations in Fig. 10-5 might occur in an actual situation.

In Figure 10-5a, we have a relatively high salvage value, with the result that the book value of the asset would decline below the estimated salvage value. While this might seem to be desirable, the U.S. Internal Revenue Service (IRS) does not permit a taxpayer to continue to deduct depreciation charges that would

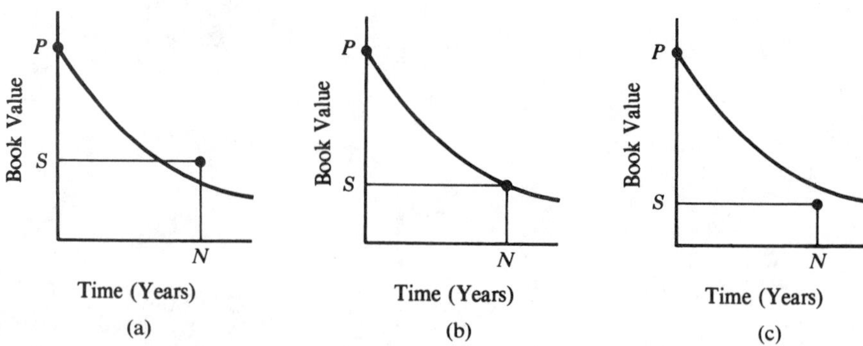

Figure 10-5 Declining balance depreciation and salvage value relationships.

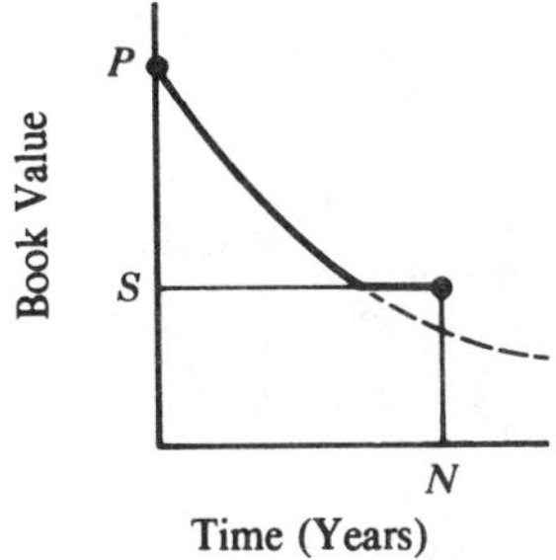

Figure 10-6 Declining balance depreciation terminated when salvage value is reached.

drop the book value below the salvage value. By applying the IRS rule, Fig. 10-5a is transformed into Fig. 10-6. No depreciation charges are made that would reduce the book value below the estimated salvage value. Depreciation charges cease when the book value equals the estimated salvage value.

Figure 10-5b presents no problem, for the declining balance depreciation schedule results in a book value exactly equal to the salvage value at the end of the useful life. Obviously, this will only happen when the salvage value happens to lie on the declining balance book-value curve.

Figure 10-5c represents the situation where the salvage value is beneath the declining balance book-value curve at the end of the depreciable life N. Since the declining balance book-value curve is independent of the estimated salvage value, this situation could easily occur. In fact, this is *always* the case when the estimated salvage value is zero.

In any given declining balance depreciation situation, the cost of the asset P, the depreciable life N, and the allowable depreciation rate ($1.50/N$ or $2/N$) will be known or determined from the facts. These data define the coordinates of the book-value curve. For example, for an asset with cost P, and depreciable life N equal to five years, the DDB book-value curve would be computed as follows:

Year		DDB *depreciation*	*Book value*
0			$1.000P$
1	$\frac{2}{5}(P - 0)$	$= 0.400P$	$0.600P$
2	$\frac{2}{5}(P - 0.400P)$	$= 0.240P$	$0.360P$
3	$\frac{2}{5}(P - 0.640P)$	$= 0.144P$	$0.216P$
4	$\frac{2}{5}(P - 0.784P)$	$= 0.086P$	$0.130P$
5	$\frac{2}{5}(P - 0.870P)$	$= 0.052P$	$0.078P$

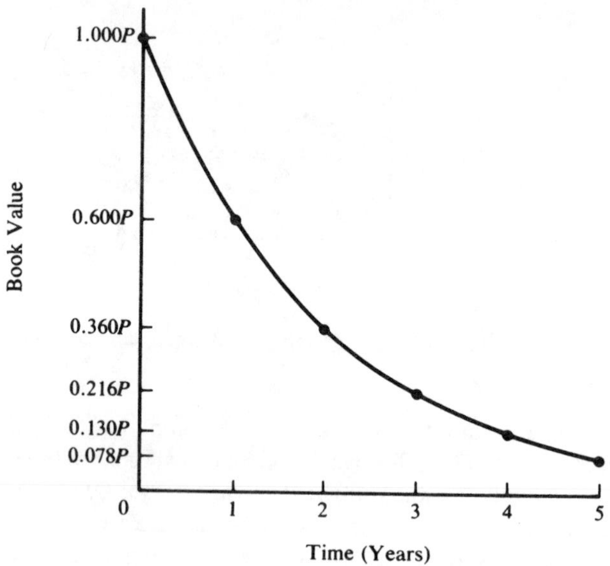

Figure 10-7 Book value curve for double declining balance depreciation and five-year depreciable life.

These data are plotted in Figure 10-7. For DDB depreciation and a five-year depreciable life, the book-value curve will decline to $0.078P$. If the estimated salvage value in this situation is less than $0.078P$, the result will look like Fig. 10-5c. This is illustrated in Ex. 10-4.

EXAMPLE 10-4

The cost of a new asset is \$900 and its useful life is five years. If the salvage value is estimated to be \$30, compute the DDB depreciation schedule.

Solution:

		DDB *depreciation*	*End-of-year book value*
1st-year depreciation	$= \frac{2}{5}(900)$	= \$360	\$540
2nd-year depreciation	$= \frac{2}{5}(900 - 360) =$	216	324
3rd-year depreciation	$= \frac{2}{5}(900 - 576) =$	130	194
4th-year depreciation	$= \frac{2}{5}(900 - 706) =$	78	116
5th-year depreciation	$= \frac{2}{5}(900 - 784) =$	46	70

The book value at the end of the useful life does not decline to the $30 salvage value. Since Fig. 10-7 is a plot of this situation (DDB depreciation and five-year depreciable life), we know that the end of useful-life book value will be 0.078*P*.

For $P = \$900$, Book value $= 0.078(900) = \$70$

Thus the situation looks like Fig. 10-5c. Two general approaches may be used to resolve the difficulty:

1. If the asset is retained in service beyond its estimated useful life (and this is probably a more frequent occurrence than one would expect), then declining balance depreciation would continue to be charged in subsequent years until either the current estimated salvage value is reached, or until the asset is disposed of. Here, the sixth-year depreciation would be $\frac{2}{5}(900 - 830) = \$28$; seventh-year depreciation would be $\frac{2}{5}(900 - 858) = \$17$. Since the seventh-year depreciation would bring the book value below the salvage value, the seventh-year depreciation is *decreased*. The resulting DDB depreciation schedule would be:

Year	*Depreciation*	*End-of-year book value*
1	$360	$540
2	216	324
3	130	194
4	78	116
5	46	70
6	28	42
7	12	30

2. A second alternative is to use a *composite* depreciation method. The rules of the IRS provide that a taxpayer may change from the declining balance method to straight line depreciation at any time during the life of an asset. In this example, we could simply use straight line depreciation in the fifth year. Since straight line depreciation in any year equals the Book value *minus* the Salvage value, all *divided by* the Remaining useful life,

$$\text{5th-year straight line depreciation} = \frac{\text{Book value} - \text{Salvage value}}{\text{Remaining useful life}} = \frac{116 - 30}{1} = \$86$$

Figure 10-8 illustrates the difficulty and the two possible ways of solving it. ◀

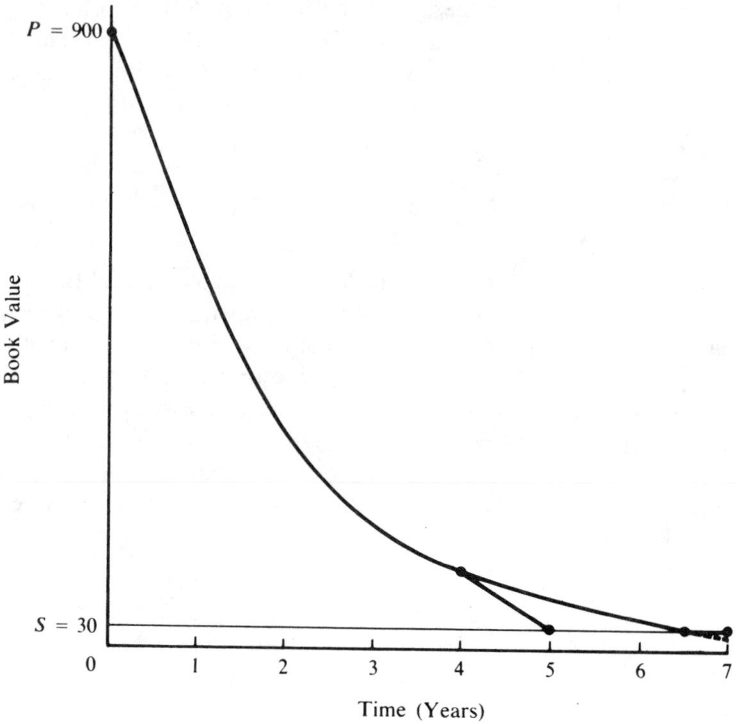

Figure 10-8 Double declining balance depreciation either converted to straight line depreciation or continued beyond the five-year estimated life.

Declining Balance Depreciation With Conversion To Straight Line Depreciation

In the previous section, we encountered a situation where declining balance depreciation was not entirely satisfactory. The depreciation method failed to achieve the desired result of allocating the cost of the asset (*minus* salvage value) over its estimated useful life. Figure 10-5c and Ex. 10-4 illustrated the problem.

We saw that one way to overcome this difficulty was to adopt a composite depreciation method. Initially, the asset is depreciated by the declining balance method; subsequently, the remainder of the depreciation is computed by the straight line method. In the composite depreciation method, one must decide when to switch from declining balance to straight line depreciation. In Example 10-4, the switch was made in the last, or fifth, year; the question arises, "Was there a better time to switch methods?"

In a subsequent section on selecting the preferred depreciation method, we will show that a taxpayer generally prefers to reduce the book value of an asset to its salvage value as quickly as possible. Thus, we would switch from declining balance to straight line whenever straight line depreciation results in larger depreciation charges and, hence, a more rapid reduction in book value of the asset.

On this basis, the choice between the alternatives in Fig. 10-8 is clear. The conversion to straight line depreciation in Year 5 is preferred over continuing the declining balance depreciation. The same criterion (reduce the book value of an asset to its salvage value as quickly as possible) will be used to determine when to switch from declining balance to straight line depreciation.

Assuming that the conversion from declining balance to straight line depreciation could take place in any of N years, there would seem to be N possibilities. Our problem is to identify the best one.

Figure 10-9 shows the results if the conversion is made at different points in time. We assume the most desirable depreciation schedule is the one that results in the most rapid decline in book value, that is, depreciates the asset as quickly as possible. Looking at Fig. 10-9, the curve that best meets this criterion is the one that converts to straight line depreciation at Point B. At this point, the slope of the declining balance curve equals the slope of the straight line. Selecting any other point for the conversion produces a curve that will be above our selected curve at some point in time. Thus, the Point B conversion is best—there is no other curve that declines as rapidly from P to S.

Actually, the declining balance curve is like stair steps rather than a smooth curve. For this reason, the year to convert is best computed as the point where

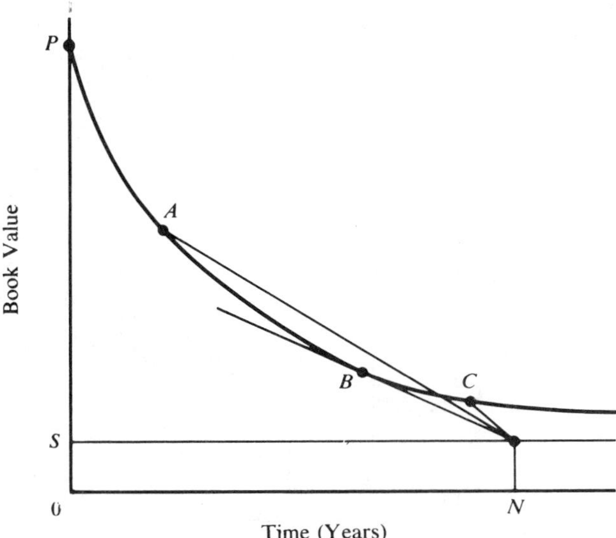

Figure 10-9 Three possible points at which to convert from double declining balance to straight line depreciation.

Table 10-1 CONVERSION FROM DOUBLE DECLINING BALANCE TO STRAIGHT LINE DEPRECIATION

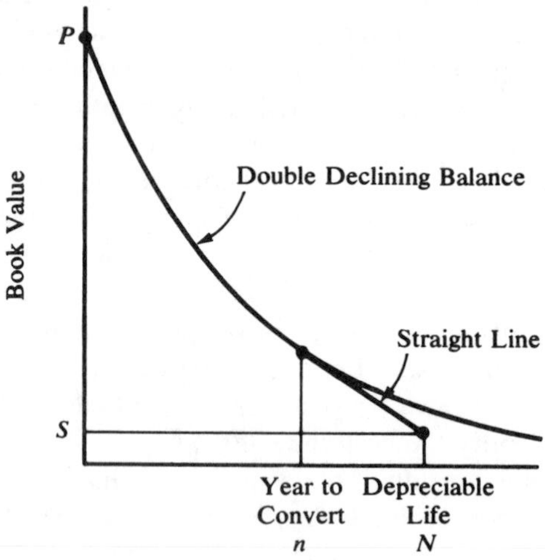

Computed year *n* to change methods

	Zero salvage value	*Salvage value ratio S/P*		
	$S = 0$	$S/P = 0.05$	$S/P = 0.10$	$S/P = 0.12$
Depreciable life of asset N (years)	*Value of n; use straight line depreciation for Year n and all subsequent years*			
3	3		***No conversion is needed***	
4	4	4		
5	4	5		
6	5	5		
7	5	6		
8	6	6	8	
9	6	7	9	
10	7	7	9	
11	7	8	10	
12	8	9	11	
13	8	9	11	
14	9	10	12	
15	9	10	13	
16	10	11	13	
17	10	11	14	
18	11	12	15	18
19	11	13	16	19
20	12	13	16	19

the conversion to straight line depreciation produces an equal or greater depreciation charge than remaining with double declining balance depreciation. Table 10-1 tabulates the conversion point for a variety of situations. Most situations covered by Table 10-1 are where the book value curve remains above the salvage value, like Fig. 10-5c, and a conversion to straight line depreciation is desirable. In situations with larger salvage values, and S/P ratios, the book value curve declines to the salvage value or below it, like Figures 10-5a and 10-5b. In this event, no conversion to straight line depreciation is needed.

EXAMPLE 10-5

Cost of the asset, P	\$900
Useful (and depreciable) life, in years, N	5
End-of-useful-life salvage value, S	30

Compute the double declining balance depreciation schedule with conversion to straight line at the most desirable time.

Solution: Table 10-1 will help locate the correct conversion point.

$$S/P = 30/900 = 0.03$$

For a depreciable life of five years and S/P of 0.05, the conversion point is the beginning of the fifth year. At $S = 0$ (and, therefore, $S/P = 0$), the conversion point is the beginning of the fourth year. Further calculations are needed to determine if the conversion point at S/P of 0.03 is at four or five years.

We will proceed by computing the DDB depreciation for each year and also the straight line depreciation if the conversion to straight line were made in that year. To obtain the most rapid decline in book value, we will use DDB depreciation until we reach a point where converting to straight line results in increased depreciation.

In Example 10-4, the DDB depreciation schedule for this case was computed as:

Year	DDB *depreciation*	*End-of-year book value*
1	\$360	\$540
2	216	324
3	130	194
4	78	116
5	46	70

We can compute the straight line depreciation for any year using Eq. 10-2:

$$\text{Straight line depreciation in any year} = \frac{\text{Book value at beginning of year} - \text{Salvage value}}{\text{Remaining useful life at beginning of year}} \qquad (10\text{-}2)$$

If convert to straight line depreciation in Year	*Straight line depreciation*	DDB *depreciation*	*Decision*
2	$\frac{540 - 30}{4} = \$127.50$	\$216	Do not convert to straight line
3	$\frac{324 - 30}{3} = 98$	130	Do not convert to straight line
4	$\frac{194 - 30}{2} = 82$	78	Convert to straight line for Year 4

The resulting depreciation schedule is:

Year	DDB *with conversion to straight line depreciation*
1	\$360
2	216
3	130
4	82
5	82
	\$870

The resulting depreciation schedule depreciates the asset to its salvage value as quickly as possible. ◀

Choosing Between SOYD and DDB Depreciation

Because they appear rather similar, it may not be readily apparent whether SOYD or DDB is the preferred method of depreciation. For a firm that pays income taxes each year, depreciation is a deduction from taxable income. The result is the greater the depreciation deduction, the less the taxable income—hence, taxes—for the year. Depreciation methods generally result in the same total depreciation deductions, so it is the *timing* of the deductions that characterize the different methods. Immediate tax savings are more valuable than tax savings some years in the future due to the time value of money. For this reason, a profitable firm prefers to depreciate its assets as rapidly as possible.

We used this criterion to determine whether to use the declining balance method or the composite method of declining balance with conversion to straight line. But the situation may not always be this simple. There are situations where one depreciation method produces larger depreciation changes (and, therefore, a lower book value) in the early years and another depreciation method is more attractive in subsequent years. In this case, we will select the method with the largest present worth of depreciation charges. Table 10-2 has been constructed in this manner to indicate the choice between SOYD and DDB depreciation.

Table 10-2 CHOOSING BETWEEN SOYD AND DDB DEPRECIATION

Methods Compared: Sum-of-years digits;
Double declining balance;
Double declining balance with conversion to straight line.

Conditions:

1. The firm is profitable and pays federal income taxes each year.
2. The income tax rate remains constant throughout the depreciable life of the asset.
3. An interest rate of 7% is used in comparing alternatives.

Note: In the non-shaded area, use double declining balance with conversion to straight line depreciation for the year indicated and all subsequent years.

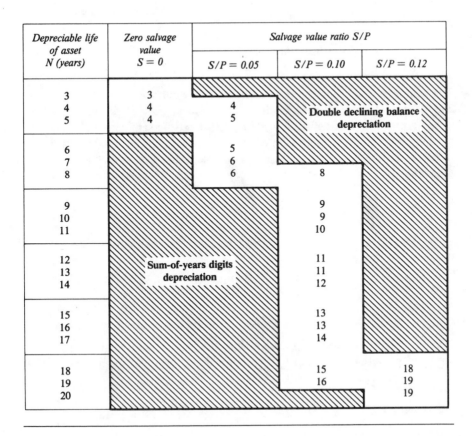

Depreciable life of asset N (years)	*Zero salvage value S = 0*	*Salvage value ratio S/P*		
		S/P = 0.05	*S/P = 0.10*	*S/P = 0.12*
3	3	Double declining balance depreciation	Double declining balance depreciation	Double declining balance depreciation
4	4	4	Double declining balance depreciation	Double declining balance depreciation
5	4	5	Double declining balance depreciation	Double declining balance depreciation
6	Sum-of-years digits depreciation	5	Double declining balance depreciation	Double declining balance depreciation
7	Sum-of-years digits depreciation	6	Double declining balance depreciation	Double declining balance depreciation
8	Sum-of-years digits depreciation	6	8	Double declining balance depreciation
9	Sum-of-years digits depreciation	Sum-of-years digits depreciation	9	Double declining balance depreciation
10	Sum-of-years digits depreciation	Sum-of-years digits depreciation	9	Double declining balance depreciation
11	Sum-of-years digits depreciation	Sum-of-years digits depreciation	10	Double declining balance depreciation
12	Sum-of-years digits depreciation	Sum-of-years digits depreciation	11	Double declining balance depreciation
13	Sum-of-years digits depreciation	Sum-of-years digits depreciation	11	Double declining balance depreciation
14	Sum-of-years digits depreciation	Sum-of-years digits depreciation	12	Double declining balance depreciation
15	Sum-of-years digits depreciation	Sum-of-years digits depreciation	13	Double declining balance depreciation
16	Sum-of-years digits depreciation	Sum-of-years digits depreciation	13	Double declining balance depreciation
17	Sum-of-years digits depreciation	Sum-of-years digits depreciation	14	Double declining balance depreciation
18	Sum-of-years digits depreciation	Sum-of-years digits depreciation	15	18
19	Sum-of-years digits depreciation	Sum-of-years digits depreciation	16	19
20	Sum-of-years digits depreciation	Sum-of-years digits depreciation	Sum-of-years digits depreciation	19

Unit Of Production Depreciation

At times, there may be situations where the recovery of depreciation on a particular asset is more closely related to use than to time. In these few situations (and they are rare), the ***Unit Of Production***—UOP—***depreciation*** in any year is:

$$\textbf{\textit{UOP depreciation in any year}} = \frac{\textbf{Production for year}}{\textbf{Total lifetime production for asset}}(P - S) \quad (10\text{-}12)$$

This method might be useful for machinery that processes natural resources if the resources will be exhausted before the machinery wears out. It is not considered an acceptable method for general use in depreciating industrial equipment.

EXAMPLE 10-6

Cost of asset, P	\$900
Salvage value, S	70

The equipment has been purchased for use in a sand and gravel pit. The sand and gravel pit will be in operation during a five-year period while a nearby airport is being reconstructed and paved. After that time, the sand and gravel pit will be shut down, and the equipment removed and sold. The airport reconstruction schedule calls for 40,000 cubic meters of sand and gravel on the following schedule:

Year	*Required sand and gravel, in* m^3
1	4,000
2	8,000
3	16,000
4	8,000
5	4,000

Compute the unit-of-production (UOP) depreciation schedule for the equipment.

Solution: The total lifetime production for the asset is 40,000 cubic meters of sand and gravel. From the airport reconstruction schedule, the first-year UOP depreciation would be:

$$\text{First-year UOP depreciation} = \frac{4000 \text{ m}^3}{40{,}000 \text{ m}^3}(\$900 - \$70) = \$83$$

Similar calculations for the subsequent four years gives the complete depreciation schedule:

Year	UOP *depreciation*
1	$ 83
2	166
3	332
4	166
5	83
	$830

It should be noted that the actual unit of production depreciation charge in any year is based on the actual production for the year rather than the scheduled production. ◀

Accelerated Cost Recovery System Depreciation

The newest depreciation method that may be used for U.S. income tax purposes is called **Accelerated Cost Recovery System**—ACRS—**depreciation**. The latest rules for its computation are in the Tax Reform Act of 1986.* Two major advantages of ACRS depreciation are: (1) that the computations are made using "property class lives" that are less than the "actual useful lives"; and, (2) that salvage values are assumed to be zero.

To understand the ACRS classes of depreciable property, a little background is needed. In 1970 the U.S. Treasury Department did an analysis of the way assets were utilized to find their actual useful lives. In 1971 they published guidelines for about 100 broad classifications of depreciable assets. For each classification there was a lower limit, midpoint, and upper limit of useful life, called the ***Asset Depreciation Range***, or ADR. The ADR midpoint lives were somewhat shorter than the actual average useful lives.

The ADR midpoint-life guidelines have been incorporated into the ACRS classification in such a way that most ACRS property classes are again shorter than the ADR midpoint lives.

The first step in ACRS depreciation is to determine the property class of the asset being depreciated. In Table 10-3, all personal property (property except real estate) falls into one of six classes. All real estate is in one of two classes.

After the property class is determined, the next step is to read the depreciation schedule from Table 10-4 for personal property or from Table 10-5 for real property.

*The tax literature calls this version MACRS (modified ACRS) to distinguish it from the prior version. For simplicity we refer to it simply as ACRS.

Table 10-3 ACCELERATED COST RECOVERY SYSTEM CLASSES OF DEPRECIABLE PROPERTY

Property class	*Personal property (all property except real estate)*
Three-Year Property	Special handling devices for food and beverage manufacture; Special tools for the manufacture of finished plastic products, fabricated metal products, and motor vehicles; Property with ADR midpoint life of 4 years or less.
Five-Year Property	Automobiles* and trucks; Aircraft (of non-air-transport companies); Equipment used in research and experimentation; Computers; Petroleum drilling equipment; Property with ADR midpoint life of more than 4 years and less than 10 years.
Seven-Year Property	All other property not assigned to another class; Office furniture, fixtures, and equipment; Property with ADR midpoint life of 10 years or more and less than 16 years.
Ten-Year Property	Assets used in petroleum refining, manufacture of tobacco products and certain food products; Railroad cars; Property with ADR midpoint life of 16 years or more and less than 20 years.
Fifteen-Year Property	Telephone distribution plants; Municipal sewage treatment plants; Property with ADR midpoint life of 20 years or more and less than 25 years.
Twenty-Year Property	Municipal sewers; Property with ADR midpoint life of 25 years and more.

Property class	*Real property (real estate)*
27.5 Years	Residential rental property (does not include hotels and motels)
31.5 Years	Nonresidential real property

*The depreciation deduction for automobiles is limited to $2660 the first tax year and further reduced in subsequent years.

Table 10-4 ACRS DEPRECIATION FOR PERSONAL PROPERTY

If the recovery year is:	The applicable percentage for the class of property is:					
	3-year class	*5-year class*	*7-year class*	*10-year class*	*15-year class*	*20-year class*
1	33.33	20.00	14.29	10.00	5.00	3.750
2	44.45	32.00	24.49	18.00	9.50	7.219
3	14.81*	19.20	17.49	14.40	8.55	6.677
4	7.41	11.52*	12.49	11.52	7.70	6.177
5		11.52	8.93*	9.22	6.93	5.713
6		5.76	8.92	7.37	6.23	5.285
7			8.93	6.55*	5.90*	4.888
8			4.46	6.55	5.90	4.522
9				6.56	5.91	4.462*
10				6.55	5.90	4.461
11				3.28	5.91	4.462
12					5.90	4.461
13					5.91	4.462
14					5.90	4.461
15					5.91	4.462
16					2.95	4.461
17						4.462
18						4.461
19						4.462
20						4.461
21						2.231

Computation Method:

The 3-, 5-, 7-, and 10-year classes are based on double declining balance (DDB) depreciation with conversion to straight line depreciation in the year with the asterisk (*) to maximize the deduction.

The 15- and 20-year classes are based on 150% declining balance depreciation with conversion to straight line depreciation in the year with the asterisk (*) to maximize the deduction.

The half-year convention applies, with all property treated as placed in service in the middle of the year. Thus a half-year of depreciation is allowed in the first recovery year and a half-year of depreciation when the property is disposed of or retired from service, or in the last recovery year.

Table 10-5 ACRS DEPRECIATION FOR REAL PROPERTY (REAL ESTATE)

	Recovery Percentages for Residential Rental Property											
Recovery	*Month placed in service*											
year	*1*	*2*	*3*	*4*	*5*	*6*	*7*	*8*	*9*	*10*	*11*	*12*
1	3.48	3.18	2.88	2.58	2.27	1.97	1.67	1.36	1.06	0.76	0.45	0.15
2–27	3.64	3.64	3.64	3.64	3.64	3.64	3.64	3.64	3.64	3.64	3.64	3.64
28	1.88	2.18	2.48	2.78	3.09	3.39	3.64	3.64	3.64	3.64	3.64	3.64
29							0.05	0.36	0.66	0.96	1.27	1.57
	Recovery Percentages for Nonresidential Real Property											
1	3.04	2.78	2.51	2.25	1.98	1.72	1.46	1.19	0.93	0.66	0.40	0.13
2–31	3.17	3.17	3.17	3.17	3.17	3.17	3.17	3.17	3.17	3.17	3.17	3.17
32	1.86	2.12	2.39	2.65	2.92	3.17	3.17	3.17	3.17	3.17	3.17	3.17
33						0.01	0.27	0.54	0.80	1.07	1.33	1.60

Useful lives are 27.5 years for residential rental property and 31.5 years for nonresidential real property. Depreciation is straight line using the mid-month convention. Thus a property placed in service in January would be allowed 11½ months depreciation for Recovery Year 1. Similarly, if the property is disposed of prior to the end of the recovery period, then the recovery percentage must take into account the number of months in the year of disposition.

The computation and use of the ACRS percentages is illustrated in the following examples.

EXAMPLE 10-7

A microcomputer costs $8000. Compute its ACRS depreciation schedule.

Solution: A microcomputer is personal property (remember, personal property is everything except real estate). From Table 10-3, we see that computers are in the five-year property class. Table 10-4 gives the applicable percentages for each year.

Year	ACRS *percentage*		*Cost*		ACRS *depreciation*
1	20.00%	×	$8000	=	$1600.00
2	32.00	×	8000	=	2560.00
3	19.20	×	8000	=	1536.00
4	11.52	×	8000	=	921.60
5	11.52	×	8000	=	921.60
6	5.76	×	8000	=	460.80
					$8000.00 ◄

EXAMPLE 10-8

Table 10-4 shows that ACRS depreciation for personal property in the three-year property class is as follows:

Year	*Depreciation*
1	33.33% of cost
2	44.44
3	14.82
4	7.41

Make the computations to verify the percentages shown.

Solution: The three-year class is based on double declining balance with conversion to straight line depreciation. One-half-year depreciation is allowed in the first recovery year and one-half-year depreciation three years later in Recovery Year 4.

Year

1 $$\frac{1}{2}\text{ year DDB deprec.} = 0.5 \times \frac{2.00}{3}(100\%\text{ cost} - 0) = 33.33\%\text{ of cost}$$

2 $$\text{DDB deprec.} = \frac{2.00}{3}(100\%\text{ cost} - 33.33\%) = 44.44\%\text{ of cost}$$

3 Compute both DDB depreciation and SL depreciation and select the larger value.

$$\text{DDB deprec.} = \frac{2.00}{3}(100\%\text{ cost} - 77.77\%) = 14.82\%\text{ of cost}$$

$$\text{SL depreciation} = \frac{\text{Book value at beginning of year} - \text{Salvage value}}{\text{Remaining useful life at beginning of year}}$$

$$= \frac{22.23\%\text{ cost} - 0}{1.5\text{ years}} = 14.82\%\text{ of cost}$$

Both methods give the same value.

4 The remaining half year of SL depreciation is:

$$= 100\%\text{ cost} - 92.59\%\text{ cost} = 7.41\%\text{ of cost}$$

These values agree with Table 10-4. ◀

EXAMPLE 10-9

A house is purchased on October 1st to be used as a rental. The $200,000 purchase price represents $150,000 for the house and $50,000 for the land. Compute the ACRS depreciation for the first two calendar years by exact calculation and compare the results with Table 10-5.

Solution: ACRS depreciation for residential rental property is based on straight line using the mid-month convention and a 27.5-year useful life. The mid-month convention assumes, in this case, that the house begins its depreciation on October 15th giving 2½ months of depreciation the first recovery year.

Year

1 $$\text{SL depreciation} = \left(\frac{2\frac{1}{2}}{12}\right)\left(\frac{150{,}000 - 0}{27.5}\right) = \$1136$$

$$\text{Table value} = 0.76\%(150{,}000) = \$1140$$

2 $$\text{SL depreciation} = \left(\frac{150{,}000 - 0}{27.5}\right) = \$5455$$

$$\text{Table value} = 3.64\%(150{,}000) = \$5460$$

The table values, being rounded to two decimal places, are consistent with the exact computations. ◀

Depletion

Depletion is the exhaustion of natural resources as a result of their *removal*. Since depletion covers such things as mineral properties, oil and gas wells, and standing timber, removal may take the form of digging up metallic or nonmetallic minerals, producing petroleum or natural gas from wells, or cutting down trees.

The reason for depletion is essentially the same as the reason for depreciation. The owner of natural resources is consuming his capital investment as the natural resources are being removed and sold. Thus a portion of the gross income should be considered a return of the capital investment. The calculation of the depletion allowance is different from depreciation as there are two distinct methods of calculating depletion: ***cost depletion*** and ***percentage depletion***. Except for standing timber and most oil and gas wells, depletion is calculated by both methods and the larger value is taken as depletion for the year. For standing timber and most oil and gas wells, only cost depletion is permissible.

Cost Depletion

In our calculation of depreciation, we generally took cost, useful life, and salvage value and used one of several methods to apportion the cost *minus* salvage value *over* the useful life. In the less frequent case where the asset is used at fluctuating rates, we might use the unit-of-production (UOP) method of depreciation. For mines, oil wells, and standing timber, fluctuating production rates are the usual situation. Thus, *cost depletion* is computed in the same manner as unit-of-production depreciation. The elements of the calculation are:

1. Cost of the property;
2. Estimation of the number of recoverable units (tons of ore, cubic meters of gravel, barrels of oil, million cubic feet of natural gas, thousand board-feet of timber, and so forth);
3. Salvage value, if any, of the property.

EXAMPLE 10-10

A small lumber company bought a tract of timber for \$35,000, of which \$5000 was the value of the land and \$30,000 was the value of the estimated 1½ million board-feet of standing timber. The first year, the company cut 100,000 board-feet of standing timber on the tract. What was the depletion allowance for the year?

Solution:

$$\text{Depletion allowance per 1000 board-ft of timber} = \frac{\$35{,}000 - \$5000}{1{,}500{,}000 \text{ board-ft}}$$
$$= \$20 \text{ per 1000 board-ft}$$

The depletion allowance for the year would be

$$100{,}000 \text{ board-ft} \times \$20 \text{ per 1000 board-ft} = \$2000$$ 

Percentage Depletion

Percentage depletion is an alternate method of calculating the depletion allowance for mineral property and some oil or gas wells. The allowance is a certain percentage of the gross income from the property during the year. This is an entirely different concept than depreciation. Unlike depreciation, which is the allocation of cost *over* useful life, ***percentage depletion*** is an annual allowance of a percentage of the gross income from the property.

Since percentage depletion is computed on the *income* rather than the cost of the property, the total depletion on a property *may exceed the cost of the property*. In computing the ***allowable percentage depletion*** on a property in any year, there is one major limitation on the amount of percentage depletion. The ***percentage depletion allowance*** in any year is limited to not more than 50% of the taxable income from the property, computed without the deduction for depletion. The percentage depletion calculations are illustrated by Ex. 10-11.

Table 10-6 PERCENTAGE DEPLETION ALLOWANCE FOR SELECTED ITEMS

Type of deposit	*Percent*
Lead, zinc, nickel, sulphur, uranium	22
Oil and gas (small producers only)	15
Gold, silver, copper, iron ore	15
Coal and sodium chloride	10
Sand, gravel, stone, clam and oyster shells, brick and tile clay	5
Most other minerals and metallic ores	14

EXAMPLE 10-11

A coal mine has a gross income of $250,000 for the year. Mining expenses equal $210,000. Compute the allowable percentage depletion deduction.

Solution: From Table 10-6, coal has a 10% depletion allowance. The percentage depletion deduction is computed from gross mining income. Then the taxable income must be computed. The allowable percentage depletion deduction is limited to the computed percentage depletion or 50% of taxable income, whichever is smaller.

Computed percentage depletion:

Gross income from mine	$250,000
Depletion percentage	× 10%
Computed percentage depletion	$ 25,000

Taxable income limitation:

Gross income from mine	$250,000
Less: expenses other than depletion	−210,000
Taxable income from mine	40,000
Deduction limitation	× 50%
Taxable income limitation	$ 20,000

Since the taxable income limitation ($20,000) is less than the computed percentage depletion ($25,000), the allowable percentage depletion deduction is $20,000. ◀

As previously stated, on mineral property and some oil and gas wells, the depletion deduction is based on either cost or percentage depletion. Each year, depletion is computed by both methods, and the allowable depletion deduction is the larger of the two amounts.

SUMMARY

To prepare for the effects of income tax on economic analysis, it is necessary to first understand the concept of depreciation. The word *depreciation* is not readily defined, but we know it means a decrease in value, and this leads us to three distinct definitions of depreciation:

1. Decline in market value of an asset.
2. Decline in value of an asset to its owner.
3. Systematic allocation of the cost of an asset *less* its salvage value *over* its useful life.

While the first two definitions are used in everyday discussions, it is the third, or accountant's, definition that is used in tax computations. For that reason, the remainder of this chapter was based on allocation of cost *over* the useful life of an asset. Book value is the remaining unallocated cost of an asset, or:

Book value = Cost − Depreciation charges made

Six depreciation methods for the systematic allocation of cost are:

1. Straight line.
2. Sum-of-years digits.
3. Declining balance.
4. Declining balance with conversion to straight line.
5. Unit of production.
6. Accelerated cost recovery system.

Of the six methods, only the declining balance method presents any computational problems. At times, the declining balance method does not achieve the desired result of allocation of the Cost *over* the Useful life of the asset. This difficulty is overcome by converting from declining balance to straight line or sum-of-years digits depreciation part way through the depreciable life of the asset.

The different depreciation methods have an impact on the taxable income of a firm in any year. A profitable firm would almost certainly prefer to depreciate assets as rapidly as possible. In this situation, the obvious choice is accelerated cost recovery system (ACRS) depreciation. Firms that are not profitable would have little or no incentive to depreciate their assets rapidly. They might choose a less rapid depreciation method like straight line depreciation.

Depletion is the exhaustion of natural resources like minerals, oil and gas wells, and standing timber. The owner of the natural resources is consuming his investment as the natural resources are removed and sold. This concept is quite similar to unit of production depreciation and results in a computation called cost depletion. For minerals and some oil and gas wells, there is an alternate calculation called percentage depletion. Percentage depletion has the unusual characteristic that in the exhaustion of the natural resources, the total allowable depletion deductions may *exceed* the invested cost.

Problems

10-1 Some special handling devices can be obtained for $12,000. At the end of four years, they can be sold for $600. Compute the depreciation schedule for the devices based on these values, by the following four methods:

a. Straight line depreciation.

b. Sum-of-years digits depreciation.

c. Double declining balance depreciation.

d. DDB depreciation with conversion to straight line depreciation.

10-2 The company treasurer is uncertain which of four depreciation methods is more desirable for the firm to use for office furniture. Its cost is $50,000, with a zero salvage value at the end of a ten-year depreciable life. Compute the depreciation schedule for the office furniture by each of the four depreciation methods listed:

a. Straight line.

b. Double declining balance.

c. Sum-of-years digits.

d. Accelerated cost recovery system.

10-3 The RX Drug Company has just purchased a capsulating machine for $76,000. The plant engineer estimates the machine has a useful life of five years and little or no salvage value. He will use zero salvage value in the computations.

a. Compute the depreciation schedule for the machine by each of the following depreciation methods:
   1. Straight line depreciation.
   2. Sum-of-years digits depreciation.
   3. Double declining balance depreciation.

b. The controller for RX Drug Co. believes that DDB with conversion to straight line depreciation may be a more desirable depreciation method. Compute the depreciation schedule for DDB depreciation with conversion to straight line depreciation at the desirable point.

10-4 A new machine tool is being purchased for $16,000 and is expected to have a zero salvage value at the end of its five-year useful life. Compute the DDB depreciation schedule for this capital asset. It may be desirable for this profitable firm to use double declining balance depreciation with conversion to straight line depreciation. If it is desirable, compute the depreciation schedule, making the conversion from DDB depreciation to straight line depreciation at the optimum time. Tabulate the resulting depreciation schedule.

10-5 A large profitable corporation purchased a small jet plane for use by the firm's executives in January. The plane cost \$1,500,000 and, for depreciation purposes, is assumed to have a zero salvage value at the end of five years.

a. Compute the depreciation schedule for DDB with conversion to sum-of-years digits depreciation at the optimum time.

b. Compute the ACRS depreciation schedule.

10-6 When a major highway was to be constructed nearby, a farmer realized that a dry streambed running through his property might be valuable as a source of sand and gravel. He shipped samples of the sand and gravel to a testing laboratory and learned that the material met the requirements for certain low-grade fill material for the highway. The farmer contacted the highway construction contractor, who offered to pay 65¢ per cubic meter for 45,000 cubic meters of sand and gravel to be scooped out of the streambed. The contractor would build a haulage road to transport the sand and gravel, and would use his own equipment to load and haul the material. All the activity would take place during a single summer.

The farmer hired an engineering student for the summer to watch the sand- and gravel-loading operation and to count the truckloads of material hauled away. For this, the student received \$2500. When the summer was over, the farmer calculated the results of his venture. He estimated that two acres of streambed had been stripped of the sand and gravel. The 640-acre farm had cost him \$300 per acre and the farmer felt the property had not changed in value. He knew that there had been no use for the sand and gravel prior to the construction of the highway, and he could foresee no future use for any of the remaining 50,000 cubic meters of sand and gravel. Determine the farmer's depletion allowance. (*Answer:* \$1462.50)

10-7 Mr. H. Salt purchased a ⅛ interest in a producing oil well for \$45,000. Recoverable oil reserves for the well were estimated at that time at 15,000 barrels, ⅛ of which represented Mr. Salt's share of the reserves. During the subsequent year, Mr. Salt received \$12,000 as his ⅛ share of the gross income from the sale of 1000 barrels of oil. From this amount, he had to pay \$3000 as his share of the expense of producing the oil. Compute Mr. Salt's depletion allowance for the year. (*Answer:* \$3000)

10-8 A heavy construction firm has been awarded a contract to build a large concrete dam. It is expected that a total of eight years will be required to complete the work. The firm will buy \$600,000 worth of special equipment for the job. During the preparation of the job cost estimate, the following utilization schedule was computed for the special equipment:

Year	*Utilization (hours/year)*	*Year*	*Utilization (hours/year)*
1	6000	5	800
2	4000	6	800
3	4000	7	2200
4	1600	8	2200

At the end of the job, it is estimated that the equipment can be sold at auction for \$60,000.

a. Compute the SOYD depreciation schedule for this equipment.

b. Compute the unit-of-production depreciation schedule.

10-9 Consider five depreciation schedules:

Year	*A*	*B*	*C*	*D*	*E*
1	$58.00	$35.00	$29.00	$58.00	$43.50
2	34.80	20.00	46.40	34.80	30.45
3	20.88	30.00	27.84	20.88	21.32
4	12.53	30.00	16.70	12.53	14.92
5	8.79	20.00	16.70	7.52	10.44
6			8.36		

They are based on the same initial cost, useful life, and salvage value. Identify each schedule as one of the following

- Straight line depreciation
- Sum-of-years digits depreciation;
- 150% declining balance depreciation
- Double declining balance depreciation
- DDB with conversion to straight line depreciation
- Unit of production depreciation
- Accelerated cost recovery system

10-10 Al Larson asked a bank to lend him money on January 1st, based on the following repayment plan: the first payment would be $2 on February 28th, with subsequent monthly payments increasing by $2 a month on an arithmetic gradient. (The March 31st payment would be $4; the April 30th payment would be $6, and so on.) The payments are to continue for eleven years, making a total of 132 payments.

a. Compute the total amount of money Al will pay the bank, based on the proposed repayment plan.

b. If the bank charges interest at 12% per year, compounded monthly, how much would it be willing to lend Al on the proposed repayment plan?

10-11 Consider a $6500 piece of machinery, with a five-year depreciable life and an estimated $1200 salvage value. The projected utilization of the machinery when it was purchased, and its actual production to date, are shown below.

Year	*Projected production, in tons*	*Actual production, in tons*
1	3500	3000
2	4000	5000
3	4500	[*Not*
4	5000	*yet*
5	5500	*known*]

Compute the machinery depreciation schedule by each of the following methods:

a. Straight line.

b. Sum-of-years digits.

c. Double declining balance.

d. Unit of production (for first two years only).

e. Accelerated cost recovery system.

10-12 A depreciable asset costs $10,000 and has an estimated salvage value of $1600 at the end of its six-year depreciable life. Compute the depreciation schedule for this asset by both SOYD depreciation and DDB depreciation.

10-13 The ACRS depreciation percentages for five-year personal property are given in Table 10-4. Make the necessary computations to determine if the percentages shown are correct.

10-14 The Acme Chemical Company purchased $45,000 of research equipment which it believes will have zero salvage value at the end of its five-year life. Compute the depreciation schedule for the equipment by each of the following methods:

a. Straight line.

b. Sum-of-years digits.

c. Double declining balance.

d. DDB depreciation with conversion to sum-of-years digits.

e. Accelerated cost recovery system.

10-15 The ACRS depreciation percentages for ten-year personal property are given in Table 10-4. Make the necessary computations to determine if the percentages shown are correct.

10-16 A $1,000,000 oil drilling rig has a six-year depreciable life and a $75,000 salvage value at the end of that time. Determine which one of the following methods: DDB, DDB with conversion to straight line, or SOYD provides the preferred depreciation schedule. Show the depreciation schedule for the preferred method.

10-17 Some equipment costs $1000, has a five-year depreciable life, and an estimated $50 salvage value at the end of that time. Ann Landers has been assigned the problem to determine whether to use:

a. DDB depreciation with conversion to straight line, or

b. SOYD depreciation.

If a 10% interest rate is appropriate, which is the preferred depreciation method for this profitable corporation? Show your computations.

10-18 The depreciation schedule for an asset, with a salvage value of $90 at the end of the recovery period, has been computed by several methods. Identify the depreciation method used for each schedule.

Depreciation Schedule

Year	*A*	*B*	*C*	*D*	*E*
1	$318.0	$212.0	$424.0	$194.0	$107.0
2	222.6	339.2	254.4	194.0	216.0
3	155.8	203.5	152.6	194.0	324.0
4	136.8	122.1	91.6	194.0	216.0
5	136.8	122.1	47.4	194.0	107.0
6		61.1			
	970.0	1060.0	970.0	970.0	970.0

10-19 TELCO Corp. has leased some industrial land near its plant. It is building a small warehouse on the site at a cost of $250,000. The building will be ready for use January 1st. The lease will expire fifteen years after the building is occupied. The warehouse will belong at that time to the landowner, with the result that there will be no salvage value to TELCO. The warehouse is to be depreciated either by ACRS or SOYD depreciation. If 10% interest is appropriate, which depreciation method should be selected?

10-20 The FOURX Corp. has purchased $12,000 of experimental equipment. The anticipated salvage value is $400 at the end of its five-year depreciable life. This profitable corporation is considering three methods of depreciation:

1. Sum-of-years digits;
2. Double declining balance;
3. DDB with conversion to straight line.

If it uses 7% interest in its comparison, which method do you recommend? Show computations to support your recommendation.

10-21 The Able Corp. is buying $10,000 of special tools for its fabricated metal products that have a four-year useful life and no salvage value. Compute the depreciation charge for the *second* year by each of the following methods:

a. DDB with conversion to straight line.
b. Sum-of-years digits.
c. Accelerated cost recovery system.

10-22 On July 1st, Nancy Regan paid $600,000 for a commercial building and an additional $150,000 for the land on which it stands. Four years later, also on July 1st, she sold the property for $850,000. Compute the accelerated cost recovery system depreciation for each of the *five* calendar years during which she had the property.

10-23 The White Swan Talc Company purchased $120,000 of mining equipment for a small talc mine. The mining engineer's report indicates the mine contains 40,000 cubic meters of commercial quality talc. The company plans to mine all the talc in the next five years as follows:

Year	*Talc production, in cubic meters*
1	15,000
2	11,000
3	4,000
4	6,000
5	4,000

At the end of five years, the mine will be exhausted and the mining equipment will be worthless. The company accountant must now decide whether to use sum-of-years digits depreciation or unit of production depreciation. The company considers 8% to be an appropriate time value of money. Compute the depreciation schedule for each of the two methods. Which method would you recommend that the company adopt? Show the computations to justify your decision.

10-24 A used small jet aircraft is purchased for $2,000,000. The salvage value is estimated to be $250,000 at the end of a five-year useful life. The plane is to be depreciated by 150% declining balance depreciation with conversion to straight line depreciation. Compute the depreciation for each of the five years.

10-25 The depreciation schedule for a microcomputer has been arrived at by several methods. The estimated salvage value of the equipment at the end of its six-year useful life is $600. Identify the resulting depreciation schedules.

	Depreciation Schedule			
Year	*A*	*B*	*C*	*D*
1	$2114	$2000	$1600	$2667
2	1762	1500	2560	1778
3	1410	1125	1536	1185
4	1057	844	922	790
5	705	633	922	527
6	352	475	460	453

10-26 Refer to Table 10-2. The table compares three methods: SOYD, DDB, and DDB with conversion to straight line depreciation.

a. No mention is made in Table 10-2 of simple Straight Line depreciation. If it were added to the list of methods compared, would the table change? If so, give a specific example.

b. Table 10-2 also makes no mention of ACRS depreciation. If it were added to the list of methods compared, would the same depreciable life (N) be applicable to both ACRS depreciation and the other depreciation methods?

10-27 A group of investors has formed Trump Corporation to purchase a small hotel. The asking price is $150,000 for the land and $850,000 for the hotel building. If the purchase takes place in June, compute the ACRS depreciation for the first three calendar years. Then assume the hotel is sold in June of the fourth year, and compute the ACRS depreciation in that year also.

10-28 Refer to Table 10-1. The point to convert from Double Declining Balance to Straight Line depreciation depends on the estimated salvage value. At some point no conversion is needed. How do you explain this?

10-29 Mr. Donald Spade purchased a computer in January to keep records on all the property he owns. The computer cost $70,000 and is to be depreciated using ACRS. Donald's accountant pointed out that under a special tax rule (the rule applies when the value of property placed in service in the last three months of the tax year exceeds 40% of all the property placed in service during the tax year), the computer and all property that year would be subject to the mid-quarter convention. The mid-quarter convention assumes that all property placed in service in any quarter year is placed in service at the midpoint of the quarter. Compute Donald's ACRS depreciation for the first year, using the mid-quarter convention.

10-30 (20-56) A company is considering buying a new piece of machinery. A 10% interest rate will be used in the computations. Two models of the machine are available.

	Machine I	*Machine II*
Initial cost	$80,000	$100,000
End of useful life salvage value, *S*	20,000	25,000
Annual operating cost	18,000	15,000 first 10 years 20,000 thereafter
Useful life	20 years	25 years

a. Determine which machine should be purchased, based on equivalent uniform annual cost.

b. What is the capitalized cost of Machine I?

c. If Machine I is purchased and a fund is set up to replace Machine I at the end of 20 years, compute the required uniform annual deposit.

d. Machine I will produce an annual saving of material of $28,000. What is the rate of return if Machine I is installed?

e. What will be the book value of Machine I after two years, based on sum-of-years digits depreciation?

f. What will be the book value of Machine II after three years, based on double declining balance depreciation?

g. Assuming Machine II is in the seven-year property class, what would be the ACRS depreciation in the third year?

CHAPTER 11

Income Taxes

As Benjamin Franklin said, two things are inevitable: death and taxes. In this chapter we will examine the structure of taxes in the United States. There is, of course, a wide variety of taxes ranging from sales taxes, gasoline taxes, property taxes, state and federal income taxes, and so forth. Here we will concentrate our attention on federal income taxes. Since income taxes are part of all real problems and have a substantial impact on many of them, no realistic analysis can ignore these income tax consequences.

First, we must understand the way in which taxes are imposed. The previous chapter concerning depreciation is an integral part of this analysis, so it is essential that the principles covered there are well understood. Then, having understood the mechanism, we will see how federal income taxes affect our economic analysis. The various analysis techniques will be used in examples of after-tax calculations.

A Partner In The Business

Probably the most straightforward way to understand the role of federal income taxes is to consider the U.S. Government as a partner in every business activity. As a partner, the Government shares in the profits from every successful venture. And in a somewhat more complex way, the Government shares in the losses of unprofitable ventures. The tax laws are complex and it is not our purpose to fully explain them.* Instead, we will examine the fundamental concepts of the federal income tax laws—and we must recognize that there are exceptions and variations to almost every statement we shall make!

*Both Prentice-Hall and Commerce Clearing House have loose-leaf income tax reporting services that fill many binders each year with detailed tax information. One or both will be found in the reference section of most libraries.

Calculation Of Taxable Income

At the mention of income taxes, one can visualize dozens of elaborate and complex calculations. And there is some truth to that vision, for there can be all sorts of complexities in the computation of income taxes. Yet some of the difficulty is removed when one defines incomes taxes as just another type of disbursement. Our economic analysis calculations in prior chapters have dealt with all sorts of disbursements: operating costs, maintenance, labor and materials, and so forth. Now we simply add one more prospective disbursement to the list—income taxes.

Taxable Income of Individuals

The amount of federal income taxes to be paid depends on taxable income and the income tax rates. Therefore, our first concern is the definition of ***taxable income***. To begin, one must compute his or her ***gross income***:

$$
\begin{aligned}
\text{Gross income} = \text{Wages, salary, etc.} &+ \text{Interest income}\\
&+ \text{Dividends}\\
&+ \text{Capital gains}\\
&+ \text{Unemployment compensation}\\
&+ \text{Other income}
\end{aligned}
$$

From gross income, we subtract any allowable retirement plan contributions and other ***adjustments***. The result is ***adjusted gross income***. From adjusted gross income, individuals may deduct two of the following items:

1. ***Personal Exemptions.*** One exemption ($2050 for 1990) is provided for each person who depends on the gross income for his or her living.
2. ***Itemized Deductions.*** Some of these are:
   a. Excessive medical and dental expenses (exceeding 7½% of adjusted gross income);
   b. State and local income, property and personal property tax;
   c. Home mortgage interest;
   d. Charitable contributions;
   e. Casualty and theft losses;
   f. Miscellaneous deductions (exceeding 2½% of adjusted gross income).
3. ***Standard Deduction.*** Each taxpayer may either itemize his or her deductions, or instead take a standard deduction as follows:
   - Single taxpayers, $3250;
   - Married taxpayers filing a joint return, $5450;

The result is ***taxable income***.

For individuals:

Adjusted gross income = Gross income − Adjustments

***Taxable income* = Adjusted gross income** (11-1)

− Personal exemption(s)

− Itemized deductions or Standard deduction

Classification of Business Expenditures

When an individual or a firm operates a business, there are three distinct types of business expenditures:

1. Expenditures for depreciable assets;
2. Expenditures for nondepreciable assets;
3. All other business expenditures.

Expenditures for depreciable assets. When facilities or productive equipment with useful lives in excess of one year are acquired, the taxpayer will recover his investment through depreciation charges.* Chapter 10 examined in great detail the several ways in which the cost of the asset could be allocated over its useful life.

Expenditures for nondepreciable assets. Land is considered a nondepreciable asset, for there is no finite life associated with it. Other nondepreciable assets are properties *not* used either in a trade, in a business, or for the production of income. An individual's home and automobile are generally nondepreciable assets. The final category of nondepreciable assets are those subject to *depletion*, rather than *depreciation*. Since business firms generally acquire assets for use in the business, their only nondepreciable assets normally are land and assets subject to depletion.

All other business expenditures. This category is probably the largest of all for it includes all the ordinary and necessary expenditures of operating a business. Labor costs, materials, all direct and indirect costs, facilities and productive equipment with a useful life of one year or less, and so forth, are part of the routine expenditures. They are charged as a business expense—*expensed*—when they occur.

Business expenditures in the first two categories—that is, for either depreciable or nondepreciable assets—are called ***capital expenditures***. In the accounting

*There is an exception. Businesses may immediately deduct (*expense*) up to $10,000 of business equipment in a year, provided their total equipment expenditure for the year does not exceed $200,000.

records of the firm, they are ***capitalized***; all ordinary and necessary expenditures in the third category are ***expensed***.

Taxable Income of Business Firms

The starting point in computing a firm's taxable income is *gross income*. All ordinary and necessary expenses to conduct the business—*except* capital expenditures—are deducted from gross income. Capital expenditures may *not* be deducted from gross income. Except for land, business capital expenditures are recovered through depreciation or depletion charges.

For business firms:

Taxable income = **Gross income** (11-2)

− **All expenditures except capital expenditures**

− **Depreciation and depletion charges**

Because of the treatment of capital expenditures for tax purposes, the taxable income of a firm may be quite different from the actual cash results.

EXAMPLE 11-1

During a three-year period a firm had the following results (in millions of dollars):

	Year 1	*Year* 2	*Year* 3
Gross income from sales	$200	$200	$200
Purchase of special tooling (useful life: 3 years)	−60	0	0
All other expenditures	−140	−140	−140
Cash results for the year	$ 0	$ 60	$ 60

Compute the taxable income for each of the three years.

Solution: The cash results for each year would suggest that Year 1 was a poor one, while Years 2 and 3 were very profitable. A closer look reveals that the firm's cash results were adversely affected in Year 1 by the purchase of special tooling. Since the special tooling has a three-year useful life, it is a capital expenditure with its cost allocated over the useful life. For straight line depreciation and no salvage value, the annual charge is:

$$\text{Annual depreciation charge} = \frac{P - S}{N} = \frac{60 - 0}{3} = \$20 \text{ million}$$

Applying Eq. 11-2,

$$\text{Taxable income} = 200 - 140 - 20 = \$40 \text{ million}$$

In each of the three years, the taxable income is $40 million. ◀

An examination of the cash results and the taxable income in Ex. 11-1 indicates that taxable income is a better indicator of the annual performance of the firm.

Income Tax Rates

Figure 11-1 illustrates that income tax rates for individuals have changed many times since 1960. The Tax Reform Act of 1986 and the 1987 technical corrections produce the lowest maximum tax rates in 55 years.

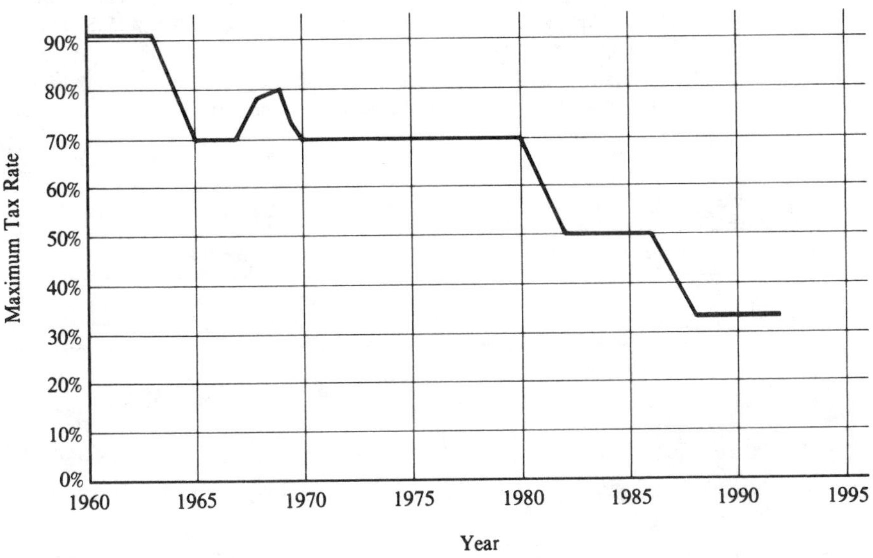

Figure 11-1 Maximum federal income tax rates for individuals.

Individual Tax Rates

There are three schedules of federal income tax rates for individuals. Single taxpayers use the Table 11-1 schedule. Married taxpayers filing a joint return use the Table 11-2 schedule. A third schedule (not shown here) is applicable to unmarried individuals with dependent relatives ("head of household").

EXAMPLE 11-2

An unmarried student earned $3500 in the summer plus another $1950 during the rest of the year. When he files an income tax return, he will be allowed one exemption (for himself). He estimates he spent $600 on allowable itemized deductions. How much income tax will he pay?

Table 11-1 1990 TAX RATES—IF YOU ARE UNMARRIED

If your taxable income is		*Your tax is*		
Over	*But not over*	*This*	*Plus following percentage*	*Over this*
$ 0	$ 19,450	$ 0	15%	$ 0
19,450	47,050	2,918	28%	19,450
47,050	97,620	10,646	33%*	47,050
97,620	109,100	27,334	33%†	97,620
Over $109,100		31,122	28%	109,100

*At $47,050 a 5% surtax is added, in effect, to phase out the benefits of the 15% tax bracket which are (28% − 15%)(19,450) = $2528. The 5% surtax is 5%(97,620 − 47,050) = $2528. Thus at $97,620, the tax is 28% of total taxable income, or $27,334.

†At $97,620 the 15% tax bracket benefits have been phased out and a second 5% surtax begins. This surtax is designed to diminish the benefits of the personal exemption. This maximum surtax is $574 for one personal exemption.

Table 11-2 1990 TAX RATES—IF TWO PEOPLE FILE A JOINT RETURN

If your taxable income is		*Your tax is*		
Over	*But not over*	*This*	*Plus following percentage*	*Over this*
$ 0	$ 32,450	$ 0	15%	$ 0
32,450	78,400	4,868	28%	32,450
78,400	162,770	17,734	33%*	78,400
162,770	185,730	45,576	33%†	162,770
Over 185,730		53,153	28%	185,730

*Above $78,400 a 5% surtax is added to phase out the benefits of the 15% tax bracket which are (28% − 15%)(32,450) = $4218. This is achieved by the surtax of 5%(162,770 − 78,400) = $4218. At $162,770 the tax is 28% of total taxable income, or $45,576.

†At $162,770 another 5% surtax begins to diminish the benefits of the personal exemption. This maximum surtax is $574 per personal exemption. For two people this is $1148, which is 5%(185,730 − 162,770). For more than two personal exemptions, the 5% surtax continues for an additional $11,480 of taxable income for each additional personal exemption.

Solution to Ex. 11-2:

$$\text{Adjusted gross income} = \$3500 + 1950 = \$5450$$

$$\begin{aligned}\text{Taxable income} &= \text{Adjusted gross income} \\ &\quad - \text{Deduction for one exemption (\$2050)} \\ &\quad - \text{Standard deduction (\$3250)} \\ &= 5450 - 2050 - 3250 = \$150\end{aligned}$$

$$\text{Federal income tax} = 15\%(150) = \$22.50 \blacktriangleleft$$

Corporate Tax Rates

Income tax for corporations is computed in a manner similar to that for individuals. The schedule is given below and is shown graphically in Fig. 11-2.

Taxable Income	*Tax Rate*	*Corporate Income Tax*
Not over $50,000	15%	15% over $0
$50,000–75,000	25%	$7,500 + 25% over $50,000
$75,000–100,00	34%	$13,750 + 34% over $75,000
$100,000–335,000	39%*	$22,250 + 39% over $100,000
Over $335,000	34%	34% of total taxable income

*Corporations with a taxable income of more than $100,000 pay an additional 5% tax on income from $100,000 to $335,000.

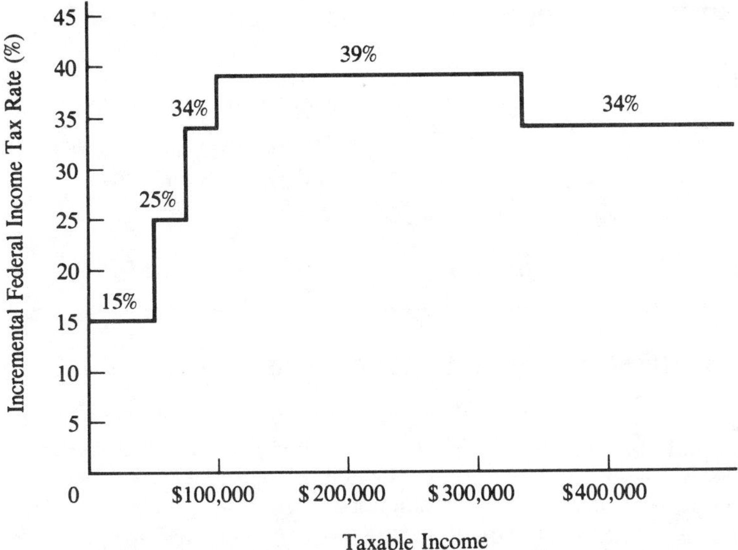

Figure 11-2 Corporation federal income tax rates (1990 rates).

EXAMPLE 11-3

The French Chemical Corporation was formed to produce household bleach. The firm bought land for $220,000, had a $900,000 factory building erected, and installed $650,000 worth of chemical and packaging equipment. The plant was completed and operations begun on April 1st. The gross income for the calendar year was $450,000. Supplies and all operating expenses, excluding the capital expenditures, were $100,000. The firm will use accelerated cost recovery system (ACRS) depreciation.

a. What is the first year depreciation charge?

b. What is the first year taxable income?

c. How much will the corporation pay in federal income taxes for the year?

Solution:

a. *ACRS depreciation:* Chemical equipment is personal property. From Table 10-3 it is probably in the "Seven-year, all other property" class.

$$\text{First-year depreciation} = \$650{,}000 \times 14.28\% = \$92{,}820$$

The building is in the 31.5-year real property class. Being placed in service April 1st, the appropriate

$$\text{First-year depreciation} = \$900{,}000 \times 2.25\% = \$20{,}250$$

The land is a nondepreciable asset.

$$\text{Total first-year ACRS depreciation} = \$92{,}820 + 20{,}250 = \$113{,}070$$

b.
$$\begin{aligned}\text{Taxable income} &= \text{Gross income} \\ &\quad - \text{All expenditures except capital expenditures} \\ &\quad - \text{Depreciation and depletion charges} \\ &= \$450{,}000 - 100{,}000 - 113{,}070 = \$236{,}930\end{aligned}$$

c.
$$\begin{aligned}\text{Federal income tax} &= \$22{,}250 + 39\%(236{,}930 - 100{,}000) \\ &= \$75{,}653 \blacktriangleleft\end{aligned}$$

Combined Federal and State Income Taxes

In addition to federal income taxes, most individuals and corporations also pay state income taxes. It would be convenient if we could derive a single tax rate to represent both the state and federal incremental tax rates. In the computation of taxable income for federal taxes, the amount of state taxes paid is one of the allowable itemized deductions. Federal income taxes are not, however, generally deductible in the computation of state taxable income. Therefore, the state income tax is applied to a *larger* taxable income than is the federal income tax rate. As a result, the combined incremental tax rate will not be the sum of two tax rates.

For an increment of income (ΔIncome),

$$\begin{aligned}\text{State income taxes} &= (\Delta\text{State tax rate})(\Delta\text{Income}) \\ \text{Federal taxable income} &= (\Delta\text{Income})(1 - \Delta\text{State tax rate}) \\ \text{Federal income taxes} &= (\Delta\text{Federal tax rate})(\Delta\text{Income}) \\ &\quad \times (1 - \Delta\text{State tax rate})\end{aligned}$$

The total of state and federal income taxes is

$$[\Delta\text{State tax rate} + (\Delta\text{Federal tax rate})(1 - \Delta\text{State tax rate})](\Delta\text{Income})$$

The terms within the left set of brackets equals the combined incremental tax rate.

Combined incremental tax rate (11-3)

$$= \Delta\text{State tax rate} + (\Delta\text{Federal tax rate})(1 - \Delta\text{State tax rate})$$

EXAMPLE 11-4

An engineer has an income that puts him in the 28% federal income tax bracket and at the 10% state incremental tax rate. He has an opportunity to earn an extra $500 by doing a small consulting job. What will be his combined state and federal income tax rate on the additional income?

Solution: Using Equation 11-3,

$$\text{Combined incremental tax rate} = 0.10 + 0.28(1 - 0.10) = 35.2\%$$ ◀

Selecting an Income Tax Rate for Economy Studies

Since income tax rates vary with the level of taxable income for both individuals and corporations, one must decide which tax rate to use in a particular situation. The simple answer is the tax rate to use is the incremental tax rate that applies to the change in taxable income projected in the economic analysis. If a married man has a taxable income of $27,700 and can increase his income by $2000, what tax rate should be used for the $2000 of incremental income? From Table 11-2, we see the $2000 falls within the 15% tax bracket.

Now suppose this individual could increase his $27,700 income by $6000. In this situation, Table 11-2 shows that the 15% incremental tax rate should be applied to the first $4750 and a 28% incremental tax rate to the last $1250 of extra income. The appropriate incremental tax rate for corporations is equally easy to determine. For larger corporations, the federal incremental tax rate is 34%. In addition, there may be up to a 12% state tax. For computational convenience, a 50% corporate tax rate is often used.

Economic Analysis Taking Income Taxes Into Account

An important step in economic analysis has been to resolve the consequences of alternatives into a cash flow. Because income taxes have been ignored, the result has been a *before-tax cash flow*. This same before-tax cash flow is an essential component in economic analysis that also considers the consequences of income tax. The principal elements in an *after-tax analysis* are:

- Before-tax cash flow;
- Depreciation;
- Taxable income (Before-tax cash flow − Depreciation);
- Income taxes (Taxable income × Incremental tax rate);
- After-tax cash flow (Before-tax cash flow − Income taxes).

These elements are usually arranged to form a *cash-flow table*. This is illustrated by Ex. 11-5.

EXAMPLE 11-5

A medium-sized profitable corporation is considering the purchase of a \$3000 used pickup truck for use by the shipping and receiving department. During the truck's five-year useful life, it is estimated the firm will save \$800 per year after paying all the costs of owning and operating the truck. Truck salvage value is estimated at \$750.

a. What is the before-tax rate of return?

b. What is the after-tax rate of return on this capital expenditure? Assume straight line depreciation.

Solution to Example 11-5a: For a before-tax rate of return, we must first compute the before-tax cash flow.

Year	*Before-tax cash flow*
0	−\$3000
1	+800
2	+800
3	+800
4	+800
5	{ +800 +750

Solve for the rate of return.

$$3000 = 800(P/A,i,5) + 750(P/F,i,5)$$

Try $i = 15\%$:

$$3000 \stackrel{?}{=} 800(3.352) + 750(0.4972) \stackrel{?}{=} 2682 + 373 = 3055$$

i is slightly low. Try $i = 18\%$:

$$3000 \stackrel{?}{=} 800(3.127) + 750(0.4371) \stackrel{?}{=} 2502 + 328 = 2830$$

$$i = 15\% + 3\%\left(\frac{3055 - 3000}{3055 - 2830}\right) = 15\% + 3\%(0.15) = 15.7\%$$

Solution to Example 11-5b: For an after-tax rate of return, we must set up the cash flow table (Table 11-3). The starting point is the before-tax cash flow. Then we will need the depreciation schedule for the truck:

$$\text{Straight line depreciation} = \frac{P - S}{N} = \frac{3000 - 750}{5} = \$450 \text{ per year}$$

Taxable income is the before-tax cash flow *minus* depreciation. For this medium-sized profitable corporation, the incremental federal income tax rate probably is 34%. Therefore income taxes are 34% of taxable income. Finally, the after-tax cash flow equals the before-tax cash flow *minus* income taxes. These data are used to compute Table 11-3.

Table 11-3 CASH FLOW TABLE FOR EXAMPLE 11-5

Year	*Before-tax cash flow*	*Straight line depreciation*	*Δ Taxable income* (a) − (b)	*34% Income taxes* −0.34(c)*	*After-tax cash flow* (a) + (d)†
	(a)	(b)	(c)	(d)	(e)
0	−$3000				−$3000
1	800	$450	$350	−$119	681
2	800	450	350	−119	681
3	800	450	350	−119	681
4	800	450	350	−119	681
5	{800 750	450	350	−119	{681 750

*Sign convention for income taxes: a minus (−) represents a disbursement of money to pay income taxes; a plus (+) represents the receipt of money by a decrease in the tax liability.

†The after-tax cash flow is the before-tax cash flow *minus* income taxes. Based on the income tax sign convention, this is accomplished by *adding* Columns **(a)** and **(d)**.

The after-tax cash flow may be solved to find the after-tax rate of return. Try $i = 10\%$:

$$3000 \stackrel{?}{=} 681(P/A,10\%,5) + 750(P/F,10\%,5)$$
$$\stackrel{?}{=} 681(3.791) + 750(0.6209) = 3047$$

i is slightly low. Try $i = 12\%$:

$$3000 \stackrel{?}{=} 681(3.605) + 750(0.5674) = 2881$$

$$i = 10\% + 2\%\left(\frac{3047 - 3000}{3047 - 2881}\right) = 10.6\% \quad ◀$$

The calculations required to compute the after-tax rate of return in Ex. 11-5 were certainly more elaborate than those for the before-tax rate of return. It must be emphasized, however, that *only* the after-tax rate of return is a meaningful value since income taxes are a major disbursement that cannot be ignored.

EXAMPLE 11-6

An analysis of a firm's sales activities indicates that a number of profitable sales are lost each year because the firm cannot deliver some of its products quickly enough. By investing an additional $20,000 in inventory it is believed that the before-tax profit of the firm will be $1000 higher the first year. The second year before-tax extra profit will be $1500. Subsequent years are expected to continue to increase on a $500 per year gradient. The investment in the additional inventory may be recovered at the end of a four-year analysis period simply by selling it and not replenishing the inventory. Compute:

a. The before-tax rate of return;

b. The after-tax rate of return assuming an incremental tax rate of 39%.

Solution: Inventory is not considered a depreciable asset, therefore, the investment in additional inventory is not depreciated. The cash flow table for the problem is presented in Table 11-4.

Table 11-4 CASH FLOW TABLE FOR EXAMPLE 11-6

Year	*Before-tax cash flow*	*Depreciation*	*Δ Taxable income* **(a) − (b)**	*39% Income taxes* **−0.39(c)**	*After-tax cash flow* **(a) + (d)**
	(a)	**(b)**	**(c)**	**(d)**	**(e)**
0	−$20,000				−$20,000
1	1,000	—	$1000	−$390	610
2	1,500	—	1500	−585	915
3	2,000	—	2000	−780	1,220
4	{2,500 20,000	—	2500	−975	{1,525 20,000

Solution to Example 11-6a: *Before-tax rate of return:*

$$20{,}000 = 1000(P/A,i,4) + 500(P/G,i,4) + 20{,}000(P/F,i,4)$$

Try $i = 8\%$:

$$20{,}000 \stackrel{?}{=} 1000(3.312) + 500(4.650) + 20{,}000(0.7350)$$

$$\stackrel{?}{=} 3312 + 2325 + 14{,}700 = 20{,}337$$

i is too low. Try $i = 10\%$:

$$20{,}000 \stackrel{?}{=} 1000(3.170) + 500(4.378) + 20{,}000(0.6830)$$
$$\stackrel{?}{=} 3170 + 2189 + 13{,}660 = 19{,}019$$

$$\text{Before-tax rate of return} = 8\% + 2\%\left(\frac{20{,}337 - 20{,}000}{20{,}337 - 19{,}019}\right) = 8.5\%$$

Solution to Example 11-6b: *After-tax rate of return:* The before-tax cash flow gradient is \$500. The resulting after-tax cash flow gradient is (1 − 0.39)(500) = \$305.

$$20{,}000 = 610(P/A,i,4) + 305(P/G,i,4) + 20{,}000(P/F,i,4)$$

Try $i = 5\%$:

$$20{,}000 \stackrel{?}{=} 610(3.546) + 305(5.103) + 20{,}000(0.8227) \stackrel{?}{=} 20{,}173$$

i is too low. Try $i = 6\%$:

$$20{,}000 \stackrel{?}{=} 610(3.465) + 300(4.4945) + 20{,}000(0.7921) \stackrel{?}{=} 19{,}304$$

$$\text{After-tax rate of return} = 5\% + 1\%\left(\frac{20{,}173 - 20{,}000}{20{,}173 - 19{,}304}\right) = 5.2\%$$ ◀

Effects Of Tax Reform Act Of 1986

Capital Gains and Losses

When a capital asset is sold or exchanged, there must be entries in the firm's accounting records to reflect the change. If the selling price of the capital asset exceeds the book value, the excess of selling price over book value is called a ***capital gain***. If the selling price is less than book value, the difference is a ***capital loss***.

$$\textit{Capital}\begin{Bmatrix}\textit{gain}\\ \textit{loss}\end{Bmatrix} = \textbf{Selling price} - \textbf{Book value}$$

There have been quite elaborate rules in the past for the tax treatment of capital gains and losses. For example, capital assets held for more than six months produced *long-term* gains or losses. Assets held less than six months produced *short-term* gains and losses.

The Tax Reform Act of 1986 eliminated the special tax treatment for long-term capital gains. Now all capital gains are simply included as another component of gross income. Nevertheless, the statutory structure for capital gains has been retained in the internal revenue code. Thus, it would be simple for Congress to reinstitute a different capital gains tax rate at some future time. The tax treatment of capital gains and losses is shown in Table 11-5.

Table 11-5 TAX TREATMENT OF CAPITAL GAINS AND LOSSES

For individuals:		
	Capital gain	Taxed as ordinary income
	Capital loss	Subtract capital losses from any capital gains; balance may be deducted from ordinary income, but not more than $3000 per year
For corporations:		
	Capital gain	Taxed as ordinary income
	Capital loss	Corporations may deduct capital losses only to the extent of capital gains. Any capital loss in the current year that exceeds capital gains is carried back three years, and, if not completely absorbed, is then carried forward for up to seven years

Investment Tax Credit

When the economy slows down and unemployment rises, the U.S. Government frequently alters its tax laws to promote greater industrial activity. One technique used to stimulate capital investments has been the ***investment tax credit***. Businesses were able to deduct from 4% to 8% of their new business equipment purchases as a *tax credit*. This meant the net cost of the equipment to the firm was reduced by the amount of the investment tax credit, yet at the same time the basis for computing depreciation remained the full cost of the equipment. The Tax Reform Act of 1986 eliminated the investment tax credit. It is likely, however, that it will reappear at some future time.

Estimating The After-Tax Rate Of Return

There is no shortcut method to compute the after-tax rate of return from the before-tax rate of return. One possible exception to this statement is in the situation of nondepreciable assets. In this special case:

After-tax rate of return **= (1 − Incremental tax rate)**
× (Before-tax rate of return)

For Ex. 11-6 we could estimate the after-tax rate of return from the before-tax rate of return as follows:

$$\text{After-tax rate of return} = (1 - 0.39)(8.5\%) = 5.2\%$$

which agrees with the value computed in Example 11-6**b**.

This relationship may be helpful for selecting a trial after-tax rate of return where the before-tax rate of return is known. It must be emphasized, however, this relationship is only a rough approximation in almost all situations.

SUMMARY

Since income taxes are part of most problems, no realistic economic analysis can ignore their consequences. Income taxes make the U.S. Government a partner in every business venture. Thus the Government benefits from all profitable ventures and shares in the losses of unprofitable ventures.

The first step in computing individual income taxes is to tabulate gross income. Any adjustments—for example, allowable taxpayer contributions to a retirement fund—are subtracted to yield adjusted gross income. Personal exemptions and either itemized deductions or the standard deduction are subtracted to find taxable income. This is used, together with a tax rate table, to compute the income tax liability for the year.

For corporations, taxable income equals gross income *minus* all ordinary and necessary expenditures (except capital expenditures) and depreciation and depletion charges. The income tax computation is relatively simple with rates ranging from 15% to 39%. The proper rate to use in an economic analysis, whether for an individual or a corporation, is the incremental tax rate applicable to the increment of taxable income being considered.

Most individuals and corporations pay state income taxes in addition to federal income taxes. Since state income taxes are an allowable deduction in computing federal taxable income, it follows that the taxable income for the federal computation is lower than the state taxable income.

$$\begin{aligned}&\text{Combined state and federal incremental tax rate}\\&\quad = \Delta\text{State tax rate} + (\Delta\text{Federal tax rate})(1 - \Delta\text{State tax rate})\end{aligned}$$

To introduce the effect of income taxes into an economic analysis, the starting point is a before-tax cash flow. Then the depreciation schedule is deducted from appropriate parts of the before-tax cash flow to obtain taxable income. Income taxes are obtained by multiplying taxable income by the proper tax rate. Before-tax cash flow less income taxes equals the after-tax cash flow.

Two income tax complexities eliminated by the Tax Reform Act of 1986 are special tax treatment of long-term capital gains and the investment tax credit. It is likely they will reappear in some future tax law.

When dealing with nondepreciable assets there is a nominal relationship between before-tax and after-tax rate of return. It is

$$\text{After-tax rate of return} = (1 - \Delta\text{Tax rate})(\text{Before-tax rate of return})$$

There is no simple relationship between before-tax and after-tax rate of return in the more usual case of investments involving depreciable assets.

Problems

11-1 An unmarried taxpayer with no dependents expects an adjusted gross income of $47,000 in a given year. His nonbusiness deductions are expected to be $3400.

a. What will his federal income tax be?

b. He is considering an additional activity expected to increase his adjusted gross income. If this increase should be $7000 and there should be no change in nonbusiness deductions or exemptions, what will be the increase in his federal income tax?

11-2 John Adams has a $32,000 adjusted gross income from Apple Corp. and allowable itemized deductions of $4000. Mary Eve has a $24,000 adjusted gross income and $2000 of allowable itemized deductions. Compute the total tax they would pay as unmarried individuals. Then compute their tax as a married couple filing a joint return. (*Answers:* $7543; $8634)

11-3 Bill Jackson worked during school and the first two months of his summer vacation. After considering his personal exemption (Bill is single) and deductions, he had a total taxable income of $1800. Bill's employer wants him to work another month during the summer, but Bill had planned to spend the month hiking. If an additional month's work would increase Bill's taxable income by $1600, how much more money would he have after paying the income tax? (*Answer:* $1360)

11-4 A prosperous businessman is considering two alternative investments in bonds. In both cases the first interest payment would be received at the end of the first year. If his personal taxable income is fixed at $40,000 and he is single, which investment produces the greater after-tax rate of return? Compute the after-tax rate of return for each bond to within ¼ of 1 percent.

Ann Arbor Municipal Bonds: A bond with a face value of $1000 pays $60 per annum. At the end of 15 years, the bond becomes due ("matures"), at which time the owner of the bond will receive $1000 plus the final $60 annual payment. The bond may be purchased for $800. Since it is a municipal bond, the annual interest is *not* subject to federal income tax. The difference between what the businessman would pay for the bond ($800) and the $1000 face value he would receive at the end of 15 years must be included in taxable income when the $1000 is received.

Southern Coal Corporation Bonds: $1000 of these bonds pay $100 per year in annual interest payments. When the bonds mature at the end of twenty years, the bondholder will receive $1000 plus the final $100 interest. The bonds may be purchased now for $1000. The income from corporation bonds must be included in federal taxable income.

11-5 Albert Chan decided to buy an old duplex as an investment. After looking for several months, he found a desirable duplex that could be bought for $93,000 cash. He decided that he would rent both sides of the duplex, and determined that the total expected income would be $800 per month. The total annual expenses for property taxes, repairs, gardening, and so forth are estimated at $600 per year. For tax purposes, Al plans to depreciate the building by the sum-of-years digits method, assuming the building has a twenty-year remaining life and no salvage value. Of the total $93,000 cost of the property, $84,000 represents the value of the building and $9000 is the value of the lot. Assume that Al is in the 38% incremental income tax bracket (combined state and federal taxes) throughout the twenty years.

In this analysis Al estimates that the income and expenses will remain constant at their present levels. If he buys and holds the property for twenty years, what after-tax rate of return can he expect to receive on his investment, using the assumptions below?

a. Al believes the building and the lot can be sold at the end of twenty years for the $9000 estimated value of the lot;

b. A more optimistic estimate of the future value of the building and the lot is that the property can be sold for $100,000 at the end of twenty years.

11-6 Mr. Sam K. Jones, a successful businessman, is considering erecting a small building on a commercial lot he owns very close to the center of town. A local furniture company is willing to lease the building for $9000 per year, paid at the end of each year. It is a net lease, which means the furniture company must also pay the property taxes, fire insurance, and all other annual costs. The furniture company will require a five-year lease with an option to buy the building and land on which it stands for $125,000 at the end of the five years. Mr. Jones could have the building constructed for $82,000 and ready for occupancy on January 5th. He could sell the commercial lot now for $30,000, the same price he paid for it. Mr. Jones is married and has an annual taxable income from other sources of $63,900. He would depreciate the commercial building by accelerated cost recovery system (ACRS) depreciation. Mr. Jones believes that at the end of the five-year lease he could easily sell the property for $125,000. What after-tax rate of return would Mr. Jones receive from this five-year venture?

11-7 A store owner, Joe Lang, believes his business has suffered from the lack of adequate automobile parking space for his customers. Thus, when he was offered an opportunity to buy an old building and lot next to his store, he was interested. He would demolish the old building and make off-street parking for twenty customer's cars. Joe estimates that the new parking would increase his business and produce an additional before-income-tax profit of $7000 per year. It would cost $2500 to demolish the old building. Mr. Lang's accountant advised that both costs (the property and demolishing the old building) would be considered the total value of the land for tax purposes, and it would not be depreciable. Mr. Lang would spend an additional $3000 right away to put a light gravel surface on the lot. This expenditure, he believes, may be charged as an operating expense immediately and need not be capitalized. To compute the tax consequences of adding the parking lot, Joe estimates that his combined state and federal incremental income tax rate will average 40%. If Joe wants a 15% after-tax rate of return from this project, how much could he pay to purchase the adjoining land with the old building? Assume that the analysis period is ten years, and that the parking lot could always be sold to recover the costs of buying the property and demolishing the old building. (*Answer:* $23,100)

11-8 Zeon, a large, profitable corporation, is considering adding some automatic equipment to its production facilities. An investment of $120,000 will produce an initial annual benefit of $29,000 but the benefits are expected to decline $3000 per year, making second-year benefits $26,000, third-year benefits $23,000, and so forth. If the firm uses sum-of-years digits depreciation, an eight-year useful life, and $12,000 salvage value, will it obtain the desired 6% after-tax rate of return? Assume that the equipment can be sold for its $12,000 salvage value at the end of the eight years. Also assume a 46% state-plus-federal income tax rate.

11-9 A group of businessmen formed a corporation to lease a piece of land for five years at the intersection of two busy streets. The corporation has invested $50,000 in car-washing equipment. They will depreciate the equipment by sum-of-years digits depreciation, assuming a $5000 salvage value at the end of the five-year useful life. The corporation is expected to have a before-tax cash flow, after meeting all expenses of operation (except depreciation), of $20,000 the first year, and declining $3000 per year in future years (second year = $17,000; third year = $14,000; and so forth). The corporation has other income, so it is taxed at a combined corporate tax rate of 20%. If the projected income is correct, and the equipment can be sold for $5000 at the end of five years, what after-tax rate of return would the corporation receive from this venture?
(*Answer:* 14%)

11-10 The effective combined tax rate in an owner-managed corporation is 40%. An outlay of $20,000 for certain new assets is under consideration. It is estimated that for the next eight years, these assets will be responsible for annual receipts of $9000 and annual disbursements (other than for income taxes) of $4000. After this time, they will be used only for stand-by purposes and no future excess of receipts over disbursements is estimated.

a. What is the prospective rate of return before income taxes?

b. What is the prospective rate of return after taxes if these assets can be written off for tax purposes in eight years using straight line depreciation?

c. What is the prospective rate of return after taxes if it is assumed that these assets must be written off for tax purposes over the next twenty years using straight line depreciation?

11-11 In January Gerald Adair bought a small house and lot for $99,700. He estimated that $9700 of this amount represented the value of the land. He rented the house for $6500 a year during the four years he owned the house. Expenses for property taxes, maintenance, and so forth were $500 per year. For tax purposes the house was depreciated by ACRS depreciation. At the end of four years the property was sold for $105,000. Gerald is married and works as an engineer. He estimates that his incremental state and federal combined tax rate is 24%. What after-tax rate of return did Gerald obtain on his investment in the property?

11-12 The management of a private hospital is considering the installation of an automatic telephone switchboard, which would replace a manual switchboard and eliminate the attendant operator's position. The class of service provided by the new equipment is estimated to be at least equal to the present method of operation. Five operators are needed to provide telephone service three shifts per day, 365 days per year. Each operator earns $14,000 per year. Company-paid benefits and overhead are 25% of wages. Money costs 8% after income taxes. Combined federal and state income taxes are 40%. Annual property taxes and maintenance are 2½% and 4% of investment, respectively. Depreciation is 15-year straight line. Disregarding inflation, how large an investment in the new equipment can be economically justified by savings obtained by eliminating the present equipment and labor costs? The existing equipment has zero salvage value.

11-13 A contractor has to choose one of the following alternatives in performing earth-moving contracts:

a. Purchase a heavy-duty truck for $13,000. Salvage value is expected to be $3000 at the end of its seven-year depreciable life. Maintenance is $1100 per year. Daily operating expenses are $35.

b. Hire a similar unit for $83 per day.

Based on a 10% after-tax rate of return, how many days per year must the truck be used to justify its purchase? Base your calculations on straight line depreciation and a 50% income tax rate. (*Answer:* 91½ days)

11-14 The Able Corporation is considering the installation of a small electronic testing device for use in conjunction with a government contract the firm has just won. The testing device will cost $20,000, and have an estimated salvage value of $5000 in five years when the government contract is finished. The firm will depreciate the instrument by the sum-of-years digits method using five years as the useful life and a $5000 salvage value. Assume Able Corp. pays 50% federal and state corporate income taxes and uses 8% *after-tax* in their economic analysis. What minimum equal annual benefit must Able obtain *before taxes* in each of the five years to justify purchasing the electronic testing device? (*Answer:* $5150)

11-15 A small business corporation is considering whether or not to replace some equipment in the plant. An analysis indicates there are five alternatives in addition to the do-nothing Alt. *A*. The alternatives have a five-year useful life with no salvage value. Straight line depreciation would be used.

Alternatives	*Cost, in thousands*	*Before-tax uniform annual benefits, in thousands*
A	$ 0	$0
B	25	7.5
C	10	3
D	5	1.7
E	15	5
F	30	8.7

The corporation has a combined federal and state income tax rate of 20%. If the corporation expects a 10% after-tax rate of return for any new investments, which alternative should be selected?

11-16 A firm is considering the following investment project:

Year	*Before-tax cash flow*	
0	−$1000	
1	+500	
2	+340	
3	+244	
4	+100	
5	+100	
	+125	Salvage value

The project has a five-year useful life with a $125 salvage value as shown. Double declining balance depreciation will be used assuming a $125 salvage value. The income tax rate is 34%. If the firm requires a 10% after-tax rate of return, should the project be undertaken?

11-17 A married couple have a combined total adjusted gross income of $75,000. They have computed that their allowable itemized deductions are $4000. Compute their federal income tax. (*Answer:* $14,108)

11-18 A major industrialized state has a state corporate tax rate of 9.6% of taxable income. (There are no steps as in the federal corporate tax rate.) If a corporation has a state taxable income of $150,000, what is the total state and federal income tax it must pay? Also, compute its combined incremental state and federal income tax rate.
(*Answers:* $50,534; 44.86%)

11-19 Jane Shay operates a management consulting business. The business has been successful and now produces a taxable income of $65,000 per year after deducting all "ordinary and necessary" expenses and depreciation. At present the business is operated as a proprietorship, that is, Jane pays personal federal income tax on the entire $65,000. For tax purposes, it is as if she had a job that pays her a $65,000 salary per year.

As an alternative, Jane is considering incorporating the business. If she does, she will pay herself a salary of $22,000 a year from the corporation. The corporation will then pay taxes on the remaining $43,000 and retain the balance of the money as a corporate asset. Thus Jane's two alternatives are to operate the business as a proprietorship or as a corporation. Jane is single and has $2500 of itemized personal deductions. Which alternative will result in a smaller total payment of taxes to the government?
(*Answer:* Now you know one of the reasons why your doctor is a corporation.)

11-20 A house and lot are for sale for $155,000. It is estimated that $45,000 is the value of the land and $110,000 is the value of the house. If purchased, the house can be rented to provide a net income of $12,000 per year after taking all expenses, except depreciation, into account. The house would be depreciated by straight line depreciation using a 27.5-year depreciable life and zero salvage value.

Mary Silva, the prospective purchaser, wants a 10% after-tax rate of return on her investment after considering both annual income taxes and a capital gain when she sells the house and lot. At what price would she have to sell the house at the end of ten years to achieve her objective? You may assume that Mary has an incremental income tax rate of 28% in each of the ten years.

11-21 Bill Alexander and his wife Valerie are both employed. Bill will have an adjusted gross income this year of $36,000. Valerie has an adjusted gross income of $2000 a month. Bill and Valerie have agreed that Valerie should continue working only until the point where the federal income tax on their joint income tax return becomes $9000. On what date should Valerie quit her job?

11-22 The Lynch Bull investment company suggests that Steven Comstock, a wealthy New York City investor (his incremental income tax rate is 50%), consider the following investment.

Buy corporate bonds on the New York Exchange with a face value (par value) of $100,000 and a 5% coupon rate (the bonds pay 5% of $100,000 which equals $5000 interest per year). These bonds can be purchased at their present market value of $75,000. At the end of each year, Steve will receive the $5000 interest, and at the end of five years, when the bonds mature, he will receive $100,000 plus the last $5000 of interest.

Steve will pay for the bonds by borrowing $50,000 at 10% interest for five years. The $5000 interest paid on the loan each year will equal the $5000 of interest income from the bonds. As a result Steve will have no net taxable income during the five years due to this bond purchase and borrowing money scheme.

At the end of five years, Steve will receive $100,000 plus $5000 interest from the bonds and will repay the $50,000 loan and pay the last $5000 interest. The net result is that he will have a $25,000 capital gain (that is, he will receive $100,000 from a $75,000 investment). *Note:* This situation represents an actual recommendation of a stock and bond brokerage firm.

a. Compute Steve's after-tax rate of return on this dual bond plus loan investment package.

b. What would be Steve's after-tax rate of return if he purchased the bonds for $75,000 cash and *did not* borrow the $50,000?

11-23 A large, profitable corporation is considering the following investment in research equipment, and has projected the benefits as follows:

Year	*Before-tax cash flow*
0	−$50,000
1	+2,000
2	+8,000
3	+17,600
4	+13,760
5	+5,760
6	+2,880

Prepare a cash flow table to determine the year-by-year after-tax cash flow assuming ACRS depreciation.

a. What is the after-tax rate of return?

b. What is the before-tax rate of return?

11-24 A major, profitable corporation is considering two alternatives:

Year	*Alt. 1* *Before-tax cash flow*	*Alt. 2* *Before-tax cash flow*
0	−$10,000	−$20,000
1–10	4,500	4,500
11–20	0	4,500

Both alternatives will be depreciated by straight line depreciation using a ten-year depreciable life and no salvage value. Neither alternative is to be replaced at the end of its useful life. If the corporation has a minimum attractive rate of return of 10% *after taxes*, which alternative should it choose? Solve the problem by:

a. Present worth analysis.

b. Annual cash flow analysis.

c. Rate of return analysis.

d. Future worth analysis.

e. Benefit–cost ratio analysis.

f. Any method you choose.

11-25 An engineer is working on the layout of a new research and experimentation facility. Two men will be required as plant operators. If, however, an additional $100,000 of instrumentation and remote controls were added, the plant could be run by a single operator. The total before-tax cost of each plant operator is projected to be $35,000 per year. The instrumentation and controls will be depreciated by accelerated cost recovery system depreciation.

If this large, profitable corporation invests in the additional instrumentation and controls, how long will it take for the after-tax benefits to equal the $100,000 cost? In other words, what is the after-tax payback period? (*Answer:* 3.24 years)

11-26 A special powertool for plastic products costs $400, has a four-year useful life, no salvage value, and a two-year before-tax payback period. Assume uniform annual end-of-year benefits.

a. Compute the before-tax rate of return.

b. Compute the after-tax rate of return, based on ACRS depreciation and a 34% corporate income tax rate.

11-27 A piece of petroleum drilling equipment costs $100,000. It will be depreciated in ten years by double declining balance depreciation with conversion to straight line depreciation at the optimal point. Assume no salvage value in the depreciation computation. The equipment is owned by Shellout, a very profitable company. Shellout will lease the equipment to others and each year receive $30,000 in rent. At the end of five years, they will sell the equipment for $35,000. (Note that this is different from the zero salvage value assumption used in computing the depreciation.) What is the after-tax rate of return Shellout will receive from this equipment investment?

11-28 The Ogi Corporation, a construction company, purchased a new pickup truck for $14,000 They used ACRS depreciation in their income tax return. During the time they had the truck they estimated that it saved the firm $5000 a year. At the end of four years, Ogi sold the truck for $3000. The combined federal and state income tax rate for Ogi is 45%. Compute their after-tax rate of return for the truck. (*Answer:* 12.5%)

11-29 A profitable wood products corporation is considering buying a parcel of land for $50,000, building a small factory building at a cost of $200,000, and equipping it with $150,000 of machinery.

If the project is undertaken ACRS depreciation will be used. Assume the plant is put in service October 1st. The before-tax net annual benefit from the project is estimated at $70,000 per year. The analysis period is to be five years and assumes the total property (land, building, and machinery) is sold at the end of five years, also on October 1st, for $328,000. Compute the after-tax cash flow based on a 34% income tax rate. If the corporation's criterion is a 15% after-tax rate of return, should it proceed with the project?

11-30 A railroad tank car was purchased by a chemical company for $55,000 and is to be depreciated by ACRS depreciation. When its requirements changed suddenly, the chemical company leased the tank car to an oil company for six years at $10,000 per year. The lease also provided that the tank car could be purchased at the end of six years by the oil company for $35,000. At the end of the six years, the oil company exercised its option and bought the tank car. The chemical company has a 34% incremental tax rate. Compute its after-tax rate of return on the tank car. (*Answer:* 9.86%)

11-31 A corporation is considering buying a medium-sized computer that will eliminate a task that must be performed three shifts per day, seven days per week, except for one eight-hour shift per week when the operation is shut down for maintenance. At present four people are needed to perform the day and night task. Thus the computer will replace four employees. Each employee costs the company $32,000 per year ($24,000 in direct wages plus $8000 per year in other company employee costs).

It will cost $18,000 per year to maintain and operate the computer. The computer will be depreciated by sum-of-years digits depreciation using a six-year depreciable life, at which time it will be assumed to have zero salvage value.

The corporation has a combined federal and state incremental tax rate of 50%. If the firm wants a 15% rate of return, after considering both state and federal income taxes, how much can it afford to pay for the computer?

11-32 A mining corporation purchased $120,000 of production machinery. They depreciated it using SOYD depreciation, a five-year depreciable life, and zero salvage value. The corporation is a profitable one that has a 34% incremental tax rate.

At the end of five years the mining company changed its method of operation and sold the production machinery for $40,000. During the five years the machinery was used, it reduced mine operating costs by $32,000 a year, before taxes. If the company MARR is 12% after taxes, was the investment in the machinery a satisfactory one?

11-33 Two mutually exclusive alternatives are being considered by a large, profitable corporation.

Year	*Alt. A Before-tax cash flow*	*Alt. B Before-tax cash flow*
0	−\$3000	−\$5000
1	1000	1000
2	1000	1200
3	1000	1400
4	1000	2600
5	1000	2800

Both alternatives have a five-year useful and depreciable life and no salvage value. Alternative *A* would be depreciated by sum-of-years digits depreciation, and Alt. *B* by straight line depreciation. If the MARR is 10% after taxes, which alternative should be selected? (*Answer:* Choose *B*.)

11-34 Xon, a large and profitable international oil company, purchased a new petroleum drilling rig for \$1,800,000. Xon will depreciate the drilling rig using ACRS depreciation. The drilling rig has been leased to a drilling company which will pay Xon \$450,000 per year for eight years. At the end of eight years the drilling rig will belong to the drilling company. If Xon has a 10% after-tax MARR, does the investment appear to be satisfactory?

11-35 An automobile manufacturer is buying some special tools for \$100,000. The tools are being depreciated by double declining balance depreciation using a four-year depreciable life and a \$6250 salvage value. It is expected the tools will actually be kept in service for six years and then sold for \$6250. The before-tax benefit of owning the tools is as follows:

Year	*Before-tax cash flow*	
1	\$30,000	
2	30,000	
3	35,000	
4	40,000	
5	10,000	
6	10,000	
	6,250	Selling price

Compute the after-tax rate of return for this investment situation, assuming a 46% incremental tax rate. (*Answer:* 11.6%)

11-36 This is the continuation of Problem 11-35. Instead of paying \$100,000 cash for the tools, the corporation will pay \$20,000 now and borrow the remaining \$80,000. The depreciation schedule will remain unchanged. The loan, at a 10% interest rate, will be repaid by four equal end-of-year payments of \$25,240.

Prepare an expanded cash flow table that takes into account both the special tools and the loan. Note that the Year 0 cash flow is −\$20,000 in this situation. *Hint:* You must determine what portion of each loan payment is interest.

a. Compute the after-tax rate of return for the tools, taking into account the $80,000 loan.

b. Explain why the rate of return obtained in Part ***a*** is different from the rate of return obtained in Problem 11-35.

11-37 A project will require the investment of $108,000 in equipment (sum-of-years-digits depreciation with a depreciable life of eight years and zero salvage value) and $25,000 in raw materials (which is not depreciable). The annual project income after paying all expenses, except depreciation, is projected to be $24,000. At the end of eight years the project will be discontinued and the $25,000 investment in raw materials will be recovered.

Assume a 34% income tax rate for this corporation. The corporation wants a 15% after-tax rate of return on its investments. Determine by present worth analysis whether or not this project should be undertaken.

11-38 A large profitable corporation is considering two mutually exclusive capital investments:

	Alt. A	*Alt. B*
Initial cost	$11,000	$33,000
Uniform annual benefit	3,000	9,000
End-of-depreciable-life salvage value	2,000	3,000
Depreciation method	SL	SOYD
Depreciable life, in years	3	4
Useful life, in years	5	5
End-of-useful-life salvage value obtained	2,000	5,000

If the firm's after-tax minimum attractive rate of return is 12% and its incremental income tax rate is 34%, which project should be selected?

11-39 A profitable incorporated business is considering an investment in equipment which has the before-tax cash flow tabulated below. The equipment will be depreciated by double declining balance depreciation with conversion, if appropriate, to straight line depreciation at the preferred time. For depreciation purposes a $700 salvage value at the end of six years is assumed. But the actual value is thought to be $1000 and it is this sum that is shown in the before-tax cash flow.

Year	*Before-tax cash flow*	
0	−$12,000	
1	1,727	
2	2,414	
3	2,872	
4	3,177	
5	3,358	
6	1,997	
	1,000	Salvage value

If the firm wants a 9% after-tax rate of return and its incremental income tax rate is 34%, determine by annual cash flow analysis whether or not the investment is desirable.

11-40 A salad oil bottling plant can either purchase caps for the glass bottles at 5 cents each, or install $500,000 worth of plastic molding equipment and manufacture the caps at the plant. The manufacturing engineer estimates the material, labor, and other costs would be 3 cents per cap.

a. If 12 million caps per year are needed and the molding equipment is installed, what is the payback period?

b. The plastic molding equipment would be depreciated by straight line depreciation using a five-year useful life and no salvage value. Assuming a 40% income tax rate, what is the after-tax payback period, and what is the after-tax rate of return?

11-41 The profitable Palmer Golf Cart Corp. is considering investing $300,000 in special tools for some of the plastic golf cart components. Executives of the company believe the present golf cart model will continue to be manufacturing and sold for five years, after which a new cart design will be needed, together with a different set of special tools.

The saving in manufacturing costs, owing to the special tools, is estimated to be $150,000 per year for five years. Assume ACRS depreciation for the special tools and a 39% income tax rate.

a. What is the after-tax payback period for this investment?

b. If the company wants a 12% after-tax rate of return, is this a desirable investment?

11-42 An individual in California with a taxable income of about $80,000 has a federal incremental tax rate of 33% and a state incremental tax rate of 9.3%. What is his combined incremental tax rate?

11-43 ARKO oil company purchased two large compressors for $125,000 each. One compressor was installed in their Texas refinery and is being depreciated by ACRS depreciation. The other compressor was placed in the Oklahoma refinery where it is being depreciated by sum-of-years digits depreciation with zero salvage value. Assume the company pays federal income taxes each year and the tax rate is constant. The corporate accounting department noted that the two compressors are being depreciated differently and wonders whether the corporation will wind up paying more income taxes over the life of the equipment as a result of this. What do you tell them?

CHAPTER **12**

Replacement Analysis

Until now we have examined problems concerning the selection of alternatives to accomplish a desired task. This has resulted in a fairly straightforward analysis of available alternatives, with the recommendation on which alternative should be adopted—typically characterized as selecting the equipment for a new plant. But economic analysis is more frequently performed in conjunction with *existing* facilities. The problem is less frequently one of building a new plant; it is keeping a present plant operating economically.

In ***replacement analysis***, we are not choosing between new ways to perform the desired task. Instead, we have equipment performing the task, and the question is, "Should the existing equipment be retained or replaced?" This adversarial situation has given rise to the terms ***defender*** and ***challenger***.* The defender is the existing equipment; the challenger, naturally, is the best available replacement equipment. In using these two terms, we are always referring to equipment.

The subject of replacement analysis can be considered as having five fundamental aspects:

1. Understanding what the comparison should be.
2. Remaining life of the defender.
3. Economic life of the challenger.
4. Replacement analysis techniques.
5. Equipment replacement models.

Chapter 12 examines these five topics.

---

*The use of the titles *defender* and *challenger* was originated by George Terborgh. He is the author of several important books on replacement economy for the Machinery and Allied Products Institute.

Defender–Challenger Comparison

The normal means of monitoring expenditures in industry, as in government, are by *annual budgets*. One important facet of a budget is the allocation of money for new capital expenditures. This may take the form of money for new facilities or for replacing and upgrading existing facilities.

Replacement analysis may therefore have as its end product a recommendation that some particular equipment be replaced and that money for the replacement be included in the capital expenditures budget. Of course, if there is not a recommendation to replace the equipment now, this recommendation may be made next year or in some subsequent year. This leads us to the first aspect of the defender–challenger comparison. The question is:

> ***Shall we replace the defender now, or shall we keep it for one or more additional years?***

Thus the question is not *whether* we are going to remove the defender: at *some* point, the existing equipment will be removed, either when the task it performs is no longer necessary to do or when the task can be performed better by different equipment. The question is not *if* it will be removed, but *when* it will be removed.

Because the defender already is in the plant, there often is a misunderstanding on what value to assign to it in an economic analysis. Example 12-1 demonstrates the problem.

EXAMPLE 12-1

An SK-30 desk calculator was purchased two years ago for $1600; it has been depreciated by straight line depreciation using a four-year life and zero salvage value. Because of recent innovations in desk calculators, the current price of the SK-30 calculators has been reduced from $1600 to $995. An office equipment supply firm has offered a trade-in allowance of $350 for the SK-30 in partial payment on a new $1200 electronic EL-40 calculator. Some discussion revealed that without a trade-in, the EL-40 can be purchased for $1050, indicating the originally quoted price of the EL-40 was overstated to allow a larger trade-in allowance. The true current market value of the SK-30 is probably only $200. In a replacement analysis, what value should be assigned to the SK-30 calculator?

Solution: In the example, five different dollar amounts relating to the SK-30 calculator have been outlined:

1. *Original cost:* the calculator cost $1600.
2. *Present cost:* the calculator now sells for $995.

3. *Book value:* the original cost less two years of depreciation is 1600 − $\frac{2}{4}$(1600 − 0) = \$800.
4. *Trade-in value:* the offer was \$350.
5. *Market value:* the estimate was \$200.

We have seen that an economy study is based on the current situation, not on the past. We referred to past costs as *sunk costs* to emphasize that, since they could not be altered, they were not relevant. (The one exception was that past costs may affect present or future income taxes.) Here the question is, "What value should be used in an economic analysis for the SK-30?" The relevant cost is the present market value for the equipment. Neither the original cost, the present cost, the book value, nor trade-in value is relevant. ◀

At first glance, the trade-in value would appear to be a suitable present value for the equipment. Often, however, the trade-in price is inflated *along with* the price for the new item. This is such a common practice in the new-car showrooms that the term *overtrade* is used to describe the excessive portion of the trade-in allowance. (The purchaser, of course, also is quoted a higher price for the new car.) This distortion of the present value of the defender, along with a distorted price for the challenger, can be serious for they do not cancel out in an economic analysis.

From Ex. 12-1 we see there are several different values that can be assigned to the defender. The appropriate one is the present market value. If a trade-in value is obtained, care should be taken to ensure that it actually represents a fair market value.

There should be less difficulty in deciding what to use for the installed cost of a challenger. Of course, there will be a past cost and a book value for the defender, but these have nothing to do with the challenger. There have been times when people added any capital loss on disposal (book value *minus* net realized salvage value) of the defender to the cost of the challenger. This is incorrect. If there is a capital loss on disposal of the defender, it will have a tax consequence, but this certainly will have no bearing on the installed cost of the challenger.

Another aspect of a defender–challenger comparison concerns which equipment *is* the challenger. If there is to be a replacement of the defender by the challenger, we would want to install the best of the available alternatives. Prior to this chapter, all our attention has been on selecting the best from two or more alternatives, so this aspect is not new. We will, however, look at the useful life of both the defender and the challenger more critically than earlier, as described in the next two sections.

Remaining Life Of The Defender

In replacement analysis, the discussion about the defender and the challenger generally amounts to a question of the old *vs.* the new. The old equipment has a relatively short remaining life compared to new equipment. Here we will examine the ***remaining life of the defender***.

How long can the defender be kept operating? Anyone who has seen—or heard!—old machinery in operation, whether it is a 300-year-old clock, a seventy-year-old automobile, or old production equipment, has probably realized that almost anything can be kept operating indefinitely, provided it receives proper repair and maintenance. However, while one might be able to keep a defender going indefinitely, the cost may prove to be excessive. So, rather than asking what the remaining *operating life* of the defender could be, we really want to know what its ***economic life*** is. This we define as the remaining useful life that results in a minimum equivalent uniform annual cost.

Economic life **= Life where EUAC is minimum**

Example 12-2 illustrates the situation.

EXAMPLE 12-2

An eleven-year-old piece of equipment is being considered for replacement. It can be sold for $2000 now and it is believed this same salvage value can also be obtained in future years. The current maintenance cost is $500 per year and is expected to increase $100 per year in future years. If the equipment is retained in service, compute the economic life that results in a minimum EUAC, based on 10% interest.

Solution: Here the salvage value is not expected to decline from its present $2000. The annual cost of this invested capital is Si = 2000(0.10) = $200. The maintenance is represented by $500 + 100G$. A year-by-year computation of EUAC is as follows:

Year n	*Age of equipment, in years*	EUAC *of invested capital* = Si	EUAC *of maintenance* = 500 + 100(A/G,10%,n)	*Total* EUAC
1	11	$200	$500	$700
2	12	200	548	748
3	13	200	594	794
4	14	200	638	838
5	15	200	681	881

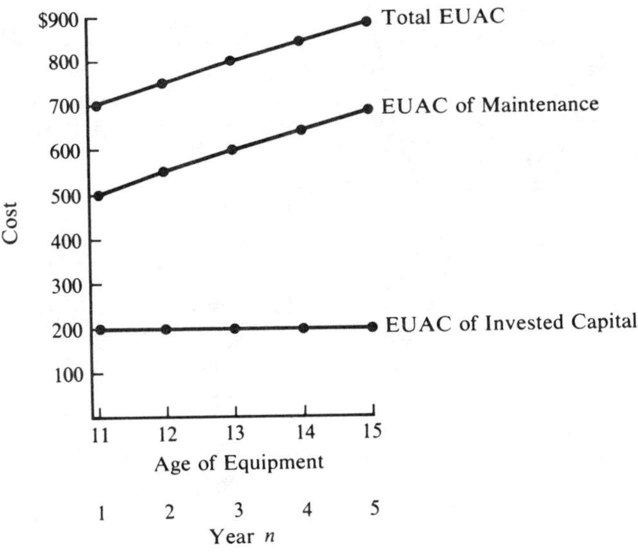

Figure 12-1 EUAC for different remaining lives.

These data are plotted in Fig. 12-1. We see that the annual cost of continuing to use the equipment is increasing. It is reasonable to assume that if the equipment is not replaced now, it will be reviewed again next year. Thus the economic life at which EUAC is a minimum is one year. ◀

Example 12-2 represents a common situation. The salvage value is stable but ***maintenance*** is increasing. The total EUAC will continue to increase as time passes. This means that an economic analysis to compare the defender at its most favorable remaining life will be based on retaining the defender one more year. This is not always the case, as is shown in Ex. 12-3.

EXAMPLE 12-3

A five-year-old machine, whose current market value is $5000, is being analyzed to determine its economic life in a replacement analysis. Compute its economic life using a 10% interest rate. Salvage value and maintenance estimates are as given.

Years of remaining life n	Estimated salvage value (S) end of Year n	Estimated maintenance cost for year	If retired at end of Year n: EUAC of capital recovery $(P - S) \times (A/P, 10\%, n) + Si$	EUAC of maintenance $100(A/G, 10\%, n)$	Total EUAC
0	$P =$ $5000				
1	4000	$ 0	$1100 + 400	$ 0	$1500
2	3500	100	864 + 350	48	1262
3	3000	200	804 + 300	94	1198
4	2500	300	789 + 250	138	1177
5	2000	400	791 + 200	181	1172
6	2000	500	689 + 200	222	1111
7	2000	600	616 + 200	262	1078
8	2000	700	562 + 200	300	1062
9	2000	800	521 + 200	337	1058 ←
10	2000	900	488 + 200	372	1060
11	2000	1000	462 + 200	406	1068

Solution: For a minimum EUAC, the machine has nine years of remaining life. ◀

From Examples 12-2 and 12-3, we see that the economic life remaining to an existing machine may be one year or it may be longer. Looking again at the two examples, we find that they could represent the same machine being examined at different points in its life. When it was five years old, Ex. 12-3 indicated the economic life remaining would be reached nine years hence, making a total of 14 years of service. This would be the point where, from Age 5 onward, the total EUAC would be a minimum. It is important to recognize that this minimum EUAC is based on the projection of future costs for the five-year-old machine, not on past or sunk costs. Therefore, the projection for the five-year-old machine (Ex. 12-3) is different from the situation when the machine is eleven years old (Ex. 12-2): the EUAC was increasing when computed from Age 11 onward.

For older equipment with a negligible or stable salvage value, it is likely that the operating and maintenance costs are increasing. Under these circumstances, the useful life at which EUAC is a minimum is one year.

Economic Life Of The Challenger

In all previous computations to determine which of several alternatives should be selected, useful lives have been assumed. But from the analysis of the economic life remaining to a defender, we recognize that a similar situation exists for the challenger. If the various costs for the challenger are known along with its year-by-year salvage values, then the economic life may be computed. Example 12-4 illustrates the computation.

EXAMPLE 12-4

A piece of machinery costs $10,000 and has no salvage value after it is installed. The manufacturer's warranty will pay all first-year maintenance and repairs. In the second year, maintenance and repairs will be $600, and they will increase on a $600 arithmetic gradient in subsequent years. If interest is 8%, compute the useful life of the equipment that results in minimum EUAC.

Solution:

	If retired at end of Year n		
Year n	EUAC *of capital recovery* $10,000(A/P,8%,n)	EUAC *of maintenance* $600(A/G,8%,n)	*Total* EUAC
1	$10800	$ 0	$10800
2	5608	289	5897
3	3880	569	4449
4	3019	842	3861
5	2505	1108	3613
6	2163	1366	3529 ←
7	1921	1616	3537
8	1740	1859	3599
9	1601	2095	3696

The total EUAC data are plotted in Fig. 12-2. From either the tabulation or the figure, we see that a useful life of six years results in a minimum EUAC.

Replacement Analysis Techniques

From the earlier sections of this chapter, we see that determining the economic life remaining to the defender and the economic life of the challenger are substantial problems in themselves. But even when this has been accomplished, we

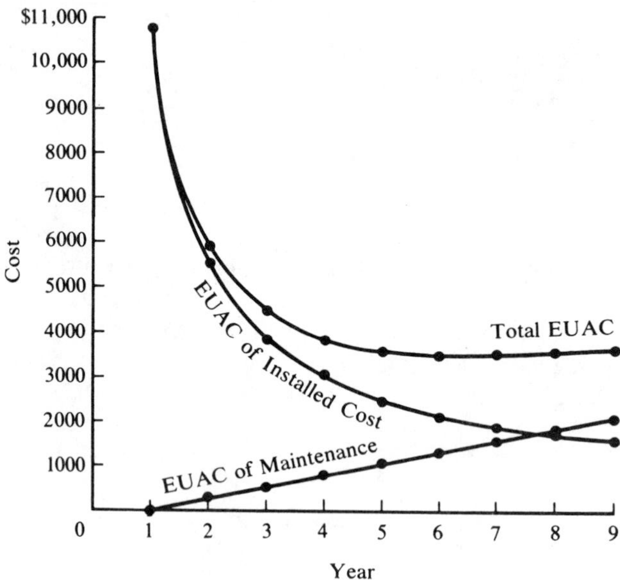

Figure 12-2 Plot of costs for Example 12-4.

have merely identified the participants in the replacement economic analysis. In this section, we will concentrate on how to proceed with the analysis.

Defender Remaining Life Equals Challenger Useful Life

When we find a situation where the remaining life of the defender equals the useful life of the challenger, we have considerable flexibility in selecting the method of analysis. As a result, these problems may be solved by present worth analysis, annual cash flow analysis, rate of return analysis, benefit–cost ratio analysis, and so forth. Examples 12-5 and 12-6 illustrate before-tax and after-tax replacement analysis for the equal-life situation.

EXAMPLE 12-5

Determine whether the SK-30 desk calculator of Ex. 12-1 should be replaced by the EL-40 electronic calculator. In addition to the data given in Ex. 12-1, the following estimates have been made:

- The SK-30 maintenance and service contract costs \$80 a year. The EL-40 will require no maintenance.
- Either calculator is expected to be used for the next five years. At the end of that time, the SK-30 will have no value, but the EL-40 probably could be sold for \$250.
- The EL-40 electronic calculator is faster and easier to use than the SK-30 mechanical calculator. This benefit is expected to save about \$120 a year by reducing the need for part-time employees.

If the MARR is 10% before taxes, should the SK-30 desk calculator be replaced by the EL-40 calculator?

Solution: Compute the EUAC for each calculator:

SK-30:

$$\text{Present market value} = \$200$$
$$\text{Future salvage value} = 0$$
$$\text{Annual maintenance} = 80 \text{ per year}$$

$$\begin{aligned} \text{EUAC} &= (200 - 0)(A/P,10\%,5) + 80 \\ &= 200(0.2638) + 80 = \$132.76 \end{aligned}$$

EL-40:

$$\text{Cash price} = \$1050$$
$$\text{Future salvage value} = 250$$
$$\text{Annual benefit of greater usefulness} = 120 \text{ per year}$$

$$\begin{aligned} \text{EUAC} &= (1050 - 250)(A/P,10\%,5) + 250(0.10) - 120 \\ &= 800(0.2638) + 25 - 120 = \$116.04 \end{aligned}$$

The EL-40 calculator has the smaller EUAC and therefore is the preferred alternative in this before-tax analysis. ◀

EXAMPLE 12-6

Solve Ex. 12-5 with a MARR equal to 8% after taxes. The EL-40 will be depreciated by straight line depreciation using a four-year depreciable life. The SK-30 is already two years old. The analysis period remains at five years. Assume a 34% corporate income tax rate.

Solution:

Alternative A: Keep the SK-30 rather than sell it. Compute the after-tax cash flow:

Year	Before-tax cash flow	Straight line depreciation	Taxable income	34% Income taxes	After-tax cash flow
0	−$200		+$600*	−$204*	−$404†
1	−80	$400	−480	+163	+83
2	−80	400	−480	+163	+83
3	−80	0	−80	+27	−53
4	−80	0	−80	+27	−53
5	{ −80 0 salvage	0	−80	+27	−53

*If sold for $200 there would be a $600 capital loss on disposal. If $600 of capital gains were offset by the loss during the year, there would be no capital gain to tax, saving 34% × $600 = $204. If the SK-30 is not sold, this loss is not realized and the resulting income taxes will be $204 higher than if it had been sold.

†This is the sum of the $200 selling price foregone *plus* the $204 income tax saving foregone.

Compute the EUAC (note that in computing a cost, the signs appear opposite to those shown in the after-tax cash flow):

$$\text{EUAC} = [404 - 83(P/A,8\%,2) + 53(P/A,8\%,3)(P/F\ 8\%,2)](A/P\ 8\%,5)$$
$$= \$93.46$$

Alternative B: Purchase an EL-40. Compute the after-tax cash flow:

Year	Before-tax cash flow	Straight line depreciation	Taxable income	34% Income taxes	After-tax cash flow
0	−$1050				−$1050
1	+120	$200	−$80	+$27	+147
2	+120	200	−80	+27	+147
3	+120	200	−80	+27	+147
4	+120	200	−80	+27	+147
5'	{ +120 +250 salvage	0	+120	−41	+329

Compute the EUAC:

$$\text{EUAC} = [1050 - 147(P/A,8\%,4) - 329(P/F,8\%,5)](A/P,8\%,5)$$
$$= \$84.97$$

Based on this after-tax analysis the EL-40 calculator is the preferred alternative. ◀

EXAMPLE 12-7

Solve Ex. 12-6 by computing the rate of return on the difference between the alternatives. In Ex. 12-6, the two alternatives were "Keep the SK-30" or "Buy an EL-40." The difference between the alternatives would be:

Buy an EL-40	rather than	Keep the SK-30
Alternative *B*	*minus*	Alternative *A*

Solution: The after-tax cash flow for the difference between the alternatives may be computed as follows:

Year	*A*	*B*	*B* − *A*
0	−$404	−$1050	−$646
1	+83	+147	+64
2	+83	+147	+64
3	−53	+147	+200
4	−53	+147	+200
5	−53	+329	+382

The rate of return on the difference between the alternatives is computed as follows:

$$\text{PW of cost} = \text{PW of benefit}$$

$$646 = 64(P/A,i,2) + 200(P/A,i,2)(P/F,i,2) + 382(P/F,i,5)$$

Try $i = 9\%$:

$$646 \stackrel{?}{=} 64(1.759) + 200(1.759)(0.8417) + 382(0.6499)$$

$$\stackrel{?}{=} 656.9$$

Try $i = 10\%$:

$$646 \stackrel{?}{=} 64(1.736) + 200(1.736)(0.8264) + 382(0.6209)$$

$$\stackrel{?}{=} 635.2$$

$$\text{The rate of return} = 9\% + \left(\frac{656.9 - 646.0}{656.9 - 635.2}\right) = 9.5\%$$

This example shows how a rate of return may be computed for an equipment replacement problem. The cash flows for both alternatives are computed and then the rate of return is computed on the difference between the alternatives. The rate of return is greater than the 8% after-tax MARR. The increment of investment is desirable. Buy the EL-40 calculator. ◀

Defender Remaining Life Different from Challenger Useful Life

When the alternative lives are equal to the analysis period, we can generally solve problems in a variety of ways. But when the alternatives (defender and challenger) have different lives, there may be difficulties. For unequal-lived alternatives, annual cash flow analysis is generally the most suitable method of analysis. In Chapter 6 it was stated that a comparison of equivalent uniform annual costs for unequal-lived alternatives is suitable only if the following assumptions are valid:

1. When an alternative has reached the end of its useful life, it is assumed to be replaced by an identical replacement (with the same costs, performance, and so forth).
2. The analysis period is a common multiple of the useful lives of the alternatives, or there is a continuing or perpetual requirement for the selected alternative.

But there are defender–challenger situations where these conditions cannot be met. It is reasonable to assume that the challenger can be replaced by an identical replacement. This is not, however, a reasonable assumption for the defender when it reaches the end of its economic life. The defender is typically an older piece of equipment with a modest current selling price. An identical replacement, even if it could be found, probably would have an installed cost far in excess of the current selling price of the defender.

Thus the assumptions we made for using an annual cash flow analysis on unequal-lived alternatives in Chapter 6 frequently cannot be met in the usual defender–challenger situation. But a careful examination of the equipment replacement problem indicates that those assumptions are not essential here. There are two alternatives in equipment replacement:

1. Replace the defender now.
2. Retain the defender for the present.

The question is not really one of selecting the defender or the challenger, but rather that of deciding if *now* is the time to replace the defender. When the defender is replaced, it will be by the challenger—the best available replacement. The challenger could be a piece of new equipment, or it could be used or reconditioned equipment. Thus in equipment replacement, an annual cash flow analysis does not assume that the defender has an identical replacement at the end of its useful life. The replacement is always by the challenger.

If the defender–challenger problem assumes a continuing requirement for the equipment, an annual cash flow analysis of the unequal-lived alternatives is proper. When there is a definite analysis period, after which the equipment will not be needed, then a careful analysis is needed to see how the analysis period affects the alternatives.

For unequal-lived alternatives, neither present worth nor rate of return is a practical method of analysis. Both methods require that the consequences of the alternatives be evaluated over the analysis period. For this reason, these methods are attempted only when there is a well-defined analysis period.

EXAMPLE 12-8

An economic analysis is to be made to determine if some existing (defender) equipment in the plant should be replaced. A \$4000 overhaul must be done now if the equipment is to be retained in service. Maintenance is estimated at \$1800 in each of the next two years, after which it is expected to increase on a \$1000 arithmetic gradient. The defender has no present or future salvage value. The equipment described in Ex. 12-4 is the challenger. Make a replacement analysis to determine whether to retain the defender or replace it by the challenger if 8% interest is used.

Solution: The first step is to determine the economic life of the defender. The pattern of overhaul and maintenance costs (see Fig. 12-3) suggests that if the overhaul is done, the remaining economic life of the equipment will be several years. The computation is as follows:

	If retired at end of Year n		
Year *n*	EUAC *of overhaul* \$4000(A/P,8%,n)	EUAC *of maintenance* \$1800 + \$1000 *gradient from Year 3 on*	*Total* EUAC
1	\$4320	\$1800	\$6120
2	2243	1800	4043
3	1552	1800 + 308*	3660 ←
4	1208	1800 + 683†	3691
5	1002	1800 + 1079	3881

*For the first 3 years, the maintenance is \$1800, \$1800, and \$2800. Thus,
EUAC = 1800 + 1000(A/F,8%,3) = 1800 + 308

†EUAC = 1800 + 1000(P/G,8%,3)(P/F,8%,1)(A/P,8%,4) = 1800 + 683

For minimum EUAC, the remaining economic life of the defender is three years. In Ex. 12-4, we determined that the economic life of the challenger is six years and that the resulting EUAC is \$3529. Since the EUAC of the challenger (\$3529) is less than the EUAC of the defender (\$3660), the challenger should be installed now to replace the defender. ◀

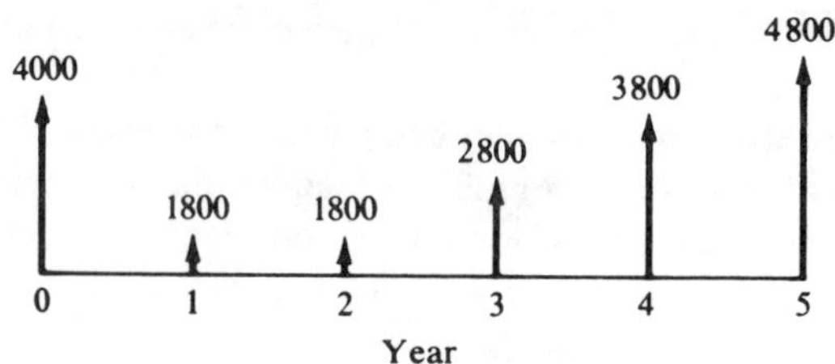

Figure 12-3 Overhaul and maintenance costs for the defender in Example 12-8.

A Closer Look at the Challenger

We defined the challenger as the best available alternative to replace the defender. But as time passes, the best available alternative can change. And given the trend in our technological society, it seems likely that future challengers will be better than the present challenger. If this is so, the prospect of improved future challengers may affect the present decision between the defender and the challenger.

Figure 12-4 illustrates two possible estimates of future challengers. In many technological areas it seems likely that the equivalent uniform annual costs associated with future challengers will decrease by a constant amount each year. There are other fields, however, where a rapidly changing technology will produce a sudden and substantially improved challenger—with decreased costs or increased benefits. The uniform decline curve of Fig. 12-4 assumes that each future challenger has a minimum EUAC that is a fixed amount less than the

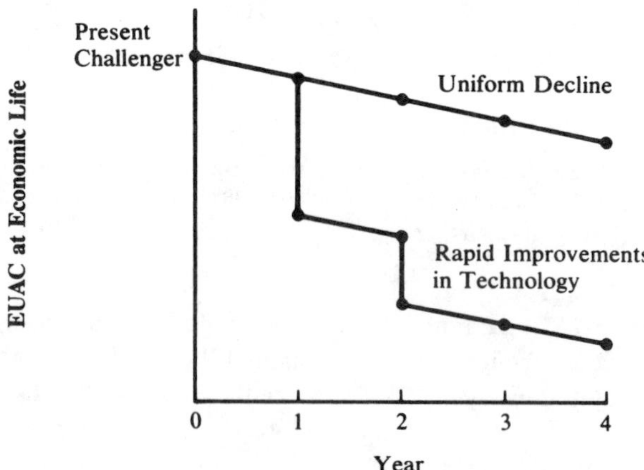

Figure 12-4 Two possible ways the EUAC of future challengers may decline.

previous year's challenger. This, of course, is only one of many possible assumptions that could be made regarding future challengers.

If future challengers will be better than the present challenger, what impact will this have on an analysis now? The prospect of better future challengers makes it more desirable to retain the defender and to reject the present challenger. By keeping the defender for now, we will be able to replace it later by a better future challenger. Or, to state it another way, the present challenger is made less desirable by the prospect of improved future challengers. Example 12-9 illustrates the situation.

EXAMPLE 12-9

Recompute Ex. 12-8. This time it is estimated that future challengers will be improved so that each year the EUAC for their six-year economic life will decline by $100. This new situation means that there are a series of challengers—the present challenger and the improved future challengers. The available alternatives are listed:

A. Keep the defender.

B. Replace the defender with the present challenger.

C. Keep the defender one year and then replace it with a better challenger next year.

D. Keep the defender two years and then replace it with a still better challenger two years hence.

E. Keep the defender three years and then replace it with a still better challenger three years hence.

F. Keep the defender four years and then replace it with a still better challenger four years hence.

And so on. At this point we must emphasize that the above alternatives are designed to provide an answer to the fundamental replacement analysis question: *Shall we replace the defender now or shall we keep it one or more additional years?* If Alt. *B* is the most economical, the best decision is to replace the defender now. If any of the other alternatives is more economical, the decision will be to retain the defender for now and to review the situation again next year.

Solution: We already have done the computations to determine the EUAC of the defender and the present challenger at their economic lives.

A. Keep the defender; from Ex. 12-8, the minimum EUAC is $3660.

B. Replace the defender with the present challenger; from Ex. 12-4, the minimum EUAC is $3529.

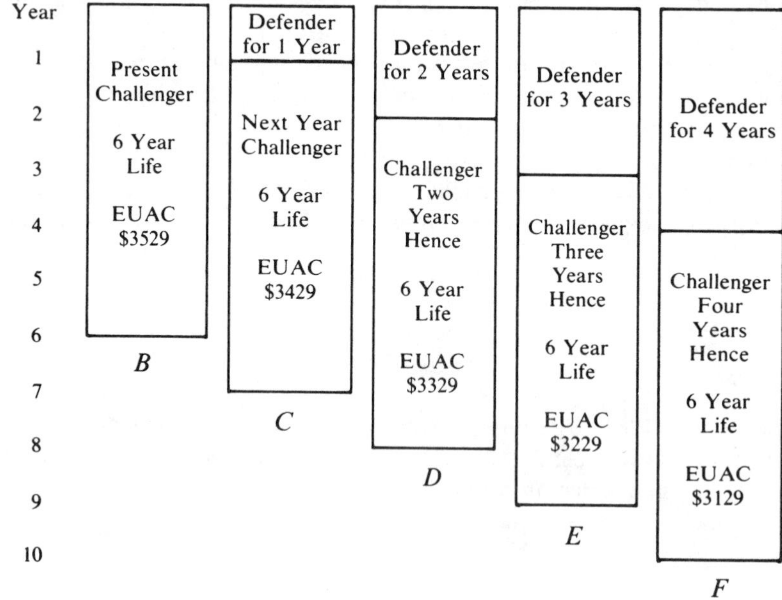

Figure 12-5 Diagrams of Alternatives *B–F*, Example 12-9.

A diagram of Alternatives *B–F* is depicted in Fig. 12-5. As stated in the problem, the EUAC for each subsequent year's challenger is $100 less than the prior year's challenger. From Ex. 12-8, the EUAC for retaining the defender was computed as follows:

Years defender retained	EUAC *for period*
1	$6120
2	4043
3	3660
4	3691

With these data, we can compute the EUAC for Alternatives *C–F*, which are the combination of retaining the defender one or more years followed by six years of a future challenger. For Alternatives *C* through *F*:

$$\text{EUAC} = [(\text{Defender EUAC})(P/A,8\%,n) + (\text{Challenger EUAC}) \times (P/A,8\%,6)(P/F,8\%,n)](A/P,8\%,n+6)$$

Alternative C:

$$\begin{aligned}\text{EUAC} &= [6120(P/A,8\%,1) + 3429(P/A,8\%,6)(P/F,8\%,1)](A/P,8\%,7) \\ &= [6120(0.926) + 3429(4.623)(0.9259)](0.1921) = \$3909\end{aligned}$$

Alternative D:

$$\text{EUAC} = [4043(P/A,8\%,2) + 3329(P/A,8\%,6)(P/F,8\%,2)](A/P,8\%,8)$$
$$= [4043(1.783) + 3329(4.623)(0.8573)](0.1740) = \$3550$$

Alternative E:

$$\text{EUAC} = [3660(P/A,8\%,3) + 3229(P/A,8\%,6)(P/F,8\%,3)](A/P,8\%,9)$$
$$= [3660(2.577) + 3229(4.623)(0.7938)](0.1601) = \$3407$$

Alternative F:

$$\text{EUAC} = [3691(P/A,8\%,4) + 3129(P/A,8\%,6)(P/F,8\%,4)](A/P,8\%,10)$$
$$= [3691(3.312) + 3129(4.623)(0.7350)](0.1490) = \$3406$$

The alternatives may be summarized as follows:

Alternative	*Description*	*Equivalent uniform annual cost*
A	Keep the defender	\$3660
B	Replace with the challenger	3529
C	Defender 1 year; better challenger 6 years	3909
D	Defender 2 years; better challenger 6 years	3550
E	Defender 3 years; better challenger 6 years	3407
F	Defender 4 years; better challenger 6 years	3406

The least cost alternative calls for keeping the defender for now. Our decision will be to retain the defender and to restudy the situation next year. ◀

In Ex. 12-8, future challengers were assumed to be the same as the present challenger. As a result, the decision was to select the present challenger rather than the defender. But in Ex. 12-9, we assumed that future challengers would be an improvement over the present challenger. This time the decision was to retain the defender. The contradictory decision in the two examples illustrates that while the present challenger is preferred over the defender, a still more attractive alternative is to delay the replacement to obtain a better future challenger.

Shortening Challenger Life to Compensate for Improved Future Challengers

We have seen that if better future challengers can be expected to become available in the future, this decreases the desirability of the present challenger. The assumption of a series of improved future challengers makes the two-alternative defender–challenger problem into a multiple-alternative problem with many more calculations. To reduce the amount of calculations, one frequently used technique is to assume that the challenger has a shortened life.

If the challenger life in Ex. 12-8 were assumed to be four years, instead of six years, the challenger EUAC would have been $3861 (as computed in Example 12-4). This would have changed the decision in Ex. 12-8 to keep the defender for the time being. Thus, the shortening of the challenger life by one-third produced the same conclusion as Ex. 12-9. While the general approach throughout this book has been to advocate accurate calculations, shortening the challenger life to reflect the prospect of better future challengers seems like a practical technique to avoid longer calculations.

Equipment Replacement Models

In the equipment replacement problems so far discussed, the assumptions have been carefully made to keep the computations to a manageable level. More complex replacement problems exist, of course, and to solve them, a variety of ***equipment replacement mathematical models*** have been developed. While these models are generally beyond the scope of this book, an introduction to the work of George Terborgh of the Machinery and Allied Products Institute—MAPI—will be given.

Several versions of the MAPI replacement analysis system exist and are described in a series of publications.* The present MAPI system, described in *Business Investment Management*, has these features:

1. Worksheets help to organize the problem by providing a format for the analysis.
2. The analysis considers both the changes in costs and the changes in benefits.
3. Future challengers are recognized as being better than the present challenger.

**Dynamic Equipment Policy* (1949); *MAPI Replacement Manual* (1950); *Business Investment Policy* (1958); *Business Investment Management* (1967). Washington, D.C.: Machinery and Allied Products Institute.

4. Standard projections are provided for certain estimates that are difficult to make.
5. The computations result in an incremental after-tax rate of return based on the difference between the proposed project and a stated alternative (like continuing with the existing situation).

The MAPI system assumes that when a piece of capital equipment is purchased, one is, in effect, buying *service values* that represent a flow of future benefits to the owner. These service values are the earnings of the equipment over the years, as compared to then-available challengers. Since future challengers are expected to be a steady improvement over the present challenger, the earnings are assumed to decline uniformly over the life of the equipment. When there is no salvage value, the projected before-tax earnings are assumed to decline to zero at the end of the service life. Figure 12-6 illustrates the earnings assumption. When there is a salvage value, the before-tax earnings are assumed to decline to an amount equal to the cost of continuing the asset in service.

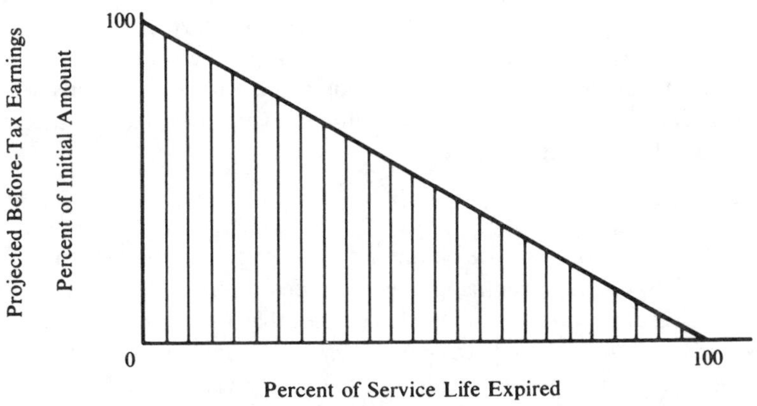

Figure 12-6 MAPI earnings projection for an asset without salvage value.

Using the service value concept, we see that the purchase of a piece of capital equipment yields the owner the present worth of all future service values *plus* the present worth of the salvage value. We therefore assume that,

Purchase price = PW of service values + PW of future salvage value

A key feature in the MAPI system is determining the ***retention value of the asset*** at the end of the comparison period. This is done by determining the year-by-year capital consumption based on the following allocation of before-tax earnings:

1. To pay income taxes (assuming a 50% tax rate) on the earnings *minus* the sum of the depreciation of the asset and interest on the debt portion of the investment.
2. To pay 3% interest on the one-quarter of the unrecovered investment that is assumed to be borrowed.
3. To provide a 10% after-tax return on the three-quarters of the unrecovered investment that is equity capital (not borrowed).
4. To provide for annual capital consumption, which is the difference between the unrecovered investment at the beginning of the year and the unrecovered investment at the end of the year. At any time, the unrecovered investment is to be equal to the present worth of the after-tax earnings *plus* the present worth of the salvage value.

The MAPI capital consumption model uses the four conditions above to devise a relationship between before-tax earnings and the initial investment. For any initial investment, there is a single before-tax earnings value that will yield the MAPI capital consumption model.

For a $10,000 investment, five-year service life, straight line depreciation, and no salvage value, the initial before-tax earnings have been computed* as $4748. Using this value the capital consumption may be computed.

Table 12-1 has been constructed to satisfy the MAPI conditions we have previously stated. A row-by-row examination of the table will aid in understanding the capital consumption model.

Row:

1. *Before-tax earnings*. The before-tax earnings follow the pattern of Figure 12-6, that is, a declining gradient series. From a MAPI formula we computed the value of the gradient series that would provide the exact amount of money needed to pay income taxes, interest on the borrowed money, a 10% after-tax return on the equity capital, and provide for the capital consumption. We must emphasize that the purpose of doing all this is to compute the capital consumption in Row **9** on the MAPI assumptions.
2. *Straight line depreciation*. For an initial cost of $10,000 and no salvage value, the annual straight line depreciation is $2000.
3. *Interest payment*. One-quarter of the unrecovered investment (Row 7) is assumed to be borrowed. Row **3** is 3% interest on the debt. For Year 1: Interest $= 0.03 \times 0.25 \times 10{,}000 =$ $75.
4. *Taxable earnings*. Taxable earnings = Before-tax earnings − Interest − Depreciation.

*From *Business Investment Management* (Pg. 328, Eq. 8).

Table 12-1 COMPUTATION OF YEAR-BY-YEAR CAPITAL CONSUMPTION USING THE MAPI ASSUMPTIONS

	Year 1	2	3	4	5	
1 Before-tax earnings	$ 4748	$3799	$2849	$1899	$ 950	
2 Straight line depreciation	2000	2000	2000	2000	2000	
3 Interest payment	75	56	38	23	10	
4 Taxable earnings	2673	1743	811	−124	−1060	
5 50% Income taxes	−1336	−872	−406	+62	+530	
6 After-tax earnings	3412	2927	2443	1961	1480	
7 Unrecovered investment	10000	7413	5098	3075	1368	0
8 10% return on equity capital	750	556	382	231	102	
9 Capital consumption	2587	2315	2023	1707	1368	= $10,000

5. *50% income taxes*. Taxes = 50% of taxable earnings.
6. *After-tax earnings*. After-tax earnings = Before-tax earnings − Income taxes.
7. *Unrecovered investment*. The initially unrecovered investment is the $10,000 initial cost. The second year, unrecovered investment is the first-year unrecovered investment ($10,000) *minus* the first-year capital consumption ($2587), and so on.
8. *10% return on equity capital*. 75% of the unrecovered investment is assumed to be equity capital. For the first year, 10% return on the equity capital = 0.10 × 0.75 × 10,000 = $750.
9. *Capital consumption*. Capital consumption = After-tax earnings − Interest payment − Return on equity capital. For the first year, Capital consumption = $3412 − 75 − 750 = $2587.

An important assumption of the MAPI capital consumption model is that the unrecovered investment at any time equals the present worth of the future after-tax earnings of the asset *plus* the present worth of any salvage value, computed at 8.25% interest. For the data tabulated above, we compute the present worth of the after-tax earnings for each year, as follows:

Year	$(P/F,8.25\%,1)$ $(P/F,8.25\%,2)$ $(P/F,8.25\%,3)$ $(P/F,8.25\%,4)$ $(P/F,8.25\%,5)$
1	3412(0.9238) + 2927(0.8534) + 2443(0.7883) + 1961(0.7283) + 1480(0.6728) = $10,000
2	2927(0.9238) + 2443(0.8534) + 1961(0.7883) + 1480(0.7283) = $7413
3	2443(0.9238) + 1961(0.8534) + 1480(0.7883) = $5098
4	1961(0.9238) + 1480(0.8534) = $3075
5	1480(0.9238) = $1368

The computed present worths of the after-tax earnings are equal to the tabulation of unrecovered investment (Row 7) in Table 12-1. The capital consumption model, therefore, meets all the MAPI assumptions. From Table 12-1, we can compute the percent retention value each year for the case of five-year service life, straight line depreciation, and no salvage value. The first year, $2587 of the $10,000 cost is consumed. The retention value is $10,000 − 2587, or $7413. This is 74.1% of the initial cost. By similar computations the percent retention values may be computed for the various comparison periods.

Comparison period, in years	*Retention value*
1	74.1%
2	50.9
3	30.7
4	13.6
5	0

Figure 12-7 is the MAPI Chart 3A of Percent Retention Values for a one-year comparison period and straight line depreciation. Our 74.1% retention value is one point on this chart. You should be able to locate the point and see that it agrees with the value on the chart. The other percent retention values may be located in Fig. 12-8 on the MAPI Chart 3B for longer comparison periods.

The MAPI model charts were computed using a 50% tax rate, 25% of the investment borrowed, 3% interest on the borrowed money, and a 10% after-tax return on equity capital, as has been described above. These values do not seem suitable at the present time. With some computer programming effort, different values could be inserted in the model formulas and a new set of MAPI charts computed to accurately reflect a specific situation.

MAPI Worksheets

A two-page worksheet is needed to solve a problem by the MAPI system. These are shown in Figures 12-9 and 12-10. On Sheet 1, the annual operating advantage is computed, assuming the project is implemented rather than some alternative (like continuing with the present situation). The comparison period should be selected as the shorter of the economic lives of the project or its alternative. As previously described, this often is one year, but it may be longer. On Sheet 2, the form is divided into three sections.

Section A. Initial Investment:

Line 26. The net cost is the Installed cost of the project *minus* any Investment tax credit.

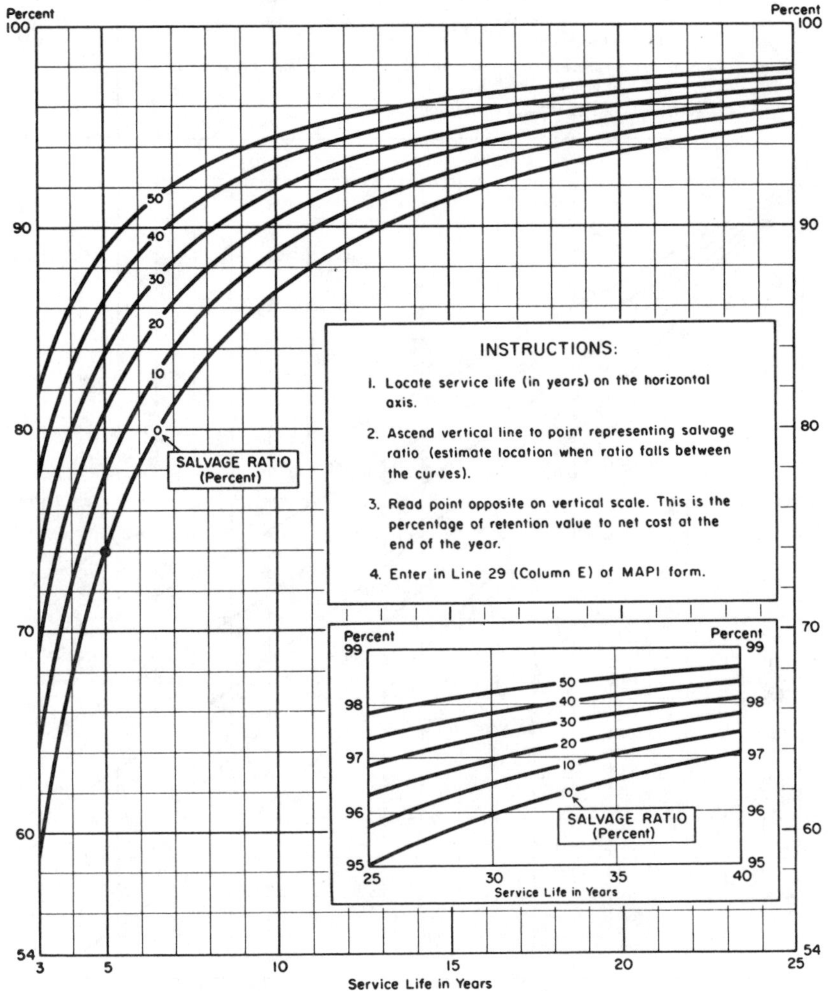
MAPI CHART No. 3A
(ONE-YEAR COMPARISON PERIOD AND STRAIGHT-LINE TAX DEPRECIATION)
Percent
Percent
SALVAGE RATIO (Percent)
INSTRUCTIONS:
1. Locate service life (in years) on the horizontal axis.
2. Ascend vertical line to point representing salvage ratio (estimate location when ratio falls between the curves).
3. Read point opposite on vertical scale. This is the percentage of retention value to net cost at the end of the year.
4. Enter in Line 29 (Column E) of MAPI form.
Service Life in Years
Service Life in Years
Copyright 1967, Machinery and Allied Products Institute and used with their permission

Figure 12-7 MAPI Chart 3A: Percent retention values for a one-year comparison period and straight line depreciation.

MAPI CHART No. 3B

(LONGER THAN ONE-YEAR COMPARISON PERIODS
AND STRAIGHT-LINE TAX DEPRECIATION)

INSTRUCTIONS:

1. Locate on horizontal axis percentage which comparison period is of service life.
2. Ascend vertical line to point representing salvage ratio (estimate location when ratio falls between the curves).
3. Read point opposite on vertical scale. This is the percentage of retention value to net cost at end of comparison period.
4. Enter in line 29 (Column E) of MAPI form.

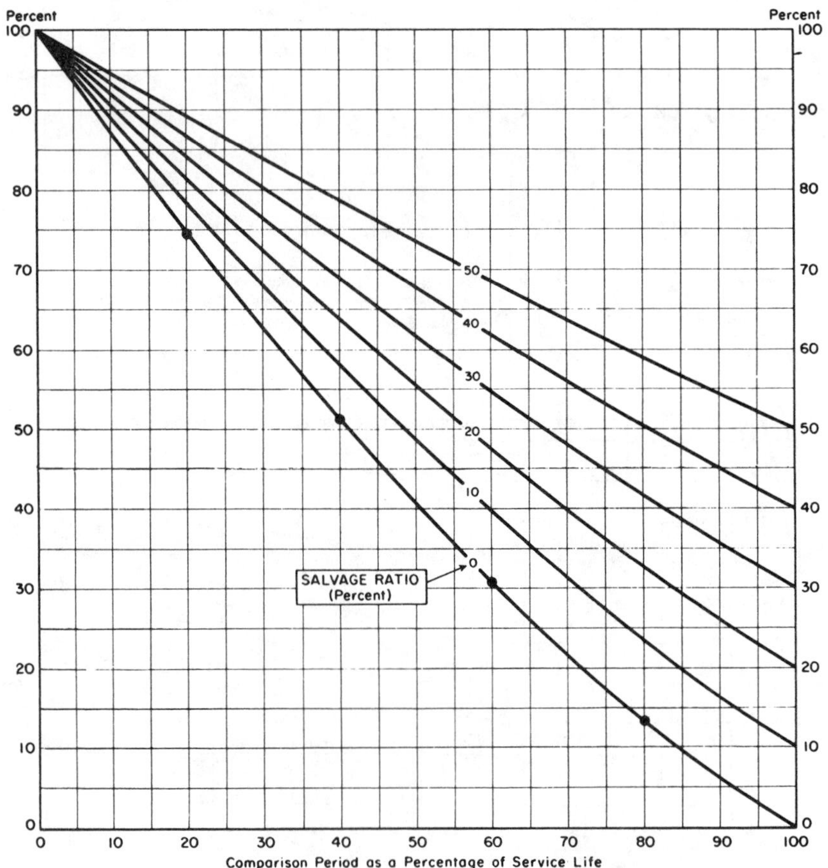

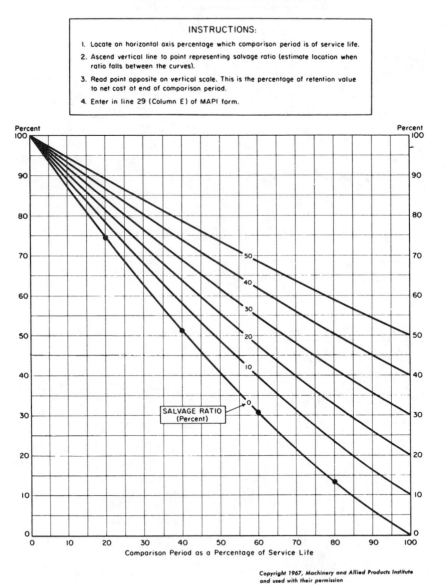

Figure 12-8 MAPI Chart 3B: Percent retention values for longer than one-year comparison periods and straight line depreciation.

PROJECT NO.__________ SHEET I

MAPI SUMMARY FORM

(AVERAGING SHORTCUT)

PROJECT____________________

ALTERNATIVE____________________

COMPARISON PERIOD (YEARS) (P)__________

ASSUMED OPERATING RATE OF PROJECT (HOURS PER YEAR) __________

I. OPERATING ADVANTAGE

(NEXT-YEAR FOR A 1-YEAR COMPARISON PERIOD,* ANNUAL AVERAGES FOR LONGER PERIODS)

A. EFFECT OF PROJECT ON REVENUE

		INCREASE	DECREASE	
1	FROM CHANGE IN QUALITY OF PRODUCTS	$	$	1
2	FROM CHANGE IN VOLUME OF OUTPUT			2
3	TOTAL	$ X	$ Y	3

B. EFFECT ON OPERATING COSTS

		INCREASE	DECREASE	
4	DIRECT LABOR	$	$	4
5	INDIRECT LABOR			5
6	FRINGE BENEFITS			6
7	MAINTENANCE			7
8	TOOLING			8
9	MATERIALS AND SUPPLIES			9
10	INSPECTION			10
11	ASSEMBLY			11
12	SCRAP AND REWORK			12
13	DOWN TIME			13
14	POWER			14
15	FLOOR SPACE			15
16	PROPERTY TAXES AND INSURANCE			16
17	SUBCONTRACTING			17
18	INVENTORY			18
19	SAFETY			19
20	FLEXIBILITY			20
21	OTHER			21
22	TOTAL	$ Y	$ X	22

C. COMBINED EFFECT

23	NET INCREASE IN REVENUE (3X−3Y)	$	23
24	NET DECREASE IN OPERATING COSTS (22X−22Y)	$	24
25	ANNUAL OPERATING ADVANTAGE (23+24)	$	25

* Next year means the first year of project operation. For projects with a significant break-in period, use performance after break-in.

Figure 12-9 MAPI summary form: Sheet 1.

SHEET 2

II. INVESTMENT AND RETURN

A. INITIAL INVESTMENT

26 INSTALLED COST OF PROJECT $ ______
MINUS INITIAL TAX BENEFIT OF $ ______ (Net Cost) $ ______ 26
27 INVESTMENT IN ALTERNATIVE
CAPITAL ADDITIONS MINUS INITIAL TAX BENEFIT $ ______
PLUS: DISPOSAL VALUE OF ASSETS RETIRED
BY PROJECT * $ ______ $ ______ 27
28 INITIAL NET INVESTMENT (26–27) $ ______ 28

B. TERMINAL INVESTMENT

29 RETENTION VALUE OF PROJECT AT END OF COMPARISON PERIOD
(ESTIMATE FOR ASSETS, IF ANY, THAT CANNOT BE DEPRECIATED OR EXPENSED. FOR OTHERS, ESTIMATE OR USE MAPI CHARTS.)

Item or Group	Installed Cost, Minus Initial Tax Benefit (Net Cost) A	Service Life (Years) B	Disposal Value, End of Life (Percent of Net Cost) C	MAPI Chart Number D	Chart Percentage E	Retention Value $\left(\frac{A \times E}{100}\right)$ F
	$					$

ESTIMATED FROM CHARTS (TOTAL OF COL. F) $ ______
PLUS: OTHERWISE ESTIMATED $ ______ $ ______ 29
30 DISPOSAL VALUE OF ALTERNATIVE AT END OF PERIOD * $ ______ 30
31 TERMINAL NET INVESTMENT (29–30) $ ______ 31

C. RETURN

32 AVERAGE NET CAPITAL CONSUMPTION $\left(\frac{28-31}{P}\right)$ $ ______ 32
33 AVERAGE NET INVESTMENT $\left(\frac{28+31}{2}\right)$ $ ______ 33
34 BEFORE-TAX RETURN $\left(\frac{25-32}{33} \times 100\right)$ % ______ 34
35 INCREASE IN DEPRECIATION AND INTEREST DEDUCTIONS $ ______ 35
36 TAXABLE OPERATING ADVANTAGE (25–35) $ ______ 36
37 INCREASE IN INCOME TAX (36×TAX RATE) $ ______ 37
38 AFTER-TAX OPERATING ADVANTAGE (25–37) $ ______ 38
39 AVAILABLE FOR RETURN ON INVESTMENT (38–32) $ ______ 39
40 AFTER-TAX RETURN $\left(\frac{39}{33} \times 100\right)$ % ______ 40

* After terminal tax adjustments.

Figure 12-10 MAPI Summary Form: Sheet 2.

Line 27. It may be that if the project is not adopted, it would be necessary to make a capital expenditure to continue with the present situation or whatever alternative is used. If so, it is added here. If, however, the project is selected, there may be alternative assets that would be retired and they would be entered here.

Line 28. The initial net investment is simply the Net cost of the project *minus* any Net capital additions that would be needed by the alternative *minus* any Assets retired by the project. In short, it is the net incremental cost of proceeding with the project rather than the alternative.

Section B. Terminal Investment:

Line 29. Our elaborate capital consumption computations were to show how the retention value is computed for use on this line.

A. The net cost is taken from Line 26.

B. Service life is estimated for the project. The model assumes better future challengers so the service life should not be reduced here.

E. The percent retention value from the chart is entered here.

Line 30. This is the salvage value of the alternative if it were retained.

Section C. Return:

Line 32. The *P* in the denominator is the comparison period on Sheet 1.

Line 35. To solve for the after-tax rate of return, it is necessary to take into account the depreciation and interest deductions that the project produces in excess of the deductions for the alternative. For a one-year comparison, it is the first-year increase; for longer comparison periods, it is the average annual increase.

The use of the MAPI system and the associated worksheets can be illustrated by an example problem.

EXAMPLE 12-10

An existing milling machine is being considered for replacement by a new milling machine. If the old milling machine is continued in service, its remaining economic life would be one year. The operating advantage of the new machine over the existing machine is tabulated on the MAPI Worksheet 1 in Fig. 12-11.

The new milling machine, which would cost $42,000, qualifies for an assumed 7% investment tax credit. It has an estimated twenty-year useful life and could be sold for 20% of its net cost at the end of that time. It will be depreciated by the straight line method. The next-year increase in depreciation and interest deductions is estimated to be $1700.

SHEET I

MAPI SUMMARY FORM
(AVERAGING SHORTCUT)

PROJECT New Milling Machine
ALTERNATIVE Continue with existing milling machine
COMPARISON PERIOD (YEARS) (P) 1 year
ASSUMED OPERATING RATE OF PROJECT (HOURS PER YEAR)

I. OPERATING ADVANTAGE
(NEXT-YEAR FOR A 1-YEAR COMPARISON PERIOD,* ANNUAL AVERAGES FOR LONGER PERIODS)

A. EFFECT OF PROJECT ON REVENUE

		INCREASE	DECREASE	
1	FROM CHANGE IN QUALITY OF PRODUCTS	$ 500	$	1
2	FROM CHANGE IN VOLUME OF OUTPUT	1000		2
3	TOTAL	$ 1500 X	$ Y	3

B. EFFECT ON OPERATING COSTS

4	DIRECT LABOR	$	$ 500	4
5	INDIRECT LABOR	200		5
6	FRINGE BENEFITS	50		6
7	MAINTENANCE		400	7
8	TOOLING			8
9	MATERIALS AND SUPPLIES		2100	9
10	INSPECTION		100	10
11	ASSEMBLY			11
12	SCRAP AND REWORK		1600	12
13	DOWN TIME			13
14	POWER	50		14
15	FLOOR SPACE			15
16	PROPERTY TAXES AND INSURANCE	100		16
17	SUBCONTRACTING			17
18	INVENTORY			18
19	SAFETY			19
20	FLEXIBILITY			20
21	OTHER			21
22	TOTAL	$ 400 Y	$ 4700 X	22

C. COMBINED EFFECT

23	NET INCREASE IN REVENUE (3X – 3Y)	$ 1500	23
24	NET DECREASE IN OPERATING COSTS (22X – 22Y)	$ 4300	24
25	ANNUAL OPERATING ADVANTAGE (23 + 24)	$ 5800	25

* Next year means the first year of project operation. For projects with a significant break-in period, use performance after break-in.

Figure 12-11a MAPI Summary Form, Sheet 1 for Example 12-10.

The existing machine could be sold for \$8500 now or about \$8000 one year hence. Other data for both milling machines are shown on the MAPI worksheets. Compute the incremental after-tax rate of return if the existing milling machine (the alternative) is replaced by the new milling machine (the project).

From the Fig. 12-11 MAPI worksheets, we see that the incremental after-tax rate of return is 7.7%.

SHEET 2

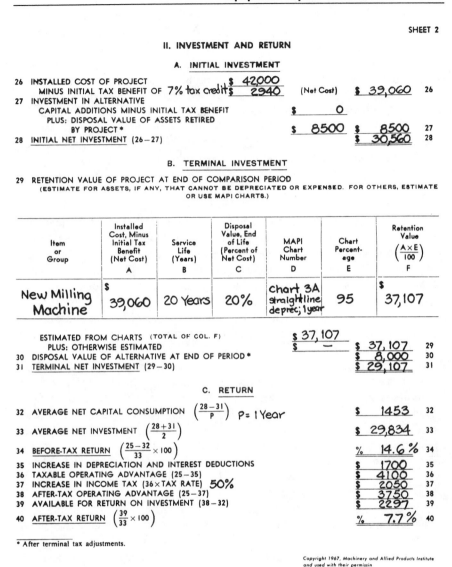

II. INVESTMENT AND RETURN

A. INITIAL INVESTMENT

26 INSTALLED COST OF PROJECT $ 42,000
MINUS INITIAL TAX BENEFIT OF 7% tax credit $ 2940 (Net Cost) $ 39,060 26
27 INVESTMENT IN ALTERNATIVE
CAPITAL ADDITIONS MINUS INITIAL TAX BENEFIT $ 0
PLUS: DISPOSAL VALUE OF ASSETS RETIRED
BY PROJECT * $ 8500 $ 8500 27
28 INITIAL NET INVESTMENT (26–27) $ 30,560 28

B. TERMINAL INVESTMENT

29 RETENTION VALUE OF PROJECT AT END OF COMPARISON PERIOD
(ESTIMATE FOR ASSETS, IF ANY, THAT CANNOT BE DEPRECIATED OR EXPENSED. FOR OTHERS, ESTIMATE OR USE MAPI CHARTS.)

Item or Group	Installed Cost, Minus Initial Tax Benefit (Net Cost) A	Service Life (Years) B	Disposal Value, End of Life (Percent of Net Cost) C	MAPI Chart Number D	Chart Percentage E	Retention Value $\left(\frac{A \times E}{100}\right)$ F
New Milling Machine	$ 39,060	20 Years	20%	Chart 3A straightline deprec; 1 year	95	$ 37,107

ESTIMATED FROM CHARTS (TOTAL OF COL. F) $ 37,107
PLUS: OTHERWISE ESTIMATED $ — $ 37,107 29
30 DISPOSAL VALUE OF ALTERNATIVE AT END OF PERIOD * $ 8,000 30
31 TERMINAL NET INVESTMENT (29–30) $ 29,107 31

C. RETURN

32 AVERAGE NET CAPITAL CONSUMPTION $\left(\frac{28-31}{P}\right)$ P = 1 Year $ 1453 32

33 AVERAGE NET INVESTMENT $\left(\frac{28+31}{2}\right)$ $ 29,834 33

34 BEFORE-TAX RETURN $\left(\frac{25-32}{33} \times 100\right)$ % 14.6 % 34

35 INCREASE IN DEPRECIATION AND INTEREST DEDUCTIONS $ 1700 35
36 TAXABLE OPERATING ADVANTAGE (25–35) $ 4100 36
37 INCREASE IN INCOME TAX (36 × TAX RATE) 50% $ 2050 37
38 AFTER-TAX OPERATING ADVANTAGE (25–37) $ 3750 38
39 AVAILABLE FOR RETURN ON INVESTMENT (38–32) $ 2297 39
40 AFTER-TAX RETURN $\left(\frac{39}{33} \times 100\right)$ % 7.7 % 40

* After terminal tax adjustments.

Figure 12-11b MAPI Summary Form, Sheet 2 for Example 12-10.

Applying Equipment Replacement Models

From the discussion of the MAPI system, it is apparent that the incremental after-tax rate of return computed by the MAPI model is based on different assumptions than our previous direct computations. Thus, our direct computations and the MAPI computations will produce slightly different results. Which is correct, if either one is, depends on how well the actual situation is portrayed in the computational models.

While we usually make simplifying assumptions in direct computations, these assumptions are readily apparent and their probable impact on the answer may be estimated. In replacement models, like the MAPI system, the complexity obscures the underlying assumptions and makes it harder to judge the sensitivity of the answer to the assumptions. A reasonable conclusion is that mathematical models may be very valuable in equipment replacement analysis, but care must be taken to ensure that the model fits the situation to which it is to be applied.

SUMMARY

In selecting equipment for a new plant the question is, "Which of the machines available on the market will be more economical?" But when one has a piece of equipment that is now performing the desired task, the analysis is made more complicated. The existing equipment (called the defender) is already in place so the question is, "Shall we replace it now, or shall we keep it for one or more years?" When a replacement is indicated, it will be by the best available replacement equipment (called the challenger). When we already have equipment, there may be a tendency to use past costs in the replacement analysis. But it is only present and future costs that are relevant.

For the defender, the economic life is the remaining useful life where the EUAC is a minimum. In situations where costs are increasing, the remaining useful life is considered to be one year. For the challenger, a similar calculation is made to determine its economic life. The replacement analysis should be based on the defender at the economic (useful) life remaining to it compared to the challenger at its economic life. When the useful lives are the same, a variety of analysis techniques (present worth, annual cash flow, rate of return, and so forth) may be used. But for different useful lives, the annual cash flow analysis is generally the most suitable method of analysis.

Until this chapter, we have assumed that future replacements will be the same as the present replacement. But it may be more realistic to assume that future challengers will be an improvement over the present challenger. The prospect of improved future challengers makes the present challenger a less desirable alternative. By continuing with the defender, we may later acquire an improved challenger. One method of adjusting a replacement analysis to account for improved future challengers is to reduce the economic life of the present challenger.

Equipment replacement models have been developed for more complex replacement problems. An example is the MAPI replacement model. The model is carefully organized to provide an accurate, incremental, after-tax rate of return based on the difference between the proposed project and a stated alternative. The difficult problem in using any replacement model is determining whether the assumptions of the model fit the situation to which the model is to be applied.

Problems

12-1 Typically there are two alternatives in a replacement analysis. One alternative is to replace the defender now. The other alternative is which one of the following?

a. keep the defender for its remaining useful life;

b. keep the defender for another year and then re-examine the situation;

c. keep the defender until there is an improved challenger that is better than the present challenger.

12-2 The economic life of the defender can be obtained if certain estimates about the defender can be made. Assuming those estimates prove to be exactly correct, one can accurately predict the year when the defender should be replaced, even if nothing is known about the challenger.

Is the above statement true or false? Explain.

12-3 A proposal has been made to replace a large heat exchanger (initial cost was $85,000 three years ago) with a new more efficient unit at a cost of $120,000. The existing heat exchanger is being depreciated by the ACRS method. Its present book value is $20,400, but it has no current value as its scrap value just equals the cost to remove it from the plant. In preparing the before-tax economic analysis to see if the existing heat exchanger should be replaced, the question arises concerning the proper treatment of the $20,400 book value. Three possibilities are that the $20,400 book value of the old heat exchanger is:

a. *added* to the cost of the new exchanger in the economic analysis;

b. *subtracted* from the cost of the new exchanger in the economic analysis;

c. *ignored* in this before-tax economic analysis.

Which of the three possibilities is correct?

12-4 A machine tool, that has been used in a plant for ten years, is being considered for replacement. It cost $9500 and was depreciated by ACRS depreciation using a five-year recovery period. An equipment dealer indicates the machine has no resale value. Maintenance on the machine tool has been something of a problem, with an $800 cost this year. Future annual maintenance costs are expected to be higher. What is the economic life of this machine tool if it is kept in service?

12-5 A new $40,000 bottling machine has just been installed in a plant. It will have no salvage value when it is removed. The plant manager has asked you to estimate the economic service life for the machine, ignoring income taxes. He estimates that the annual maintenance cost will be constant at $2500 per year. What service life will result in the least equivalent uniform annual cost?

12-6 Which one of the following is the proper dollar value of defender equipment to use in replacement analysis?

a. original cost;

b. present market value;

c. present trade-in value;

d. present book value;

e. present replacement cost, if different from original cost.

12-7 The Ajax Corporation purchased a railroad tank car eight years ago for $60,000. It is being depreciated by SOYD depreciation, assuming a ten-year depreciable life and a $7000 salvage value. The tank car needs to be reconditioned now at a cost of $35,000. If this is done, it is estimated the equipment will last for ten more years and have a $10,000 salvage value at the end of the ten years.

On the other hand, the existing tank car could be sold now for $10,000 and a new tank car purchased for $85,000. The new tank car would be depreciated by ACRS depreciation. Its estimated actual salvage value would be $15,000. In addition, the new tank car would save $7000 per year in maintenance costs, compared to the reconditioned tank car.

Based on a 15% before-tax rate of return, determine whether the existing tank car should be reconditioned, or a new one purchased. *Note:* The problem statement provides more data than are needed. This is typical of realistic problems.

(*Answer:* Recondition the old tank car.)

12-8 The Clap Chemical Company needs a large insulated stainless steel tank for the expansion of their plant. Clap has located one at a brewery that has just been closed. The brewery offers to sell the tank for $15,000 delivered to the chemical plant. The price is so low that Clap believes it can sell the tank at any future time and recover its $15,000 investment.

The outside of the tank is lined with heavy insulation that requires considerable maintenance. It is estimated the maintenance costs will be as follows:

Year	*Insulation maintenance cost*
0	$2000
1	500
2	1000
3	1500
4	2000
5	2500

a. Based on a 15% before-tax MARR, what is the economic life of the insulated tank?

b. Is it likely that the insulated tank will be replaced by another tank at the end of its computed economic life? Explain.

12-9 The plant manager has just purchased a piece of unusual machinery for $10,000. Its resale value at the end of one year is estimated to be $3000 and is rising at the rate of $500 per year, because it is sought by antique collectors.

The maintenance cost is expected to be $300 per year for each of the first three years, and then it is expected to double each year after that. Thus the fourth-year maintenance will be $600; the fifth-year maintenance, $1200; and so on. Based on a 15% before-tax MARR, what is the economic life of this machinery?

12-10 A firm manufactures padded shipping bags. One hundred bags are packed in a cardboard carton. At present, machine operators fill the cardboard cartons by eye, that is, when the cardboard carton looks full, it is assumed to contain 100 shipping bags. Actual inspection reveals that the cardboard carton may contain anywhere from 98 to 123 bags with an average quantity of 105.5 bags.

The management has never received complaints from its customers about cartons containing less than 100 bags. Nevertheless, management realizes that they are giving away 5½% of their output by overfilling the cartons. One solution would be to count the shipping bags to ensure that 100 are packed in each carton.

Another solution would be to weigh each filled shipping carton. Underweight cartons would have additional shipping bags added, and overweight cartons would have some shipping bags removed. This would not be a perfect solution as the actual weight of the shipping bags varies slightly. If the weighing is done, it is believed that the average quantity of bags per carton could be reduced to 102, with almost no cartons containing less than 100 bags.

The weighing equipment would cost $18,600. The equipment would be depreciated by straight line depreciation using a ten-year depreciable life and a $3700 salvage value at the end of ten years. The $18,600 worth of equipment qualifies for a 10% investment tax credit. One person, hired at a cost of $16,000 per year, would be required to operate the weighing equipment and to add or remove padded bags from the cardboard cartons. 200,000 cartons will be checked on the weighing equipment each year, with an average removal of 3.5 padded bags per carton with a manufacturing cost of 3 cents per bag.

This large profitable corporation has a 50% combined federal-plus-state incremental tax rate. Assume a six-year comparison period for the analysis. Compute the MAPI after-tax return.

12-11 The Quick Manufacturing Co., a large profitable corporation, is considering the replacement of a production machine tool. A new machine would cost $3700, have a four-year useful and depreciable life, and have no salvage value. For tax purposes, sum-of-years digits depreciation would be used. The existing machine tool was purchased four years ago at a cost of $4000, and has been depreciated by straight line depreciation assuming an eight-year life and no salvage value. It could be sold now to a used equipment dealer for $1000 or be kept in service for another four years. It would then have no salvage value. The new machine would save about $900 per year in operating costs compared to the existing machine. Assume a 40% combined state and federal tax rate.

- ***a.*** Compute the before-tax rate of return on the replacement proposal of installing the new machine rather than keeping the existing machine.
- ***b.*** Compute the after-tax rate of return on this replacement proposal.

(*Answer:* ***a.*** 12.6%)

12-12 The Plant Department of the local telephone company purchased four special pole hole diggers eight years ago for $14,000 each. They have been in constant use to the present time. Due to an increased workload, it is considered that additional machines will soon be required. Recently it was announced that an improved model of the digger has been put on the market. The new machines have a higher production rate and lower maintenance expense than the old machines, but their cost will be $32,000 each. The service life of the new machines is estimated to be eight years with a salvage estimated at $750 each. The four original diggers have an immediate salvage of $2000 each and an estimated salvage of $500 each eight years hence. The estimated average annual maintenance expense associated with the old machines is approximately $1500 each, compared to $600 each for the new machines.

A field study and trial indicates that the workload would require three additional new machines if the old machines are continued in service. However, if the old machines are all retired from service, the present workload plus the estimated increased load could be carried by six new machines with an annual savings of $12,000 in operator costs. Because the new machines employ a new principle of operation, it is contemplated that a special personnel training program will be necessary before the machines can be placed in operation. It is estimated that this training program will cost about $700 per new machine. If the MARR is 9% before taxes, what should the company do?

12-13 Fifteen years ago the Acme Manufacturing Company bought a propane powered forklift truck for $4800. They depreciated it using straight line depreciation, a twelve-year life, and zero salvage value. Over the years, the forklift has been a good piece of equipment, but lately the maintenance cost has risen sharply. Estimated end-of-year maintenance costs for the next ten years are as follows:

Year	*Maintenance cost*
1	$ 400
2	600
3	800
4	1000
5–10	1400/year

The old forklift has no present or future net salvage value as its scrap metal value just equals the cost to haul it away. A replacement is now being considered for the old forklift. A modern unit can be purchased for $6500. It has an economic life equal to its ten-year depreciable life. Straight line depreciation will be employed with zero salvage value at the end of the ten-year depreciable life. At any time the new forklift can be sold for its book value. Maintenance on the new forklift is estimated to be a constant $50 per year for the next ten years. After that maintenance is expected to increase sharply. Should Acme Manufacturing keep its old forklift truck for the present, or replace it now with a new forklift truck? The firm expects an 8% after-tax rate of return on its investments. Assume a 40% combined state-and-federal tax rate.

(*Answer:* Keep the old forklift truck.)

12-14 A firm is concerned about the condition of some of its plant machinery. Bill James, a newly hired engineer, was assigned the task of looking into the situation and determining what alternatives are available. After a careful analysis, Bill reports that there are five feasible, mutually exclusive alternatives.

Alternative A. Spend $44,000 now repairing various items. The $44,000 can be charged as a current operating expense (rather than capitalized) and deducted from other taxable income immediately. These repairs are anticipated to keep the plant functioning for the next seven years with operating costs remaining at present levels.

Alternative B. Purchase $49,000 of general purpose equipment. Depreciation would be straight line, with the depreciable life equal to the seven-year useful life of the equipment. The equipment will have no end-of-useful-life salvage value. The new equipment will reduce operating costs $6000 per year below their present levels.

Alternative C. Purchase $56,000 of new specialized equipment. This equipment would be depreciated by sum-of-years digits depreciation over its seven-year useful life. This equipment would reduce operating costs $12,000 per year below their present levels. It will have no end-of-useful-life salvage value.

Alternative D. This alternative is the same as Alt. *B*, except that this particular equipment would reduce operating costs $7000 per year below their present levels.

Alternative E. This is the "do-nothing" alternative. If nothing is done, future annual operating costs are expected to be $8000 above the present level.

This profitable firm pays 40% corporate income taxes. In their economic analysis, they require a 10% after-tax rate of return. Which of the five alternatives should the firm adopt?

12-15 In a replacement analysis problem, the following facts are known:

- Initial cost: $12,000
- Annual Maintenance: None for the first three years;
  $2000 at the end of the fourth year;
  $2000 at the end of the fifth year;
  Increasing $2500 per year after the fifth year ($4500 at the end of the sixth year, $7000 at the end of the seventh year, and so forth).

Actual salvage value in any year is zero. Assume a 10% interest rate and ignore income taxes. Compute the best useful life for this challenger. (*Answer:* Five years)

12-16 (20-58) Machine *A* has been completely overhauled for $9000 and is expected to last another 12 years. The $9000 was treated as an expense for tax purposes last year. It can be sold now for $30,000 net after selling expenses, but will have no salvage value 12 years hence. It was bought new nine years ago for $54,000 and has been depreciated since then by straight line depreciation using a 12-year depreciable life.

Because less output is now required, Machine *A* can now be replaced with a smaller Machine *B*. Machine *B* costs $42,000, has an anticipated life of 12 years, and would reduce operating costs $2500 per year. It would be depreciated by straight line depreciation with a 12-year depreciable life and no salvage value.

Both the income tax and capital gains tax rates are 40%. Compare the after-tax annual cost of the two machines and decide whether Machine *A* should be retained or replaced by Machine *B*. Use a 10% after-tax rate of return in the calculations.

CHAPTER 13

Inflation And Deflation

Until now we have assumed a stable economic situation, that is, where prices are relatively unchanged over substantial periods of time. Unfortunately, this is not always a realistic assumption. Chapter 13 describes how either inflation or deflation may be incorporated into an economic analysis. If all costs and benefits are changing at equal rates, then inflation has no net effect on before-tax economic analyses. Even in this idealized situation, both inflation and deflation have an impact on the after-tax results. Because it occurs far more frequently, the emphasis of this chapter is on inflation.

Inflation And Deflation Defined

As all readers will recognize, ***inflation*** is the situation where prices of goods and services are increasing, that is, apartment rents, textbook prices, haircuts, groceries—and almost everything else—tend to increase as time passes.

This general upward movement of prices in an inflationary period does not necessarily produce a single rate of inflation. The prices of different items change at different rates, due to the complexities of a free competitive economy (plus monopolistic and other not-so-competitive influences).

If inflation means prices are rising, what about the value of money? Ten years ago the wholesale price of a box of oranges was $4.75, while recently it was $10.50. In other words $4.75 will not buy today what it would buy ten years ago. The purchasing power, or value, of money has declined. ***Inflation makes future dollars less valuable than present dollars.***

EXAMPLE 13-1

Ten years ago the owner of an orange grove went to the bank and borrowed $4750. This represented an amount of money equal to 1000 boxes of oranges. The loan was for ten years at 7% interest. At $10.50 per box, how many boxes of oranges now represent repayment of the debt?

Solution: The orchardist promised to pay

$$4750(F/P,7\%,10) = 4750(1.967) = \$9343$$

at the end of ten years. He had computed that he would have to sell (9343 ÷ 4.75) = 1791 boxes of oranges to repay the debt.

Now oranges sell for \$10.50 per box. To repay the debt, the orchardist has to sell (9343 ÷ 10.50) = 890 boxes of oranges. Although the orchardist borrowed money equivalent to 1000 boxes of oranges, the increase in their selling price means that he is repaying the equivalent of fewer oranges than he borrowed. ◀

Inflation benefits long-term borrowers of money—this is because they will repay their debt in dollars with reduced purchasing power. If inflation helps long-term borrowers, it just as clearly is unfavorable for long-term lenders of money.

Deflation is just the reverse of inflation. Prices that tend to decline result in future dollars that have more purchasing power than present dollars. Here the lenders are benefited as they lend present dollars which must be subsequently repaid in future dollars that have greater purchasing power.

Price Changes

Price changes may or may not need to be considered in an economic analysis. We know that costs and benefits must be computed in comparable units. It would make no sense to compute costs in Year-1992 dollars and then measure benefits in Year-1997 dollars, if 1997 dollars are a different unit of measure than 1992 dollars. It would be like measuring costs in apples and benefits in oranges.

On the other hand, one might have a situation where future benefits fluctuate *along with* the inflation or deflation of the period. This could result in a situation where the future benefits are constant when measured in Year-0 dollars. If this had been the original assumption (constant benefits in Year-0 dollars), the subsequent inflation or deflation would have no effect on a before-tax economic analysis. We would need to be concerned only with any differential inflation or deflation effects which distort the cash flow when it is measured in comparable units.

In after-tax economy studies, the income tax consequences are computed on the taxable portion of the cash flow in a given year. To compute the 1992 taxable income, one must use 1992 dollars and 1992 data. At the same time, the 1992 depreciation deduction would be based on the original cost of the asset. Therefore, in after-tax economic analysis, the impact of inflation or deflation cannot be eliminated by simply adjusting future before-tax benefits. Initially, the effects of inflation or deflation will be presented without considering income taxes. Then the more difficult after-tax situation will be considered in the last part of this chapter.

Measuring the Rate of Price Change

To introduce the effect of price changes into our computations, it would be helpful if we could determine the ***rate of price change***. In a period of increasing prices, the rate of price change could simply be called the ***inflation rate*** and designated f. Unless otherwise noted, the inflation rate f is assumed to be compounded annually.

A look at the cost to mail a letter in the United States can help explain the situation. In the ten-year period 1979–1989 (see Fig. 13-1), for example, the cost to mail a letter increased from 15¢ to 25¢. Using these values, we can compute the inflation rate f for the period:

$$F = P(F/P,f,n \text{ years})$$

$$25 = 15(F/P,f,10) \qquad (F/P,f,10) = 1.67$$

Interpolating from the Compound Interest Tables, $f = 5.2\%$. Thus, the inflation rate in the cost of mailing a letter has been 5.2%.

To obtain a broad measure of the inflation rate in the United States, the U.S. Government gathers statistics and computes many indices. The best known one is the ***Consumer Price Index*** (CPI) which attempts to measure the cost of food, other commodities, and services in the proportions appropriate to urban consumers. The result is a measure of the *change in the cost of living*. Table 13-1 tabulates values of the Consumer Price Index. The CPI may be used to compute an inflation rate for the ten-year period 1977–1987.

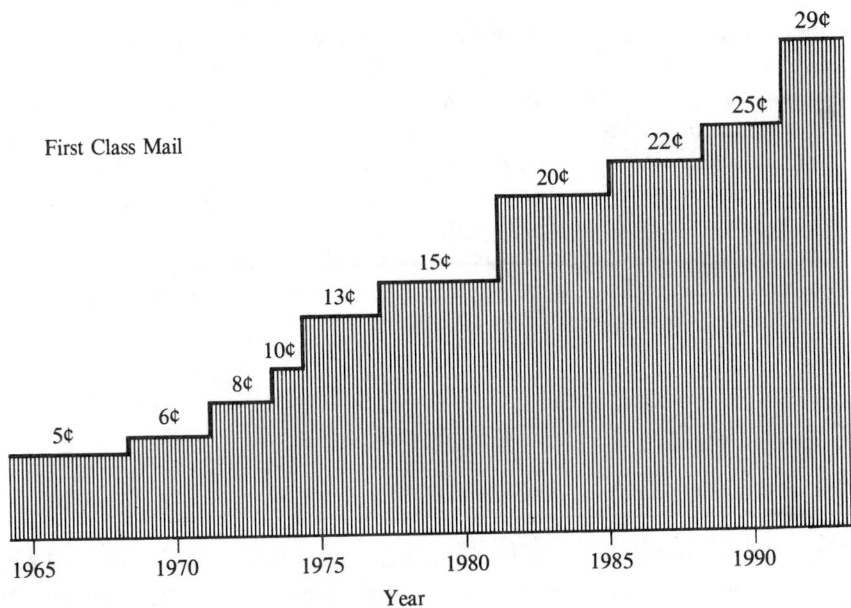

Figure 13-1 Cost to mail a first-class letter.

Table 13-1 INFLATION AS MEASURED BY THE CONSUMER PRICE INDEX

Year	Consumer Price Index*	Inflation rate f	Year	Consumer Price Index*	Inflation rate f
1973	44.3	6.0%	1983	99.6	3.2%
1974	49.3	11.3	1984	103.9	4.3
1975	53.8	9.1	1985	107.6	3.6
1976	56.9	5.8	1986	109.6	1.9
1977	60.6	6.5	1987	113.6	3.6
1978	65.2	7.6	1988	118.3	4.1
1979	72.6	11.3	1989	124.0	4.8
1980	82.4	13.5	1990	130.7	5.4
1981	90.9	10.3			
1982	96.5	6.2			

*Reference base: 1982–84 = 100.

$$(F/P,f,10) = \frac{113.6}{60.6} = 1.87 \qquad f = 6.5\%$$

The computed inflation rates tabulated in Table 13-1 show the large variation in the annual inflation during the period.

Are Price Indexes Suitable for Estimating Future Prices?

Often one must project the future price of some specific item, like a piece of machinery or a raw-material item. Rather than use some general index like the Consumer Price Index or the ***Producer Price Index*** (PPI, the cost of commodities and services to producers), it would be better to consider the specific components of an item whose cost is being projected. It may be, for example, that although producer prices are rising, a specific item may be reasonably expected to remain unchanged, or even decline. There could be many reasons for such a situation: improved technology, expanded production facilities, increased price competition, and so forth. Thus, general price indexes may—or may not—be suitable for predicting future prices of specific items.

Considering Price Changes In Before-Tax Calculations

Estimates of price changes in the future that reflect increased or decreased purchasing power of money certainly should be included in economy studies. In this situation, a disbursement at one point in time no longer yields the same *amount* of an item as the same disbursement at another point in time. Thus, dollars at one point in time represent more or less buying power than dollars at another point in time.

This variation in buying power must be adjusted. The obvious way is to convert all components of a cash flow to equivalent dollars with equal buying power. The standard dollar could be dollars with 1992 buying power, or 1997 buying power, or what have you. The essential item is that a standard be adopted. Economic analyses typically are analyses *now* concerning alternative *future* actions. Since the buying power of *now* dollars is both known and a reasonable standard, the usual practice is to adjust the cash flow to equivalent dollars *now*—Year 0. This may be accomplished by making a year-by-year computation to adjust the cash flow.

EXAMPLE 13-2

For the cash flow it is expected that the future benefits will be received in inflated dollars. What are the equivalent Year-0 dollars if the inflation rate f is +4% per year?

Year	*Cash flow*
0	−\$50
1	+10
2	+25
3	+20
4	+20

Solution:

Year	*Unadjusted cash flow*	*Multiplied by*		*Cash flow in Year-0 dollars*
0	−\$50	1	=	−\$50
1	+10	$(1 + 0.04)^{-1}$	=	+9.6
2	+25	$(1 + 0.04)^{-2}$	=	+23.1
3	+20	$(1 + 0.04)^{-3}$	=	+17.8
4	+20	$(1 + 0.04)^{-4}$	=	+17.1

The general form of the multiplier is $(1 + f)^{-n}$. When f is positive (indicating increasing prices), the multiplier may be read from compound interest tables as $(P/F,f,n)$. A negative f indicates prices decreasing (deflation). ◀

If the cash flow fluctuates as in Example 13-2, each year must be individually adjusted. Then to compute the present worth of the future benefits, for example, a series of single payment present worth factors are used.

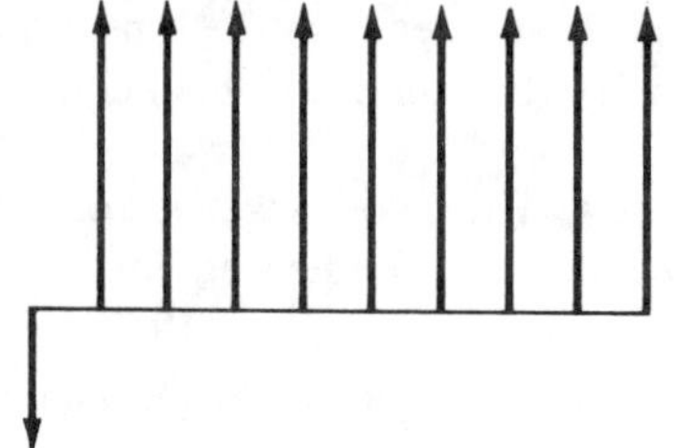

Figure 13-2 Actual cash flow for a loan or annuity.

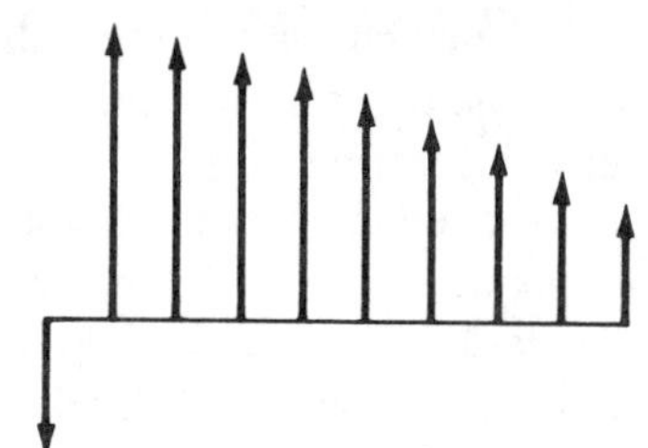

Figure 13-3 Cash flow measured in Year-0 dollars for a loan or annuity with inflation.

Impact of Price Changes on Constant Future Receipts or Disbursements

In situations of loans, annuities, and so forth, there is a commitment to a series of uniform end-of-period payments or receipts. These payments or receipts are unaffected by price changes. Yet the value of these future cash flow payments, computed in Year-0 dollars, is greatly affected by the price change rate. For equal future payments or receipts, the actual cash flow is represented by Figure 13-2. If there were inflation (for example, $f = +5\%$), the actual cash flow, measured in Year-0 dollars, would look like Fig. 13-3. The present worth of these future payments or receipts may be computed in Year-0 dollars by the method demonstrated in Example 13-3.

EXAMPLE 13-3

Mr. Dale Hart is considering the purchase of an annuity that will pay $1000 per year for ten years, beginning at the end of the first year. A friend tells him he believes there will be inflation of 6% per year for the next 10 years. If Mr. Hart decides that he should obtain a 5% rate of return, after considering the effect of the 6% inflation, how much would he be willing to pay now for the annuity?

Solution: The individual $1000 annuity payments are adjusted by the inflation rate f to obtain their equivalent values in Year-0 dollars. The inflation rate may be used to convert an actual amount of money F_{actual} in any year to the equivalent sum at the same point in time but in Year-0 dollars, $F_{\$0}$.

$$F_{\$0} = F_{\text{actual}}(1 + f)^{-n}$$

If this computation were carried out, the changed form of the cash flow diagram would be like Fig. 13-4.

But the computation need not be made. If we designate the interest rate without inflation as i', then we can write the equation for the present worth of F in Year-0 dollars:

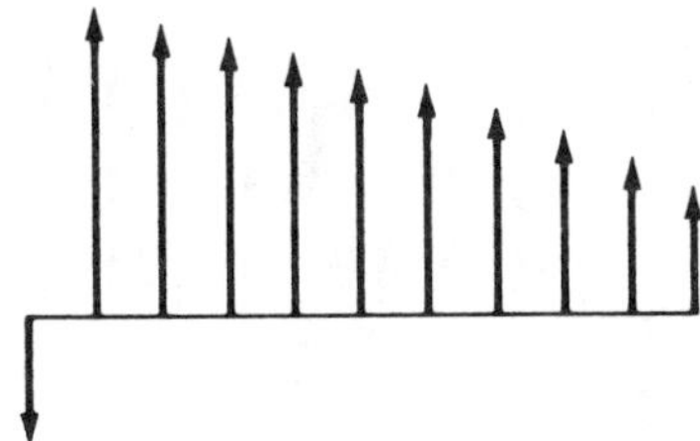

Figure 13-4 Cash flow for Example 13-3 in Year-0 dollars.

$$\text{PW of } F \text{ in Year-0 dollars} = F_{\text{actual}}\frac{(1+f)^{-n}}{(1+i')^n}$$

$$= F_{\text{actual}}\frac{1}{(1+i')^n(1+f)^n}$$

This is similar to the single payment present worth factor: $1/(1+i)^n$. The present worth factor may be applied here if we substitute for i:

$$\textbf{Equivalent } \boldsymbol{i = (1+i')(1+f) - 1 = i' + f + i'f} \qquad (13\text{-}1)$$

where i = interest rate

i' = interest rate without inflation

f = inflation rate

This equivalent value of i may be used with either single payment or uniform series present worth factors. In this problem:

$$\text{Equivalent } i = [0.05 + 0.06 + (0.05)(0.06)] = 0.113 = 11.3\%$$

The amount Mr. Hart would be willing to pay:

$$\text{PW} = 1000(P/A,11.3\%,10) = 1000(5.816) = 5816 \blacktriangleleft$$

Impact of Price Changes on a Fixed Quantity of Future Goods or Services

Future payments or receipts may be associated with a fixed quantity of goods. In the situation of inflation, for example, more dollars would be required to obtain a fixed quantity of goods. Of course it takes more or less actual dollars, depending on whether there is inflation or deflation. Figure 13-5 is the actual cash flow for 5% inflation.

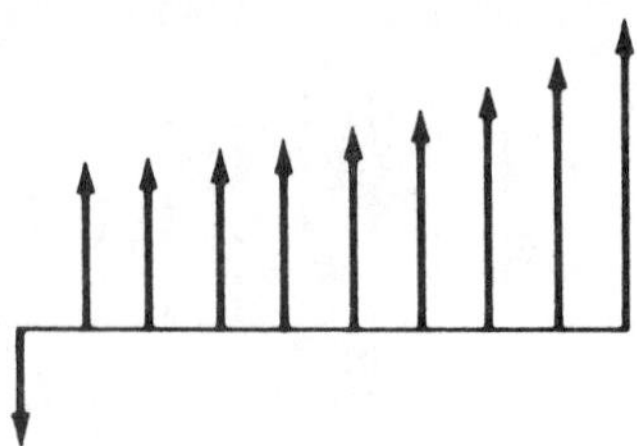

Figure 13-5 Actual cash flow for a fixed quantity of goods or services with inflation.

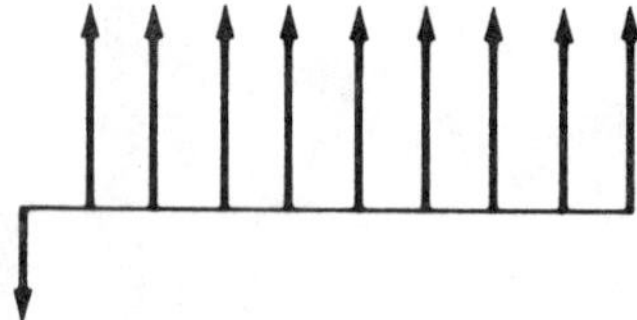

Figure 13-6 Cash flow measured in Year-0 dollars for a fixed quantity of goods or services.

Figure 13-6 shows that since a fixed quantity of goods or services is being priced, the cash flow measured in Year-0 dollars is constant. Thus present worth of cost of these goods or services in Year-0 dollars would simply be

$$\text{PW of cost} = A(P/A, i, n) \tag{13-2}$$

Inflation Effect On After-Tax Calculations

In the previous sections, we have noted the impact of inflation on before-tax calculations. We found that if the subsequent benefits brought constant quantities of dollars, then inflation will diminish the true value of the future benefits and, hence, the real rate of return. If, however, the future benefits keep up with the rate of inflation, the rate of return will not be adversely affected due to the inflation. Unfortunately, we are not so lucky when we consider a situation with income taxes. This is illustrated by Example 13-4.

EXAMPLE 13-4

A $12,000 investment will return annual benefits for six years with no salvage value at the end of six years. Assume straight line depreciation and a 46% income tax rate. The problem is to be solved for both before- and after-tax rates of return, the latter for two situations:

1. *No inflation:* the annual benefits are constant at $2918 per year.
2. *Inflation equal to 5%:* the benefits from the investment increase at this same rate, so that they continue to be the equivalent of $2918 in Year-0 dollars.

The benefit schedule for the two situations:

Year	*Annual benefit for both situations, in Year-0 dollars*	*No inflation, actual dollars received*	*5% inflation factor**	*5% inflation, actual dollars received*
1	$2918	$2918	$(1.05)^1$	$3064
2	2918	2918	$(1.05)^2$	3217
3	2918	2918	$(1.05)^3$	3378
4	2918	2918	$(1.05)^4$	3547
5	2918	2918	$(1.05)^5$	3724
6	2918	2918	$(1.05)^6$	3910

*May be read from the 5% Compound Interest Table as $(F/P,5\%,n)$.

Solutions:

Before-tax rate of return: Since both situations (no inflation and 5% inflation) have an annual benefit, stated in Year-0 dollars of $2918, they have the same before-tax rate of return.

$$\text{PW of cost} = \text{PW of benefit}$$
$$12{,}000 = 2918(P/A,i,6)$$
$$(P/A,i,6) = \frac{12{,}000}{2918} = 4.11$$

From Interest Tables: Before-tax rate of return equals 12%.

After-tax rate of return, no inflation:

Year	*Before-tax cash flow*	*Straight line depreciation*	*Taxable income*	*46% income taxes*	*Actual dollars and Year-0 $ after-tax cash flow*
0	−$12,000				−$12,000
1–6	+2,918	$2000	$918	−$422	+2,496

$$\text{PW of cost} = \text{PW of benefit}$$
$$12{,}000 = 2496(P/A,i,6)$$
$$(P/A,i,6) = \frac{12{,}000}{2496} = 4.81$$

From Interest Tables: After-tax rate of return equals 6.7%.

After-tax rate of return, 5% inflation:

Year	*Before-tax cash flow*	*Straight line depreciation*	*Taxable income*	*46% income taxes*	*Actual dollars, after-tax cash flow*
0	−$12000				−$12000
1	+3064	$2000	$1064	−$489	+2575
2	+3217	2000	1217	−560	+2657
3	+3378	2000	1378	−634	+2744
4	+3547	2000	1547	−712	+2835
5	+3724	2000	1724	−793	+2931
6	+3910	2000	1910	−879	+3031

Converting to Year-0 dollars and solving for the rate of return:

Year	*Actual dollars, After-tax cash flow*	*Conversion factor*		*Year-0 dollars, after-tax cash flow*	*Present worth at 5%*	*Present worth at 4%*
0	−$12000			−$12000	−$12000	−$12000
1	+2575	$\times (1.05)^{-1}$	=	+2452	+2335	+2358
2	+2657	$\times (1.05)^{-2}$	=	+2410	+2186	+2228
3	+2744	$\times (1.05)^{-3}$	=	+2370	+2047	+2107
4	+2835	$\times (1.05)^{-4}$	=	+2332	+1919	+1993
5	+2931	$\times (1.05)^{-5}$	=	+2297	+1800	+1888
6	+3031	$\times (1.05)^{-6}$	=	+2262	+1688	+1788
					−25	+362

Linear interpolation between 4% and 5%:

After-tax rate of return = 4% + 1%[362/(362 + 25)] = 4.9% ◀

From Example 13-4, we see that the before-tax rate of return for both situations (no inflation and 5% inflation) is the same. This was expected because the benefits in the inflation situation increased in proportion to the inflation. This shows that where future benefits fluctuate with changes in inflation or deflation, the effects do not alter the Year-0 dollar estimates. Thus, no special calculations are needed in before-tax calculations when future benefits are expected to respond to inflation or deflation rates.

The after-tax calculations illustrate a different result. The two situations, with equal before-tax rates of return, do not produce equal after-tax rates of return:

Situation	***Before-tax rate of return***	***After-tax rate of return***
No inflation	**12%**	**6.7%**
5% inflation	**12%**	**4.9%**

Thus, 5% inflation resulted in a smaller after-tax rate of return even though the benefits increased at the same rate as the inflation. A review of the cash flow table reveals that while benefits increased, the depreciation schedule did not. Thus the inflation resulted in increased taxable income and, hence, larger income tax payments; but there were not sufficient increases in benefits to offset these additional disbursements.

The result was that while the after-tax cash flow in actual dollars was increased, it was not large enough to offset *both* inflation and increased income taxes. This is readily apparent when the equivalent Year-0 dollar after-tax cash flow is examined. With inflation, the Year-0 dollar after-tax cash flow is smaller than the Year-0 dollar after-tax cash flow without it. Of course, inflation might cause equipment to have a salvage value that was not forecast, or a larger one than had been projected. This would tend to reduce the unfavorable effect of inflation on the after-tax rate of return.

SUMMARY

Inflation is characterized by rising prices for goods and services, while deflation produces a fall in prices. An inflationary trend makes future dollars have less purchasing power than present dollars. This helps long-term borrowers of money for they may repay a loan of present dollars in the future with dollars of reduced buying power. The help to borrowers is at the expense of lenders.

Deflation has the opposite effect from inflation. If money is borrowed at one point in time, and then a deflationary period occurs, the borrower has to repay his loan with dollars of greater purchasing power than he borrowed. This would be advantageous to lenders at the expense of borrowers.

While price changes occur in a variety of ways, one method of stating a price change is as a uniform rate of price change per year, or a compound price change. For this method the proper single payment multiplier to adjust a future cash flow to Year-0 buying power is

$$\textbf{Year-0 dollars} = [(1 + f)^{-n}]\textbf{Year-}n\textbf{ dollars}$$

When there is a uniform future cash flow series, the Year-0 buying power has the same general form as the compound interest factor. An equivalent interest rate,

$$i_{\text{equivalent}} = i' + f + i'f$$

may be computed and then used in conjunction with the present worth factor to compute the present worth of the future cash flow in Year-0 dollars.

The effect of inflation on the computed rate of return for an investment depends on how future benefits respond to the inflation. If benefits produce constant dollars, which are not increased by inflation, the effect of inflation is to reduce the before-tax rate of return on the investment. If, on the other hand, the dollar benefits increase to keep up with the inflation, the before-tax rate of return will not be adversely affected by the inflation.

This is not true when an after-tax analysis is made. Even if the future benefits increase to match the inflation rate, the allowable depreciation schedule does not increase. The result will be increased taxable income and income tax payments. This reduces the available after-tax benefits and, therefore, the after-tax rate of return. The important conclusion is that estimates of future inflation or deflation may be important in evaluating capital expenditure proposals.

Problems

13-1 One economist has predicted that there will be a 7% per year inflation of prices during the next ten years. If this proves to be correct, an item that presently sells for $10 would sell for what price ten years hence? (*Answer:* $19.67)

13-2 A man bought a 5% tax-free municipal bond. It cost $1000 and will pay $50 interest each year for twenty years. The bond will mature at the end of the twenty years and return the original $1000. If there is 2% annual inflation during this period, what rate of return will the investor receive after considering the effect of inflation?

13-3 A firm is having a large piece of equipment overhauled. It anticipates the machine will be needed for the next twelve years. The firm has an 8% minimum attractive rate of return. The contractor has suggested three alternatives:

a. A complete overhaul for $6000 that should permit twelve years of operation.

b. A major overhaul for $4500 that can be expected to provide eight years of service. At the end of eight years, a minor overhaul would be needed.

c. A minor overhaul now. At the end of four and eight years, additional minor overhauls would be needed.

If minor overhauls cost $2500, which alternative should the firm select? If minor overhauls, which now cost $2500, increase in cost at +5% per year, but other costs remain unchanged, which alternative should the firm select? (*Answers:* Alt. c; Alt. **a**)

13-4 A man wishes to set aside some money for his daughter's college education. His goal is to have a bank savings account containing an amount equivalent to $20,000 with today's purchasing power of the dollar, at the girl's 18th birthday. The estimated inflation rate is 8%. If the bank pays 5% compounded annually, what lump sum of money should he deposit in the bank savings account on his daughter's fourth birthday? (*Answer:* $29,670)

13-5 One economist has predicted that for the next five years, the United States will have an 8% annual inflation rate, followed by five years at a 6% inflation rate. This is equivalent to what average price change per year for the entire ten-year period?

13-6 A homebuilder's advertising has the caption, "Inflation to Continue for Many Years." The advertisement continues with the explanation that if one bought a home now for $97,000, and inflation continued at a 7% annual rate, the home would be worth

$268,000 in 15 years. This means, the advertisement says, a profit of $171,000 from the purchase of a home now. Do you agree with the homebuilder's logic? Explain.

13-7 Sam Johnson inherited $85,000 from his father. Sam is considering investing the money in a house which he will then rent to tenants. The $85,000 cost of the property consists of $17,500 for the land, and $67,500 for the house. Sam believes he can rent the house and have $8000 a year net income left after paying the property taxes and other expenses. The house will be depreciated by straight line depreciation using a 45-year depreciable life.

a. If the property is sold at the end of five years for its book value at that time, what after-tax rate of return will Sam receive? Assume that his incremental personal income tax rate is 34%.

b. Now assume there is 7% per year inflation, compounded annually. The tenants will have their rent increased 7% per year to match the inflation rate, so that after considering increased taxes and other expenses, the annual net income will go up 7% per year. Assume Sam's incremental income tax rate remains at 34% for all ordinary taxable income related to the property. The value of the property is now projected to increase from its present $85,000 at a rate of 10% per year, compounded annually.

If the property is sold at the end of five years, compute the rate of return on the after-tax cash flow in actual dollars. Also compute the rate of return on the after-tax cash flow in Year-0 dollars.

13-8 Tom Ward put $10,000 in a five-year certificate of deposit that pays 12% interest per year. At the end of the five years the certificate will mature and he will receive his $10,000 back. Tom has substantial income from other sources and estimates that his incremental income tax rate is 42%. If the inflation rate is 7% per year, what is his:

a. Before-tax rate of return, ignoring inflation?

b. After-tax rate of return, ignoring inflation?

c. After-tax rate of return, after taking inflation into account?

13-9 Linda Lovelace bought a lot at the Salty Sea for $18,000 cash. She does not plan to build on the lot, but instead will hold it as an investment for ten years. She wants a 10% after-tax rate of return after taking the 6% annual inflation rate into account. If income taxes amount to 15% of the capital gain, at what price must she sell the lot at the end of the ten years? (*Answer:* $95,188)

13-10 A newspaper reports that in the last five years, prices have increased a total of 50%. This is equivalent to what annual inflation rate, compounded annually? (*Answer:* 8.45%)

13-11 A South American country has had a great amount of inflation. Recently, its exchange rate was 15 cruzados per dollar, that is, one dollar will buy 15 cruzados in the foreign exchange market.

It is likely the country will continue to experience a 25% inflation rate, and the United States will continue at a 7% inflation rate. Assume that the exchange rate will vary the same as the inflation. In this situation, one dollar will buy how many cruzados five years from now? (*Answer:* 32.6)

13-12 A group of students decided they would lease and run a gasoline service station. The lease is for ten years. Almost immediately the students were confronted with the need to alter the gasoline pumps to read in liters. The Dayton Co. has a conversion kit available for $900 that may be expected to last ten years. As an alternative, they offer a $500 conversion kit that has a five-year useful life. The students believe that any money not invested in the conversion kits may be invested elsewhere at a 10% interest rate. Income tax consequences are to be ignored in this problem.

a. Assuming that future replacement kits cost the same as today, which alternative should be selected?

b. If one assumes a 7% inflation rate, which alternative should be selected?

13-13 An automobile manufacturer has an automobile that gets ten kilometers per liter of gasoline. It is estimated that gasoline prices will increase at a 12% per year rate, compounded annually, for the next eight years. This manufacturer feels that the automobile fuel consumption for its new automobiles should decline as fuel prices increase, so that the fuel cost will remain constant. To achieve this, what must be the fuel rating, in km/l, of his automobiles eight years hence?

13-14 Pollution control equipment must be purchased to remove the suspended organic material from liquid being discharged from a vegetable packing plant. Two alternative pieces of equipment are available that would accomplish the task. A Filterco unit presently costs $7000 and has a five-year useful life. A Duro unit, on the other hand, now costs $10,000 but will have a ten-year useful life.

Equipment costs are rising at 8% per year, compounded annually, due to inflation, so when the Filterco unit would be replaced, the cost would be much more than $7000. Based on a ten-year analysis period, and a 20% minimum attractive rate of return, before taxes, which piece of pollution control equipment should be purchased?

13-15 Dick Bernhard and his wife have a total taxable income of $60,000 this year and file a joint federal income tax return. If inflation continues for the next twenty years at a 7% rate, compounded annually, Dick wonders what their taxable income must be in the future to provide them the same purchasing power, after taxes, as their present taxable income. Assuming the federal income tax rate table is unchanged, what must their taxable income be twenty years from now?

13-16 One economist has predicted that for the next six years prices in the United States will increase 55%. After that he expects a further increase of 25% in the subsequent four years, so that prices at the end of ten years will have increased to 180% of the present level. Compute the inflation rate, f, for the entire ten-year period.

13-17 The U.S. tax laws provide that depreciation on equipment is based on its original cost. Yet due to substantial inflation, the replacement cost of equipment is often much greater than the original cost. What effect, if any, does this have on a firm's ability to buy new equipment to replace old equipment?

13-18 The City of Columbia is trying to attract a new manufacturing business to the area. It has offered to install and operate a water pumping plant to provide service to the proposed plant site. This would cost $50,000 now, plus $5000 per year in operating costs for the next ten years, all measured in Year-0 dollars.

To reimburse the city, the new business must pay a fixed uniform annual fee, A, at the end of each year for ten years. In addition, they are to pay the city $50,000 at the end of ten years. It has been agreed that the computations should be based on the city's receiving a 3% rate of return, after taking an inflation rate, f, of 7% into account.

The city council has directed the city engineer to determine the amount of the uniform annual fee, A. She has assigned the task to you. (*Answer:* $12,100)

13-19 A small research device is purchased for $10,000 and depreciated by ACRS depreciation. The net benefits from the device, before deducting depreciation, are $2000 at the end of the first year, and increasing $1000 per year after that (second year equals $2000, third year equals $3000, and so on), until the device is hauled to the junkyard at the end of seven years. During the seven-year period there is an inflation rate f of 7%.

This profitable corporation has a 50% combined federal and state income tax rate. If it requires a 12% after-tax rate of return on its investment, after taking inflation into account, should the device have been purchased?

13-20 Sally Johnson loaned a friend $10,000 at 15% interest, compounded annually. She is to repay the loan in five equal end-of-year payments. Sally estimates the inflation rate during this period is 12%. After taking inflation into account, what rate of return is Sally receiving on the loan? Compute your answer to the nearest 0.1%. (*Answer:* 2.7%)

13-21 Sam purchased a home for $150,000 with some creative financing. The bank agreed to lend Sam $120,000 for six years at 15% interest. It received a first mortgage on the house. The Joneses, who sold Sam the house, agreed to lend Sam the remaining $30,000 for six years at 12% interest. They received a second mortgage on the house. Thus Sam became the owner without putting up any cash. Sam pays $1500 a month on the first mortgage and $300 a month on the second mortgage. In both cases these are "interest only" loans, and the balance due on the loans does not diminish.

Sam rented the house, but after paying the taxes, insurance, and so on, he only had $800 left, so he was forced to put up $1000 a month of his own money to make the monthly payments on the mortgages. At the end of three years, Sam sold the house for $205,000. After paying off the two loans and the real estate broker, he had $40,365 left. After taking an 8% inflation rate into account, what was his before-tax rate of return?

13-22 Dale saw that the campus bookstore is having a special sale on pads of computation paper. The normal price is $3.00 a pad, but it is on sale at $2.50 a pad. This is unusual and Dale assumes it will not be put on sale again. On the other hand, Dale expects that there will be no increase in the $3.00 regular price, even though the inflation rate is 2% every three months. Dale believes that competition in the paper industry will keep wholesale and retail prices constant. He uses a pad of computation paper every three months.

Dale considers 19.25% a suitable minimum attractive rate of return. Dale will buy one pad of paper for his immediate needs. How many extra pads of computation paper should he buy? (*Answer:* 4)

13-23 When there is little or no inflation, a homeowner can expect to rent an unfurnished home for 12 percent of the market value of the property (home and land) per year. About 1/8 of the rental income is paid out for property taxes, insurance, and other operating expenses. Thus the net annual income to the owner is 10.5% of the market value of the property. Since prices are relatively stable, the future selling price of the property often equals the original price paid by the owner.

For a $150,000 property (where the land is estimated at $46,500 of the $150,000), compute the after-tax rate of return, assuming the selling price 59 months later (in December) equals the original purchase price. Use accelerated cost recovery system depreciation beginning January 1st. Also, assume a 35% income tax rate.
(*Answer:* 6.84%)

13-24 (This is a continuation of Problem 13-23.) As inflation has increased throughout the world, the rental income of homes has decreased and a net annual rental income of 8% of the market value of the property is common. On the other hand, the market value of homes tends to rise about 2% per year more than the inflation rate. As a result, both annual net rental income, and the resale value of the property rise faster than the inflation rate. Consider the following situation.

A $150,000 property (with the house valued at $103,500 and the land at $46,500) is purchased for cash in Year 0. Use ACRS depreciation, beginning January 1st. The market value of the property increases at a 12% annual rate. The annual rental income is 8% of the beginning-of-year market value of the property. Thus the rental income also increases each year. The general inflation rate f is 10%.

The individual who purchased the property has an average income tax rate of 35%.

a. Compute the actual dollar after-tax rate of return for the owner, assuming he sells the property 59 months later (in December).

b. Compute the after-tax rate of return for the owner, after taking the general inflation rate into account, assuming he sells the property 59 months later.

13-25 An investor wants a real rate of return i' (rate of return without inflation) of 10% per year on any projects in which he invests. If the expected annual inflation rate for the next several years is 6%, what interest rate i should be used in project analysis calculations?

13-26 In Chapters 5 (Present Worth Analysis) and 6 (Annual Cash Flow Analysis) it is assumed that prices are stable and a machine purchased today for $5000 can be replaced for the same amount many years hence. In fact, prices have generally been rising, so the stable price assumption is incorrect. Under what circumstances is it appropriate to use the "stable price" assumption when prices actually are changing?

13-27

a. Compute the equivalent annual inflation rate, based on the consumer price index, for the period from 1981 to 1986.

b. Using the equivalent annual inflation rate computed in part ***a***, estimate the consumer price index in 1996, working from the 1987 consumer price index.

CHAPTER **14**

Estimation Of Future Events

Economic analysis, if it is to be of any value, must concern itself with present and future consequences. We know that a post audit is desirable to see how well prior estimates were actually achieved, but that is not the central thrust of economic analysis. Our task is to take the present situation and our appraisal of the future and make sound decisions based upon them. This is probably a lot easier said than done. It may be relatively easy to determine the present situation; but it is not at all easy to look into the future and appraise it accurately.

One ironic aspect of an engineering education is that engineers are well trained to evaluate a situation analytically; yet there is very little attention given to learning how to predict the future. This seems strange when we recognize that engineering inevitably deals with both the present and the future. Engineering—including economic analysis—inevitably *requires* the estimation of future events.

In this chapter we consider the problem of evaluating the future. The easiest way to begin is to make a careful estimate. Then we examine the possibility of predicting a range of possible outcomes. Finally, we consider the situation where the probabilities of the various outcomes are known or may be estimated.

Precise Estimates

In an economic analysis, we need to evaluate the future consequences of an alternative. While that cannot be easy, it must be done. In practically every chapter of this book, there are cash flow tables where the costs and benefits for future years are precisely described. Do we really believe that we can exactly foretell a future cost? No one believes he can predict the future with certainty. Instead, the goal is to select a single value which represents the *best* estimate that can be made of the future.

Once our best estimates are made of the various future consequences, we must put them into the economic analysis. Yet the way they are used is really quite different from the way they were determined.

In estimating the future consequences, we recognize that they are not precise and that the actual values will be somewhat different from our estimates. Once these estimates are entered into the economic analysis itself, however, it is likely that we will have proceeded on the tacit assumption that these estimates *are* correct. We know the estimates will not always turn out to be correct; yet we treat them like facts once they are *in* the economic analysis. This can lead to trouble. If actual costs and benefits are different from the estimates, an undesirable alternative could be selected. This is because the variability of the future consequences is concealed by assuming that the best estimates will actually occur. The problem is illustrated by Example 14-1.

EXAMPLE 14-1

Two alternatives are being considered. The best estimates for the various consequences are as follows:

	A	*B*
Cost	$1000	$2000
Net annual benefit	150	250
Useful life, in years	10	10
End-of-useful-life salvage value	100	400

If interest is 3½%, which alternative should be selected?

Solution:
Alternative A:

$$\begin{aligned} \text{NPW} &= -1000 + 150(P/A,3^1/2\%,10) + 100(P/F,3^1/2\%,10) \\ &= -1000 + 150(8.317) + 100(0.7089) \\ &= -1000 + 1248 + 71 = +\$319 \end{aligned}$$

Alternative B:

$$\begin{aligned} \text{NPW} &= -2000 + 250(P/A,3^1/2\%,10) + 400(P/F,3^1/2\%,10) \\ &= -2000 + 250(8.317) + 400(0.7089) \\ &= -2000 + 2079 + 284 = +\$363 \end{aligned}$$

Alternative *B*, with its larger NPW, would be selected.

Alternate Formation of Ex. 14-1. Suppose that at the end of ten years, the actual salvage value turns out to be $300 instead of the $400 best estimate. If all the other estimates were correct, is *B* still the preferred alternative?

Solution: *Corrected B:*

$$\begin{aligned} \text{NPW} &= -2000 + 250(P/A,3^1/2\%,10) + 300(P/F,3^1/2\%,10) \\ &= -2000 + 250(8.317) + 300(0.7089) \\ &= -2000 + 2079 + 213 = +\$292 \end{aligned}$$

Under these circumstances, *A* is now the preferred alternative. ◀

Example 14-1 shows that the change in the salvage value of Alternative *B* actually results in a change of preferred alternative.

EXAMPLE 14-2

Using Example 14-1 data, compute the sensitivity of the decision to the Alt. *B* salvage value. For Alt. *A*, NPW = +319. For breakeven between the alternatives,

$$\text{NPW}_A = \text{NPW}_B$$

$$+319 = -2000 + 250(P/A,3\tfrac{1}{2}\%,10) + \text{Salvage value}_B(P/F,3\tfrac{1}{2}\%,10)$$

$$= -2000 + 250(8.317) + \text{Salvage value}_B(0.7089)$$

At the breakeven point

$$\text{Salvage value}_B = \frac{319 + 2000 - 2079}{0.7089} = \frac{240}{0.7089} = \$339$$

For Alt. *B* salvage value > \$339, *B* is preferred;
salvage value < \$339, *A* is preferred. ◀

Breakeven and sensitivity analysis provide one means of examining the impact of the variability of some estimate on the outcome. It helps by answering the question, "How much variability can a parameter have before the decision will be affected?" This approach does not, however, answer the basic problem of how to take the inherent variability of parameters into account in an economic analysis. This will be considered next.

A Range Of Estimates

Realistically, the *true* situation is that there is a range of possible values for most parameters. One could, for example, construct values for the ***optimistic*** estimate, the ***most likely*** estimate, and the ***pessimistic*** estimate. Then the economic analysis could be performed on each set of data to determine if the decision is sensitive to the range of projected values.

EXAMPLE 14-3

A firm is considering an investment. Three estimates for the various parameters are as follows:

	Optimistic value	*Most likely value*	*Pessimistic value*
Cost	\$1000	\$1000	\$1052
Net annual benefit	200	198	190
Useful life, in years	12	12	9
End-of-useful-life salvage value	100	0	0

If a 10% before-tax minimum attractive rate of return is required, is the investment justified under all three estimates? Compute the rate of return for each estimate.

Solutions to Ex. 14-3.

Optimistic estimate:

$$\text{PW of cost} = \text{PW of benefit}$$
$$\$1000 = 200(P/A,i,12) + 100(P/F,i,12)$$

Try $i = 18\%$:

$$\$1000 \stackrel{?}{=} 200(4.793) + 100(0.1372) = 973$$

18% is too high. $i \approx 17\%$.

Most likely estimate:

$$\$1000 = 198(P/A,i,12)$$
$$(P/A,i,12) = \frac{1000}{198} = 5.05$$

From Compound Interest Tables, $i = 16.7\%$.

Pessimistic estimate:

$$\$1052 = 190(P/A,i,9)$$
$$(P/A,i,9) = \frac{1052}{190} = 5.54$$

From Compound Interest Tables, $i = 11\%$.

From the calculations we conclude that the rate of return for this investment is most likely to be 16.7%, but might range from 11% to 17%. The investment meets the 10% MARR criterion for all estimates. ◀

Example 14-3 required that three separate calculations be made for the investment, one each for the optimistic values, the most likely values, and the pessimistic values. This approach emphasizes the unlikely situations where all parameters prove to be very favorable or very unfavorable. Neither is likely to happen. Rather, there is likely to be a blend of results, with the parameters assuming values near to the most likely estimate, but with due consideration for the possible range of values. One way to accomplish this is to estimate an average or mean* value for each parameter, based on the following weighting factors:†

$$\textit{Mean value} = \frac{\textbf{Optimistic value} + \textbf{4(Most likely value)} + \textbf{Pessimistic value}}{\textbf{6}}$$

This approach is illustrated in Ex. 14-4.

EXAMPLE 14-4

Solve Ex. 14-3 by applying the weighting factors given above. Compute the resulting mean rate of return.

Solution:

$$\text{Mean cost} = \frac{1000 + 4(1000) + 1052}{6} = 1009$$

$$\text{Mean net annual benefit} = \frac{200 + 4(198) + 190}{6} = 197$$

$$\text{Mean useful life} = \frac{12 + 4(12) + 9}{6} = 11.5 \text{ years}$$

$$\text{End-of-useful-life salvage value} = \frac{100}{6} = 17$$

Compute the mean rate of return:

$$\text{PW of cost} = \text{PW of benefit}$$
$$\$1009 = 197(P/A,i,11.5) + 17(P/F,i,11.5)$$

From Interest Tables, the mean rate of return is approximately 16%. ◀

Example 14-4 gave a mean rate of return (16%) that was different from the most likely rate of return (16.7%) computed in Example 14-3. The immediate question is, "Why are these values different?" The reason for the difference can be seen from the way the two values were calculated. One rate of return was based exclusively on the most likely values. The mean rate of return, on the other hand, took into account not only the most likely values, but also the variability of the parameters.

---

*The mean value is defined as the sum of all the values divided by the number of values.

†If you are interested, these weighting factors represent an approximation of the beta distribution.

In examining the data we see that the pessimistic values are further away from the most likely values than are the optimistic values. This causes the resulting weighted mean values to be less favorable than the most likely values. As a result, the mean rate of return, in this example, is less than the rate of return based on the most likely values.

Probability And Risk

Probability can be considered to be the long-run relative frequency of occurrence of an outcome. There are just two possible outcomes from flipping a coin (a Head or a Tail). If, for example, a coin is flipped over and over, we can expect in the long run that half the time Heads will appear and half the time Tails. We would say the probability of flipping a Head is 0.50 and of flipping a Tail is 0.50. Since probabilities are defined so that the sum of probabilities for all possible outcomes is 1, the situation is

$$\begin{aligned} \text{Probability of flipping a Head} &= 0.50 \\ \text{Probability of flipping a Tail} &= \underline{0.50} \\ \text{Sum of all possible outcomes} &= 1.00 \end{aligned}$$

A more complex situation is given in the following example.

EXAMPLE 14-5

If one were to roll one die (that is, one-half of a pair of dice), what is the probability that either a 1 or a 6 would result?

Solution: Since a die is a perfect six-sided cube, the probability of any side appearing is 1/6.

$$\begin{aligned} \text{Probability of rolling a } 1 &= P(1) = 1/6 \\ 2 &= P(2) = 1/6 \\ 3 &= P(3) = 1/6 \\ 4 &= P(4) = 1/6 \\ 5 &= P(5) = 1/6 \\ 6 &= P(6) = \underline{1/6} \\ \text{Sum of all possible outcomes} &= 6/6 = 1 \end{aligned}$$

The probability of rolling either a 1 or a 6 $= 1/6 + 1/6 = 1/3$. ◀

In the two examples, the probability of each outcome was the same. This need not be the case.

EXAMPLE 14-6

In the game of Blackjack, a perfect hand is a Ten or Facecard plus an Ace. What is the probability of being dealt a Ten or a Facecard from a newly shuffled deck of 52 cards? What is the probability of being dealt an Ace in this same situation?

Solution: The three outcomes being examined are to be dealt a Ten or a Facecard, an Ace, or some other card. Every card in the deck represents one of these three possible outcomes. There are 4 Aces; 16 Tens, Jacks, Queens, and Kings; and 32 other cards.

The probability of being dealt a Ten or a Facecard = 16/52 = 0.31
The probability of being dealt an Ace = 4/52 = 0.08
The probability of being dealt some other card = 32/52 = 0.61
1.00 ◀

The term ***risk*** has a special meaning when it is used in statistics. It is defined as a situation where *there are two or more possible outcomes and the probability associated with each outcome is known.*

In the two previous examples there is a risk situation. We could not know in advance what playing card would be dealt or what number would be rolled by the die. However, since the various probabilities could be computed, our definition of risk has been satisfied.

Probability and risk are not restricted to gambling games. For example:

In a particular engineering course, a student has computed the probability for each of the letter grades he might receive as follows:

Outcome		
Grade	*Grade point*	*Probability P(grade)*
A	4.0	0.10
B	3.0	0.30
C	2.0	0.25
D	1.0	0.20
F	0.0	0.15
		1.00

From the table we see that the grade with the highest probability is B. This, therefore, is the most likely grade.

We see from the table that there is a substantial probability that some grade other than B will be received. And the probabilities indicate that if a B is not received, the grade will probably be something less than a B. But in saying the most likely grade is a B, these other outcomes are ignored. In the next section we will show that a composite statistic may be computed using all the data.

Expected Value

In the example just discussed, we saw that the most likely grade of B in an engineering class had a probability of 0.30. That is not a very high probability. In some other course, like a math class, we might estimate a probability of 0.65 of obtaining a B, making the B the most likely grade. While a B is most likely in both the classes, it is more certain in the math class.

In the early part of this chapter, we computed a weighted mean to give a better understanding of the total situation as represented by various possible outcomes. We can do the same thing here. An obvious selection of the weighting factors is to use the probabilities for the various outcomes. Then, since the sum of the probabilities equals 1, the computation is:

$$\text{Weighted mean} = \frac{\text{Outcome}_A \times P(A) + \text{Outcome}_B \times P(B) + \cdots}{1}$$

When the probabilities are used as the weighting factors, we call the result the ***expected value*** and write the equation as:

$$\textbf{\textit{Expected value}} = \textbf{Outcome}_A \times \boldsymbol{P}\textbf{(A)} + \textbf{Outcome}_B \times \boldsymbol{P}\textbf{(B)} + \cdots$$

EXAMPLE 14-7

Compute the student's expected grade in the engineering course using the probabilities given in the text.

Solution:

Grade	*Grade point*	*P(grade)*	*Grade point × P(grade)*
A	4.0	0.10	0.40
B	3.0	0.30	0.90
C	2.0	0.25	0.50
D	1.0	0.20	0.20
F	0.0	0.15	0
		Expected (GP) =	2.00

The expected grade point (GP) of 2.00 indicates a grade of C. ◀

From the calculations in Ex. 14-7, we find that for a given set of probabilities, the most likely grade is B and the expected grade is C. How can we resolve these conflicting results? First, the results are correct. The most likely grade *is* B and the expected grade *is* C.

The two values tell us different things about the probabilities of the outcomes.

For example, suppose 1000 students took courses in which they each believed the distribution of probabilities in Ex. 14-7 was correct. Each person could correctly state that his most likely grade would be B, with an expected grade of C. If the projected probabilities proved to be correct, suppose the 1000 students received grades as follows:

Grade	*Number of students*
A	100
B	300
C	250
D	200
F	150
	1000

We would note immediately that only 300 students received B grades; most students received some other grade. If the average grade were computed, what would it be?

To average A, B, C, D, and F, we would assign the numerical values 4, 3, 2, 1, and 0. The computation is:

Grade	*Number of students*		*Grade × students*
A = 4.0	100		400
B = 3.0	300		900
C = 2.0	250		500
D = 1.0	200		200
F = 0.0	150		0
	Sum	=	2000

$$\text{Average grade} = \frac{\text{Sum}}{\text{Number of students}} = \frac{2000}{1000} = 2.0$$

The average grade is C.

We recognize that the average grade is exactly the expected grade. This helps us to understand expected value. If a situation were to occur over and over again, then the accumulated results will approach the expected value. Nevertheless, B remains the most likely grade to be received.

EXAMPLE 14-8

Just before a horse race is about to begin, a spectator decides that the situation on the four-horse race is as follows:

Horse	*Probability of winning*	*Outcome of a $10 bet if horse wins**
1	0.15	$48.00
2	0.15	58.00
3	0.50	16.50
4	0.20	42.00
	1.00	

*In horserace betting, a ticket is purchased for $10. The outcome of a winning ticket represents the refund of the $10 bet plus the amount won.

What is the expected value of the ticket if he bets $10 on Horse #3 to win?

Solution:

$$\text{Expected value} = \text{Outcome if \#3 wins} \times P(\text{Win}_3) + \text{Outcome if \#3 loses} \times P(\text{Loss}_3)$$

$$= \$16.50(0.50) + \$0(0.50) = \$8.25$$

Thus a ticket purchased for $10 has an expected value of $8.25. (Is this the way to get rich?) One also notes that there is no way for the bettor to actually win the expected value. He must either win $16.50 or nothing. There are no other possibilities. Figure 14-1 illustrates the situation. In this example, we find that betting $10 on the favorite horse (the one with the highest probability of winning) could be expected to result in a net loss to the bettor.

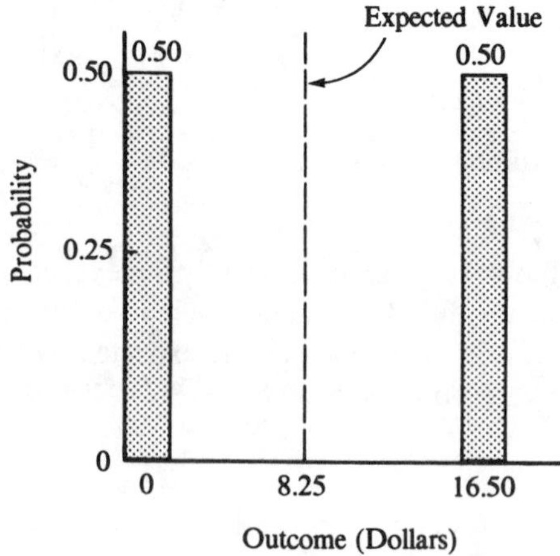

Figure 14-1 Example 14-8 situation.

Alternate Formation of Ex. 14-8: An alternate strategy would be to bet \$10 on each of the horses to win. In this way one is certain to have a winning ticket. What will be the expected value of the four tickets for this betting scheme?

Solution:

$$\begin{aligned}\text{Expected value} &= \$48.00(0.15) + \$58.00(0.15) + \$16.50(0.50) \\ &\quad + \$42.00(0.20) \\ &= \$7.20 + 8.70 + 8.25 + 8.40 \\ &= \$32.55\end{aligned}$$

Figure 14-2 is a plot of this betting plan.

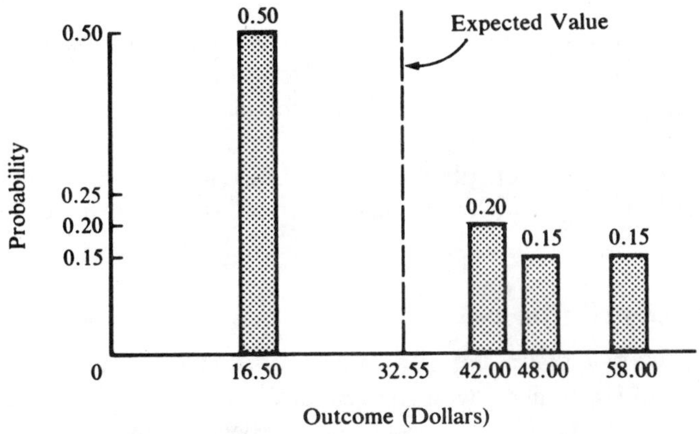

Figure 14-2 Example 14-8, alternate formation.

Regarding the two strategies offered in this example, which betting scheme is better?

$$\text{Expected result from one \$10 bet} = \$8.25 - \$10.00 = \$1.75 \text{ loss}$$
$$\text{Expected result from four \$10 bets} = \$32.55 - \$40.00 = \$7.45 \text{ loss}$$

On this basis, the single \$10 bet is preferred. Of course, it is clear that the best decision would be to make no bet at all. (Or would it?) ◀

Expected value may be a useful analysis tool in certain kinds of situations, for it represents the long-term outcome if the particular situation occurs over and over again. If one were to put money into a slot machine and play it over and over for an hour or two, the expected value may be a rather accurate estimate of the results from playing the slot machine a couple of hundred times. But would it be useful in estimating the results from playing the machine once? Obviously not. The most likely result would be the loss of the coin. Much less frequently, one might receive three or more coins back.

Thus the expected value (at possibly 0.75 coin returned per play) is not very useful when we are trying to evaluate a situation that is not repeated. Example 14-8 represented a one-time event. A particular horserace will be run once. Thus the expected value cannot tell us much about the outcome from a particular race; it does say that people who continue to bet on horseraces (or play slot machines) must expect to lose money over the long term.

Distribution Of Outcomes

In the three previous sections, we have considered ways of treating situations where the outcomes vary. There was no discussion of the distributions of the outcomes as in Figures 14-1 and 14-2. In this section, we will examine two specific distributions (uniform and normal) and describe how to randomly sample from these or any other distributions. The ability to randomly sample from a distribution is prerequisite for the discussion of simulation in the following section.

Uniform Distribution

In Ex. 14-5 we saw that a die, being a perfect six-sided cube, can be rolled to provide each of six numbers with an equal probability of 1/6. The distribution is discrete since only integers appear and values in between are not possible. When the outcomes have an equal probability, this is called a ***uniform distribution*** and is illustrated by Fig. 14-3.

When we would like to know what the outcome is from the roll of a die, the easiest way is simply to roll the die. On the other hand, we could simulate rolling the die in a variety of ways. One way would be to make up six slips of paper and assign the numbers 1 through 6 to them. Then shake them in a hat

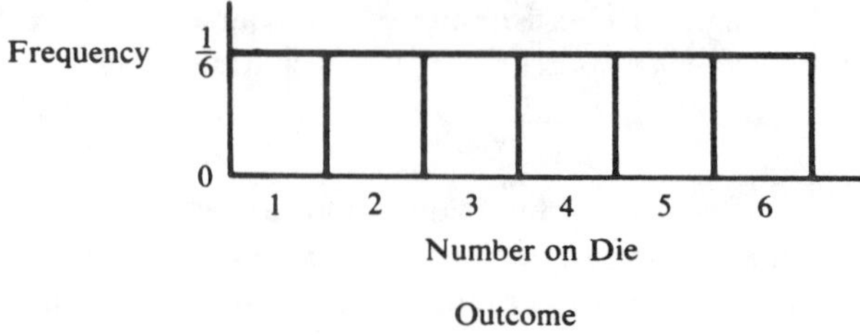

Figure 14-3 Uniform distribution—outcomes of a die.

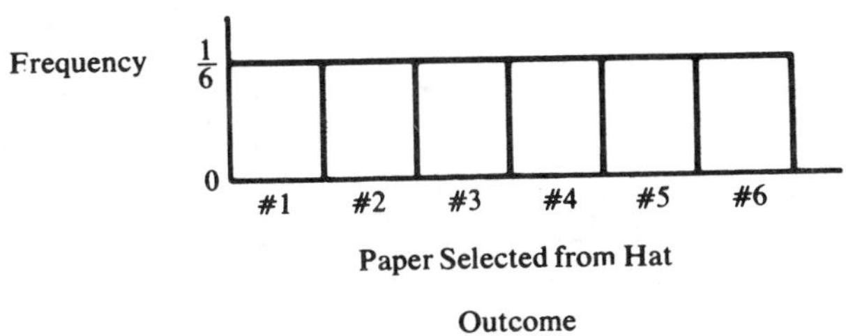

Figure 14-4 Uniform distribution—papers in a hat.

and choose one of them at random. In this manner, we would have simulated the rolling of a die. Since there are six papers in the hat and they are shaken up so each piece of paper is equally likely to be selected, we really have created another uniform distribution—Figure 14-4.

It is easy to use six pieces of paper and create a uniform distribution with outcomes of 1 through 6. A more general approach to simulating the uniform distribution can now be devised.

Suppose someone puts each of the possible two-digit numbers—00 through 99—on slips of paper and put the 100 slips in a hat. Then after careful mixing, one is chosen at random and the two-digit value written down. The slip of paper is returned to the hat and the process repeated. Table 14-1 is a sample of random two-digit numbers. A careful examination of Table 14-1 shows that some two-digit numbers appear several times and some do not appear at all. This is, of course, possible for the slips of paper were replaced in the hat after they were drawn.

The table of random numbers represents random sampling from a uniform distribution of values from 00 through 99 as depicted by Fig. 14-5. We can use the table of random numbers to simulate the outcome of a die by assigning 1/6 of the random numbers to a 1 on the die, another 1/6 of the random numbers to

Table 14-1 150 TWO-DIGIT RANDOM NUMBERS

77	41	71	94	66	14	16	77	50	17	65	61	85	23	15
06	40	75	90	34	75	45	34	96	74	34	92	24	52	99
46	19	39	60	20	50	05	80	16	33	79	03	27	22	37
02	66	67	29	92	03	24	58	08	05	56	57	47	72	02
68	40	65	20	78	69	76	30	39	21	00	22	40	61	19
43	10	23	08	48	55	14	45	52	22	90	71	15	89	04
80	87	71	81	45	19	30	72	88	08	44	24	18	41	97
35	17	82	18	84	00	77	87	38	83	42	38	55	17	31
87	21	94	49	66	74	96	10	70	09	76	34	21	06	55
15	69	32	47	88	87	14	99	19	27	41	61	40	53	03

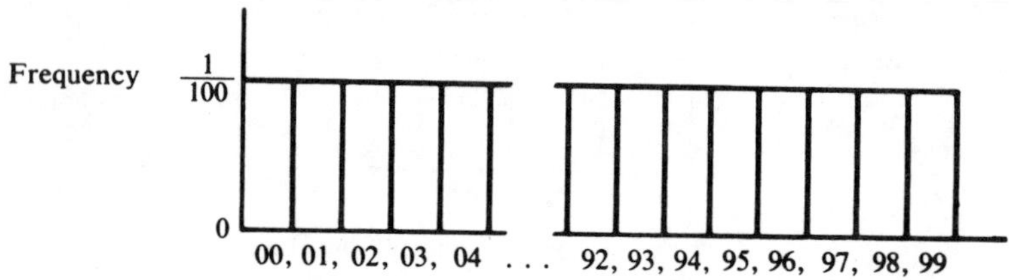

Figure 14-5 Uniform distribution—two-digit random numbers.

a 2 on the die, and so forth. Since 100/6 equals 16.6, it will not come out even. We will let 16 random numbers represent each face of the die. The assignment of numbers is as follows:

Random numbers	*Outcome on the die*
00–15	1
16–31	2
32–47	3
48–63	4
64–79	5
80–95	6
96–99	Random numbers not used

Graphically, the situation would look like Fig. 14-6.

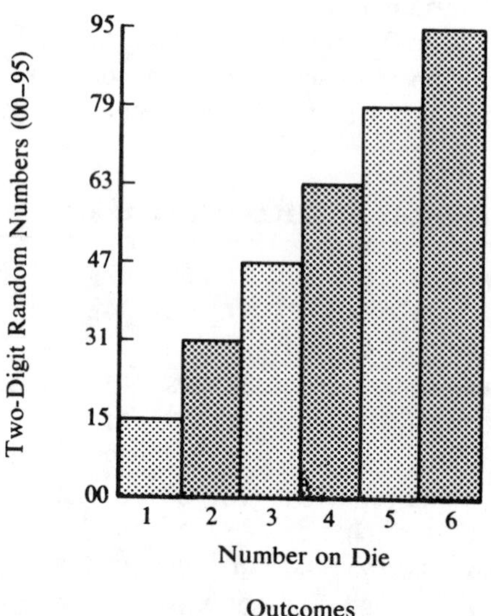

Figure 14-6 Plot of outcome on die *vs*. random numbers.

EXAMPLE 14-9

Using the table of two-digit random numbers, Table 14-1, and Figure 14-6, simulate the rolling of a die three times.

Solution: We first obtain two-digit random numbers from Table 14-1. The proper way to use a table of random numbers is to enter the table at some random point and read in a consistent manner. Following this method, we might read the numbers 50, 96, and 16 from the ninth column in the table. Enter the Figure 14-6 graph with 50 as the random number. Read from the y-axis across to the stair-step curve and read down to 4 on the x-axis. This indicates that the first roll of the die is a 4.

The second random number, 96, is not being used in the simulation. One should discard the number and go on to the next random number selected. A random number of 16 corresponds to a 2 on the die. Another random number must be selected from Table 14-1. The selected 08 corresponds to 1 on the die. Using a table of random numbers we have simulated the rolling of a die with the numbers 4, 2, and 1 having been selected. ◀

With the technique illustrated in Ex. 14-9, any uniform distribution may be simulated with a table of random numbers.

Normal Distribution

Possibly the best known of all distributions is the normal distribution. It is defined by the equation:

$$y(x) = \frac{1}{\sigma\sqrt{2\pi}} \exp -\frac{1}{2}\left(\frac{x-\mu}{\sigma}\right)^2$$

where μ = *mean*, σ = *standard deviation*

Since all values are possible, the normal distribution is a continuous distribution. We see that a particular normal distribution is defined by its mean μ and standard deviation σ. Different values of μ and σ give normal distributions that look different from one another. Two normal distributions are shown in Fig. 14-7.

Because the equation defining the normal distribution is cumbersome to use, data are usually obtained from a table of the distribution with mean μ equal to 0 and standard deviation σ equal to 1. For continuous distributions the area under the curve equals the probability. As the sum of the probabilities for all possible outcomes is 1, the total area under the normal curve is 1. We can find the area under portions of the distribution from Fig. 14-8. The figure shows that 68.3% of the area under the normal curve lies between $(\mu - 1.0\sigma)$ and $(\mu + 1.0\sigma)$. Thus in any situation, we could expect a normally distributed variable to be within this range about 68.3% of the time.

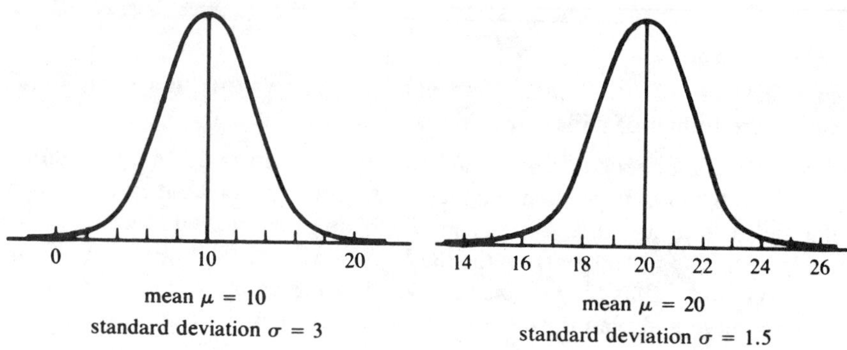

Figure 14-7 Two normal distributions with different means and standard deviations.

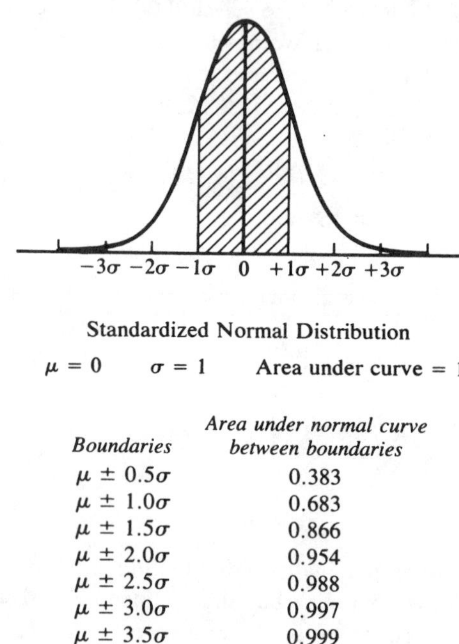

Boundaries	*Area under normal curve between boundaries*
$\mu \pm 0.5\sigma$	0.383
$\mu \pm 1.0\sigma$	0.683
$\mu \pm 1.5\sigma$	0.866
$\mu \pm 2.0\sigma$	0.954
$\mu \pm 2.5\sigma$	0.988
$\mu \pm 3.0\sigma$	0.997
$\mu \pm 3.5\sigma$	0.999

Figure 14-8 Area under the normal distribution curve.

A comment about the notation is in order. When one is referring to the normal distribution of some population (like the ages of *all* college students) we say the mean is μ and the standard deviation σ. But when we are referring to a *sample* of ages of college students, we say the mean is $\bar{x}$ and the standard deviation is s. Thus

	All college students	*Sample of college students*
Mean	μ	$\bar{x}$
Standard deviation	σ	s

EXAMPLE 14-10

The ages of twenty college students are as follows:

21 22 24 23 25 23 22 22 26 22
23 21 23 24 25 23 24 23 22 24

For this group of people, it is believed their ages are normally distributed. Compute the mean and standard deviation for the twenty students.

Solution: If all the ages are added, their sum $\Sigma x = 462$. The mean of this sample of twenty students is

$$\textbf{Mean } \bar{x} = \frac{\Sigma x}{n} = \frac{462}{20} = 23.1 \text{ years}$$

The standard deviation was once called the *root-mean-square deviation*, for this is one of the methods of its calculation. The ***standard deviation*** is the square root of the sum of the square of the deviations about the mean, divided by the sample size minus 1, or

$$\textit{Standard deviation } s = \sqrt{\frac{\Sigma(x - \bar{x})^2}{n - 1}}$$

The twenty ages may be grouped and this equation used to find the standard deviation.

Number in age group, n	*Age group, x*	$(x - \bar{x})$	$(x - \bar{x})^2$	$n(x - \bar{x})^2$
2	21	−2.1	4.4	8.8
5	22	−1.1	1.2	6.0
6	23	−0.1	0.0	0.0
4	24	+0.9	0.8	3.2
2	25	+1.9	3.6	7.2
1	26	+2.9	8.4	8.4
20			$\Sigma(x - \bar{x})^2 =$	33.6

$$s = \sqrt{\frac{\Sigma(x - \bar{x})^2}{n - 1}} = \sqrt{\frac{33.6}{19}} = 1.33$$

It is frequently easier to solve for the standard deviation when the equation is rewritten as follows:

$$\textbf{\textit{Standard deviation s}} = \sqrt{\frac{\Sigma x^2}{n-1} - \frac{(\Sigma x)^2}{n(n-1)}}$$

In this problem,

$$\Sigma x^2 = 21^2 + 23^2 + 22^2 + 21^2 + 24^2 + \cdots = 10{,}706$$

$$\Sigma x = 462$$

$$s = \sqrt{\frac{10{,}706}{19} - \frac{462^2}{20(19)}} = \sqrt{563.47 - 561.69} = \sqrt{1.78}$$

$$= 1.33$$

We have computed for the twenty college students $\bar{x} = 23.1$, $s = 1.33$. The sample mean and standard deviation are our best estimates of the population. We therefore estimate that $\mu = 23.1$ and $\sigma = 1.33$. ◀

To define a particular normal distribution it is necessary to specify the mean μ and standard deviation σ. Suppose, for example, you believed that the useful life of a particular type of equipment was normally distributed with a mean life of 15 years and a standard deviation of 2.4 years. How could you obtain random samples from this distribution?

A convenient method is based on the standardized normal distribution. Figure 14-9 shows both the standardized normal distribution and the particular useful life normal distribution we want. The deviation of any value of x from the mean of the distribution may be expressed in number of standard deviations.

$$z = \frac{x - \mu}{\sigma} \qquad \text{or} \qquad x = z\sigma + \mu$$

From this equation, we see that any Point x on a specific distribution (with μ and σ) has an equivalent Point z on the standardized normal distribution. This relationship allows us to relate the standardized normal distribution to any other normal distribution.

In our useful life example, two standard deviations above the mean would be at $x = 15 + 2(2.4) = 19.8$ years on the useful life distribution, Fig. 14-9b. The equivalent point on the standardized normal distribution is

$$z = \frac{x - \mu}{\sigma} = \frac{19.8 - 15}{2.4} = +2.0$$

This interrelationship means that if we randomly sample from the standardized normal distribution, we can relate this to an equivalent random sample for any normal distribution.

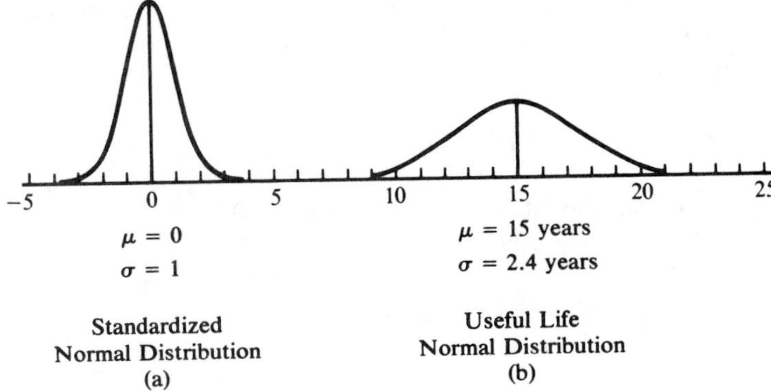

Figure 14-9a & b Standardized normal distribution and a specific useful life normal distribution.

A point on the standardized normal distribution ($\mu = 0$, $\sigma = 1$) is fully defined by specifying the number of standard deviations the point is to the left (negative) or to the right (positive) of the mean. Thus the value +1.02 would indicate the point is 1.02 standard deviations to the right of the mean.

Table 14-2 represents random samples from the standardized normal distribution. The use of random normal numbers is illustrated in Ex. 14-11.

EXAMPLE 14-11

On a particular portion of a highway, observations indicate that the speed of automobiles is normally distributed with mean $\mu = 104$ kilometers per hour and standard deviation $\sigma = 11$ kilometers per hour. Obtain a random sample of five automobiles on the highway.

Table 14-2 100 RANDOM NORMAL NUMBERS (z)

−0.22	−0.87	2.32	−0.94	0.63	0.81	0.74	1.08	−1.82	0.07
−0.89	−0.39	0.29	−0.27	1.06	−0.42	2.26	−0.35	1.09	−2.55
0.06	1.28	−1.74	2.47	0.58	0.69	1.41	−1.19	2.37	−0.06
−0.01	−0.40	0.64	−2.22	1.10	0.47	−0.09	−0.35	−0.72	0.30
−0.87	−1.34	0.85	0.27	−1.35	0.58	−1.72	1.88	−0.45	0.82
0.24	0.40	0.50	1.41	−1.95	−0.02	−1.00	−0.20	−1.08	−0.78
−1.05	−0.06	0.27	−0.04	0.99	−0.78	0.46	−1.18	0.37	1.07
−0.57	0.24	−1.02	0.86	0.78	−1.69	−0.17	−0.23	−0.87	−0.45
−1.24	−0.63	0.03	−0.83	0.25	−0.89	−0.77	0.90	−0.27	0.94
2.11	0.78	−1.69	−0.17	0.21	0.48	−2.82	−0.86	1.40	−1.20

Solution: From Table 14-2, randomly select a value of random normal number (z). The value of z selected is -0.94. This value of z represents a vehicle on the highway travelling at a speed of

$$x = z\sigma + \mu = -0.94(11) + 104 = 93.7 \text{ km/hr}$$

Four other random normal numbers, 1.06, 0.69, −0.09, and 1.88, are selected. The indicated automobile speeds are

$$x = 1.06(11) + 104 = 115.7 \text{ km/hr}$$
$$x = 0.69(11) + 104 = 111.6$$
$$x = -0.09(11) + 104 = 103.0$$
$$x = 1.88(11) + 104 = 124.7$$

Through the use of random normal numbers, the speed of five random automobiles has been computed. ◀

Sampling from Any Distribution Using Random Numbers

In the two previous sections we saw how to obtain a random sample from the uniform distribution and from the normal distribution. While these are two important situations, there are times when we want to obtain a random sample from some other distribution. We might, for example, wish to obtain a random sample from the grade distribution of Figure 14-10.

The procedure is to replot the data to represent a cumulative distribution of grades. This has been done in Figure 14-11.

Where the x-axis represents all possible outcomes, as in this case, the cumulative probability on the y-axis must vary from 0 to 1.00. To facilitate random sampling, random numbers may be assigned to each segment of the ordinate in proportion to its probability. Figure 14-12 shows the data of Fig. 14-11 with two-digit random numbers assigned to each segment of the y-axis.

To obtain a random sample from the distribution of grades (Figure 14-10), the first step is to randomly select a two-digit number from the table of random numbers, Table 14-1. This number is then entered as the ordinate in Figure 14-12. Read across from the ordinate to the curve and then read the corresponding value on the x-axis. The random number 35, for example, corresponds to a grade of C on the x-axis.

The procedure described for sampling from the grade distribution can be used for any discrete or continuous distribution. In fact, looking back at the discussion of the uniform distribution reveals that we were actually using a cumulative distribution (Fig. 14-6) to randomly sample from the uniform distribution.

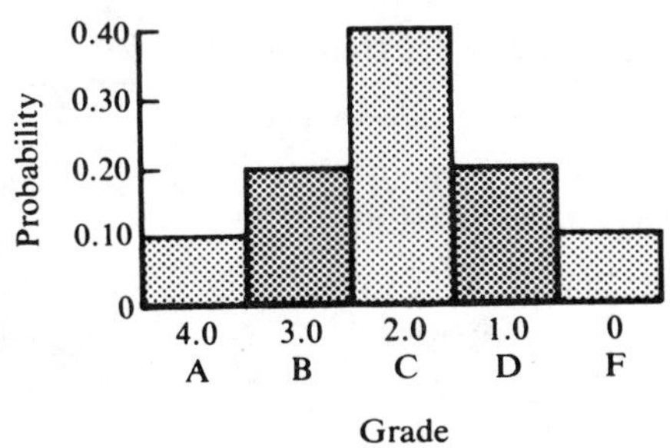

Grade	*Grade point average*	*Probability*
A	4.0	0.10
B	3.0	0.20
C	2.0	0.40
D	1.0	0.20
F	0.0	0.10
		1.00

Figure 14-10 Distribution of grades.

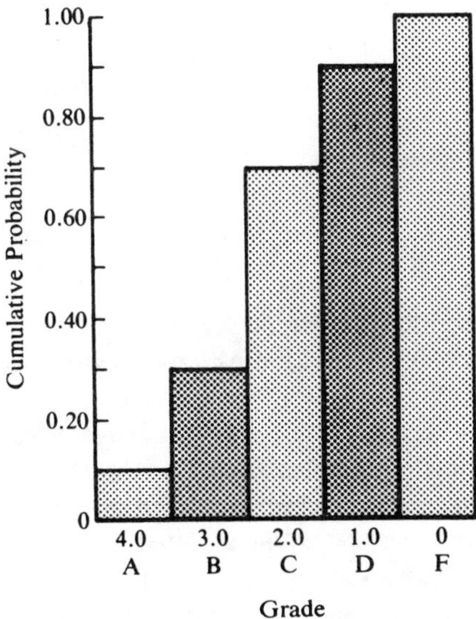

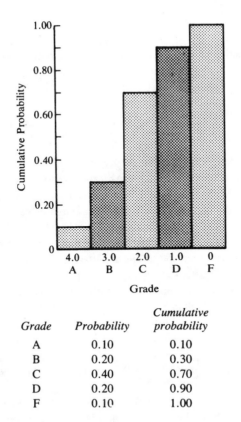

Grade	*Probability*	*Cumulative probability*
A	0.10	0.10
B	0.20	0.30
C	0.40	0.70
D	0.20	0.90
F	0.10	1.00

Figure 14-11 Cumulative distribution of grades.

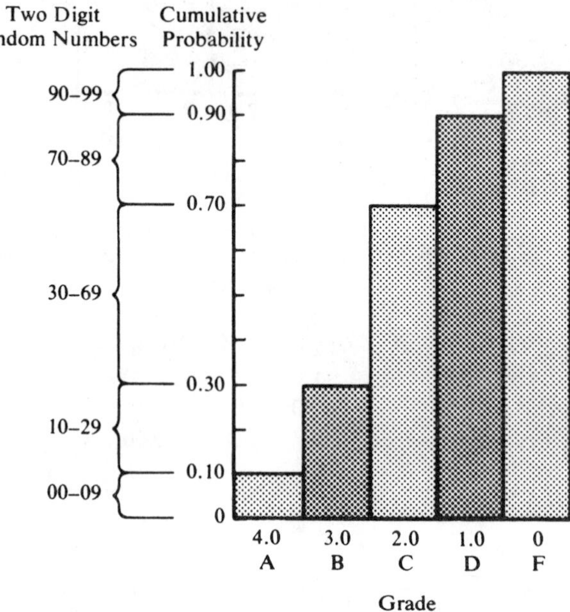

Figure 14-12 Cumulative distribution of grades with two-digit random numbers assigned to the *y*-axis.

Simulation

Simulation may be described as the repetitive analysis of a mathematical model. When we can accurately predict all the various consequences of an investment project with precise estimates, a single computation will give us the results. Sometimes future events are stated in terms of a range of values like optimistic, most likely, and pessimistic estimates. In this circumstance, one approach is to compute the results (like rate of return) for each of the three estimates. Here, for the first time, multiple computations are used to describe the results.

When values for an economic analysis are stated in terms of one or more probability distributions, the analysis is further complicated. The number of different combinations of values quickly becomes so large—often infinite—that one can no longer consider evaluating each possibility. The only practical solution is to randomly sample from each of the variables and compute the results. Is a single random sample adequate to portray the situation? The answer is, "Yes and no." Surely we know more about the situation than if we had no random sample, but not much more.

We could compare this with a random sample of a piece of chicken from a bucket full of pieces of chicken. If the one piece is a leg, does that mean the bucket consists only of chicken legs, or that all pieces are as good as the chicken leg? We simply cannot say. The random sample that produced the chicken leg proves at least that there was one in the bucket, but it says very little about the rest of the contents.

To obtain greater information about an economic analysis the procedure is to continue to take additional random samples for evaluation. This technique is simulation. In Ex. 14-9 we simulated the rolling of a die three times. We now see that this was a very simple simulation. Example 14-12 illustrates an ***economic analysis simulation***.

EXAMPLE 14-12

If a more accurate scales is installed on a production line, it will reduce the error in computing postage charges and save $250 a year. The useful life of the scales is believed to be uniformly distributed and range from 12 to 16 years. The initial cost of the scales is estimated to be normally distributed with a mean of $1500 and a standard deviation of $150.

Simulate 25 random samples of the problem and compute the rate of return for each sample. Construct a graph of rate of return *vs*. frequency of occurrence.

Solution: The useful life is uniformly distributed between 12 and 16 years. This is illustrated in Fig. 14-13. The data are replotted as a cumulative distribution function in Fig. 14-14.

We can assign 20 two-digit random numbers to each of the five possible useful lives as follows:

Random numbers	*Useful life, in years*
00–19	12
20–39	13
40–59	14
60–79	15
80–99	16

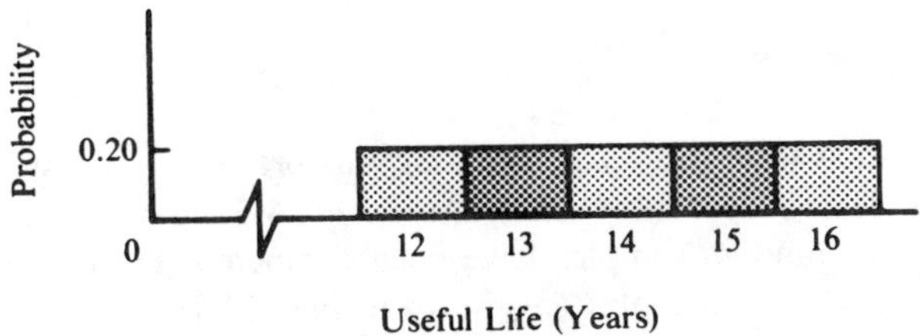

Figure 14-13 Useful life distribution for Example 14-12.

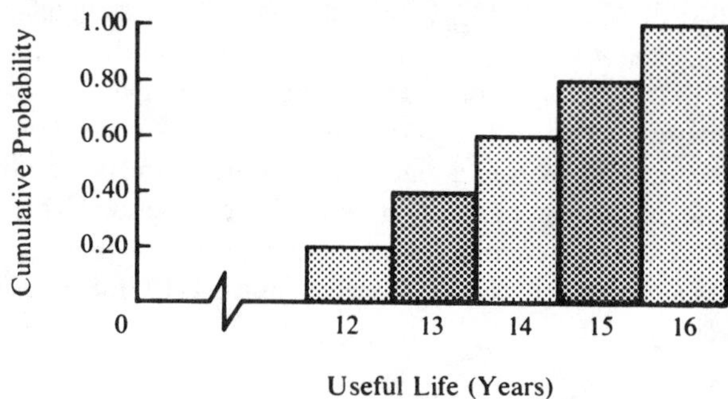

Figure 14-14 Useful life distribution for Example 14-12 replotted as a cumulative distribution function.

For the first random sample, we randomly enter the random number table (Table 14-1) and read the number 82. We previously assigned this number to represent a useful life of 16 years.

Next, to sample from the normal distribution, we read a random normal number of −1.82 from Table 14-2. To obtain the cost of the scales, multiply this value by the standard deviation and add the mean.

$$\text{Cost of scales} = -1.82(\$150) + \$1500 = \$1227$$

With these values computed, the cash flow is:

Year	*Cash flow*
0	−$1227
1–16	250

For the cash flow, the rate of return may be computed as follows:

$$\text{PW of cost} = \text{PW of benefit}$$
$$1227 = 250(P/A, i, 16) \qquad i = 19\%$$

Table 14-3 shows the results of repeating this process to obtain 25 values of rate of return. The rates of return shown in Table 14-3 are plotted in Fig. 14-15. An additional 75 random samples have been computed (but not detailed here) and the data for all 100 samples are shown in Fig. 14-16. ◀

Table 14-3 RESULTS OF 25 RANDOM SAMPLES FOR EXAMPLE 14-12

Random number	*Useful life, in years*	*Random normal number*	*Cost*	*Computed rate of return*
82	16	−1.82	$1227	19%
17	12	1.08	1662	11%
35	13	0.74	1611	12%
87	16	0.81	1622	13%
21	13	−0.42	1437	14%
94	16	0.69	1604	14%
49	14	0.47	1570	13%
66	15	1.10	1665	12%
74	15	−2.22	1167	20%
96	16	0.64	1596	14%
10	12	−0.40	1440	14%
70	15	−0.01	1498	14%
09	12	0.06	1509	13%
76	15	1.28	1692	12%
34	13	−0.39	1442	14%
21	13	−0.87	1370	15%
06	12	2.32	1848	8%
55	14	0.29	1544	13%
03	12	−0.27	1460	13%
53	14	1.06	1659	12%
40	14	−0.42	1437	15%
61	15	2.26	1839	11%
41	14	1.41	1712	11%
27	13	−0.09	1486	14%
19	12	−1.72	1242	17%

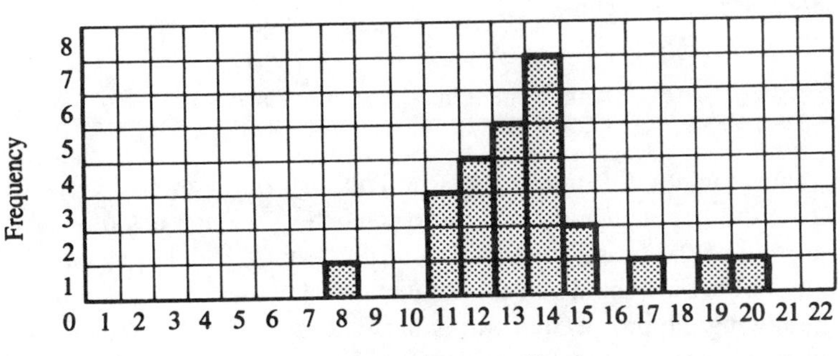

Figure 14-15 Graph of rate of return *vs.* frequency for 25 random samples in Example 14-12.

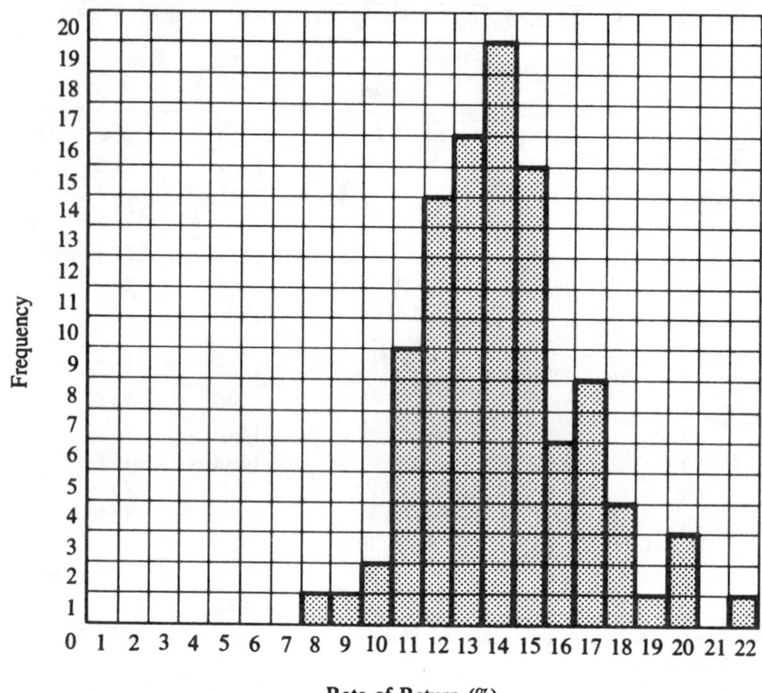

Figure 14-16 Graph of rate of return *vs*. frequency for 100 random samples in Example 14-12.

Instead of recognizing the variability of the cost of the scales in Ex. 14-12, or its useful life, we might have used the single best estimates for them. The problem would have become:

Cost of scales	$1500
Annual benefit	250
Useful life, in years	14

These values yield a 14% rate of return. Thus, all our elaborate computations in this case did not change the results very much. The computations, however, give a picture of the possible variability of the results. Without recognizing the variability of the data and doing the simulation computations we would have no way of knowing the prospective variation in the rate of return. Simulation may add a dimension to our knowledge of a problem. With it we can learn about a projected result and also the likelihood that the result or some other one will occur.

From Example 14-12, it is apparent that if one goes from a situation of best estimates to one of probability distributions, the computations are multiplied about a hundred times. Fortunately, most of these calculations may be performed on a computer. In Chapter 19, a microcomputer program is presented to solve simulation problems. This makes simulation a practical tool.

SUMMARY

Estimation of the future is an important element of an engineer's work. In economic analysis we have several ways of describing the future. Precise estimates will not be exactly correct, but they are considered to be the best single values to represent what we think will happen.

A simple way to represent variability is through a range of estimates for future events. Frequently this is done by making three estimates: optimistic, most likely, and pessimistic. If the problem is solved three times, using first the optimistic values, followed by the most likely and pessimistic values, the full range of prospective results may be examined. The disadvantage of this approach is that it is extremely unlikely that either the optimistic results (based on optimistic estimates for all the components of the problem) or the pessimistic results will occur.

A variation to the solution of the range of estimates is to assign relative weights to the various estimates and solve the problem using the weighted values. The weights suggested are:

Estimate	*Relative weight*
Optimistic	1
Most likely	4
Pessimistic	1

When the probability of a future event is known or may be reasonably predicted, the technique of *expected value* may be used. Here the probabilities are applied as the relative weights.

$$\text{Expected value} = \text{Outcome}_A \times \text{Probability}_A + \text{Outcome}_B \times \text{Probability}_B + \cdots$$

Expected value is a useful technique in projecting the long term results when a situation occurs over and over again. For situations where the event will only occur once or a few times, expected value gives little insight into the infrequent event.

Two probability distributions are described. The uniform distribution is where each outcome has an equal probability of occurrence. The 52 different cards in a deck of cards, when shuffled, represent a uniform distribution. For computations, two-digit random numbers are a practical representation of the uniform distribution.

The normal distribution is possibly the best known of all distributions. A normal distribution is defined by its two parameters, the mean (a measure of central tendency) and standard deviation (a measure of the variability of the distribution). A random sample from any normal distribution may be obtained through the use of random normal numbers. A random sample may also be obtained from other distributions if the cumulative distribution function can be plotted.

Where the elements of an economic analysis are stated in terms of probability distributions, a repetitive analysis of a random sample is often done. This simulation is based on the premise that a random sampling of increasing size becomes a better and better estimate of the possible outcomes. The large number of computations limit the usefulness of the technique when hand methods are used.

Problems

14-1 Two instructors announced that they "grade on the curve," that is, give a fixed percentage of each of the various letter grades to each of their classes. Their curves are:

Grade	*Instructor A*	*Instructor B*
A	10%	15%
B	15%	15%
C	45%	30%
D	15%	20%
F	15%	20%

If a student came to you and said that his object was to enroll in the class where his expected grade point average would be greater, which instructor would you recommend?

(*Answer:* Instructor *A*)

14-2 A man wants to determine whether or not to invest $1000 in a friend's speculative venture. He will do so if he thinks he can get his money back. He believes the probabilities of the various outcomes at the end of one year are:

Result	*Probability*
$2000 (double his money)	0.3
1500	0.1
1000	0.2
500	0.3
0 (lose everything)	0.1

What would be his expected outcome if he invests the $1000?

14-3 The M.S.U. football team has ten games scheduled for next season. The business manager wishes to estimate how much money the team can be expected to have left over after paying the season's expenses, including any post-season "bowl game" expenses. From records for the past season and estimates by informed people, the business manager has assembled the following data:

Situation	*Probability*	*Situation*	*Net income*
Regular season:		Regular season:	
Win 3 games	0.10	Win 5 or fewer	
Win 4 games	0.15	games	$250,000
Win 5 games	0.20		
Win 6 games	0.15	Win 6 to 8	
Win 7 games	0.15	games	400,000
Win 8 games	0.10		
Win 9 games	0.07	Win 9 or 10	
Win 10 games	0.03	games	600,000
Post-season:		Post-season:	Additional income
Bowl game	0.10	Bowl game	of $100,000

Based on the business manager's data, what is the expected net income for the team next season? (*Answer:* $355,000)

14-4 Telephone poles are an example of items that have varying useful lives. Telephone poles, once installed in a location, remain in useful service until one of a variety of events occur.

a. Name three reasons why a telephone pole might be removed from useful service at a particular location.

b. You are to estimate the total useful life of telephone poles. If the pole is removed from an original location while it is still serviceable, it will be installed elsewhere. Estimate the optimistic life, most likely life, and pessimistic life for telephone poles. What percentage of all telephone poles would you expect to have a total useful life greater than your estimated optimistic life?

14-5 In the New Jersey and Nevada gaming casinos, the crap table is a popular gambling game. One of the many bets available is the "Hard-way 8." A $1 bet in this fashion will win the player $4 if in the game the pair of dice come up 4 and 4 prior to one of the other ways of totaling 8. For a $1 bet, what is the expected result? (*Answer:* 80¢)

14-6 When a pair of dice are tossed the results may be any number from 2 through 12. In the game of craps one can win by tossing either a 7 or an 11 on the first roll. What is the probability of doing this? *Hint:* There are 36 ways that a pair of six-sided dice can be tossed. What portion of them result in either a 7 or an 11? (*Answer:* 8/36)

14-7 Your grade point average for the current school term probably cannot yet be determined with certainty.

a. Estimate the probability of obtaining a grade point average in each of the five categories below.

Grade point average (A = 4.00)	*Probability of obtaining the GPA*
0.00–0.80	
0.81–1.60	
1.61–2.40	
2.41–3.20	
3.21–4.00	

b. Plot the data from Part ***a*** with the grade point average as the x-axis and cumulative probability (of obtaining the GPA category or some lower one) as the y-axis.

c. Assign the values between 00 and 99 to represent the cumulative probability 0.00–1.00 on the y-axis. Obtain a random number from Table 14-1 and determine the matching cumulative probability. Read and record the corresponding value of the GPA on the x-axis. In this manner, obtain 25 values of GPA. Graph the results as a bar graph with the x-axis as GPA and the y-axis as frequency. How does this bar graph compare with the table in Part ***a***?

14-8 A decision has been made to perform certain repairs on the outlet works of a small dam. For a particular 36-inch gate valve, there are three available alternatives:

1. leave the valve as it is;

2. repair the valve; or

3. replace the valve.

If the valve is left as it is, the probability of a failure of the valve seats, over the life of the project, is 60%; the probability of failure of the valve stem is 50%; and of failure of the valve body is 40%.

If the valve is repaired, the probability of a failure of the seats, over the life of the project, is 40%; of failure of the stem is 30%; and of failure of the body is 20%. If the valve is replaced, the probability of a failure of the seats, over the life of the project, is 30%; of failure of the stem is 20%; and of failure of the body is 10%.

The present worth of cost of future repairs and service disruption of a failure of the seats is $10,000; the present worth of cost of a failure of the stem is $20,000; the present worth of cost of a failure of the body is $30,000. The cost of repairing the valve now is $10,000; and of replacing it is $20,000. If the criterion is to minimize expected costs, which alternative is best?

14-9 A man went to Atlantic City with $500 and placed 100 bets of $5 each on a number on the roulette wheel. There are 38 numbers on the wheel and the gaming casino pays 35 times the amount bet if the ball drops into the bettor's numbered slot in the roulette wheel. In addition, the bettor receives back the original $5 bet. Estimate how much money the man is expected to win or lose in Atlantic City.

14-10 A heat exchanger is being installed as part of a plant modernization program. It costs $80,000, including installation, and is expected to reduce the overall plant fuel cost by $20,000 per year. Estimates of the useful life of the heat exchanger range from an optimistic 12 years to a pessimistic 4 years. The most likely value is 5 years. Using the range of estimates to compute the mean life, determine the estimated before-tax rate of return. Assume the heat exchanger has no salvage value at the end of its useful life.

14-11 A factory building is located in an area subject to occasional flooding by a nearby river. You have been brought in as a consultant to determine whether or not floodproofing of the building is economically justified. The alternatives are as follows:

a. Do nothing. Damage in a moderate flood is $10,000 and in a severe flood, $25,000.

b. Alter the factory building at a cost of $15,000 to withstand moderate flooding without damage and to withstand severe flooding with $10,000 damages.

c. Alter the factory building at a cost of $20,000 to withstand a severe flood without damage.

In any year the probability of flooding is as follows: 0.70—no flooding of the river; 0.20—moderate flooding; and 0.10—severe flooding. If interest is 15% and a 15-year analysis period is used, what do you recommend?

14-12 Al took a midterm examination in physics and received a score of 65 in the exam. The mean was 60 and the standard deviation was 20. Bill received a score of 14 in mathematics where the exam mean was 12 and the standard deviation was 4. Which student ranked higher in his class? Explain.

14-13 An engineer decided to make a careful analysis of the cost of fire insurance for his $200,000 home. From a fire rating bureau he found the following risk of fire loss in any year.

Outcome	*Probability*
No fire loss	0.986
$ 10,000 fire loss	0.010
40,000 fire loss	0.003
200,000 fire loss	0.001

a. Compute his expected fire loss in any year.

b. He finds that the expected fire loss in any year is less than the $550 annual cost of fire insurance. In fact, an insurance agent explains that this is always true. Nevertheless, the engineer buys fire insurance. Explain why this is or is not a logical decision.

14-14 An industrial park is being planned for a tract of land near the river. To protect the industrial buildings that will be built on this low-lying land from flood damage, an earthen embankment can be constructed. The height of the embankment will be determined by an economic analysis of the costs and benefits. The following data have been gathered.

Embankment height above roadway (in meters)	*Initial cost*
2.0	$100,000
2.5	165,000
3.0	300,000
3.5	400,000
4.0	550,000

Flood level above roadway (in meters)	*Average frequency that flood level will exceed height in Col. 1*
2.0	Once in 3 years
2.5	Once in 8 years
3.0	Once in 25 years
3.5	Once in 50 years
4.0	Once in 100 years

The embankment can be expected to last 50 years and will require no maintenance. Any time the flood water flows over the embankment, $100,000 of damage occurs. Should the embankment be built? If so, to which of the five heights above the roadway? A 12% rate of return is required.

14-15 (20-59) Five years ago a dam was constructed to impound irrigation water and to provide flood protection for the area below the dam. Last winter a 100-year flood caused extensive damage both to the dam and to the surrounding area. This was not surprising since the dam was designed for a 50-year flood.

The cost to repair the dam now will be $250,000. Damage in the valley below amounts to $750,000. If the spillway is redesigned at a cost of $250,000 and the dam is repaired for another $250,000, the dam may be expected to withstand a 100-year flood without sustaining damage. However, the storage capacity of the dam will not be increased and the probability of damage to the surrounding area below the dam will be unchanged. A second dam can be constructed up the river from the existing dam for $1 million. The capacity of the second dam would be more than adequate to provide the desired flood protection. If the second dam is built, redesign of the existing dam spillway will not be necessary, but the $250,000 of repairs must be done.

The development in the area below the dam is expected to be complete in ten years. A new 100-year flood in the meantime will cause a $1 million loss. After ten years the loss would be $2 million. In addition there would be $250,000 of spillway damage if the spillway is not redesigned. A 50-year flood is also likely to cause about $200,000 of damage, but the spillway would be adequate. Similarly, a 25-year flood would cause about $50,000 of damage.

The three alternatives are (1) Repair the existing dam for $250,000 but make no other alterations. (2) Repair the existing dam ($250,000) and redesign the spillway to take a 100-year flood ($250,000). (3) Repair the existing dam ($250,000) and build the second dam ($1 million). Based on an annual cash flow analysis, and a 7% interest rate, which alternative should be selected?

CHAPTER 15

Selection Of A Minimum Attractive Rate Of Return

The preceding chapters have said very little about what interest rate or minimum attractive rate of return is suitable for use in a particular situation. Since this problem is quite complex, there is no single answer that is always appropriate. A discussion of a suitable interest rate to use must inevitably begin with an examination of the sources of capital, followed by a look at the prospective investment opportunities. Only in this way can an intelligent decision be made on the choice of an interest rate or minimum attractive rate of return.

Sources Of Capital

In broad terms there are four sources of capital available to a firm. They are: money generated from the operation of the firm, borrowed money, sale of mortgage bonds, and sale of capital stock.

Money Generated from the Operation of the Firm

A major source of capital investment money is through the retention of profits resulting from the operation of the firm. Since only about half of the profits of industrial firms are paid out to stockholders, the half that is retained is an important source of funds for all purposes, including capital investments. In addition to profit, there is money generated in the business equal to the annual depreciation charges on existing capital assets. In other words, a profitable firm will generate money equal to its depreciation charges *plus* its retained profits. Even a firm that earns zero profit will still generate money from operations equal to its depreciation charges. (A firm with a loss, of course, will have still less funds.)

External Sources of Money

When a firm requires money for a few weeks or months, it typically borrows the money from banks. Longer term unsecured loans (of, say, 1–4 years) may also be arranged through banks. While banks undoubtedly finance a lot of capital expenditures, regular bank loans cannot be considered a source of permanent financing.

Longer term secured loans may be obtained from banks, insurance companies, pension funds, or even the public. The security for the loan is frequently a mortgage on specific property of the firm. When sold to the public, this financing is by mortgage bonds. The sale of stock in the firm is still another source of money. While bank loans and bonds represent debt that has a maturity date, stock is a permanent addition to the ownership of the firm.

Choice of Source of Funds

Choosing the source of funds for capital expenditures is a decision for the board of directors; sometimes it also requires approval of the stockholders. In situations where internal operations generate adequate funds for the desired capital expenditures, external sources of money probably are seldom used. But when the internal sources are inadequate, external sources must be employed or the capital expenditures will have to be deferred or cancelled.

Cost Of Funds

Cost of Borrowed Money

A first step in deciding on a minimum attractive rate of return might be to determine the interest rate at which money can be borrowed. Longer term secured loans may be obtained from banks, insurance companies, or the variety of places where substantial amounts of money accumulates (for example, Japan or the oil-producing nations).

A large, profitable corporation might be able to borrow money at the ***prime rate***, that is, the interest rate that banks charge their best and most sought after customers. All other firms are charged an interest rate that is higher by anywhere from one-half to several percent. In addition to the financial strength of the borrower and his ability to repay the loan, the interest rate will vary depending on the duration of the loan.

Cost of Capital

Another relevant interest rate is the ***cost of capital***. The general assumption concerning the cost of capital is that all the money the firm uses for investments is drawn from all the components of the overall capitalization of the firm. The mechanics of the computation is given in Ex. 15-1.

EXAMPLE 15-1

For a particular firm, the purchasers of common stock require an 11% rate of return, mortgage bonds are sold at a 7% interest rate and bank loans are available at 9%. Compute the cost of capital for the following capital structure:

		Rate of return	*Annual amount*
$ 20 million	Bank loan	9%	$1.8 million
20 million	Mortgage bonds	7%	1.4 million
60 million	Common stock and retained earnings	11%	6.6 million
$100 million			$9.8 million

Solution: Interest payments on debt, like bank loans and mortgage bonds, are tax deductible business expenses. Thus:

$$\text{After-tax interest cost} = \text{Before-tax interest cost}(1 - \text{tax rate})$$

If we assume the firm pays 40% income taxes, the computations become:

$$\text{Bank loan:} \quad \text{After-tax interest cost} = 9\%(1 - 0.40) = 5.4\%$$
$$\text{Mortgage bonds:} \quad \text{After-tax interest cost} = 7\%(1 - 0.40) = 4.2\%$$

Dividends paid on the ownership in the firm (common stock + retained earnings) are not tax deductible. Combining the three components, the after-tax interest cost for the $100 million of capital is:

$$\$20 \text{ million}(5.4\%) + \$20 \text{ million}(4.2\%) + \$60 \text{ million}(11\%) = \$8.52 \text{ million}$$

$$\text{Cost of capital} = \frac{\$8.52 \text{ million}}{\$100 \text{ million}} = 8.52\% \blacktriangleleft$$

In an actual situation, the cost of capital is quite difficult to compute. The fluctuation in the price of common stock, for example, makes it difficult to pick a cost, and the fluctuating prospects of the firm makes it even more difficult to estimate the future benefits the purchasers of the stock might expect to receive. Given the fluctuating costs and prospects of future benefits, what rate of return do stockholders require? There is no precise answer, but we can obtain an approximate answer. Similar assumptions must be made for the other components of a firm's capitalization.

Investment Opportunities

An industrial firm has many more places in which it can invest its money than does an individual. A firm has larger amounts of money and this alone makes certain kinds of investment possible that are unavailable to individual investors, with their more limited investment funds. The U.S. Government, for example, borrows money for short terms of 90 or 180 days by issuing certificates called Treasury Bills that frequently yield a greater interest rate than savings accounts. The customary minimum purchase is $25,000.

More important, however, is the fact that a firm conducts a business and this business offers many investment opportunities. While exceptions can be found, a good generalization is that the opportunities for investment of money within the firm are superior to the investment opportunities outside the firm. Consider the following situation: a tabulation of the available investment opportunities for a particular firm is outlined in Table 15-1. A plot of these projects by rate of return *vs*. investment is shown in Fig. 15-1. The cumulative investment required for all projects at or above a given rate of return is given in Fig. 15-2.

The two figures illustrate that a firm may have a broad range of investment opportunities available at varying rates of return. It may take some study and searching to identify the better investment projects available to a firm. If this is done, the available projects will almost certainly exceed the money the firm budgets for capital investment projects.

Opportunity Cost

We see that there are two aspects of investing that are basically independent. One factor is the source and quantity of money available for capital investment

Table 15-1 A FIRM'S AVAILABLE INVESTMENT OPPORTUNITIES

Project number	*Project*	*Cost* ($\times 10^3$)	*Estimated rate of return*
Investment Related to Current Operations:			
1	New equipment to reduce labor costs	$150	30%
2	New equipment to reduce labor costs	50	45%
3	Overhaul particular machine to reduce material costs	50	38%
4	New test equipment to reduce defective products produced	100	40%
New Operations:			
5	Manufacture parts that previously had been purchased	200	35%
6	Further processing of products previously sold in semi-finished form	100	28%
7	Further processing of other products	200	18%
New Production Facilities:			
8	Relocate production to new plant	250	25%
External Investments:			
9	Investment in a different industry	300	20%
10	Investment in a different industry	300	10%
11	Overseas investment	400	15%
12	Buy Treasury Bills	Unlimited	8%

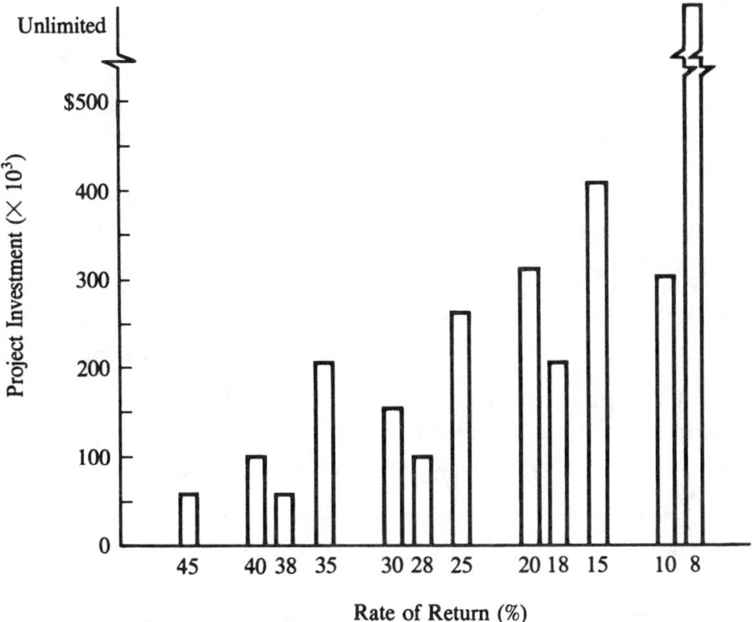

Figure 15-1 Rate of return *vs*. project investment.

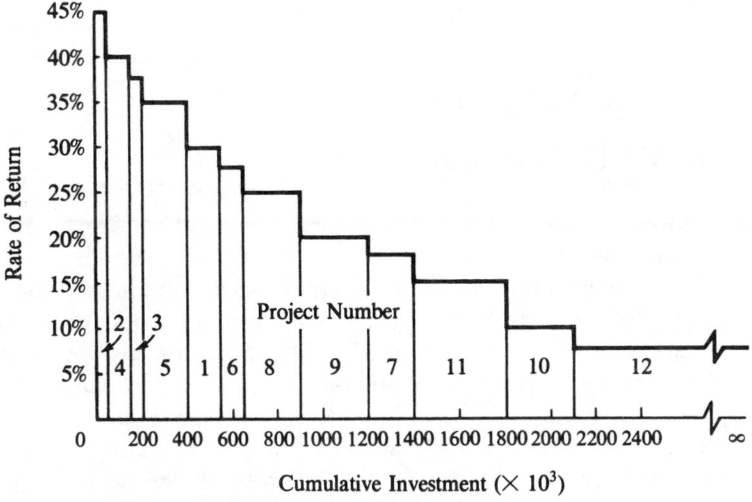

Figure 15-2 Cumulative investment required for all projects at or above a given rate of return.

projects. The other aspect is the investment opportunities themselves that are available to the firm.

These two situations are typically out of balance, with investment opportunities exceeding the available money supply. Thus some investment opportunities can be selected and many must be rejected. Obviously, we want to ensure that *all the selected projects are better than the best rejected project*. To do this, we must know something about the rate of return on the best rejected project. The best rejected project is the best opportunity foregone, and this in turn is called the ***opportunity cost***.

Opportunity cost **= Cost of the best opportunity foregone**

= Rate of return on the best rejected project

If one could predict in advance what the opportunity cost would be for some future period (like the next twelve months), this rate of return could be one way to judge whether to accept or reject any proposed capital expenditure.

EXAMPLE 15-2

Consider the situation represented by Figures 15-1 and 15-2. For a capital expenditure budget of \$1200 ($\times 10^3$), what is the opportunity cost?

Solution: From Fig. 15-2 we see that the eight projects with a rate of return of 20% or more require a cumulative investment of \$1200 ($\times 10^3$). We would take on these projects and reject the other four (7, 11, 10, and 12) with rates of return of 18% or less. The best rejected project is #7 and it has an 18% rate of return. This indicates the opportunity cost is 18%. ◀

Selecting A Minimum Attractive Rate Of Return

Using the three concepts on the cost of money (the cost of borrowed money, the cost of capital, and opportunity cost), which, if any, of these values should be used as the minimum attractive rate of return (MARR) in economic analyses?

Fundamentally, we know that unless the benefits of a project exceed the cost of the project, we cannot add to the profitability of the firm. A lower boundary for the minimum attractive rate of return must be the cost of the money invested in the project. It would be unwise, for example, to borrow money at 8% and invest it in a project yielding a 6% rate of return.

Further, we know that no firm has an unlimited ability to borrow money. Bankers—and others who evaluate the limits of a firm's ability to borrow money—look at both the profitability of the firm and the relationship between the components in the firm's capital structure. This means that continued borrowing of money will require that additional stock must be sold to maintain an acceptable ratio between ***ownership*** and ***debt***. In other words, borrowing for a particular investment project is only a block of money from the overall capital structure of the firm. This suggests that the MARR should not be less than the cost of capital. Finally, we know that the MARR should not be less than the rate of return on the best opportunity foregone. Stated simply,

> **Minimum attractive rate of return should be equal to the highest one of the following: cost of borrowed money; cost of capital; or opportunity cost.**

Adjusting MARR To Account For Risk And Uncertainty

We know from our previous study of estimating the future that what actually occurs is often different from the estimate. When we are fortunate enough to be able to assign probabilities to a set of possible future outcomes, we call this a ***risk*** situation. We saw in Chapter 14 that techniques like expected value and simulation may be used when the probabilities are known.

Uncertainty is the term used to describe the condition when the probabilities are *not* known. Thus, if the probabilities of future outcomes are known we have *risk*, and if they are unknown we have *uncertainty*.

One way to reduce the likelihood of undertaking projects that do not produce satisfactory results is to pass up marginal projects. In other words, no matter what projects are undertaken, some will turn out better than anticipated and some worse. Some undesirable results can be prevented by selecting only the best projects and avoiding those whose expected results are closer to a minimum standard: then (in theory, at least) the selected projects will provide results *above* the minimum standard even if they do considerably worse than anticipated.

In projects where there is normal business risk and uncertainty, the MARR is used without adjustment. For projects with greater risk or uncertainty, the MARR is increased. This is certainly not the best way to handle conditions of risk. A preferable way is to deal explicitly with the probabilities using one of the techniques from Chapter 14. This may be more acceptable as an adjustment for uncertainty. When the interest rate (MARR) used in economic analysis calculations is raised to adjust for risk or uncertainty, greater emphasis is placed on immediate or short term results and less emphasis on longer term results.

EXAMPLE 15-3

Consider the two following alternatives: the MARR has been raised from 10% to 15% to take into account the greater risk and uncertainty that Alt. *B*'s results may not be as favorable as indicated. What is the impact of this change of MARR on the decision?

Year	*Alt. A*	*Alt. B*
0	−80	−80
1–10	10	13.86
11–20	20	10

Solution:

Year	*Alt. A*	NPW *at* 14.05%	NPW *at* 10%	NPW *at* 15%
0	−80	−80.00	−80.00	−80.00
1–10	10	52.05	61.45	50.19
11–20	20	27.95	47.38	24.81
		0	+28.83	−5.00

Year	*Alt. B*	NPW *at* 15.48%	NPW *at* 10%	NPW *at* 15%
0	−80	−80.00	−80.00	−80.00
1–10	13.86	68.31	85.14	69.56
11–20	10	11.99	23.69	12.41
		0	+28.83	+1.97

Computations at* MARR *of* 10% *ignoring risk and uncertainty. Both alternatives have the same positive NPW (+28.83) at a MARR of 10%. Also, the differences in the benefits schedules (*A* − *B*) produce a 10% incremental rate of return. (The calculations are not shown here.) This must be true if NPW for the two alternatives is to remain constant at a MARR of 10%.

***Considering risk and uncertainty with* MARR *of* 10%.** At 10%, both alternatives are equally desirable. Since Alt. *B* is believed to have greater risk and uncertainty, a logical conclusion is to select Alt. *A* rather than *B*.

***Increase* MARR *to* 15%.** At a MARR of 15%, Alt. *A* has a negative NPW and Alt. *B* has a positive NPW. Alternative *B* is preferred under these circumstances.

Conclusion. Based on a business-risk MARR of 10%, the two alternatives are equivalent. Recognizing some greater risk of failure for Alt. *B* makes *A* the preferred alternative. If the MARR is increased to 15%, to add a margin of safety against risk and uncertainty, the computed decision is to select *B*. Since Alt. *B* has been shown to be less desirable than *A*, the decision, based on a MARR of 15%, may be an unfortunate one. The difficulty is that the same risk adjustment (increase the MARR by 5%) is applied to both alternatives even though they have different amounts of risk. ◀

The conclusion to be drawn from Ex. 15-3 is that increasing the MARR to compensate for risk and uncertainty is only an approximate technique and may not always achieve the desired result. Nevertheless, adjusting the MARR upward for increased risk and uncertainty is commonly done in industry.

Inflation and the Cost of Borrowed Money

As inflation has varied, what is its effect on the cost of borrowed money? A widely held view has been that interest rates on long term borrowing, like twenty-year Treasury bonds, will be about 3% more than the inflation rate. For borrowers this is the real—that is, after-inflation—cost of money, and for lenders the real return on loans. If inflation rates increase, it would follow that borrowing rates would also increase. All this suggests a rational and orderly situation, about as we might expect.

Unfortunately, things have not worked out this way. Figure 15-3 shows that the real interest rate has not been 3% in recent times and, in fact, there have been long periods when the real interest rate was negative. Can this be possible? Would anyone invest money at an interest rate several percent less than the inflation rate? Well, consider this: when the U.S. inflation rate was 12%, savings banks were paying 5½% on regular passbook deposits. And there was a lot of money in those accounts. While there must be relationship between interest rates and inflation, Figure 15-3 suggests that it is complex.

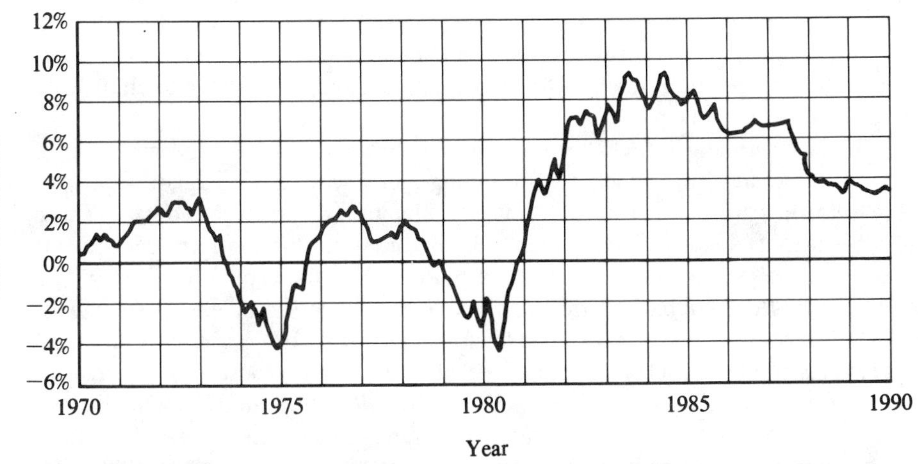

Figure 15-3 *The real interest rate.* The interest rate on twenty-year Treasury bonds *minus* the Inflation rate, f, as measured by changes in the Consumer Price Index.

Representative Values Of MARR Used In Industry

We saw that the minimum attractive rate of return should be established at the highest one of the following: cost of borrowed money, cost of capital, or the opportunity cost.

The cost of borrowed money will vary from enterprise to enterprise, with the lowest rate being the prime interest rate. The prime rate may change several times in a year; it is widely reported in newspapers and business publications. As we pointed out, the interest rate for firms that do not qualify for the prime interest rate may be ½% to several percent higher.

The cost of capital of a firm is an elusive value. There is no widely accepted way to compute it; we know that as a *composite value* for the capital structure of the firm, it conventionally is higher than the cost of borrowed money. The cost of capital must consider the market valuation of the shares (common stock, and so forth) of the firm, which may fluctuate widely, depending on future earnings prospects of the firm. We cannot generalize on representative costs of capital.

Somewhat related to cost of capital is the computation of the return on total capital (long term debt, capital stock, and retained earnings) actually achieved by firms. *Fortune* magazine, among others, does an annual analysis of the rate of return on total capital. The after-tax rate of return on total capital for individual firms ranges from 0% to about 40% and averages 8%. *Business Week* magazine does a periodic survey of corporate performance. It reports an after-tax rate of return on common stock and retained earnings. We would expect the values to be higher than the rate of return on total capital, and this is the case. The after-tax return on common stock and retained earnings ranges from 0% to about 65% with an average of 14%.

When discussing MARR, firms can usually be divided into two general groups. First, there are firms which are struggling along with an inadequate supply of investment capital, or are in an unstable situation or unstable industry. These firms cannot or do not invest money in anything but the most critical projects with very high rates of return and a rapid return of the capital invested. Often these firms use payback period and establish a criterion of one year or less, before income taxes. For an investment project with a five-year life, this corresponds to about a 60% after-tax rate of return. When these firms do rate of return analysis, they reduce the MARR to possibly 25% to 30% after income taxes. There is potentially a substantial difference between a one-year before-tax payback period and a 30% after-tax MARR, but this apparently does not disturb firms that specify this type of dual criteria.

The second group of firms represents the bulk of all enterprises. They are in a more stable situation and take a longer range view of capital investments. Their greater money supply enables them to invest in capital investment projects that firms in the first group will reject. Like the first group, this group of firms also uses payback and rate of return analysis. When small capital investments (of about $500 or less) are considered, payback period is often the only analysis technique used. The criterion for accepting a proposal may be a before-tax payback period not exceeding one or two years. Larger investment projects are analyzed by rate of return. Where there is a normal level of business risk, an after-tax MARR of 12% to 15% appears to be widely used. The MARR is increased when there is greater risk involved.

In Chapter 9 we saw that payback period is not a proper method for the economic analysis of proposals. Thus, industrial use of payback criteria is *not* recommended. Fortunately, the trend in industry is toward greater use of accurate methods and less use of payback period.

Note that the values of MARR given above are approximations. But the values quoted appear to be opportunity costs, rather than cost of borrowed money or cost of capital. This indicates that firms cannot or do not obtain money to fund projects whose rates of return are nearer to the cost of borrowed money or cost of capital. While one could make a case that good projects are needlessly being rejected, there may be practical business reasons why firms operate as they do.

One cannot leave this section without noting that the MARR used by enterprises is so much higher than can be obtained by individuals. (Where can you get a 30% after-tax rate of return without excessive risk?) The reason appears to be that businesses are not faced with the intensively competitive situation that confronts an individual. There might be thousands of people in any region seeking a place to invest $2000 with safety; but how many people could—or would—want to invest $500,000 in a business? This diminished competition, combined with a higher risk, appears to explain at least some of the difference.

SUMMARY

There are four general sources of capital available to an enterprise. The most important one is money generated from the operation of the firm. This has two components: there is the portion of profit that is retained in the business; in addition, a profitable firm generates funds equal to its depreciation charges that are available for reinvestment.

The three other sources of capital are from outside the operation of the enterprise:

1. Borrowed money from banks, insurance companies, and so forth.
2. Longer term borrowing from a lending institution or from the public in the form of mortgage bonds.
3. Sale of equity securities like common or preferred stock.

Retained profits and cash equal to depreciation charges are the primary sources of investment capital for most firms, and the only sources for many enterprises.

In selecting a value of MARR, three values are frequently considered:

1. *Cost of borrowed money.*
2. *Cost of capital.* This is a composite cost of the components of the overall capitalization of the enterprise.
3. *Opportunity cost.* This refers to the cost of the opportunity foregone; stated more simply, opportunity cost is the rate of return on the best investment project that is rejected.

The MARR should be equal to the highest one of these three values.

When there is a risk aspect to the problem (probabilities are known), this can be handled by techniques like expected value and simulation. Where there is uncertainty (probabilities of the various outcomes are not known), there are analytical techniques, but they are less satisfactory. A method commonly used to adjust for risk and uncertainty is to increase the MARR. This method has the effect of distorting the time-value-of-money relationship. The effect is to discount longer term consequences more heavily compared to short term consequences, which may or may not be desirable.

Problems

15-1 Examine the financial pages of your newspaper (or *The Wall St. Journal*) and determine the current interest rate on the following securities:

a. U.S. Treasury bond due in five years.

b. General obligation bond of a municipal district, city, or a state due in twenty years.

c. Corporate debenture bond of a U.S. industrial firm due in twenty years.

Explain why the interest rates are different for these different bonds.

15-2 Consider four mutually exclusive alternatives:

	A	*B*	*C*	*D*
Initial cost	0	100	50	25
Uniform annual benefit	0	16.27	9.96	5.96
Computed rate of return	0%	10%	15%	20%

Each alternative has a ten-year useful life and no salvage value. Over what range of interest rates is *C* the preferred alternative? (*Answer:* $4.5\% < i \leq 9.6\%$)

15-3 Frequently we read in the newspaper that one should lease an automobile rather than buying it. For a typical 24-month lease on a car costing $9400, the monthly lease charge is about $267. At the end of the 24 months, the car is returned to the lease company (which owns the car). As an alternative, the same car could be bought with no down payment and 24 equal monthly payments, with interest at a 12% nominal annual percentage rate. At the end of 24 months the car is fully paid for. The car would then be worth about half of its original cost.

a. Over what range of nominal before-tax interest rates is leasing the preferred alternative?

b. What are some of the reasons that would make leasing more desirable than is indicated in ***a***?

15-4 Assume you have $2000 available for investment for a five-year period. You wish to *invest* the money—not just spend it on fun things. There are obviously many alternatives available. You should be willing to assume a modest amount of risk of loss of some or all of the money if this is necessary, but not a great amount of risk (no investments in poker games or at horse races). How would you invest the money? What is your minimum attractive rate of return? Explain.

15-5 There are many venture capital syndicates that consist of a few (say, eight or ten) wealthy people who combine to make investments in small and (hopefully) growing businesses. Typically, the investors hire a young investment manager (often an engineer with an MBA) who seeks and analyzes investment opportunities for the group. Would you estimate that the MARR sought by this group is more or less than 12%? Explain.

CHAPTER **16**

Economic Analysis In Government

So far we have considered economic analysis where a firm pays all the investment costs and receives all the benefits. But when we examine the activities of government we find a quite different situation. They receive revenue through various forms of taxation and are expected to do things "in the public interest." Thus the government pays, but it receives few if any of the benefits. The original requirement for economic analysis of federal flood control projects contained what has become a famous line for describing benefits in federal projects,* " . . . the benefits to whomsoever they may accrue . . . " Unlike industrial firms, the federal government pays the costs to obtain benefits for whomsoever might be fortunate enough to receive them.

Perhaps it is desirable to construct a canal across Florida. The costs will be paid by the federal government. But it is clear that the federal government will not receive the benefits of the canal. Instead the benefits will be received by landowners in the area and the people who visit and enjoy the area being enhanced by the canal.

This situation can present all sorts of problems. For one, it means that the intended beneficiaries of a federal project will be very anxious to get the project approved and funded. A second problem concerns the measurement of the benefits. It is obviously more difficult to measure actual benefits when they are so widely disseminated. Other difficulties include the selection of an interest rate and choosing the correct viewpoint from which the analysis should be made. Finally, in benefit–cost ratio analysis, there may be a problem deciding what goes in the numerator and what goes in the denominator of the ratio.

---

*Flood Control Act of 1936.

From Whose Viewpoint Should The Analysis Be Made?

When governmental bodies do economic analysis, an important question concerns the proper viewpoint of the analysis. A look at industry will help to explain the problem. Industrial economic analysis must also be based on a viewpoint, but in this case there is an obvious answer—a firm pays the costs and counts *its* benefits. Thus, both the costs and benefits are measured from the point of view of the firm. Costs and benefits that occur outside of the firm generally are ignored.

In years past, the consequences (costs and benefits) of a firm's actions that occurred outside the firm were called external consequences and ignored. Ask anyone who has lived near a cement plant, a slaughterhouse, or a steel mill about external consequences! More recently, government has forced industry to reduce pollution and other undesirable external consequences (see Fig. 16-1). The result has been to force firms to take a larger, or community-oriented, viewpoint in evaluating the consequences of their actions.

The councilmembers of a small town that levies taxes can be expected to take the "viewpoint of the town" in making decisions: unless it can be shown that the money from taxes can be used *effectively*, it is unlikely the town council will spend it. But what happens when the money is contributed to the town by the federal government as in "revenue sharing" or some other federal grant? Often the federal government pays a share of project costs varying from 10% to 90%. Example 16-1 illustrates the viewpoint problem that is created.

EXAMPLE 16-1

A municipal project will cost $1 million. The federal government will pay 50% of the cost if the project is undertaken. Although the original economic analysis showed the PW of benefits was $1.5 million, a subsequent detailed analysis by

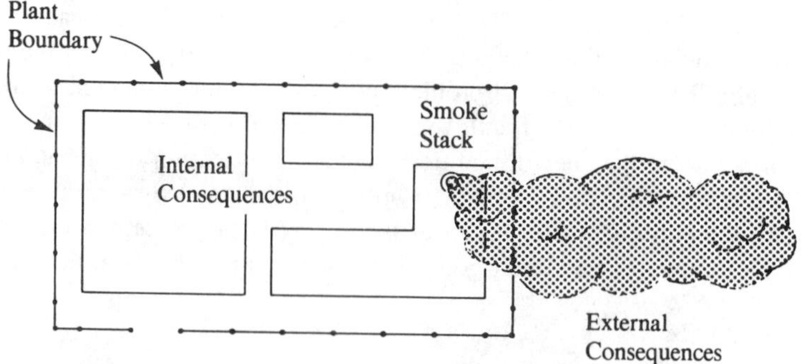

Figure 16-1 Internal and external consequences for an industrial plant.

the town engineer indicates a more realistic estimate of the PW of benefits is $750,000. The town council must decide whether or not to proceed with the project. What would you advise?

Solution: From the viewpoint of the town, the project is still a good one. If the town puts up half the cost ($500,000) it will receive all the benefits ($750,000). On the other hand, from an *overall* viewpoint, the revised estimate of $750,000 of benefits does not justify the $1 million expenditure. This illustrates the dilemma caused by varying viewpoints. For economic efficiency, one does not want to encourage the expenditure of money, regardless of the source, unless the benefits at least equal the costs. ◀

Some possible viewpoints are: an individual, a business firm or corporation, a town or city district, a city, a state, a nation, or a group of nations. To avoid suboptimizing, the proper approach is to *take a viewpoint at least as broad as those who pay the costs and those who receive the benefits*. When the costs and benefits are totally confined to a town, for example, that seems an appropriate viewpoint. But when the money or the benefits go beyond the proposed viewpoint, then the viewpoint should be enlarged to this broader view.

What Interest Rate?

Another thorny problem in economic analysis for government concerns the appropriate interest rate to use in an analysis. For industrial economic analyses, some of the alternatives were to use the cost of borrowed money, the cost of capital, or the opportunity cost.

In government, most of the money is obtained by taxation and spent about as quickly as it is obtained. There is little time delay between collecting the money from taxpayers and spending it. (Remember, the federal government and many states collect taxes from every paycheck in the form of withholding tax.) The collection of taxes and its disbursement, although based on an annual budget, is actually a continuing process. Using this line of reasoning, some people would argue that there is no interest rate involved because there is little or no time lag between collecting and spending the money. They would advocate a 0% interest rate.

Another approach to the *What interest rate?* question is to recognize that in addition to collecting taxes, most levels of government (federal, state, or local) borrow money for capital expenditures. Where money is borrowed for a specific project, one line of reasoning is to use an interest rate equal to the cost of borrowed money. This is less valid an argument for state and local governments than for industrial firms. The reason is that the federal government, through the income tax laws, subsidizes state and local bonded debt. If a state or one of its political subdivisions (like a county, city, or special assessment district) raises money for governmental purposes through the sale of bonds, interest paid on the bonds is generally not taxable income to the individual who owns them. This is an important benefit of these bonds (typically called "municipal bonds") with the result that people will naturally require a lower interest rate on these bonds than on similar bonds where the interest is fully taxable. As a rough estimate, when fully taxed bonds yield a 9% interest rate, municipal bonds or other tax-free bonds might have only a 7% interest rate. The difference of 2% represents the effect of the preferred treatment for federal income tax purposes and, hence, a form of hidden federal subsidy on tax-free bonds.

Opportunity cost, which is the interest rate on the best opportunity foregone, may take two forms in governmental economic analysis. It may be the opportunity cost within government or the opportunity cost of the taxpayers from whom the money was obtained. We know that for economic efficiency one should select the best projects from among all the prospective projects. We would expect the rate of return on all accepted projects to be higher than the rate of return on any of the rejected projects. This is accomplished by setting the interest rate for use in economic analyses equal to the opportunity cost. If the interest rate is determined solely on alternative governmental use of money, this might be called the ***governmental opportunity cost***.

Another question is whether or not the money should be taken from the taxpayer for governmental use. Individual taxpayers have varying individual opportunity costs, but the opportunity cost for taxpayers as a group can be approximated. It is economically undesirable to take money from a taxpayer with a 9% opportunity cost, for example, and spend it on a governmental project yielding 6%.

From the foregoing discussion, a reasonable conclusion is that in governmental economic analyses the interest rate should be the *largest* one of the following: cost of borrowed money (plus subsidy on tax-free bonds); governmental opportunity cost; or taxpayers' opportunity cost.

The tendency in government has been to use relatively low interest rates for economy studies. The structure of typical government projects may partly explain the reason. A highway project, for example, represents a large initial expenditure with only a small continuing maintenance expenditure. The benefits increase over the years as the use of the highway is projected to increase. The present worth of benefits is very sensitive to the interest rate used in the analysis. This is illustrated in Ex. 16-2.

EXAMPLE 16-2

On a public project the annual benefits will be $50,000 beginning two years after start of construction and increasing each year by $50,000 on an arithmetic gradient. Assume a thirty-year analysis period. What capital investment can be justified as of the beginning of construction, **a.** at a 3% interest rate? **b.** at a 9% interest rate?

Solution to Ex. 16-2a: For a benefit–cost ratio of one,

$$\begin{aligned}\text{Cost (equals PW of cost)} &= \text{PW of benefits} \\ &= \$50{,}000(P/G,3\%,30) \\ &= \$50{,}000(241.4) = \$12{,}070{,}000\end{aligned}$$

Solution to Ex. 16-2b: At a 9% interest rate,

$$\begin{aligned}\text{Cost} &= \$50{,}000(P/G,9\%,30) \\ &= \$50{,}000(89.0) = \$4{,}450{,}000\end{aligned}$$

For a benefit–cost ratio of 1 or more, the construction cost may be as much as $12,070,000 for a 3% interest rate, but only $4,450,000 for a 9% interest rate. ◀

Benefit–Cost Ratio Analysis

Benefit–cost ratio analysis requires that all the various consequences of a proposed project be classified and placed into either the numerator or the denominator of the ratio. At first glance, this would appear simply a matter of sorting out the consequences into benefits (for the numerator) or costs (for the denominator). This works satisfactorily when applied to the projects of a firm or an individual. In governmental projects, however, there may be difficulties deciding whether to classify various consequences as items for the numerator or for the denominator, as shown by Ex. 16-3.

EXAMPLE 16-3

On a proposed governmental project, the following consequences have been identified:

- Initial cost of project to be paid by government is 100.
- Present worth of future maintenance to be paid by government is 40.
- Present worth of benefits to the public is 300.
- Present worth of additional public user costs is 60.

Show the various ways of computing the benefit–cost ratio.

Solution: Putting the benefits in the numerator and all the costs in the denominator gives:

$$\text{Benefit–cost ratio} = \frac{\text{All benefits}}{\text{All costs}} = \frac{300}{100 + 40 + 60} = \frac{300}{200} = 1.5$$

An alternate computation is to consider user costs a *dis*benefit and to subtract them in the numerator rather than adding them in the denominator:

$$\text{Benefit–cost ratio} = \frac{\text{Public benefits} - \text{Public costs}}{\text{Governmental costs}}$$

$$= \frac{300 - 60}{100 + 40} = \frac{240}{140} = 1.7$$

Still another variation would be to consider maintenance costs as a disbenefit:

$$\text{Benefit–cost ratio} = \frac{300 - 60 - 40}{100} = \frac{200}{100} = 2.0$$

It should be noted that, while three different benefit–cost ratios may be computed, the value of NPW does not change:

$$\text{NPW} = \text{PW of benefits} - \text{PW of costs} = 300 - 60 - 40 - 100 = 100$$ ◀

There is no inherently correct way to compute the benefit–cost ratio. For highway projects, the authoritative guide is published by AASHTO.* In it the ratio is:

$$\text{Benefit–cost ratio} = \frac{\text{Net public benefits}}{\text{Government costs}}$$

which corresponds to the second of the three ratios in Example 16-3. Some analysts use the third of the ratios in Ex. 16-3. This is:

$$\text{Benefit–cost ratio} = \frac{\text{Net public benefits} - \text{Governmental maintenance costs}}{\text{Governmental investment cost}}$$

The alternate methods of computing the benefit–cost ratio do not change desirable projects ($B/C > 1$) into undesirable projects ($B/C < 1$) or *vice versa*. The danger is that by changing the method of its computation, one might infer that the resulting higher benefit–cost ratio represents a better project.

Other Aspects

Economic analyses made to determine how to spend public monies bring together the full range of analysis difficulties. Since the people who will receive the benefits may pay none of the costs directly (one way or the other, of course,

---

*American Association of State Highway and Transportation Officials, *A Manual on User Benefit Analysis of Highway and Bus-Transit Improvements*, Washington, D.C., 1978.

the public *does* pay for public expenditures), they may represent a strong pressure group to obtain the approval and funding of a particular project. Many times a second, opposing group of people effectively challenge the local, state, or federal government on a particular project. The struggle often concerns consequences that are *not* converted to money, thus, are not incorporated in the analysis. As a result of the countervailing power of the two groups, it is likely that better decision making is taking place.

From our examination of benefit–cost ratio analysis in Chapter 9, we know that incremental analysis is necessary in analyzing multiple alternatives. Unfortunately, some people who use benefit–cost ratio analysis are not aware of the need for incremental analysis.* If one goes to the trouble to make an analysis, it should be a fair and reasonably accurate picture of the situation.

SUMMARY

Governmental economic analyses use the benefit–cost ratio method, largely because of federal legislation in the 1930's. It has become as familiar to people in government as rate of return is to people in industry.

The viewpoint to take in an analysis is a more difficult problem in government, possibly because of the varied and sometimes remote sources of money. When some or all the money is state or federal funds, the appropriate viewpoint should be at this same level. The frequently localized nature of the benefits produces a local viewpoint that considers money from "outside" (either Washington, D.C., or the state capital) as "free" because it need not be repaid from the local benefits. From the local viewpoint this seems to be true.

People are not in agreement on *what interest rate* to use. Theory suggests it should be the *largest* one of the following: cost of borrowed money (plus subsidy on tax-free bonds); governmental opportunity cost; or taxpayers opportunity cost. We would expect the opportunity cost to be substantially higher than the cost of borrowed money and, therefore, recommend that opportunity cost is the proper choice in this situation.

The interest rates in actual use (one example is 6⅝%) appear to better reflect the cost of borrowed money than the opportunity's cost. The typical situation of "Cost now and benefits later" makes projects appear more desirable when lower interest rates are used. This is not unique to public projects, but the effect is probably exaggerated due to the relatively long analysis periods.

Aside from all the other difficulties with benefit–cost ratio analysis, we find that, like payback period, there is more than one way by which it may be computed. The classification of some costs as disbenefits moves them from the denominator to the numerator. The result will be to increase the value of the ratio. While this has no real effect, it may create problems in deciding which of several projects is best, or otherwise mislead people.

*The author once almost lost a struggle with a federal agency over whether or not incremental benefit–cost ratio analysis was required in a particular multiple-alternative situation.

Problems

16-1 Consider the following investment opportunity:

Initial cost	\$100,000
Additional cost at end of Year 1	50,000
Benefit at end of Year 1	0
Annual benefits at end of Years 2–10, per year	20,000

With interest at 7%, what is the benefit–cost ratio for this project? (*Answer:* 0.83)

16-2 A government agency has estimated that a flood control project has costs and benefits that are parabolic, according to the equation:

$$(\text{Present worth of benefits})^2 - 22(\text{Present worth of cost}) + 44 = 0$$

where both benefits and costs are stated in millions of dollars. What is the present worth of cost for the optimal size project?

16-3 The Highridge Water District needs an additional supply of water from Steep Creek. The engineer has selected two plans for comparison:

Gravity plan: Divert water at a point ten miles up Steep Creek and carry it through a pipeline by gravity to the district.

Pumping plan: Divert water at a point near the district and pump it through two miles of pipeline to the district. The pumping plant can be built in two stages, with half capacity installed initially and the other half ten years later.

Use a forty-year analysis period and 8% interest. Salvage values can be ignored. During the first ten years, the average use of water will be less than during the remaining thirty years. Using present worth analysis, select the more economical plan.

	Gravity	*Pumping*
Initial investment	\$2,800,000	\$1,400,000
Additional investment in tenth year	0	200,000
Operation, maintenance, replacements, per year	10,000	25,000
Power cost (average first 10 years), per year	0	50,000
Power cost (average next 30 years), per year	0	100,000

(*Answer:* Pumping plan)

16-4 The federal government proposes to construct a multi-purpose water project. This project will provide water for irrigation and for municipal uses. In addition, there will be flood control benefits and recreation benefits. The estimated project benefits computed for ten-year periods for the next fifty years are given in the table.

Purpose	*1st decade*	*2nd decade*	*3rd decade*	*4th decade*	*5th decade*
Municipal	\$ 40,000	\$ 50,000	\$ 60,000	\$ 70,000	\$110,000
Irrigation	350,000	370,000	370,000	360,000	350,000
Flood control	150,000	150,000	150,000	150,000	150,000
Recreation	60,000	70,000	80,000	80,000	90,000
Totals:	\$600,000	\$640,000	\$660,000	\$660,000	\$700,000

The annual benefits may be assumed to be one-tenth of the decade benefits. The operation and maintenance cost of the project is estimated to be $15,000 per year. Assume a fifty-year analysis period with no net project salvage value.

a. If an interest rate of 5% is used, and a benefit–cost ratio of unity, what capital expenditure can be justified to build the water project now?

b. If the interest rate is changed to 8%, how does this change the justified capital expenditure?

16-5 The city engineer has prepared two plans for the construction and maintenance of roads in the city park. Both plans are designed to provide the anticipated road and road maintenance requirements for the next forty years. The minimum attractive rate of return used by the city is 7%.

Plan *A* is a three-stage development program. $300,000 is to be spent immediately, followed by $250,000 at the end of 15 years and $300,000 at the end of thirty years. Maintenance will be $75,000 per year for the first 15 years, $125,000 per year for the next 15 years, and $250,000 per year for the final ten years.

Plan *B* is a two-stage program. $450,000 is required immediately (including money for some special equipment), followed by $50,000 at the end of 15 years. Maintenance will be $100,000 per year for the first 15 years and $125,000 for each of the subsequent years. At the end of forty years, it is believed the equipment may be sold for $150,000.

a. Determine which plan should be chosen, using benefit–cost ratio analysis.

b. If you favored Plan *B*, what value of MARR would you want to use in the computations? Explain.

16-6 The state is considering the elimination of a railroad grade crossing by building an overpass. The new structure, together with the needed land, would cost $1,800,000. The analysis period is assumed to be thirty years on the theory that either the railroad or the highway above it will be relocated by then. Salvage value of the bridge (actually, the net value of the land on either side of the railroad tracks) thirty years hence is estimated to be $100,000. A 6% interest rate is to be used.

At present, about 1000 vehicles per day are delayed due to trains at the grade crossing. Trucks represent 40% and 60% are other vehicles. Time for truck drivers is valued at $18 per hour and for other drivers at $5 per hour. Average time saving per vehicle will be two minutes if the overpass is built. No time saving occurs for the railroad.

The installation will save the railroad an annual expense of $48,000 now spent for crossing guards. During the preceding ten-year period, the railroad has paid out $600,000 in settling lawsuits and accident cases related to the grade crossing. The proposed project will entirely eliminate both these expenses. The state estimates that the new overpass will save it about $6000 per year in expenses directly due to the accidents. The overpass, if built, will belong to the state.

Should the overpass be built? If the overpass is built, how much should the railroad be asked to contribute to the state as its share of the $1,800,000 construction cost?

16-7 An existing two-lane highway between two cities is to be converted to a four-lane divided freeway. The distance between them is ten miles. The average daily traffic

(ADT) on the new freeway is forecast to average 20,000 vehicles per day over the next twenty years. Trucks represent 5% of the total traffic. Annual maintenance on the existing highway is $1500 per lane-mile. The existing accident rate is 4.58 per million vehicle miles (MVM). Three alternate plans of improvement are now under consideration.

Plan A: Improve along the existing development by adding two lanes adjacent to the existing lanes at a cost of $450,000 per mile. It is estimated that this plan will reduce auto travel time by two minutes, and will reduce truck travel time by one minute when compared to the existing highway. The Plan *A* estimated accident rate is 2.50 per MVM. Annual maintenance is estimated to be $1250 per lane-mile.

Plan B: Improve along the existing alignment with grade improvements at a cost of $650,000 per mile. Plan *B* would add two additional lanes, and it is estimated that this plan would reduce auto and truck travel time by three minutes each, when compared to the existing facility. The accident rate on this improved road is estimated to be 2.40 per MVM. Annual maintenance is estimated to be $1000 per lane-mile.

Plan C: Construct a new freeway on new alignment at a cost of $800,000 per mile. It is estimated that this plan would reduce auto travel time by five minutes and truck travel time by four minutes when compared to the existing highway. Plan *C* is 0.3 miles longer than *A* or *B*. The estimated accident rate for *C* is 2.30 per MVM. Annual maintenance is estimated to be $1000 per lane-mile. Plan *C* includes the abandonment of the existing highway with no salvage value.

Useful data: Incremental	operating cost—autos:	6¢ per mile
	—trucks:	18¢ per mile
Time saving	—autos:	3¢ per minute
	—trucks:	15¢ per minute
	Average accident cost:	$1200

If a 5% interest rate is used, which of the three proposed plans should be adopted?
(*Answer:* Plan *C*)

16-8 The local highway department is preparing an economic analysis to see if reconstruction of the pavement on a mountain road is justified. The number of vehicles travelling on the road increases each year, hence the benefits to the motoring public of the pavement reconstruction also increases. Based on a traffic count, the benefits are projected as follows:

Year	*End-of-year benefit*
19*X*1	$10,000
19*X*2	12,000
19*X*3	14,000
19*X*4	16,000
19*X*5	18,000
19*X*6	20,000
	and so on, increasing $2000 per year

The reconstructed pavement will cost $275,000 when it is installed and will have a 15-year useful life. The construction period is short, hence a beginning-of-year reconstruction will result in the end-of-year benefits listed in the table. Assume a 6% interest rate. The reconstruction, if done at all, must be done not later than 19*X*6. Should it be done, and if so, in what year?

16-9 A section of road in the state highway system needs repair at a cost of $150,000. At present, the volume of traffic on the road is low, with the result that few motorists will benefit from the work. Future traffic is expected to increase with resulting increased motorist benefits. The repair work will produce benefits for ten years after it is completed. The highway planning department is examining five mutually exclusive alternatives on whether or not to repair the road, and if so, the timing of the repair.

Year	*Do not repair*	*Repair now*	*Repair 2 years hence*	*Repair 4 years hence*	*Repair 5 years hence*
0	0	−$150,000			
1	0	5,000			
2	0	10,000	−$150,000		
3	0	20,000	20,000		
4	0	30,000	30,000	−$150,000	
5	0	40,000	40,000	40,000	−$150,000
6	0	50,000	50,000	50,000	50,000
7	0	50,000	50,000	50,000	50,000
8	0	50,000	50,000	50,000	50,000
9	0	50,000	50,000	50,000	50,000
10	0	50,000	50,000	50,000	50,000
11	0	0	50,000	50,000	50,000
12	0	0	50,000	50,000	50,000
13	0	0	0	50,000	50,000
14	0	0	0	50,000	50,000
15	0	0	0	0	50,000

Should the road be repaired and, if so, when should the work be done? Use a 15% MARR.

16-10 A fifty-meter tunnel must be constructed as part of a new aqueduct system for a city. Two alternatives are being considered. One is to build a full-capacity tunnel now for $500,000. The other alternative is to build a half-capacity tunnel now for $300,000 and then to build a second parallel half-capacity tunnel twenty years hence for $400,000. The cost of repair of the tunnel lining at the end of every ten years is estimated to be $20,000 for the full-capacity tunnel and $16,000 for each half-capacity tunnel.

Determine whether the full-capacity tunnel or the half-capacity tunnel should be constructed now. Solve the problem by benefit–cost ratio analysis, using a 5% interest rate, and a fifty-year analysis period. There will be no tunnel lining repair at the end of the fifty years.

CHAPTER **17**

Rationing Capital Among Competing Projects

We have until now dealt with situations where, at some interest rate, there is an ample amount of money to make all desired capital investments. But the concept of ***scarcity of resources*** is fundamental to a free market economy. It is through this mechanism that more economically attractive activities are encouraged at the expense of less desirable activities. For industrial firms, there are often more ways of spending money than there is money that is available. The result is that we must select from available alternatives the more attractive projects and reject—or, at least, delay—the less attractive projects.

This problem of rationing capital among competing projects is one part of a two-part problem called ***capital budgeting***. In planning its capital expenditures, an industrial firm is faced with two questions: "Where will money for capital expenditures come from?" and, "How shall we allocate available money among the various competing projects?" In Chapter 15 we discussed the sources of money for capital expenditures as one aspect in deciding on an appropriate interest rate for economic analysis calculations. Thus, the first problem has been treated.

Throughout this book, we have examined for any given project two or more feasible alternatives. We have, therefore, sought to identify in each project the most attractive alternative. For the sake of simplicity, we have looked at these projects in an isolated setting—almost as if a firm had just one project it was considering. In the business world, we know that this is rarely the case. A firm will find that there are a great many projects that are economically attractive. This situation raises two problems not previously considered:

1. How do you rank projects to show their order of economic attractiveness?
2. What do you do if there is not enough money to pay the costs of all economically attractive projects?

In this chapter we will look at the typical situation faced by a firm: multiple attractive projects, with an inadequate money supply to fund all of them. To do this, we must review our concepts of capital expenditure situations and available alternatives. Then we can summarize the various techniques that have been presented for determining if an alternative is economically attractive, first by screening all alternatives to find those that merit further consideration. Following this, we will select the best alternative from each project, assuming there is no shortage of money. The next step will be the addition of a budget constraint.

When there is not enough money to fund the best alternative from each project, we will have to do what we can with the limited amount of money available. It will become important that we have a technique for accurately ranking the various competing projects in their order of economic attractiveness. All this is designed to answer the question, "How shall we allocate available money among the various competing projects?"

Capital Expenditure Project Proposals

At the beginning of the book, we described decision making as the process of selecting the best alternative to achieve the desired objective in a given situation or problem. By carefully defining our objective, the model, and the choice of criteria, the given situation is reduced to one of selecting the best from the feasible alternatives. In this chapter we call the engineering decision-making process for a given situation or problem a ***project proposal***. Associated with various project proposals are their particular available alternatives. For a firm with many project proposals, the following situation may result:

Capital Expenditure Proposals.

Project 1—Additional manufacturing facility:

Alternative *A*. Lease an existing building.
B. Construct a new building.
C. Contract for the manufacturing to be done overseas.

Project 2—Replace old grinding machine:

Alternative *A*. Purchase semi-automatic machine.
B. Purchase automatic machine.

Project 3—Production of parts for the assembly line:

Alternative *A*. Make the parts in the plant.
B. Buy the parts from a subcontractor.

Our task is to apply economic analysis techniques to this more complex problem.

Mutually Exclusive Alternatives and Single Project Proposals

Until now we have dealt with mutually exclusive alternatives, that is, where selecting one alternative results in rejecting the other alternatives being considered. Even in the simplest problems encountered, the question was one of selection *between* alternatives. Should, for example, Machine *A or* Machine *B* be purchased to perform the necessary task? Clearly, the purchase of one of the machines meant that the other one would not be purchased. Since either machine would perform the task, the selection of one precludes the possibility of selecting the other one as well.

Even in the case of multiple alternatives, we have been considering mutually exclusive alternatives. A typical example was: "What size pipeline should be installed to supply water to a remote construction site?" Only one alternative is to be selected. This is different from the situation for single proposals where only one course of action is outlined. Consider Example 17-1.

EXAMPLE 17-1

The general manager of a manufacturing plant has received the following project proposals from the various operating departments:

1. The foundry wishes to purchase a new ladle to speed up their casting operation.
2. The machine shop has asked for some new inspection equipment.
3. The painting department reports they must make improvements to the spray booth to conform with new air pollution standards.
4. The office manager wants to buy a larger, more modern safe.

Each project consists of a single course of action. Note that the single project proposals are also independent, for there is no interrelationship or interdependence among them. The general manager can decide to allocate money for none, some, or all of the various project proposals.

Solution: *Do-Nothing Alternative*. The four project proposals above each have a single course of action. The general manager could, for example, buy the office manager a new safe and buy the inspection equipment for the machine shop. But he could also decide to *not* buy the office manager a safe or the equipment for the machine shop. There is, then, an alternative to buying the safe for the office manager: not to buy him the safe—to do nothing. Similarly, he could decide to do nothing about the request for the machine shop inspection equipment. Naturally, there are do-nothing alternatives for each of the four single project proposals:

1*A*. Purchase the foundry a new ladle.
1*B*. Do nothing. (Do not purchase a new ladle.)

2*A*. Obtain the inspection equipment for the machine shop.
2*B*. Do nothing. (Do not obtain the inspection equipment.)

3*A*. Make improvements to the spray booth in the painting department.
3*B*. Do nothing. (Make no improvements.)

4*A*. Buy a new safe for the office manager.
4*B*. Do nothing. (Let him use the old safe!)

One can adopt Alt. 1*A* (buy the ladle) or 1*B* (do not buy the ladle), but not both. We find that what we considered to be a single course of action is really a pair of mutually exclusive alternatives. Even Alt. 3 is in this category. The originally stated single proposal was:

> "The painting department reports that they must make improvements to the spray booth to conform with new air pollution standards."

Since the painting department reports they *must* make the improvements, is there actually another alternative? Although at first glance we might not think so, the company does have one or more alternatives available. It may be possible to change the paint, or the spray equipment, and thereby solve the air pollution problem without any improvements to the spray booth. In this situation, there does not seem to be a practical do-nothing alternative, for failure to comply with the air pollution standards might result in large fines or even shutting down the plant. But if there is not a practical do-nothing alternative, there might be a number of do-something-else alternatives.

We conclude that all project proposals may be considered to have mutually exclusive alternatives. ◀

Identifying and Rejecting Unattractive Alternatives

It is clear that no matter what the circumstances may be, we want to eliminate from further consideration any alternative that fails to meet the minimum level of economic attractiveness, provided one of the other alternatives does meet the criterion. Table 17-1 summarizes five techniques that may be used.

At first glance it appears that many calculations are required, but this is not the situation. *Any* of the five techniques listed in Table 17-1 may be used to determine whether or not to reject an alternative. Each will produce the same decision regarding *Reject–Don't reject*.

Selecting the Best Alternative from Each Project Proposal

The task of selecting the best alternative from among two or more mutually exclusive alternatives has been a primary subject of this book. Since a project proposal is this same form of problem, we may use any of the several methods discussed in Chapters 5 through 9. The criteria are summarized in Table 17-2.

Table 17-1 CRITERIA FOR REJECTING UNATTRACTIVE ALTERNATIVES

For each alternative compute	*Reject alternative when*	*Do not reject alternative when*
Rate of return, i	$i <$ MARR	$i \geq$ MARR
Present worth, PW	PW of benefits < PW of costs	PW of benefits ≥ PW of costs
Annual cost, EUAC Annual benefit, EUAB	EUAC > EUAB	EUAC ≤ EUAB
Benefit–cost ratio, B/C	B/C < 1	B/C ≥ 1
Net present worth, NPW	NPW < 0	NPW ≥ 0

Table 17-2 CRITERIA FOR CHOOSING THE BEST ALTERNATIVE FROM AMONG MUTUALLY EXCLUSIVE ALTERNATIVES

	Situation		
Analysis method	*Fixed input* (*The cost of each alternative is the same*)	*Fixed output* (*The benefits from each alternative are the same*)	*Neither input nor output fixed* (*Neither the costs nor the benefits for each alternative are the same*)
Present worth	**Maximize present worth of benefits**	**Minimize present worth of cost**	**Maximize net present worth**
Annual cash flow	**Maximize equivalent uniform annual benefits**	**Minimize equivalent uniform annual cost**	**Maximize (EUAB − EUAC)**
Benefit–cost ratio	**Maximize benefit–cost ratio**	**Maximize benefit–cost ratio**	**Incremental benefit–cost ratio analysis is required**
Rate of return	**Incremental rate of return analysis is required**		

Rationing Capital By Rate Of Return

One way of looking at the capital rationing problem is through the use of rate of return. The technique for selecting from among independent projects may be illustrated by an example.

EXAMPLE 17-2

Nine independent projects are being considered. Figure 17-1 may be prepared from the following data.

Project	*Cost*	*Uniform annual benefit*	*Useful life, in years*	*Salvage value*	*Computed rate of return*
1	$100	$23.85	10	$ 0	20%
2	200	39.85	10	0	15%
3	50	34.72	2	0	25%
4	100	20.00	6	100	20%
5	100	20.00	10	100	20%
6	100	18.00	10	100	18%
7	300	94.64	4	0	10%
8	300	47.40	10	100	12%
9	50	7.00	10	50	14%

If a capital budget of $650 is available, which projects should be selected?

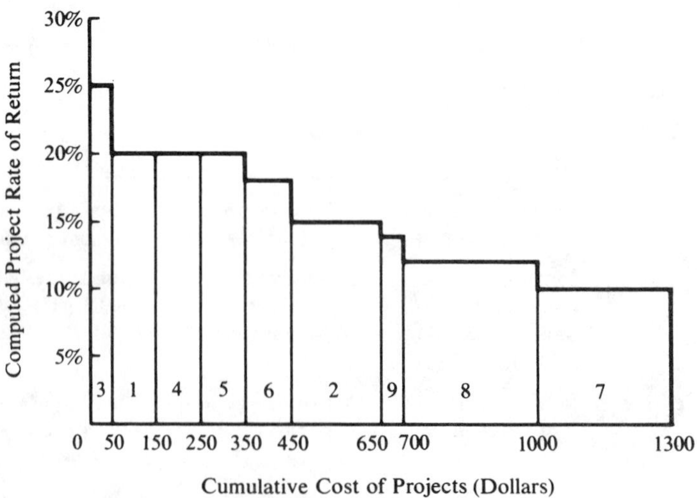

Figure 17-1 Cumulative cost of projects *vs.* rate of return.

Solution: Looking at the nine projects, we see that some are expected to produce a larger rate of return than others. It is natural that if we are to select from among them, we will pick those with a higher rate of return. When the projects are arrayed by rate of return, as in Fig. 17-1, the choice of Projects 3, 1, 4, 5, 6, and 2 is readily apparent, and is a correct decision. ◀

In Example 17-2, the rate of return was computed for each project and then the projects were arranged in order of decreasing rate of return. For a fixed amount of money in the capital budget, the projects are selected by going down the list until the money is exhausted. Using this procedure, the point where the money runs out is where we cut off approving projects. This point is called the ***cutoff rate of return***. Figure 17-2 illustrates the general situation.

For any set of ranked projects and any capital budget, the rate of return at which the budget is exhausted is the cutoff rate of return. In Fig. 17-2 the cost of each individual project is small compared to the capital budget. The cumulative cost curve is a relatively smooth curve producing a specific cutoff rate of return. Looking back at Fig. 17-1, we see the curve is actually a step function. For Example 17-2, the cutoff rate of return is between 14% and 15% for a capital budget of $650.

Significance of the Cutoff Rate of Return

Cutoff rate of return is determined by the comparison of an established capital budget and the available projects. One must examine all the projects and all the money for some period of time (like an annual budget) to compute the cutoff rate of return. It is a computation relating known projects with a known money

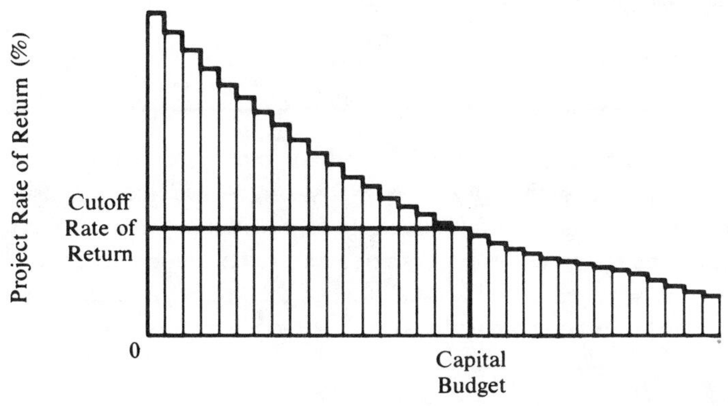

Figure 17-2 Location of the cutoff rate of return.

supply. For this period of time, the cutoff rate of return is the opportunity cost (rate of return on the opportunity or project foregone) and also the minimum attractive rate of return. In other words, the minimum attractive rate of return to get a project accomplished *is* the cutoff rate of return.

MARR = Cutoff rate of return = Opportunity cost

We generally use the minimum attractive rate of return to decide whether or not to approve an individual project even though we do not know exactly what other projects will be proposed during the year. In this situation, we cannot know if the MARR is equal to the cutoff rate of return. When the MARR is different from the cutoff rate of return, incorrect decisions may occur. This will be illustrated in the next section.

Rationing Capital By Present Worth Methods

Throughout this book we have chosen from among project alternatives to maximize net present worth. If we can do the same thing for a group of projects, and we do not exceed the available money supply, then the capital budgeting problem is solved.

But more frequently, the capital budgeting problem is one where we will be unable to accept all desirable projects. We, therefore, have a task not previously encountered. We must choose the best from the larger group of acceptable projects.

Lorie and Savage* showed that the proper technique is to use a multiplier, p, to decrease the attractiveness of an alternative in proportion to its use of the scarce supply of money. The revised criterion is

NPW − *p*(PW of cost) (17-1)

where p is a multiplier computed by trial and error

If a value of p were selected (say, 0.1), then some alternatives with a positive NPW will have a negative [NPW − p(PW of cost)]. This new criterion will reduce the number of favorable alternatives and thereby reduce the combined cost of the projects meeting this more severe criterion. By trial and error, the multiplier p is adjusted until the total cost of the projects meeting the [NPW − p(PW of cost)] criterion equals the available money supply—the capital budget.

*Lorie, J. and L. Savage, "Three Problems in Rationing Capital," *Journal of Business*, October, 1955, pp. 229–239.

EXAMPLE 17-3

Using the present worth method, determine which of the nine independent projects of Ex. 17-2 should be included in a capital budget of $650. The minimum attractive rate of return has been set at 8%.

Project	*Cost*	*Uniform annual benefit*	*Useful life, in years*	*Salvage value*	*Computed* NPW
1	$100	$23.85	10	$ 0	$60.04
2	200	39.85	10	0	67.40
3	50	34.72	2	0	11.91
4	100	20.00	6	100	55.48
5	100	20.00	10	100	80.52
6	100	18.00	10	100	67.10
7	300	94.64	4	0	13.46
8	300	47.40	10	100	64.38
9	50	7.00	10	50	20.13

Solution: Locating a value of p in [NPW − p(PW of cost)] by trial and error:

			Trial p = 0.20		*Trial p* = 0.25	
Project	*Cost*	*Computed* NPW	[NPW − *p*(PW *of cost*)]	Cost	[NPW − *p*(PW *of cost*)]	Cost
1	$ 100	$60.04	$40.04	$ 100	$35.04	$100
2	200	67.40	27.40	200	17.40	200
3	50	11.91	1.91	50	−0.59	
4	100	55.48	35.48	100	30.48	100
5	100	80.52	60.52	100	55.52	100
6	100	67.10	47.10	100	42.10	100
7	300	13.46	−46.54		−61.54	
8	300	64.38	4.38	300	−10.62	
9	50	20.13	10.13	50	7.63	50
	$1300			$1000		$650

For a value of p equal to 0.25, the best selection is computed to be Projects 1, 2, 4, 5, 6, and 9.

Alternate Formation of Example 17-3: This answer does not agree with the solution obtained in Ex. 17-2. The difficulty is that the interest rate used in the present worth calculations is not equal to the computed cutoff rate of return. In Ex. 17-2 the cutoff rate of return was between 14% and 15%, say 14.5%. We will recompute the present worth solution using MARR = 14.5%.

Project	*Cost*	*Computed* NPW *at 14.5%*	*Cost of projects with positive* NPW
1	$100	$22.01	$100
2	200	3.87	200
3	50	6.81	50
4	100	21.10	100
5	100	28.14	100
6	100	17.91	100
7	300	−27.05	
8	300	−31.69	
9	50	−1.28	
			$650

Solution: At a MARR of 14.5% the best set of projects is the same as computed in Ex. 17-2, namely, Projects 1, 2, 3, 4, 5, and 6, and their cost equals the capital budget. One can see that only projects with a rate of return greater than MARR can have a positive NPW at this interest rate. With MARR equal to the cutoff rate of return, we *must* obtain the same solution by either the rate of return or present worth methods. ◀

Figure 17-3 outlines the present worth method for the more elaborate case where there are independent projects each with mutually exclusive alternatives.

EXAMPLE 17-4

A company is preparing its capital budget for next year. The amount has been set at $250 by the Board of Directors. The MARR of 8% is believed to be close to the cutoff rate of return. The following project proposals are being considered.

Project proposals	*Cost*	*Uniform annual benefit*	*Salvage value*	*Useful life, in years*	*Computed* NPW
Proposal 1					
Alt. *A*	$100	$23.85	$0	10	$60.04
B	150	32.20	0	10	66.06
C	200	39.85	0	10	67.40
D	0	0			0
Proposal 2					
Alt. *A*	50	14.92	0	5	9.57
B	0	0			0
Proposal 3					
Alt. *A*	100	18.69	25	10	36.99
B	150	19.42	125	10	38.21
C	0	0			0

Which project alternatives should be selected, based on present worth methods?

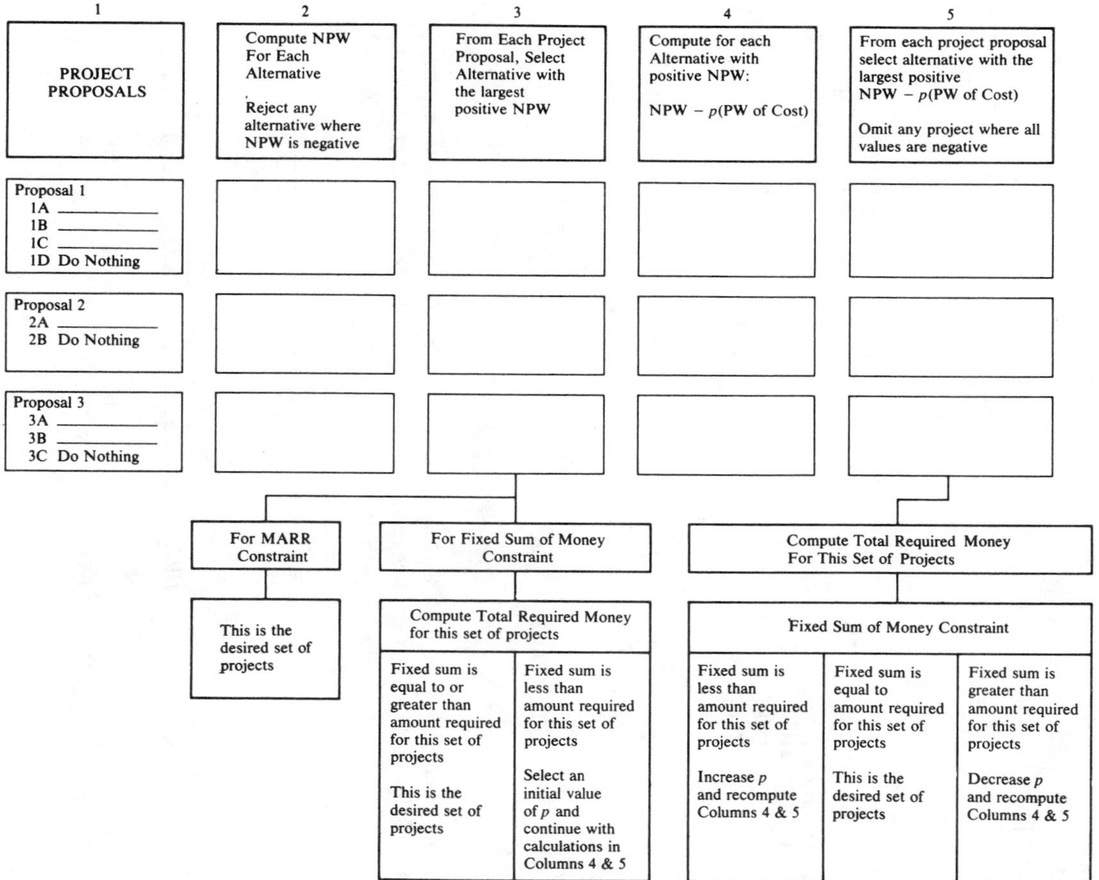

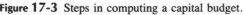

Figure 17-3 Steps in computing a capital budget.

Solution: The tabulation below shows that to maximize NPW, we would choose Alternatives 1*C*, 2*A*, and 3*B*. The total cost of these three projects is $400. Since the capital budget is only $250, we cannot fund these projects. To penalize all projects in proportion to their cost, we will use Equation 17-1 with its multiplier, *p*. As a first trial, a value of $p = 0.10$ is selected and the alternatives with the largest [NPW − *p*(PW of cost)] selected.

					$p = 0.10$		
			Alternative with largest positive NPW			*Alternative with largest positive* [NPW − *p*(PW *of cost*)]	
Project proposals	*Cost*	NPW	*Alt.*	*Cost*	[NPW − *p*(PW *of cost*)]	*Alt.*	*Cost*
Proposal 1							
Alt. *A*	$100	$60.04			$50.04		
B	150	66.06			51.06	1*B*	$150
C	200	67.40	1*C*	$200	47.40		
D	0	0			0		
Proposal 2							
Alt. *A*	50	9.57	2*A*	50	4.57	2*A*	50
B	0	0			0		
Proposal 3							
Alt. *A*	100	36.99			26.99	3*A*	100
B	150	38.21	3*B*	150	23.21		
C	0	0			0		
				$400			$300

The first trial with $p = 0.10$ selects Alternatives 1*B*, 2*A*, and 3*A* with a total cost of $300. This still is greater than the $250 capital budget. Another trial is needed with a larger value of *p*. Select $p = 0.15$ and recompute.

			$p = 0.15$		
				Alternative with largest positive [NPW − *p*(PW *of cost*)]	
Project proposals	*Cost*	NPW	[NPW − *p*(PW *of cost*)]	*Alt.*	*Cost*
Proposal 1					
Alt. *A*	$100	$60.04	$45.04	1*A*	$100
B	150	66.06	43.56		
C	200	67.40	37.40		
D	0	0	0		
Proposal 2					
Alt. *A*	50	9.57	2.07	2*A*	50
B	0	0	0		
Proposal 3					
Alt. *A*	100	36.99	21.99	3*A*	100
B	150	38.21	15.71		
C	0	0	0		
					$250

The second trial, with $p = 0.15$, points to Alternatives 1*A*, 2*A*, and 3*A* for a total cost of $250. This equals the capital budget, hence is the desired set of projects. ◀

EXAMPLE 17-5

Solve Ex. 17-4 by the rate of return method. For project proposals with two or more alternatives, incremental rate of return analysis is required. The data from Ex. 17-4 and the computed rate of return for each alternative and each increment of investment is shown in the tabulation below.

Solution:

					Incremental analysis			
	Cost	*Uniform annual benefit*	*Salvage value*	*Computed rate of return*	*Cost*	*Uniform annual benefit*	*Salvage value*	*Computed rate of return*
Proposal 1								
A	$100	$23.85	$ 0	20.0%				
B − *A*					$50	$8.35	$ 0	10.6%
B	150	32.20	0	17.0%				
C − *B*					50	7.65	0	8.6%
C − *A*					100	16.00	0	9.6%
C	200	39.85	0	15.0%				
D	0	0	0	0%				
Proposal 2								
A	50	14.92	0	15.0%				
B	0	0	0	0%				
Proposal 3								
A	100	18.69	25	15.0%				
B − *A*					50	0.73	100	8.3%
B	150	19.42	125	12.0%				
C	0	0	0	0%				

The various separable increments of investment may be ranked by rate of return. They are plotted in a cumulative cost *vs*. rate of return graph in Fig. 17-4. The ranking of projects by rate of return gives the following:

Project

1*A*
2*A*
3*A*
1*B* in place of 1*A*
1*C* in place of 1*B*
3*B* in place of 3*A*

For a budget of $250, the selected projects are 1*A*, 2*A*, and 3*A*. Note that if a budget of $300 were available, 1*B* would replace 1*A*, making the proper set of

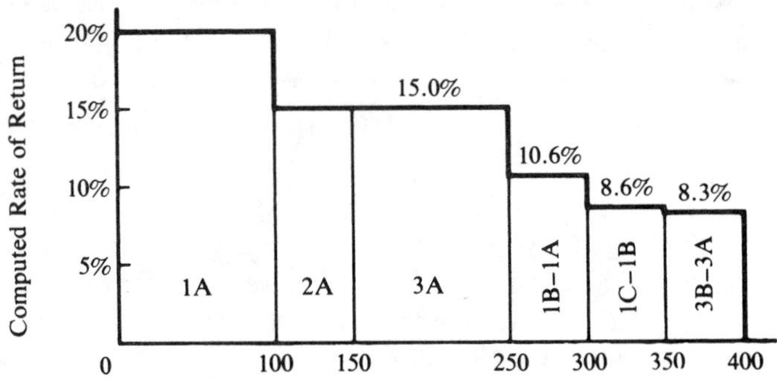

Figure 17-4 Cumulative cost *vs.* incremental rate of return.

projects 1*B*, 2*A*, and 3*A*. At a budget of $400, 1*C* would replace 1*B*; and 3*B* would replace 3*A*, making the selected projects 1*C*, 2*A*, and 3*B*. These answers agree with the computations in Ex. 17-4. ◀

Ranking Project Proposals

Closely related to the problem of capital budgeting is the matter of ranking project proposals. We will first examine a method of ranking by present worth methods and then show that project rate of return is not a suitable method of ranking projects.

Anyone who has ever bought firecrackers probably used the practical ranking criterion of "biggest bang for the buck" in selecting the fireworks. This same criterion—stated more eloquently—may be used to correctly rank independent projects.

> **Rank independent projects according to their value of net present worth divided by the present worth of cost. The appropriate interest rate is MARR (as a reasonable estimate of the cutoff rate of return).**

Example 17-6 illustrates the method of computation.

EXAMPLE 17-6

Rank the following nine independent projects in their order of desirability, based on a 14.5% minimum attractive rate of return. (To facilitate matters the necessary computations are included in the tabulation.)

Project	*Cost*	*Uniform annual benefit*	*Useful life, in years*	*Salvage value*	*Computed rate of return*	*Computed NPW at 14.5%*	*Computed NPW/Cost*
1	$100	$23.85	10	$0	20%	$22.01	0.2201
2	200	39.85	10	0	15%	3.87	0.0194
3	50	34.72	2	0	25%	6.81	0.1362
4	100	20.00	6	100	20%	21.10	0.2110
5	100	20.00	10	100	20%	28.14	0.2814
6	100	18.00	10	100	18%	17.91	0.1791
7	300	94.64	4	0	10%	−27.05	−0.0902
8	300	47.40	10	100	12%	−31.69	−0.1056
9	50	7.00	10	50	14%	−1.28	−0.0256

Solution: Ranked by NPW/PW of cost, the projects are listed:

Project	$\frac{\text{NPW}}{\text{PW of cost}}$	*Rate of return, %*
5	0.2814	20
1	0.2201	20
4	0.2110	20
6	0.1791	18
3	0.1362	25
2	0.0194	15
9	−0.0256	14
7	−0.0902	10
8	−0.1056	12

The rate of return tabulation illustrates that it is *not* a satisfactory ranking criterion and would have given a different ranking from the present worth criterion. ◀

In Example 17-6, the projects are ranked according to the ratio NPW/PW of cost. In Fig. 17-3, the criterion used is [NPW − p(PW of cost)]. If one were to compute the value of p at which [NPW − p(PW of cost)] = 0, we would obtain p = (NPW/PW of cost). Thus the multiplier p is the ranking criterion at the point where [NPW − p(PW of cost)] = 0.

If independent projects can be ranked in their order of desirability, then the selection of projects to be included in a capital budget is a simple task. One may proceed down the list of ranked projects until the capital budget is exhausted. The only difficulty with this scheme occasionally occurs when the capital budget is more than enough for n projects, but too little for $(n + 1)$ projects.

In Example 17-6, a capital budget of $300 is just right to fund the top three projects. But a capital budget of $550 is more than enough for the top five projects (sum = $450) but not enough for the top six projects (sum = $650).

When we have this lumpiness problem, one cannot always say with certainty that the best use of a capital budget of $550 is to fund the top five projects. There may be some other set of projects that makes better use of the available $550. While some trial and error computations may indicate the proper set of projects, more elaborate techniques are needed to prove optimality.

As a practical matter, a capital budget probably has some flexibility. If in Ex. 17-6 the tentative capital budget is $550, then a careful examination of Project 2 will dictate whether to expand the capital budget to $650 (to be able to include Project 2) or to drop back to $450 (and leave Project 2 out of the capital budget).

SUMMARY

Prior to this chapter we have assumed that all worthwhile projects are approved and implemented. But industrial firms, like individuals and governments, are typically faced with more good projects than there is money available. The task is to select the best projects and reject, or at least delay, the rest.

Mutually exclusive alternatives are those where the acceptance of one alternative effectively prevents the adoption of the other alternatives. This could be because the alternatives perform the same function (like Pump *A* *vs*. Pump *B*) or occupy the same physical location (like a gas station *vs*. a hamburger stand). If a project has a single alternative of doing something, we know there is likely to be a mutually exclusive alternative of doing nothing—or possibly doing something else. A project proposal may be thought of as having two or more mutually exclusive alternatives. Projects are assumed in this chapter to be independent.

Capital may be rationed among competing investment opportunities by either rate of return or present worth methods. The results may not always be the same for these two methods in many practical situations.

If projects are ranked by rate of return, a proper procedure is to go down the list until the capital budget is exhausted. The rate of return at this point is the cutoff rate of return. This procedure gives the best group of projects, but does not necessarily have them in the proper priority order.

Maximizing NPW is an appropriate present worth selection criterion where the available projects do not exhaust the money supply. But if the amount of money required for the best alternative from each project exceeds the available money, a more severe criterion is imposed: adopt only those alternatives and projects that have a positive [NPW $-$ p(PW of cost)]. The value of the multiplier p is chosen by trial and error until the alternatives and projects meeting the criterion just equal the available capital budget money.

It has been shown in earlier chapters that the usual business objective is to maximize NPW, and this is not necessarily the same as maximizing rate of return. One suitable procedure is to use the ratio (NPW/PW of cost) to rank the projects.

This present worth ranking method will order the projects so that, for a limited capital budget, NPW will be maximized. We know that MARR must be adjusted from time to time to reasonably balance the cost of the projects that meet the MARR criterion and the available supply of money. This adjustment of the MARR to equal the cutoff rate of return is essential for the rate of return and present worth methods to yield compatible results.

Another way of ranking is by incremental rate of return analysis. Once a ranking has been made, we can go down the list and accept the projects until the money runs out. There is a theoretical difficulty if the capital budget contains more money than is required for n projects, but not enough for one more, or $(n + 1)$ projects. As a practical matter, capital budgets are seldom inflexible, with the result that some additional money may be allocated if the $(n + 1)$ project looks like it should be included.

Problems

17-1 The following ten independent projects each have a ten-year life and no salvage value.

Project	*Cost, in thousands*	*Uniform annual benefit, in thousands*	*Computed rate of return, %*
1	$ 5	$1.03	16
2	15	3.22	17
3	10	1.77	12
4	30	4.88	10
5	5	1.19	20
6	20	3.83	14
7	5	1.00	15
8	20	3.69	13
9	5	1.15	19
10	10	2.23	18

The projects have been proposed by the staff of the Ace Card Company. The MARR of Ace has been 12% for several years.

a. If there is ample money available, what projects should Ace approve?

b. Rank order all the acceptable projects in their order of desirability.

c. If only $55,000 is available, which projects should be approved?

17-2 At Miami Products, four project proposals (three with mutually exclusive alternatives) are being considered. All the alternatives have a ten-year useful life and no salvage value.

Project proposal	*Cost, in thousands*	*Uniform annual benefit, in thousands*	*Computed rate of return, %*
Project 1			
Alt. *A*	$25	$4.61	13
B	50	9.96	15
C	10	2.39	20
Project 2			
Alt. *A*	20	4.14	16
B	35	6.71	14
Project 3			
Alt. *A*	25	5.56	18
B	10	2.15	17
Project 4	10	1.70	11

a. Using rate of return methods, determine which set of projects should be undertaken if the MARR is 10%.

b. Using rate of return methods, which set of projects should be undertaken if the capital budget is limited to $100,000?

c. For a budget of $100,000, what interest rate should be used in rationing capital by present worth methods? (Limit your answer to a value for which there is a Compound Interest Table available at the back of the book.)

d. Using the interest rate determined in **c**, rank order the eight different investment opportunities by the present worth method.

e. For a budget of $100,000 and the ranking in ***d***, which of the investment opportunities should be selected?

17-3 Al Dale is planning his Christmas shopping as he must buy gifts for seven people. To quantify how much the various people would enjoy receiving a list of prospective gifts, Al has assigned appropriateness units (called "ohs") for each gift if given to each of the seven people. A rating of five ohs represents a gift that the recipient would really like. A rating of four ohs indicates the recipient would like it four-fifths as much; three ohs, three-fifths as much, and so forth. A zero rating indicates an inappropriate gift that cannot be given to that person. These data are tabulated below.

		"Oh" rating of gift if given to various family members						
	Prospective gift	*Father*	*Mother*	*Sister*	*Brother*	*Aunt*	*Uncle*	*Cousin*
1.	$20 box of candy	4	4	2	1	5	2	3
2.	$12 box of cigars	3	0	0	1	0	1	2
3.	$16 necktie	2	0	0	3	0	3	2
4.	$20 shirt or blouse	5	3	4	4	4	1	4
5.	$24 sweater	3	4	5	4	3	4	2
6.	$30 camera	1	5	2	5	1	2	0
7.	$ 6 calendar	0	0	1	0	1	0	1
8.	$16 magazine subscription	4	3	4	4	3	1	3
9.	$18 book	3	4	2	3	4	0	3
10.	$16 game	2	2	3	2	2	1	2

The objective is to select the most appropriate set of gifts for the seven people (that is, maximize total ohs) that can be obtained with the selected budget.

a. How much will it cost to buy the seven gifts the people would like best, if there is ample money for Christmas shopping?

b. If the Christmas shopping budget is set at $112, which gifts should be purchased, and what is their total appropriateness rating in ohs?

c. If the Christmas shopping budget must be cut to $90, which gifts should be purchased, and what is their total appropriateness rating in ohs?

(*Answer:* **a.** $168)

The following facts are to be used in solving Problems 17-4 through 17-7: In assembling data for the Peabody Company annual capital budget, five independent projects are being considered. Detailed examination by the staff has resulted in the identification of from three to six mutually exclusive do-something alternatives for each project. In addition, each project has a do-nothing alternative. The projects and their alternatives are listed below.

Project proposal	*Cost, in thousands*	*Uniform annual benefit, in thousands*	*Useful life, in years*	*End-of-useful-life salvage value, in thousands*	*Computed rate of return, %*
Project 1					
Alt. *A*	$40	$13.52	2	$20	10
B	10	1.87	16	5	18
C	55	18.11	4	0	12
D	30	6.69	8	0	15
E	15	3.75	2	15	25
Project 2					
Alt. *A*	10	1.91	16	2	18
B	5	1.30	8	0	20
C	5	0.97	8	2	15
D	15	5.58	4	0	18
Project 3					
Alt. *A*	20	2.63	16	10	12
B	5	0.84	16	0	15
C	10	1.28	16	0	10
D	15	2.52	16	0	15
E	10	3.50	4	0	15
F	15	2.25	16	15	15
Project 4					
Alt. *A*	10	2.61	8	0	20
B	5	0.97	16	0	18
C	5	0.90	16	5	18
D	15	3.34	8	0	15
Project 5					
Alt. *A*	5	0.75	8	5	15
B	10	3.50	4	0	15
C	15	2.61	8	5	12

Each project concerns operations at the St. Louis brewery. The plant was leased from another firm many years ago and the lease expires 16 years from now. For this reason, the analysis period for all projects is 16 years. Peabody considers 12% to be the minimum attractive rate of return.

In solving the Peabody Co. problems, an important assumption concerns the situation at the end of the useful life of an alternative when the alternative has a useful life less than the 16-year analysis period. Two replacement possibilities are listed.

Assumption 1: When an alternative has a useful life less than 16 years, it will be replaced by a new alternative with the same useful life as the original. This may need to occur more than once. The new alternative will have a 12% computed rate of return and, hence, a NPW = 0 at 12%.

Assumption 2: When an alternative has a useful life less than 16 years, it will be replaced at the end of its useful life by an identical alternative (one with the same cost, uniform annual benefit, useful life, and salvage value as the original alternative).

17-4 For an unlimited supply of money, and replacement Assumption 1, which project alternatives should Peabody select? Solve the problem by present worth methods.
(*Answer:* Project Alternatives *1B*, *2A*, *3F*, *4A*, and *5A*)

17-5 For an unlimited supply of money, and replacement Assumption 2, which project alternatives should Peabody select? Solve the problem by present worth methods.

17-6 For an unlimited supply of money, and replacement Assumption 2, which project alternatives should Peabody select? Solve the problem by rate of return methods. (*Hint:* By careful inspection of the alternatives, you should be able to reject about half of them. Even then the problem requires lengthy calculations.)

17-7 For a capital budget of $55,000, and replacement Assumption 2, which project alternatives should Peabody select?
(*Answer:* Project Alternatives *1E*, *2A*, *3F*, *4A*, and *5A*)

17-8 A financier has a staff of three people whose job it is to examine possible business ventures for him. Periodically they present him their findings concerning business opportunities. On a particular occasion, they presented the following investment opportunities:
Project A: This is a project for the utilization of the commercial land the financier already owns. Three mutually exclusive alternatives are:

Project A1. Sell the land for $500,000.

Project A2. Lease the property for a car-washing business. An annual income, after all costs, like property taxes, and so on, of $98,700 would be received at the end of each year for twenty years. At the end of the twenty years, it is believed the property could be sold for $750,000.

Project A3. Construct an office building on the land. The building will cost $4,500,000 to construct and will not produce any net income for the first two years. The probabilities of various levels of rental income, after all expenses, for the subsequent 18 years are as follows:

Annual rental income	*Probability*
$1,000,000	0.1
1,100,000	0.3
1,200,000	0.4
1,900,000	0.2

The property (building and land) probably can be sold for $3 million at the end of twenty years.

Project B: An insurance company is seeking to borrow money for ninety days. They offer to pay 13¾% per annum, compounded continuously.

Project C: The financier owns a manufacturing company. The firm desires additional working capital to allow it to increase its inventories of raw materials and finished products. An investment of $2,000,000 will allow the company to obtain sales that in the past the company had to forgo. The additional capital will increase company profits by $500,000 a year. The financier can recover this additional investment by ordering the company to reduce its inventories and to return the $2,000,000. For planning purposes, assume the additional investment will be returned at the end of ten years.

Project D: The owners of *Sunrise* magazine are seeking a loan of $500,000 for ten years at a 16% interest rate.

Project E: The Galveston Bank has indicated they are willing to accept a deposit of any sum of money over $100,000, for any desired duration, at a 14.06% interest rate, compounded monthly. It seems likely that this interest rate will be available from Galveston, or some other bank, for the next several years.

Project F: A car rental company is seeking a loan of $2,000,000 to expand their fleet of automobiles. They offer to repay the loan by paying $1,000,000 at the end of one year, and $1,604,800 at the end of two years.

a. If there is $4 million available for investment now (or $4.5 million if the Project *A* land is sold), which projects should be selected? What is the MARR in this situation?

b. If there is $9 million available for investment now (or $9.5 million if the Project *A* land is sold), which projects should be selected?

17-9 The Raleigh Soap Company has been offered a five-year contract to manufacture and package a leading brand of soap for Taker Bros. It is understood the contract will not be extended past the five years as Taker Bros. plans to build their own plant nearby. The contract calls for 10,000 metric tons (one metric ton equals 1000 kilograms) of soap a year. Raleigh normally produces 12,000 metric tons of soap a year, so production for the five-year period would be increased to 22,000 metric tons. Raleigh must decide what changes, if any, to make to accommodate this increased production. Five projects are under consideration.

Project 1: Increase liquid storage capacity.

At present, Raleigh has been forced to buy caustic soda in tank truck quantities due to inadequate storage capacity. If another liquid caustic soda tank is installed to hold 1000 cubic meters, the caustic soda may be purchased in railroad tank car quantities at a more favorable price. The result would be a saving of 0.1 cent per kilogram of soap. The tank, which would cost $83,400, has no net salvage value.

Project 2: Another sulfonation unit.

The present capacity of the plant is limited by the sulfonation unit. The additional 12,000 metric tons of soap cannot be produced without an additional sulfonation unit. Another unit can be installed for $320,000.

Project 3: Packaging department expansion.

With the new contract, the packaging department must either work two 8-hour shifts, or have another packaging line installed. If the two-shift operation is used, a 20% wage premium must be paid for the second shift. This premium would amount to $35,000 a year. The second packaging line could be installed for $150,000. It would have a $42,000 salvage value at the end of five years.

Project 4: New warehouse.

The existing warehouse will be inadequate for the greater production. It is estimated that 400 square meters of additional warehouse is needed. A new warehouse can be built on a lot beside the existing warehouse for $225,000, including the land. The annual taxes, insurance, and other ownership costs would be $5000 a year. It is believed the warehouse could be sold at the end of five years for $200,000.

Project 5: Lease a warehouse.

An alternative to building an additional warehouse would be to lease warehouse space. A suitable warehouse one mile away could be leased for $15,000 per year. The $15,000 includes taxes, insurance, and so forth. The annual cost of moving materials to this more remote warehouse would be $34,000 a year.

The contract offered by Taker Bros. is a favorable one which Raleigh Soap plans to accept. Raleigh management has set a 15% before-tax minimum attractive rate of return as the criterion for any of the projects. Which projects should be undertaken?

17-10 Ten capital spending proposals have been made to the budget committee as they prepare the annual budget for their firm. The independent projects each have a 5-year life and no salvage value.

Project	*Initial cost in thousands*	*Uniform annual benefit in thousands*	*Computed rate of return*
A	$10	$2.98	15%
B	15	5.58	25
C	5	1.53	16
D	20	5.55	12
E	15	4.37	14
F	30	9.81	19
G	25	7.81	17
H	10	3.49	22
I	5	1.67	20
J	10	3.20	18

a. Based on a MARR of 14%, which projects should be approved?

b. Rank order all the projects in their order of desirability.

c. If only $85,000 is available, which projects should be approved?

CHAPTER 18

A Further Look At Rate Of Return

In Chapter 7, the rate of return is defined for an *investment* as follows:

> ***Rate of return* is the interest rate earned on the unrecovered investment such that the payment schedule makes the unrecovered investment equal to zero at the end of the life of the investment.**

In a *borrowing* situation, the definition is:

> ***Rate of return* is the interest rate paid on the unpaid balance of a loan such that the payment schedule makes the unpaid loan balance equal to zero when the final payment is made.**

In the actual calculation of rate of return, we wrote one equation relating costs and benefits (for example, Present worth of cost equals Present worth of benefits) and solved the equation for the unknown rate of return. This works fine if the situation represents either a pure investment or a pure borrowing situation and there is a single positive rate of return.

Unfortunately, there are times when these two conditions are not met. A remedy was suggested in Chapter 7A: by application of an external interest rate, the cash flow was adjusted until the number of sign changes was reduced to 1. The *cash flow rule of signs* then tells us there is either none or one positive rate of return. And with only one sign change in the cash flow, the situation must be one of either pure investment or of pure borrowing. This remedy works. But, as we will find in this chapter, we may be adjusting the cash flow when no adjustment is necessary, and even when an adjustment is required, we may be making too large an adjustment. The resulting computed rate of return is thus affected by adjustments which are either unnecessary or too much.

Cash Flow Situations

A cash flow may represent any kind of situation. It may be a pure borrowing, pure investment, or a mixture of the two. This is illustrated by Table 18-1. In Case *A*, the typical investment situation is represented by the investment of $50 at Year 0, followed by the return of the resulting benefits in Years 1 through 4. Case *B* represents a borrowing situation. Fifty dollars is received in Year 0, followed by four payments of $15 each in Years 1–4 to repay the loan. Note that *A* and *B* are mirror images of one another. That is, changing the sign of all the cash flows in *A* gives us *B*, and *vice versa*.

Case *C* represents a mixed situation. Initially there is a receipt of benefits (like a borrowing situation), followed by investments in Years 1 and 2. Finally, there are benefits in Years 3 and 4. The result is a mixed situation with Years 0, 1, and 2 looking like a borrowing situation and, at the same time, Years 1, 2, 3, and 4 look like an investment situation.

In computing the rate of return for a cash flow, one must carefully decide what one wishes the number to represent. Is it to be the internal rate of return earned on invested money, the external rate of return paid on borrowed money, or what? As this is a book on capital expenditure analysis, the view here is that we want to determine the rate of return earned on the money invested in the project while it is actually in the project. Money that is invested outside of the project will generally be assumed to earn some established external rate of return.

Analysis Of A Cash Flow As An Investment Situation

A good deal can be learned about the desirability of a cash flow as an investment situation. The goal is to produce a single value that accurately portrays the profitability of the investment opportunity reflected by the cash flow. We do not

Table 18-1 EXAMPLES OF DIFFERENT CASH FLOW SITUATIONS

	Case A	*Case B*	*Case C*
	pure investment	*pure borrowing*	*mixed borrowing and investment*
Year	*Cash flow*	*Cash flow*	*Cash flow*
0	−$50	+$50	+$50
1	+15	−15	−30
2	+15	−15	−30
3	+15	−15	+15
4	+15	−15	+15

want multiple rates of return. (After all, how can an investment have both a 20% and a 60% rate of return at the same time?) And we do not want a rate of return that assumes that one must temporarily invest money outside of the investment project at an unrealistic interest rate. If a suitable external investment rate is 6%, for example, then the computations must not be based on some other rate. In short, we want a single, realistic value representing the rate of return (note that this could also be called the *profitability rate*) on the investment.

There are four tests that will help us to understand the investment situation represented by any cash flow. They are:

1. Cash flow rule of signs.
2. Accumulated cash flow sign test.
3. Algebraic sum of the cash flow.
4. Net investment conditions.

These will be examined one by one.

Cash Flow Rule of Signs

There may be as many positive rates of return as there are sign changes in the cash flow.

If we let a_i represent the cash flow in Year i, then the entire cash flow could be represented as follows:

Year	*Cash flow*
0	a_0
1	a_1
2	a_2
3	a_3
.	.
.	.
.	.
n	a_n

As described in detail in Chapter 7A, a *sign change* is where successive terms in the cash flow (ignoring zeros) have different signs.

Number of sign changes in cash flow	*Number of positive rates of return*
0	0 (or ROR = ∞)
1	1, or 0
2	2, 1, or 0
3	3, 2, 1, or 0
.	.
.	.
.	.

Accumulated Cash Flow Sign Test

The accumulated cash flow is the algebraic sum of the cash flow to that point in time. If we let a_i represent the cash flow in a year and A_i to represent the accumulated cash flow, the situation is:

Year	*Cash flow*	*Accumulated cash flow*
0	a_0	$A_0 = a_0$
1	a_1	$A_1 = a_0 + a_1$
2	a_2	$A_2 = a_0 + a_1 + a_2$
3	a_3	$A_3 = a_0 + a_1 + a_2 + a_3$
.	.	.
.	.	.
.	.	.
n	a_n	$A_n = a_0 + a_1 + a_2 + a_3 + \cdots + a_n$

The sequence of accumulated cash flows ($A_0, A_1, A_2, A_3, \ldots, A_n$) is examined to determine the number of sign changes. As before, a sign change is where successive terms in the accumulated cash flow (ignoring zeros) have different signs.

Norstrøm† proved that sufficient (but not necessary) conditions for a single positive rate of return are:

1. The accumulated cash flow in Year n is greater than zero ($A_n > 0$).
2. There is exactly one sign change in the sequence of accumulated cash flows.

Algebraic Sum of the Cash Flow

$$\text{Algebraic sum} = \sum_{i=0}^{n} a_i$$

or looking at the accumulated cash flow,

$$\text{Algebraic sum} = A_n$$

A positive algebraic sum ($A_n > 0$) suggests a positive rate of return and an algebraic sum equal to zero ($A_n = 0$) suggests a 0% rate of return. In either case, however, there may be multiple positive rates of return.

A negative algebraic sum means the costs exceed the benefits of the project. This would seem to immediately mean that a positive rate of return would be

†Norstrøm, Carl J., "A Sufficient Condition for a Unique Nonnegative Internal Rate of Return," *Journal of Financial and Quantitative Analysis*, VII (June, 1972), pp. 1835–9.

impossible in an investment situation, which is usually true. It has been shown by Merrett and Sykes,† however, that where there are two or more sign changes in the cash flow, and a substantial outlay near the end of the life of a project, the result may be one or more positive rates of return.

Net Investment Conditions

Two conditions have been shown‡ to be sufficient (but not necessary) to establish that a computed positive rate of return, i^*, is the only positive rate of return. The conditions are as follows:

1. The cash flow in the nth year must be positive ($a_n > 0$).
2. Given that a positive rate of return i^* has been computed, there is a net investment throughout the life of the project until the end of the nth year when the net investment becomes zero. Mathematically, this is:

$$\text{Net investment in any Year } k = \sum_{j=0}^{n} a_j(1 + i^*)^{k-j}$$

$$\text{for } k = 0, 1, 2, \ldots, n$$

When either of these two conditions is not met, we know a net investment does not exist throughout the life of the project. This means there are one or more periods when the project has a net outflow of money *which will later be required to be returned to the project*. This money can be put into an external investment until such time as it is needed in the project. The interest rate of the external investment (e^*) will be the interest rate at which the money can in fact be invested outside the project. The external interest rate (e^*) is unrelated to the internal rate of return (i^*) on the project. If there is no external investment, then no value of e^* is required in the computations.

Application of the Four Tests of a Cash Flow

The four tests can be used to produce a chart for computing the rate of return for any investment project. Figure 18-1 is such a chart. The required computations may be done by hand or on a microcomputer. A microcomputer program called RORiPC has been written to perform all the computations. It is described in Chapter 19. The computation of the rate of return for any investment project may be illustrated by the following example problems.

---

†Merrett, A. J. and Allen Sykes. *The Finance and Analysis of Capital Projects*, 2nd ed. London: Longman, 1973, p. 135.

‡Soper, C. S., "The Marginal Efficiency of Capital: A Further Note," *Economic Journal* LXIX, pp. 174–7.

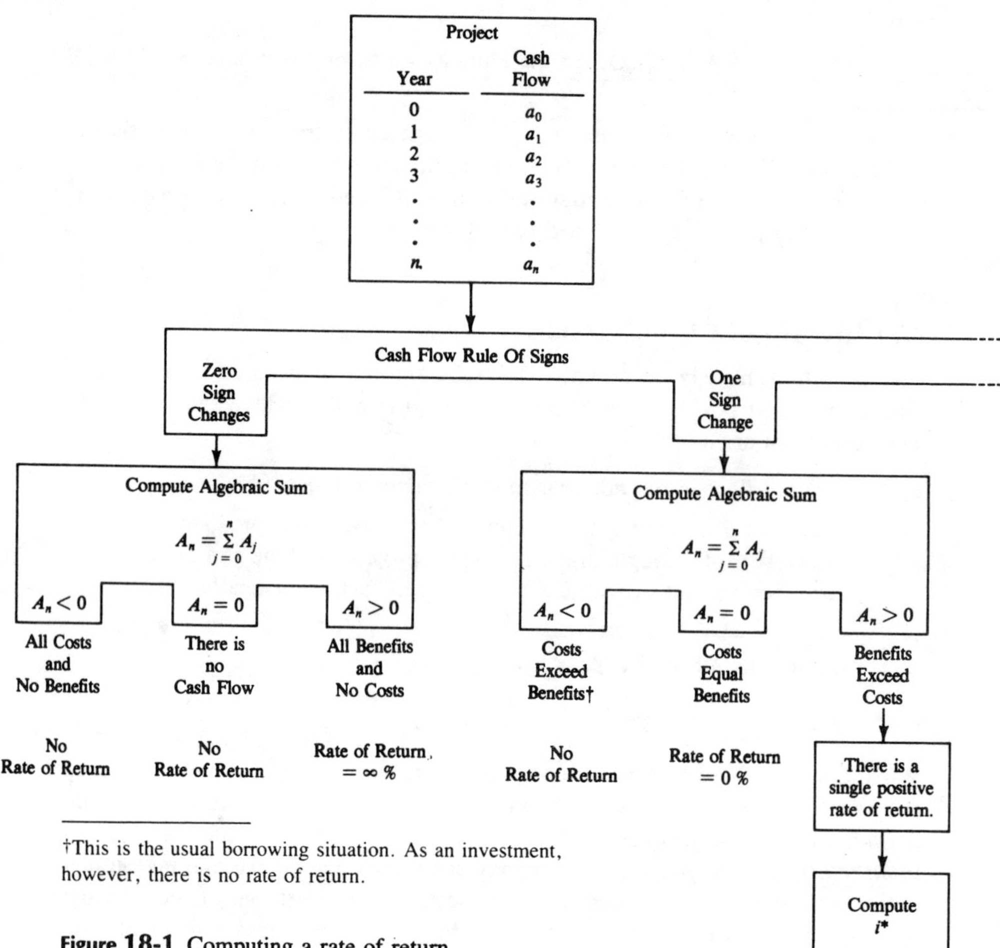

Figure 18-1 Computing a rate of return.

EXAMPLE 18-1

Given the following cash flow, compute the internal rate of return i^*. If needed, use an external interest rate e^* of 6%.

Year	*Cash flow*
0	−\$2000
1	+1200
2	+400
3	+400
4	−200
5	+400

First, we will write out the cash flow and accumulated cash flow and then determine the sign changes in both, along with the algebraic sum of the cash flow.

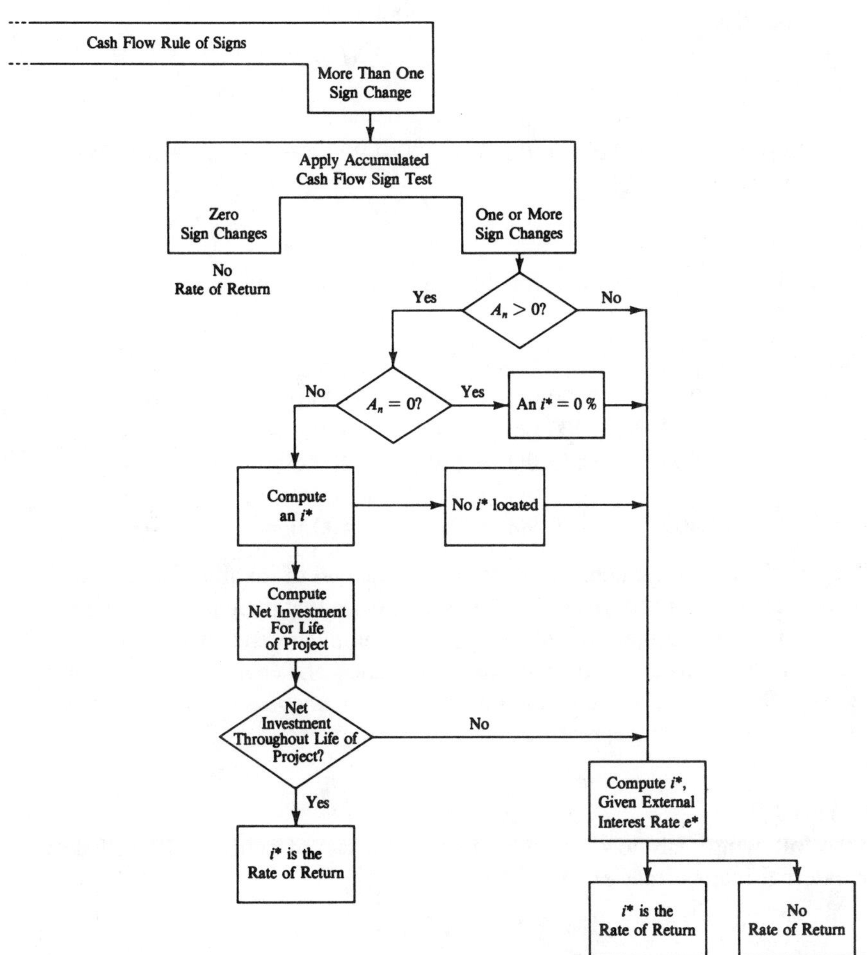

Solution:

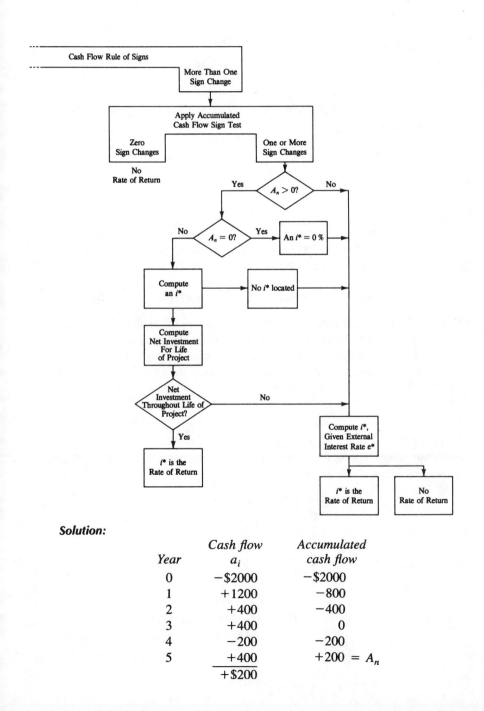

Year	*Cash flow* a_i	*Accumulated cash flow*
0	−$2000	−$2000
1	+1200	−800
2	+400	−400
3	+400	0
4	−200	−200
5	+400	+200 = A_n
	+$200	

The cash flow rule of signs indicates there may be as many as three positive rates of return. Since there is one sign change in the accumulated cash flow and $A_n > 0$, these are sufficient conditions for a single positive rate of return. It may be computed in the usual manner.

$$\text{PW of cost} = \text{PW of benefits}$$

$$2000 + 200(P/F,i,4) = 1200(P/F,i,1) + 400[(P/F,i,2) + (P/F,i,3) + (P/F,i,5)]$$

Try $i = 5\%$:

$$2000 + 200(0.8227) = 1200(0.9524) + 400(0.9070 + 0.8638 + 0.7835)$$

$$2165 = 2165$$

Thus, $i^* = 5\%$.

Also: Compute the net investment for the project at each time period:

Year	*Cash flow*		*Net investment*
0	−$2000		−$2000.0
1	+1200	−2000.0(1 + 0.05) + 1200.0 =	−900.0
2	+400	−900.0(1 + 0.05) + 400.0 =	−545.0
3	+400	−545.0(1 + 0.05) + 400.0 =	−172.3
4	−200	−172.3(1 + 0.05) − 200.0 =	−380.9
5	+400	−380.9(1 + 0.05) + 400.0 =	0

The fact that the net investment becomes zero at the end of the fifth year proves that 5% is a rate of return for the cash flow. We also see that there is a continuing net investment in the project for all time periods until the end of the project. This means there is no external investment of money and, hence, no need for an external interest rate. We conclude that $i^* = 5\%$ is a correct measure of the profitability of the project represented by the cash flow. ◀

EXAMPLE 18-2

Given the following cash flow, compute the internal rate of return i^*. If needed, use an external interest rate e^* of 6%.

Year	*Cash flow*
0	−$100
1	+50
2	+50
3	+50
4	+50
5	−73.25

Solution: As in the previous example, we first determine the sign changes in the cash flow and the accumulated cash flow and compute the sum of the cash flow.

Year	*Cash flow* a_i	*Accumulated cash flow* A_i
0	−\$100	−\$100
1	+50	−50
2	+50	0
3	+50	+50
4	+50	+100
5	−73.25	+26.75 = A_n
	+\$26.75	
Sign changes	= 2	= 1

With one sign change in the accumulated cash flow and $A_n > 0$, we know there is a single positive rate of return. The problem was devised with a 20% interest rate as is shown:

$$\text{PW of cost} = \text{PW of benefits}$$
$$100 + 73.25(P/F,20\%,5) = 50(P/A,20\%,4)$$
$$100 + 73.25(0.4019) = 50(2.589)$$
$$129.45 = 129.45$$

Thus, $i^* = 20\%$. Check net investment:

Year	*Cash flow*		*Net investment*
0	−\$100		−\$100
1	+50	−100.00(1 + 0.20) + 50.00 =	−70
2	+50	−70.00(1 + 0.20) + 50.00 =	−34
3	+50	−34.00(1 + 0.20) + 50.00 =	+9.20
4	+50	+9.20(1 + 0.20) + 50.00 =	+61.04
5	−73.25	+61.04(1 + 0.20) − 73.25 =	0

Note that the fact that net investment becomes zero at the end of the last year confirms that 20% is a rate of return for the cash flow. It is also clear that there is not net investment throughout the life of the project. At the end of Year 3, there is in fact +9.20 that must be invested outside of the project represented by the cash flow. And in Year 4 the amount for external investment has increased to +61.04. The money invested externally must be returned to the project in Year 5 to provide the needed final disbursement of 73.25. In the table above, the money has been assumed to be invested at a 20% interest rate. More importantly, this same assumption of a 20% external investment is implicit in the PW of cost = PW of benefits calculation that also indicated that $i^* = 20\%$.

The problem statement specifies that external investment may be expected to earn a 6% interest rate. Since this is less than the 20% in the calculations above, this means that a benefit—the external interest earned—of the cash flow is reduced. For $e^* = 6\%$, we can see that i^* is less than 20%. The calculation must be done by trial and error. Try $i = 15\%$:

Year	*Cash flow*		*Net investment*
0	−$100		−$100.00
1	+50	−100.00(1 + 0.15) + 50.00 =	−65.00
2	+50	−65.00(1 + 0.15) + 50.00 =	−24.75
3	+50	−24.75(1 + 0.15) + 50.00 =	+21.54
4	+50	+21.54(1 + 0.06) + 50.00 =	+72.83
5	−73.25	+72.83(1 + 0.06) − 73.25 =	+3.95

The net investment does not equal zero at the end of the project. This indicates our trial $i = 15\%$ is in error. The remaining positive net investment signals that i should be increased. Further trials would guide us to $i = 16.5\%$. For $e^* = 6\%$ and $i = 16.5\%$ we have:

Year	*Cash flow*		*Net investment*
0	−$100		−$100.00
1	+50	−100.00(1 + 0.165) + 50.00 =	−66.50
2	+50	−66.50(1 + 0.165) + 50.00 =	−27.47
3	+50	−27.47(1 + 0.165) + 50.00 =	+17.99
4	+50	+17.99(1 + 0.060) + 50.00 =	+69.07
5	−73.25	+69.07(1 + 0.060) − 73.25 =	−0.03

We see that with $e^* = 6\%$, i^* is very close to 16.5%. ◀

The results of Example 18-2 are important. The two sign changes in the cash flow warned us that there might be as many as two positive rates of return. From the accumulated cash flow sign test we proved that in reality there is one positive rate of return. We then showed that 20% is the single positive rate of return. One would be tempted to stop at this point, satisfied that $i^* = 20\%$ is a proper measure of the profitability of the project represented by the cash flow.

Examination of the net investment throughout the life of the project reveals that external investments occur at the end of Years 3 and 4. Using an external interest rate e^*, we found the true profitability of the project is $i^* = 16.5\%$. The surprising discovery is that where there are multiple sign changes in an investment project cash flow, there may be a unique positive rate of return, but the existence of external investment may mean that the unique positive rate of return is *not* a suitable measure of project profitability.

EXAMPLE 18-3

Given the following cash flow, compute the internal rate of return i^*. If needed, use an external interest rate e^* of 6%.

Year	*Cash flow*
0	−$100
1	+330
2	−362
3	+132

Solution: We will begin the solution by writing the accumulated cash flow and checking the sign changes.

Year	*Cash flow*	*Accumulated cash flow*
0	−$100	−$100
1	+330	+230
2	−362	−132
3	+132	0 = A_n
	0	
Sign changes	= 3	= 2

Careful examination of these data reveal the following:

1. There may be as many as three positive rates of return.
2. One rate of return is 0%.
3. The accumulated cash flow does not meet the sufficient conditions for a single positive rate of return.
4. At $i^* = 0\%$ there is external investment at the end of Year 1. (The accumulated cash flow sequence becomes the net investment sequence at 0%.)

At this point we have two choices. We can search to find out how many positive rates of return there are and their numerical values. As an alternative we can seek i^* for the given $e^* = 6\%$.

This particular cash flow was devised by picking the desired roots of 0%, 10%, and 20% for a third order polynomial as follows: let $x = 1 + i$. For $i =$ 0.00, 0.10, and 0.20,

$$(x - 1.0)(x - 1.1)(x - 1.2) = 0$$

Multiplying we obtain:

$$x^3 - 3.3x^2 + 3.62x - 1.32 = 0$$

Multiplying by −100 gives:

$$-100x^3 + 330x^2 - 362x + 132 = 0$$

$$-100(1 + i)^3 + 300(1 + i)^2 - 362(1 + i) + 132 = 0$$

This represents the future worth of the following cash flow.

Year	*Cash flow*
0	−$100
1	+330
2	−362
3	+132

Thus the three positive rates of return for the cash flow are 0%, 10%, and 20%. Knowing that there are three positive rates of return, and their values, has not helped us in our search for a true measure of project profitability.

A trial and error search is made for i^*, given that $e^* = 6\%$. For a trial $i =$ 5.98%:

Year	*Cash flow*		*Net investment*
0	−$100		−$100.00
1	+330	$-100.00(1 + 0.0598) + 330 =$	+224.02
2	−362	$-224.02(1 + 0.0600) - 362 =$	−124.54
3	+132	$-124.54(1 + 0.0598) + 132 =$	+0.01

For $e^* = 6\%$, i^* is about 5.98%. We see that the exterior investment is more desirable than the internal project. ◀

EXAMPLE 18-4

Given the following cash flow, compute the internal rate of return i^*. If needed, use an external interest rate e^* of 6%.

Year	*Cash flow*	*Accumulated cash flow*
0	−$200	−$200
1	+150	−50
2	+50	0
3	+50	+50
4	+50	+100
5	+50	+150
6	+50	+200
7	+50	+250
8	−285	$-35 = A_n$
	−$35	
Sign changes	= 2	= 2

Solution: Although the algebraic sum of the cash flow is negative, this is one of the cash flows where there are two positive rates of return (at about 8.7% and 17.4%). The accumulated cash flow shows that an important benefit is the substantial external investment of money for relatively long periods of time. If we compute the external investment at $e^* = 6\%$ (rather than 8.7% or 17.4%), the benefits will be substantially reduced and the computed internal rate of return will decline. The internal rate of return is computed by trial and error. Try $i = 4\%$:

Year	Cash flow		Net investment
0	−\$200		−\$200.0
1	+150	−200.0(1.04) + 150 =	−58.0
2	+50	−58.0(1.04) + 50 =	−10.3
3	+50	−10.3(1.04) + 50 =	+39.3
4	+50	+39.3(1.06) + 50 =	+91.7
5	+50	+91.7(1.06) + 50 =	+147.2
6	+50	+147.2(1.06) + 50 =	+206.0
7	+50	+206.0(1.06) + 50 =	+268.4
8	−285	+268.4(1.06) − 285 =	−0.5

With $e^* = 6\%$, i^* is about 4%. ◀

EXAMPLE 18-5

Consider the following project cash flow. The external interest rate $e^* = 6\%$.

Year	Cash flow
0	−\$500
1	+300
2	+300
3	+300
4	−600
5	+200
6	+135.66

a. Solve for the project internal rate of return (i^*) using the methods described in Chapter 7A.

b. Solve for the project internal rate of return (i^*) using the methods described in this chapter.

c. Explain the difference in the results obtained in Parts **a** and **b**.

Solution to Example 18-5a: There are three sign changes in the cash flow. We wish to reduce this to one sign change. We will eliminate the −600 at the end of Year 4 by accumulating money at the external interest rate ($e^* = 6\%$) to equal the required \$600. The computations are as follows:

Year	Cash flow	External investment			Transformed cash flow
0	−\$500				−\$500
1	+300				+300
2	+300		$x(1.06)^2$	$x = 250.98$	+49.02
3	+300	+300(1.06) →	↓		0
4	−600	+318	+282		0
5	+200				+200
6	+135.66				+135.66

As described in the computations above, all the money at Year 3 is assumed to be invested externally at 6%. The +\$300 increases to +\$318 at the end of one year and can be brought back to the cash flow at the end of Year 4. This makes the internal investment cash flow have a zero for Year 3 and −600 + 318 = −\$282 for Year 4.

A further alteration of the cash flow is needed to reduce the number of sign changes to 1. Part of the Year 2 cash flow must be set aside in an external investment so that its accumulated sum at the end of two years will be +282. This can be brought back to the cash flow at the end of Year 4 with the result that the internal investment cash flow will have a zero for Year 4. The amount to invest externally in Year 2 is $x = +282/(1.06)^2 = +250.98$. The result is the transformed cash flow. The transformed cash flow has one sign change and, therefore, cannot have more than one positive rate of return. Using the technique described in Chapter 7A, we will see that i^* is very close to 11.1%.

Year	*Transformed cash flow*	*Present worth at 11%*	*Present worth at 12%*	*Present worth at 11.1%*
0	−\$500	−\$500	−\$500	−\$500
1	+300	+270.27	+267.86	+270.03
2	+49.02	+39.79	+39.08	+39.71
3	0	0	0	0
4	0	0	0	0
5	+200	+118.69	+113.49	+118.16
6	+135.66	+72.53	+68.73	+72.14
	+\$184.68	+\$1.28	−\$10.86	+\$0.04

Solution to Example 18-5b: The procedure will be to solve the cash flow for a rate of return, check net investment, and proceed using e^*, if necessary.

Some preliminary computations, not shown here, lead us to a 15% rate of return. This may be verified, and net investment computed for $i^* = 15\%$.

Year	*Cash flow*	*Present worth at 15%*		*Net investment at 15%*
0	−\$500	−\$500		−\$500.00
1	+300	260.87	−500.00(1.15) + 300.00 =	−275.00
2	+300	+226.84	−275.00(1.15) + 300.00 =	−16.25
3	+300	+197.25	−16.25(1.15) + 300.00 =	+281.31
4	−600	−343.05	+281.31(1.15) − 600.00 =	−276.49
5	+200	+99.44	−276.49(1.15) + 200.00 =	−117.96
6	+135.66	+58.65	−117.96(1.15) + 135.66 =	0
	+\$135.66	\$0		

The computation of net investment at 15% indicates there is not net investment throughout the life of the project. At the end of Year 3 there is an external investment of $281.31. The computation is based on $e^* = i^* = 15\%$. For $e^* = 6\%$, we must compute i^* by trial and error.

For $e^* = 6\%$, and Trial $i = 13\%$:

Year	*Cash flow*		*Net investment*
0	−$500		−$500.00
1	+300	−500.00(1.13) + 300.00 =	−265.00
2	+300	−265.00(1.13) + 300.00 =	+0.55
3	+300	+0.55(1.06) + 300.00 =	+300.58
4	−600	+300.58(1.06) − 600.00 =	−281.38
5	+200	−281.38(1.13) + 200.00 =	−117.96
6	+135.66	−117.96(1.13) + 135.66 =	+2.36

The end of Year 6 net investment is not zero, indicating that i^* is not equal to our 13% Trial i. Select 13.2% as another Trial i and repeat the computation:

Year	*Cash flow*		*Net investment*
0	−$500		−$500.00
1	+300	−500.00(1.132) + 300.00 =	−266.00
2	+300	−266.00(1.132) + 300.00 =	−1.11
3	+300	−1.11(1.132) + 300.00 =	+298.74
4	−600	+298.74(1.06) − 600.00 =	−283.33
5	+200	−283.33(1.132) + 200.00 =	−120.73
6	+135.66	−120.73(1.132) + 135.66 =	−1.01

Results of the two trials:

Trial i	*End of Year 6 net investment*
13.0%	+2.36
13.2%	−1.01

We conclude that for $e^* = 6\%$, i^* is approximately 13.13%.

At 13.13% the net investment is:

Year	*Cash flow*		*Net investment*
0	+$500		−$500.00
1	+300	−500.00(1.1313) + 300.00 =	−265.65
2	+300	−265.65(1.1313) + 300.00 =	−0.53
3	+300	−0.53(1.1313) + 300.00 =	+299.40
4	−600	+299.40(1.06) − 600.00 =	−282.64
5	+200	−282.64(1.1313) + 200.00 =	−119.75
6	+135.66	−119.75(1.1313) + 135.66 =	+0.19

Solution to Example 18-5c: The two computation methods transform the project cash flow into different amounts of internal and external investment. This is illustrated by the following tabulation:

		Part **a**		*Part* **b**	
Year	*Cash flow*	*Internal investment*	*External investment*	*Internal investment*	*External investment*
0	−\$500	−\$500.00		−\$500.00	
1	+300	+300		+300.00	
2	+300	+49.02	+\$250.98	+300.00	
3	+300		+300.00	+0.60	+\$299.40
4	−600		−600.00	−282.64	−317.36
5	+200	+200.00		+200.00	
6	+135.66	+135.66		+135.66	

The tabulation shows that the methods outlined in this chapter do not necessarily assume as much external investment as is required by the methods of Chapter 7A. We now see clearly that the criterion on how much to alter a cash flow is properly based on maintaining a net investment—not on reducing the number of sign changes in the cash flow.

Based on net investment a smaller portion of the benefits are assumed in an external investment at $e^* = 6\%$. For values of i^* greater than e^*, we will find Part **b**'s i^* to be larger than Part **a**'s i^*. ◀

From what has been said, plus the five example problems, there are some important conclusions to be noted. When there are multiple positive rates of return for a cash flow, in general, none of them are a suitable measure of project profitability. Even in the situation where there is only one positive rate of return, that value still may not be a good indicator of project profitability. The critical question is whether the rate of return is exclusively the return on funds invested in the project or is the return on the combination of funds in the project and funds temporarily invested outside the project. If the project cash flow reflects both internal and external investments, the one or more positive rates of return assume that the internal rate of return i^* equals the external rate of return e^*. This is seldom a valid assumption. Where it is not, an external rate of return e^* should be selected and i^* computed, given e^*.

In a situation where an initial investment is made, followed by subsequent benefits (only one sign change in the cash flow), there is no temporary external investment and no difficulty in computing a suitable i^*. There will be other situations where in the later years of a cash flow an additional net investment is required. This may or may not result in multiple positive rates of return. It will, however, mean that an external interest rate e^* is needed to compute a suitable i^*.

SUMMARY

We can now see that the computation of rate of return for an investment project is far more complex than was stated in Chapter 7.

The cash flow rule of signs is the simple test of the number of changes of sign in the project cash flow. Zero sign changes is the unusual case of all benefits or all disbursements. This is immediately seen as either good or a disaster. One sign change is the conventional situation. This is the case that has been discussed throughout most of this book. It generally leads to a single positive rate of return that is a valid measure of project profitability. Multiple sign changes are a warning sign that a valid measure of project profitability may be difficult to locate.

The accumulated cash flow sign test is a test of the accumulated cash flows ($A_0, A_1, A_2, \ldots, A_n$). The algebraic sum of the cash flow is simply the sum of the project cash flow and equals the accumulated cash flow, A_n. If there is one sign change in the sequence and $A_n > 0$, these are sufficient (but not necessary) conditions for a single positive rate of return. When there is more than one sign change in the cash flow, the existence of a single positive rate of return does not necessarily give us a suitable measure of project profitability. The critical test is net investment.

Net investment is the computation to determine whether or not there is a continuing net investment throughout the life of the project for a value of i^*. Thus the first step in computing net investment is to locate a value of i^*. Then net investment is computed year-by-year with the net investment earning i^* each year. A cash flow will either have net investment throughout the life of the project, or it will not have it. A cash flow that does not have net investment is one where there is a net outflow of money from the project which will later be required to be returned to the project. At this point we need to invest the outflow of money someplace else (at external interest rate e^*) until it is required back in the project.

If there is net investment throughout the life of the project, we know this is sufficient (but not necessary) for a single positive rate of return, and this i^* is a suitable measure of project profitability. When there is not net investment throughout the life of the project, another rate of return i^* must be computed, with an external interest rate e^* assumed for project money temporarily invested externally. This revised i^* is an appropriate measure of project profitability. As we have seen, the calculation of i^*, using an external interest rate e^*, can be lengthy. The microcomputer program RORiPC, described in the next chapter, performs the necessary computations.

Problems

18-1 Consider the following situation:

Year	*Cash flow*
0	−$500
1	+2000
2	−1200
3	−300

a. For the cash flow, what information can be learned from—
1. Cash flow rule of signs?
2. Accumulated cash flow sign test?
3. Algebraic sum of the cash flow?
4. Net investment conditions?

b. Compute the internal rate of return for the cash flow. If there is external investment, assume it is made at the same interest rate as the internal rate of return.

c. Compute the internal rate of return for the cash flow. This time assume a 6% external interest rate.

18-2 Repeat Problem 18-1 for the following cash flow:

Year	*Cash flow*
0	−$500
1	+200
2	−500
3	+1200

(*Answers:* ***b.*** 21.1%; **c.** 21.1%)

18-3 Repeat Problem 18-1 for the following cash flow:

Year	*Cash flow*
0	−$500
1	+200
2	−500
3	+200

18-4 Repeat Problem 18-1 for the following cash flow:

Year	*Cash flow*
0	−$100
1	+360
2	−570
3	+360

18-5 Given the following cash flow:

Year	*Cash flow*
0	−$200
1	+100
2	+100
3	+100
4	−300
5	+100
6	+200
7	+200
8	−124.5

Compute the internal rate of return assuming—

a. External interest rate equals the internal rate of return.

b. External interest rate equals 6%.

(*Answer:* **a.** 20%; **b.** 18.9%)

18-6 In Examples 7A-1 and 7A-2 (in Chapter 7A), an analysis was made of the proposed sale of an airplane by Going Aircraft Co. to Interair. In Ex. 7A-2, the rate of return was computed to be 8.4%, given a 6% external interest rate. Using the methods described in this chapter, compute i^*, given $e^* = 6\%$. Explain why your answer is or is not different from the 8.4% computed in Ex. 7A-2.

18-7 Refer to Ex. 5-9 in Chapter 5.

a. Construct a plot of NPW *vs.* i for the cash flow.

b. Compute the rate of return for the investment project, assuming if necessary a 10% interest rate on external investments.

c. Would the method described in Chapter 7A produce a different answer from that obtained in **b**? Explain.

18-8 Consider the following situation:

Year	*Cash flow*	*Present worth at 70.7%*
0	−$200	−$200.00
1	+400	+234.33
2	−100	−34.32
		+$0.01

a. What is i^* if e^* equals 70.7%?

b. What is i^* if e^* equals 0%?

18-9 An investor is considering two mutually exclusive projects in which to invest his money. He can obtain a 6% before-tax rate of return on external investments, but he requires a minimum attractive rate of return of 7% for these projects. Use a ten-year analysis period.

	Project A: *Build drive-up photo shop*	*Project B:* *Buy land in Hawaii*
Initial capital investment	$58,500	$ 48,500
Net uniform annual income	6,648	0
Salvage value ten years hence	30,000	138,000
Computed rate of return	8%	11%

Compute the incremental rate of return from investing in Project *A* rather than Project *B*.

CHAPTER **19**

Microcomputer Programs

Simple economic problems can be readily solved using compound interest tables and a hand calculator. But when working on more difficult problems, manual computation can become rather tiring. When doing things like simulation, hand figuring becomes simply too cumbersome to be practical.

Whether dealing with simple or complex problems, it is desirable to be able to check one's computations. If you have access to an IBM-PC or -XT (or compatible) microcomputer, you are in luck. A computer program, EEAi, written by Dr. Jan Wolski, New Mexico Institute of Mining and Technology, solves a variety of engineering economic analysis problems. The main program consists of six component programs.

Engineering Economic Analysis Microcomputer Program EEAi

EAiPC — ***Elementary Economic Analysis.***
The easiest program to run. Input a cash flow. The program allows the computation of P or F and A and i.

EAiGR — ***Graph of* NPW *vs. Interest Rate.***
For use in conjunction with EAiPC.

MORTGiPC — ***Mortgage Repayment.***
The program computes the uniform payment schedule for a mortgage or a loan. It can be used in other situations where A is known or is to be calculated.

RORiPC ***Rate of Return.***

Solves any cash flow for a rate of return up to 100% in a sophisticated manner as described in Chapter 18. Handles all complexities caused by multiple sign changes, multiple rates of return, and an external interest rate.

EQANiPC ***After-Tax Equipment Selection and Replacement Analysis.***

An elaborate program with a full range of tax and inflation options.

CEAPiPC ***Capital Expenditure Analysis Program: Simulation.***

Handles complex problems and after-tax simulation where input values may be probability distributions.

Directory Of Diskette, Version 3.1

README	883	EATST3.DAT	495
EEA31.EXE	189244	ROR1.DAT	993
EEA31.HLP	14919	RORTST.DAT	693
VIEW.EXE	9840	EQ1.DAT	2982
EA_DEMO.DAT	415	EQAN1.DAT	3034
EA1.DAT	290	EQTST.DAT	3013
EA72.DAT	327	CEAP1.DAT	775
EA81.DAT	951	CEAP2.DAT	1001
EA816.DAT	615	CEAP3.DAT	971
EA87.DAT	596	PLTDAT.DAT	653

Running EEAi

The EEAi program (where "i" is the current version number) can be run on any IBM compatible PC computer with the DOS operating system. A graphics card is needed to get the interest curves. Since the diskette is write protected, you must copy its content to another disk (hard or floppy) before you can use the program. Files EEAi.EXE, EEAi.HLP, and VIEW.EXE must be in the same directory as the data files you want to use. To get the information related to the current version of the program, at the DOS prompt type **README** and press the *Enter* key. After you have read this file, you are ready to start the program.

Type **EEA31** and press *Enter*. The first page of the program will appear.

```
PROGRAM FOR ENGINEERING ECONOMIC ANALYSIS

(c) Copyright 1990 Engineering Press, Inc.
    All rights reserved.
    Version 3.1
    See README for details

            Written by Jan Wolski
```

```
┌── Select ───┐
│ 1  EA3PC    │
│ 2  EA3GR    │
│ 3  MORTG3PC │
│ 4  ROR3PC   │
│ 5  EQAN3PC  │
│ 6  CEAP3PC  │
│ 7  Exit     │
└─────────────┘
```

To select a desired program, type the appropriate number or move the cursor using the up or down arrow keys (hereafter ↑ and ↓ ; these are usually located on the numeric keypad at the right side of the keyboard) and press *Enter*. The "Exit" option allows you to get back to the DOS prompt. If you select a program, you will see the next display:

```
PROGRAM programname
─────────────────────

┌── Select ────┐
│ Run          │
│ Display file │
│ Print file   │
│ Exit         │
└──────────────┘
```

Select "Run", "Display file", "Print file" or "Exit" using the ↑ or ↓ keys and press *Enter*.

"Run"	This command allows you to input data from the keyboard, or to edit and execute the existing data files.
"Display file"	This command allows you to see the list of existing data files and to examine their content.
"Print file"	This command allows you to print a selected data or output file. You can also print a selected page displayed on the screen by simultaneously pressing the *Shift* and *PrtSc* (Print Screen) keys. This is particularly useful for printing graphs, but to do so you must have a graphics printer.
"Exit"	This command allows you to get back to the main menu.

Normally you use the "Run" option (press *Enter* when "Run" option is highlighted). You will see:

```
┌─────────────── Y-Yes or N-No ───────────────┐
│ Do you want to use your input data file ?   │
└─────────────────────────────────────────────┘
```

(An exception is MORTGiPC: only data from the keyboard are accepted.)

If you type **Y**, a list of the existing data files will be displayed. Note that the first two letters are specific to each program. For example, EA is specific to all EAiPC program data files. Each data file has an extension .DAT. Select a data file using the ↑ or ↓ keys and press *Enter*. The title of your data file will be displayed with a reversed field (dark symbols on a light background) over it and the cursor over the first symbol of the name in the lower window entitled "Edit".

Type **N** if you want to input data from the keyboard. You will be guided through the data input process specific to each program. In both situations, the data can be modified (edited) and saved.

Data Editing System

Every data window can be modified (edited). An example is as follows:

```
PROGRAM EA3PC  -  ELEMENTARY ECONOMIC ANALYSIS

================ Problem description ================
      EXAMPLE #1 - TEST CASH FLOW
======================== Edit =======================
     EXAMPLE #1 - TEST CASH FLOW
```

At this point you can modify the title using the following keys:

- Left or right arrow keys to move cursor along the line;
- *Ins* (Insert) or *Del* (Delete) keys to insert or to delete a character.

To save the changes, press *Enter*. You will see the modified title in the upper field, and the reversed field is now located over the second line of the title (if one exists). Again, this line can be modified. If there are no changes to make, press *Enter*. The new page will display your data file which can be modified in the same manner as described above, or press *Enter* to go on.

Use the ↑ or ↓ keys to move from one data to another, even if the actual move is done along a horizontal line. *Remember that the left and right arrows are reserved for the "Edit" field.* Please note that this file editing system is a bit different from those used in most other programs. Here, the editing is performed in the special field and is saved only after pressing *Enter*. This way you avoid any confusion about which are old or new data.

After all pages of data have been displayed, the computer does the calculations, and the first page of output is shown on the screen. You can use:

- ↑ or ↓ keys to move the text line-by-line;
- *PgDn* or *PgUp* keys to move down or up by page;
- simultaneously press *Ctrl–PgDn* or *Ctrl–PgUp* to move to the end or to the beginning of the file.

Use the *Esc* (Escape) key to get back to the menu. Then new values can be input using the same data file, a new data file can be initiated, and so on, depending on the program you are using. The *Esc* key will always return you to the previous menu.

An output file is created every time you run one of the programs. For example, EA.OUT file is produced if you run the EAiPC program. This means that every time you run the program, you lose the previous output file. To save the old content, you may copy the standard output file to a specific file (for example, at the DOS prompt, type: **COPY EA.OUT EA1.OUT**) or print it before running the program again.

You accept a given data window by pressing *Enter*. Therefore, never press *Enter* twice in a row during data editing! To move from one data field to another, use the ↑ or ↓ keys.

Program For Elementary Economic Analysis

A program called EAiPC (where "i" is the current version number) is available to analyze a cash flow and to compute the following unknown economic factors:

- If the interest rate INT is given, then the present or future value PF is calculated for a desired moment of time NO and the equivalent uniform annual benefit A in a selected period of time NF through NL is calculated.
- If the interest rate is not given (0 or blank in the input), then only the rate of return is calculated.
- Plotting the relation between the present or future worth and the interest rate is optional; if the plot is desired, then an input file PLTDAT.DAT is created for the plotting program EAiGR.

Two data options are available: existing data files can be used or data can be input directly from the terminal and saved for future use if desired. The previous section describes how to create or edit existing data files. Of course, a new data file should not be given an existing file name. In such a case, the old file would be overwritten and lost. Just skip over a particular parameter using the up or down arrow keys if it is not given as part of the input.

Each line of the data block represents one cash flow period and contains three values:

1. PERIOD (IN) = Period number (between −120 and 120);
2. CONSTANT VALUE (VC) = Value which remains constant during stated period and following periods until a new value is input;
3. SINGLE VALUE (VS) = Single value.

To change a non-zero VC value to 0, type **0.01**, which is the lowest greater-than-zero value which can be stored in the data file. The period numbers can be negative, zero or positive. They can be repeated several times, and they can be placed in any order. The lines are sorted by the program, and the next time you look at your data file, you will see the lines of data arranged in ascending period numbers.

It is not necessary to write lines representing periods where nothing new happens. For example, a $100.00 investment at the beginning of the first period, which brings a profit of $20.00 for the first five years, $15.00 for the next five years, and has a salvage value of $10, would require the following input dat block:

```
PERIOD    CONT.VALUE    SINGLE VALUE
  Test Data Block:
  0           0            -100
  1          20               0
  6          15               0
 10           0              10
```

The same cash flow graphically is:

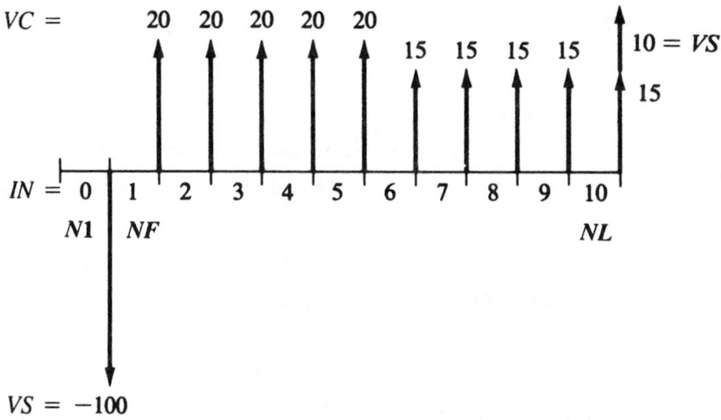

As can be seen, Periods 2–5 and 7–9 are omitted in the data file since nothing new happens through their duration. VC value in Period 10 is $15.00 but 0 value can be written instead, since "zero" means lack of data. Next a new set of data may be written or the data file may end. EA1.DAT is the input file for this problem and is used in Example 19-1.

After the data file is completed, you are asked to input values for the interest rate, period number for the present/future worth calculation, and first and last period for the equivalent uniform amount. Input a very small value (for example, 0.001) for a zero interest rate.

The main advantage of the EAiPC program is its simplicity. Any cash flow can be solved for the present or future value, equivalent uniform annual benefit or internal rate of return with the minimum information required and maximum input simplicity.

The limitation of this program is that it finds only the first lowest value of the internal rate of return. Of course, there may be more than one rate of return if there is more than one change in sign of the cash flow during the time span under consideration. If this is the case, it is better to use the RORiPC program rather than EAiPC.

The program EAiGR produces a plot of the relationship between the present or future value and interest rate for all alternatives under consideration. A PC computer with graphics capabilities is needed to run this program. If plotting is desired, an output file PLTDAT.DAT is created by program EAiPC and this file is used as the input file for the plotting program. The graph is produced on the display and can be transferred to the printer. Note that GRAPHICS must be set at the DOS prompt prior to execution of EAiGR to get the graph on a printer with graphics capabilities.

Normally, the program EAiGR is called from EAiPC if a plot is requested. Also, it can be run separately using the existing data file PLTDAT.DAT. Even without using the graphics program, the relationship between present worth and interest rate can be plotted manually using the rate-of-return convergence control data displayed during each execution of the program EAiPC.

Variables

A = Equivalent uniform amount between NF and NL

INT = Interest, % (input 0.001 for 0 value)

IN = Period number (between −120 and 120)

NO = Period for which past, present, or future value is calculated (can be negative)

N1 = Period preceding first amount A (N1 = NF − 1)

NF = Period of first amount A

NL = Period of last amount A

PF = Present or future worth

P1 = Present worth at N1 period

ROR = Rate of return, %

VC = Values which continue until a new value is input (input 0.01 for new 0 values)

VS = Single value, in any of IN periods

The following data files for EAiPC are on the diskette:

EA_DEMO.DAT	Demonstration problem
EA1.DAT	Example 19-1
EA81.DAT	Problem 8-1
EA87.DAT	Problem 8-7
EA816.DAT	Problem 8-16
EA72.DAT	Problem 7-2
EATST3.DAT	Sample Problem

EXAMPLE 19-1

Solve the following cash flow using EAiPC. Data of this example is stored in the data file EA1.DAT.

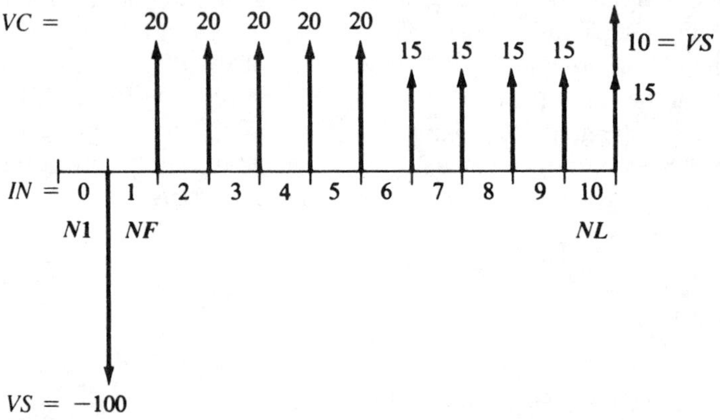

Solution: Start the EEA31 program. Select the EA3PC option. You will see the following display:

```
PROGRAM EA3PC - ELEMENTARY ECONOMIC ANALYSIS

Select
Run
Display file
Print file
Exit
```

Select "Run". Next you will see:

```
Y-Yes or N-No
Do you want to use your input data file ?
```

Type **Y**. A list of existing data files will appear.

```
┌─── Select ───┐
│ EA_DEMO.DAT  │
│ EA1.DAT      │
│ EA72.DAT     │
│ EA81.DAT     │
│ EA816.DAT    │
│ EA87.DAT     │
│ EATST3.DAT   │
└──────────────┘
```

The reversed field will cover the first data file. Move to EA1.DAT using the down arrow key. Press *Enter* to get this file. You will see:

```
PROGRAM EA3PC  -  ELEMENTARY ECONOMIC ANALYSIS

──────────────── Problem description ────────────────
        EXAMPLE #1 - TEST CASH FLOW
──────────────────────── Edit ───────────────────────
        EXAMPLE #1 - TEST CASH FLOW
```

Press *Enter* to go the next screen:

```
───────────────────────────────────── Data ──────────────────────────────────────
TITLE: ALTERNATIVE A
  PERIOD  CONT.VALUE  SINGLE VALUE        PERIOD  CONT.VALUE  SINGLE VALUE
----------------------------------    ----------------------------------
 1:   0      0.00       -100.00        16:
 2:   1     20.00          0.00        17:
 3:   6     15.00          0.00        18:
 4:  10      0.00         10.00        19:
 5:                                    20:
 6:                                    21:
 7:                                    22:
 8:                                    23:
 9:                                    24:
10:                                    25:
11:                                    26:
12:                                    27:
13:                                    28:
14:                                    29:
15:                                    30:
───────────────────────────────────── Edit ──────────────────────────────────────
 ALTERNATIVE A
```

As can be seen, a maximum of 30 periods can be introduced. Press *Enter* to continue.

```
═══════════ Y-Yes or N-No ═══════════
 Do you want to plot NPW/INTEREST curve ?
```

Press **N** and you are asked to input the following values:

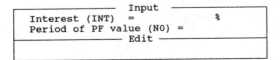

```
──────────── Input ────────────
 Interest (INT)  =           %
 Period of PF value (NO) =
──────────── Edit ─────────────
```

Enter **10** and **0**. Next window:

```
        Equivalent Uniform Amount  (A):
 Period of first amount (NF) =
 Period of last amount (NL)  =
                   Edit
```

Enter **1** and **10**. After introduction of the above values, the calculations will be performed, and the output will be displayed on the screen and written in the output file EA.OUT. Printed, it will look like this:

```
                    EA.OUT file display
THE DATA ARE:

          EXAMPLE #1 - TEST CASH FLOW

         ALTERNATIVE A
 PERIOD  CONT.VALUE  SINGLE VALUE
---------------------------------
     0        0.00       -100.00
     1       20.00          0.00
     6       15.00          0.00
    10        0.00         10.00


ROR Calculation, convergence control:
      INT           PF
        0.00        85.00000
        8.84        20.73870
       11.82         6.75197
       13.26         0.91498
       13.48         0.04736
       13.50         0.00038


THE RESULTS ARE:
 Present or future worth at NO =   0 and
 Interest rate INT =   10.00%              PF =      14.98

 Equivalent uniform amount for the period
 from NF =   1 through NL =  10             A =       2.44

 Rate of return                           ROR =      13.50 %
```

EXAMPLE 19-2

Solve Example 7-2 using EAiPC, EAiGR, and data file EA72.DAT. Compute the rate of return for the following cash flow:

Year	*Cash flow*
0	−$700
1	+100
2	+175
3	+250
4	+325

Solution: Select EA3PC from the main menu, and "Run" from the second one. Answer **Y** to the question "Do you want to use your data file?" Select EA72.DAT from the list of the data files. You will see:

```
                    Problem description
        EXAMPLE   #7-2
                                 Edit
        EXAMPLE   #7-2
```

Press *Enter* to go to the next screen:

```
                                        Data
TITLE:            CALCULATION OF THE RATE OF RETURN
  PERIOD  CONT.VALUE  SINGLE VALUE       PERIOD  CONT.VALUE  SINGLE VALUE
----------------------------------    ----------------------------------
 1:   0        0.00        -700.00      16:
 2:   1        0.00         100.00      17:
 3:   2        0.00         175.00      18:
 4:   3        0.00         250.00      19:
 5:   4        0.00         325.00      20:
 6:                                     21:
 7:                                     22:
 8:                                     23:
 9:                                     24:
10:                                     25:
11:                                     26:
12:                                     27:
13:                                     28:
14:                                     29:
15:                                     30:
                                       Edit
              CALCULATION OF THE RATE OF RETURN
```

Press *Enter* to continue.

```
                  Y-Yes or N-No
  Do you want to plot NPW/INTEREST curve ?
```

Press **N**. Leave the interest field blank (using the down arrow key), and input zero for the period of PF value:

```
                  Input
  Interest (INT)   =              %
  Period of PF value (NO) =    0
                  Edit
```

After introduction of the above values, the calculations will be performed, and the output displayed on the screen and written in an output file EA.OUT. Printed, it will look like this:

```
                              EA.OUT file display
THE DATA ARE:

        EXAMPLE   #7-2

        CALCULATION OF THE RATE OF RETURN
   PERIOD  CONT.VALUE  SINGLE VALUE
 ------------------------------------------
   0          0.00        -700.00
   1          0.00         100.00
   2          0.00         175.00
   3          0.00         250.00
   4          0.00         325.00


ROR Calculation, convergence control:
      INT           PF
        0.00       150.00000
        6.13        14.90913
        6.82         1.69255
        6.91         0.02626


THE RESULTS ARE:
 Rate of return                                 ROR =       6.91 %  ◀
```

Press *Esc* and you will see:

```
                 Y-Yes or N-No
Do you want to try new values using same data file ?
```

Press **N**.

```
                    Y-Yes or N-No
Do you want to Do you want to analyze another set of data ?
```

Press **N** again.

```
        Y-Yes or N-No
Do you want to get a graph ?
```

Press **N**.

EXAMPLE 19-3

Find the rate of return and produce the NPW vs. Interest Rate graph for cash flows *A* and *B* stored in the datafile EA_DEMO.DAT. Enter EA_DEMO.DAT file. You will see successive windows with the data. Normally you could edit the existing data or you could input data from the keyboard, but for now press *Enter* after reviewing each data window.

Also press *Enter* when you see the "input" window, since this time you do not need to input values for the interest and for the period PF values. You will see a large window entitled EA.OUT with the results. Since EA_DEMO.DAT file contains two subjects, press *Esc* to see the second part. Type **N** to the question "Do you want to try new values using same data file?" and answer **Y** to "Do you want to analyze another set of data?" You will see windows with data for the alternative *B*. Press *Enter* at each of them. The last EA.OUT window will display the first page of output for both subjects. To see subsequent pages, press the down arrow key multiple times, or use *PgDn* or *PgUp* keys to move more quickly up and down, as desired. Press *Esc* after you finish reviewing the output.

Since there are no more data in the EA_DEMO.DAT file, answer **N** to the two following questions. Next you see "Do you want to get a graph?" Press **Y**. Program EAiPC switches automatically to EAiGR. You then see the following questions.

Are you satisfied with DELTY = 25.114?	Press **N**
(NPW axis would look awkward if you press **Y**.)	
Input your own "DELTY" value (greater than 25.114)	Type **30**
Are you satisfied with DELTX = 2.0?	Press **Y**
Do you want a grid?	Press **Y** or **N**

Press **S** to start the graphics. After reviewing the graph, press **P** to get a printout of the graph (if a printer is connected and ON), and/or press *Esc* to return to the program menu. Press **E** to return to the main menu. The graph for this DEMO example is shown below.

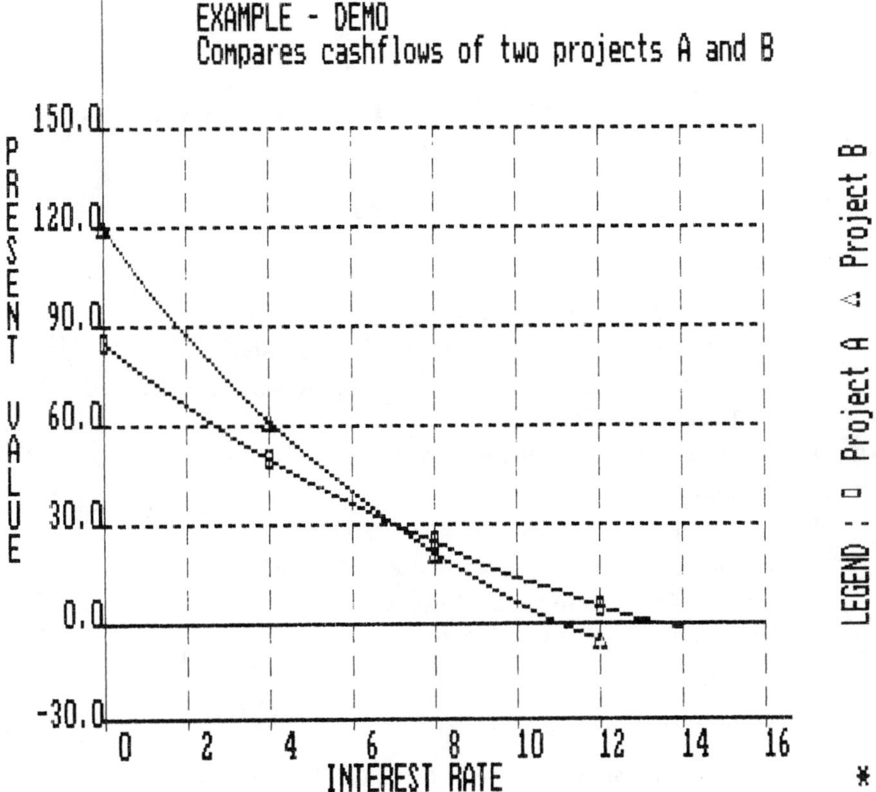

Program For Uniform Series Calculations

A program called MORTGiPC (where "i" indicates the current version number) is available to analyze uniform series cash flows. Annuity and mortgage payments are examples of this type of cash flow.

The MORTG program is structured in a way similar to the general-purpose EA program. The various economic parameters can be calculated depending on which data are given. The program is interactive and all data are input from the terminal. The space for a value which is not given must be left blank. (Skip it using the ↓ key.) The cash flow model, used in this program, is shown below.

The present or future value (PF) at the given interest rate (INT) can be calculated for any period of time NO using the uniform series (A) which starts at NF and is of duration N. NO can be chosen before, during or after the period of uniform series payments.

Monthly, continuous, or other compounding systems can be used. In addition, the program accepts a nominal or an effective interest rate per year.

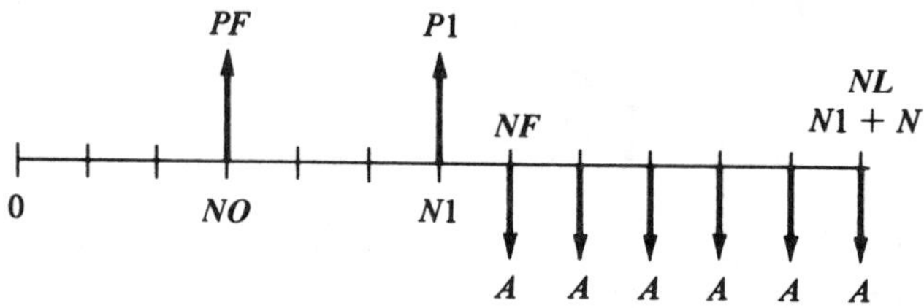

Variables

A = Equivalent uniform amount between NF and NL

INT = Interest, % per year (input −999 for 0 value)

NO = Period for which past, present, or future value is calculated (can be negative)

N1 = Period preceding first amount A (N1 = NF − 1)

NF = Period of first amount A

NL = Period of last amount A (NL = N1 + N)

N = Number of periods of amount A (N = NL − N1)

PF = Present or future worth at Period NO

P1 = Present worth at Period N1

ROR = Rate of return, %

The following options are available:

- PF can be found at NO if A, NF, N, and INT are given;

- A can be found if PF, NO, NF, N, and INT are given;
- INT can be found if PF, A, NO, NF, and N are given;
- N can be found if PF, A, INT, NO, and NF are given; in this case the value of the last payment (AN) is also calculated.

Upon request, the program also calculates the actual principal and interest, and the cumulative principal and interest for each period. The results are presented in a tabular form.

A period is considered to be a month. Principal and interest payments per month and per year are displayed in the table. Also, other than a month basic period can be used if one remembers to convert the input interest rate. For example, Example 4-21 can be solved using a year as a basic period. Then input an interest rate 8% × 12 = 96%, and 4 × 12 = 48 compounding periods.

Program MORTGiPC is easier to use than EAiPC where there are uniform series; in addition, it displays more detailed information.

To execute the program, begin at the main menu and move the highlighted area (with ↓ key) to MORTGiPC. Press *Enter* twice. You will see the single data window. In this program, all data are input from the keyboard. Leave blank the variable to be calculated. (Skip the variable with the ↓ key). After you input all data, press *Enter*. The output file MORTG.OUT is displayed. Use *PgDn* key to see subsequent pages. Press *Esc* and **E** to exit and to return to the main menu.

EXAMPLE 19-4

A $1000 loan is to be repaid in twelve equal monthly payments. The nominal interest rate is 12% per year, compounded monthly. Find the monthly payment, and produce a table of monthly principal and interest payments.

Solution: Select MORTG3PC from the main menu. The following table will be displayed on the screen:

```
          PROGRAM MORTG3PC - MORTGAGE REPAYMENT

 ---------------- MORT input data --------------------
| Present or Future Worth             (PF)  =         |
| Time Period of PF (in months)       (NO)  =         |
| Interest Rate        type=         (INT)  =         |
| (Type: Nominal=N, Effective=E)                      |
| Number of Repayment Periods          (N)  =         |
| Time Period of First Repayment      (NF)  =         |
| Uniform Repayment Amount             (A)  =         |
 ------------------------ Edit -----------------------
|                                                     |
 -----------------------------------------------------
```

Input PF = **1000**, NO = **0**, Type = **N**, INT = **12**, N = **12**, and NF = **1**. Leave the A field blank, since this value is unknown (is to be calculated). Press *Enter* and a new window will be displayed:

```
─────────── Compounding system ───────────
 Monthly
 Continuous
 Other (input number of periods per year)
```

Select "Monthly" and press *Enter*. The following will be displayed:

```
─────────── Y-Yes or N-No ───────────
 Do you want a table of principal and interest ?
```

Press **Y**. The following output will be displayed:

```
─────────── MORTG.OUT file display ───────────
DATA ARE:

Present or Future worth              (PF)  =   1000.00
Time Period of PF (in months)        (NO)  =         0
Interest Rate       type=N           (INT) =     12.00
(Type: Nominal=N, Effective=E)
Number of Repayment Periods          (N)   =        12
Time Period of First Repayment       (NF)  =         1
Uniform Repayment Amount             (A)   = Not given

Monthly Compounding System:

RESULTS:

Capital recovery factor is:              0.088849
Uniform repayment amount is:            88.85     ◀


  TABLE OF PRINCIPAL AND INTEREST PAYMENTS
                                                  CUMULATIVE
 PERIOD       PRINCIPAL        INTEREST      PRINCIPAL        INTEREST
      1           78.85           10.00          78.85           10.00
      2           79.64            9.21         158.49           19.21
      3           80.43            8.42         238.92           27.63
      4           81.24            7.61         320.16           35.24
      5           82.05            6.80         402.21           42.04
      6           82.87            5.98         485.08           48.01
      7           83.70            5.15         568.78           53.16
      8           84.54            4.31         653.32           57.48
      9           85.38            3.47         738.70           60.94
     10           86.24            2.61         824.93           63.56
     11           87.10            1.75         912.03           65.31
     12           87.97            0.88        1000.00           66.19
----------------------------------------------------------------------
YEAR  1         1000.00           66.19        1000.00           66.19

 Effective interest rate per month is      1.000%
 Nominal interest rate per year is        12.0000%
 Effective interest rate per year is      12.6825%
```

The output is saved in the file MORTG.OUT. Press *Esc* to return to the menu. Then you can review this file or print it.

EXAMPLE 19-5

Solve Problem 4-32 using MORTGiPC: A $150 bicycle was purchased December 1st with a $15 down payment. The balance is to be paid at the rate of $10 at the end of each month with the first payment due on December 31st. The last payment may be some amount less than $10. If interest on the unpaid balance is computed at 1½% per month, how many payments will there be? What is the amount of the final payment?

Solution:

```
DATA ARE:

Present or Future Worth              (PF)   =      135.00
Time Period of PF (in months)        (NO)   =         0
Interest Rate       Type = N         (INT)  =       18.00
(Type: Nominal=N, Effective=E)
Number of Repayment Periods          (N)    = Not given
Time Period of First Repayment       (NF)   =         1
Uniform Repayment Amount             (A)    =       10.00

Monthly Compounding System

RESULTS:

 TABLE OF PRINCIPAL AND INTEREST PAYMENTS
                                                  CUMULATIVE
 PERIOD        PRINCIPAL        INTEREST       PRINCIPAL        INTEREST
      1             7.97            2.03            7.97            2.03
      2             8.09            1.91           16.07            3.93
      3             8.22            1.78           24.29            5.71
      4             8.34            1.66           32.62            7.38
      5             8.46            1.54           41.09            8.91
      6             8.59            1.41           49.68           10.32
      7             8.72            1.28           58.40           11.60
      8             8.85            1.15           67.25           12.75
      9             8.98            1.02           76.24           13.76
     10             9.12            0.88           85.35           14.65
     11             9.26            0.74           94.61           15.39
     12             9.39            0.61          104.00           16.00
---------------------------------------------------------------------------
YEAR   1          104.00           16.00          104.00           16.00


                                                  CUMULATIVE
 PERIOD        PRINCIPAL        INTEREST       PRINCIPAL        INTEREST
     13             9.54            0.46          113.54           16.46
     14             9.68            0.32          123.22           16.78
     15             9.82            0.18          133.04           16.96
     16             1.96            0.03          135.00           16.99
---------------------------------------------------------------------------
YEAR   2           31.00            0.99          135.00           16.99

 Repayment period N is: 16   last payment:             1.99  ◄

 Effective interest rate per month is          1.50%
 Nominal interest rate per year is            18.00%
 Effective interest rate per year is          19.56%
```

Program For Rate of Return

The program ROR was written by D. G. Newnan and modified and adapted to the PC by Dr. Jan Wolski. The program is designed to analyze a cash flow and compute a rate of return on the internal investment using, if necessary, a preselected external interest rate. The computations are based on the method described in Chapter 18. Use this program instead of EA or MORTG in situations where there is more than one sign change in the cash flow during the project life.

An input file can be used or data may be input directly from the terminal during program execution. The structure of the data file is as follows:

TXT(3) = Three lines of text (not used by the program);
BLTXT = Title of one data block;
EXTR = External interest rate, % (EXTR = 999 if the external rate is to be equal to the internal rate of return);
CASH(i) = Cash flows at the end of periods i = 0, N.

The output file ROR.OUT is produced by the program. The output may be presented in an abbreviated form or with a full description of the calculation process and complete tables for the first and last rate of return calculations.

EXAMPLE 19-6

Solve Example 18-2 using RORiPC. Given the following cash flow, compute the internal rate of return i^*. If needed, use an external interest rate e^* of 6%.

Year	*Cash flow*
0	−$100
1	+50
2	+50
3	+50
4	+50
5	−73.25

Solution: Select RORiPC from the main menu. Answer **N** to the question "Do you want comments and calculation tables?" You have to create a new data file, so answer **N** to the question "Do you want to use your input data file?"

The Entry and Edit windows will be displayed. Then input the title: **18-2**. Next input **6** for the external rate of return, and **−100**, **50**, **50**, **50**, **50**, and **−73.25** for the cash flows. Press *Enter* after the last value, and you will be asked whether you want to save the above data in a data file. If you answer **Y**, then you will be asked to input the file name. Its name automatically starts with ROR and ends with .DAT. Thus, if you want the file to be ROR18_2.DAT, just input 18_2. After pressing *Enter*, you will see the output. It is saved in file

ROR.OUT. If you press *Esc*, you will get back to the menu, and you can review or print this file. If you want to see the above data file, go to "Review file" and select ROR18_2.DAT using the down arrow. You will see the following:

```
DATA FILE: ROR18_2.DAT
EXTR   CASH(I),I=0,N
************************************************************
 18-2
  6.00, -100,  50,  50,  50,  50,  -73.25 *
************************************************************

VARIABLES:

EXTR     = EXTERNAL INTEREST RATE, %.  IT IS EQUAL TO INTERNAL
           RATE OF RETURN IF EXTR = 999
CASH(I)  = CASH FLOW IN PERIOD I
N        = NUMBER OF PERIODS TO BE ANALYSED

NOTE: THREE LINES OF TEXT MUST PRECEDE DATA BLOCKS.
      DATA BLOCK: LINE 1 - TITLE, LINE 2 - EXTR, CASH(I),I=0,N.
      PLACE COMMA AFTER EACH DATA. PLACE '*' AFTER LAST VALUE.
```

(The above note is important only if you create a data file using a text editor, and not from within the EEAi program.)

Output file:

```
              RATE OF RETURN COMPUTATION
     FROM NEWNAN - ENGINEERING ECONOMIC ANALYSIS
    ============================================

 18-2

CASH(I):
     -100.00
       50.00
       50.00
       50.00
       50.00
      -73.25

     18-2
                                       ACCUMULATED
         YEAR        CASH FLOW          CASH FLOW
           0          -100.00            -100.00
           1            50.00             -50.00
           2            50.00               0.00
           3            50.00              50.00
           4            50.00             100.00
           5           -73.25              26.75

           SUM =          26.75

    NUMBER OF SIGN CHANGES IN CASH FLOW =     2
                  IN ACCUMULATED CASH FLOW =     1

      THERE IS A SINGLE POSITIVE RATE OF RETURN
 ONE SIGN CHANGE IN ACCUMULATED CASH FLOW
 LAST TERM IN CASH FLOW NEGATIVE (OR ZERO)
 RESULTS OF RATE OF RETURN COMPUTATION WITH EXTERNAL INTEREST
      RATE =    6.00 % :
      EXTERNAL INTEREST RATE NEEDED IN COMPUTATION.
      COMPUTED INTERNAL RATE OF RETURN EQUALS  16.49 %    ◄
```

EXAMPLE 19-7

Solve Example 18-5**b** for the internal rate of return using RORiPC. The external interest rate $e^* = 6\%$.

Year	*Cash flow*
0	−\$500
1	+300
2	+300
3	+300
4	−600
5	+200
6	+136.66

Solution: Select RORiPC from the main menu. Answer **Y** to the question "Do you want comments and calculation tables?" You have to create a new data file, so answer **N** to the question "Do you want to use your input data file?"

The Entry and Edit windows will be displayed. Then input the title:

EXAMPLE PROBLEM 18-5b SOLVED USING RORiPC

Next input **6** for the external rate of return, and **−500**, **300**, **300**, **300**, **−600**, **200** and **135.66** for the cash flows. Press *Enter* after the last value, and you will be asked whether you want to save the above data in a data file. If you answer **Y**, then you will be asked to input the file name. Next you will see the output.

Output file:

```
              RATE OF RETURN COMPUTATION
     FROM NEWNAN - ENGINEERING ECONOMIC ANALYSIS
     ==========================================

EXAMPLE PROBLEM 18-5b SOLVED USING RORiPC

 CASH(I):
      -500.00
       300.00
       300.00
       300.00
      -600.00
       200.00
       135.66


     EXAMPLE PROBLEM 18-5b SOLVED USING RORiPC
                                        ACCUMULATED
            YEAR         CASH FLOW       CASH FLOW
             0            -500.00          -500.00
             1             300.00          -200.00
             2             300.00           100.00
             3             300.00           400.00
             4            -600.00          -200.00
             5             200.00             0.00
             6             135.66           135.66

              SUM =          135.66
```

```
NUMBER OF SIGN CHANGES IN CASH FLOW =    3
           IN ACCUMULATED CASH FLOW =    3

CASH FLOW RULE OF SIGNS:
      THERE MAY BE AS MANY POSITIVE RATES OF RETURN AS THERE
     ARE SIGN CHANGES IN THE CASH FLOW.  FOR THIS CASH FLOW
     POSSIBLE NUMBER OF POSITIVE RATES RETURN
     IS BETWEEN 0 AND   3

SIGN CHANGES IN ACCUMULATED CASH FLOW:
     TWO CONDITIONS THAT ARE SUFFICIENT (BUT NOT NECESSARY)
     FOR A SINGLE POSITIVE RATE OF RETURN ARE
     1) THE ACCUMULATED CASH FLOW IN YEAR N IS NOT EQUAL TO 0.
     2) THERE IS ONE SIGN CHANGE IN THE ACCUMULATED CASH FLOW.

FOR THIS CASH FLOW
     ACCUMULATED CASH FLOW IN YEAR N =  6 is        135.66
     AND NUMBER OF SIGN CHANGES =   3
     NO CONCLUSIONS CAN BE REACHED FROM THIS CALCULATION.

ALGEBRAIC SUM OF CASH FLOW:
     A NEGATIVE SUM INDICATES THAT COSTS EXCEED BENEFITS.
     GENERALLY AN UNSATISFACTORY INVESTMENT SITUATION.
     POSITIVE SUM INDICATES ONE OR MORE POSITIVE RATES OF RETURN.
     A ZERO SUM FREQUENTLY INDICATES A ZERO PERCENT RATE OF
     RETURN. FOR THIS CASH FLOW SUM =        135.66

EXTERNAL INTEREST RATE IS EQUAL =   6.00 %

USING ABOVE DATA, PROCEED WITH ANALYSIS.

TRIAL NUMBER  1 : INTERNAL RATE OF RETURN =  14.52 %
                   EXTERNAL INTEREST RATE =   6.00 %

       CASH                                                 INTERNAL  EXTERNAL
YEAR   FLOW              COMPUTATION                        INVESTM.  INVESTM.
 0   -500.00                                                -500.00      0.00
 1    300.00   -500.00*(1+0.145)+   300.00 =   -272.61      300.00      0.00
 2    300.00   -272.61*(1+0.145)+   300.00 =    -12.20      300.00      0.00
 3    300.00    -12.20*(1+0.145)+   300.00 =    286.03       13.97    286.03
 4   -600.00    286.03*(1+0.060)+  -600.00 =   -296.81     -296.81   -303.19
 5    200.00   -296.81*(1+0.145)+   200.00 =   -139.91      200.00      0.00
 6    135.66   -139.91*(1+0.145)+   135.66 =    -24.57      135.66      0.00

TRIAL NUMBER 13 : INTERNAL RATE OF RETURN =  13.14 %
                   EXTERNAL INTEREST RATE =   6.00 %

       CASH                                                 INTERNAL  EXTERNAL
YEAR   FLOW              COMPUTATION                        INVESTM.  INVESTM.
 0   -500.00                                                -500.00      0.00
 1    300.00   -500.00*(1+0.131)+   300.00 =   -265.71      300.00      0.00
 2    300.00   -265.71*(1+0.131)+   300.00 =     -0.64      300.00      0.00
 3    300.00     -0.64*(1+0.131)+   300.00 =    299.28        0.72    299.28
 4   -600.00    299.28*(1+0.060)+  -600.00 =   -282.76     -282.76   -317.24
 5    200.00   -282.76*(1+0.131)+   200.00 =   -119.93      200.00      0.00
 6    135.66   -119.93*(1+0.131)+   135.66 =     -0.03      135.66      0.00

AFTER 13 TRIALS, THE TRUE INTERNAL RATE OF RETURN WAS FOUND
TO BE VERY CLOSE TO  13.143 %. THIS VALUE IS CONSIDERED THE ANSWER.


RESULTS OF RATE OF RETURN COMPUTATION WITH EXTERNAL INTEREST
     RATE =    6.00 % :
     EXTERNAL INTEREST RATE NEEDED IN COMPUTATION.
     COMPUTED INTERNAL RATE OF RETURN EQUALS  13.14 %          ◄
```

Program For After-Tax Cost Analysis

A program called EQANiPC ("i" indicates the version number of the program) has been written by Dr. Jan Wolski to calculate the present cost and the equivalent uniform annual cost of a project. Various depreciation methods—including accelerated cost recovery system (MACRS), the investment tax credit and its recapture, salvage value handling, income tax, and cost inflation—are included. Also various project financing methods can be selected. The detailed cash flow table can be printed on request for up to 25 years of the project life.

The description of main variables appearing in the program can be found below. Only a limited number of data are required in the data file. They are:

AI = Capitalized cost of project in present dollar value, k$
NT = Project life, years
OT = Operating time, hours/year

All remaining data are assumed internally. For example, the income tax rate is assumed to be 34%, and the salvage value of the project is zero if not given in the data file. Test data files EQ1.DAT, EQAN1.DAT and EQTST3.DAT are provided on the diskette.

Program EQAN is very useful in situations where the benefits of the project are impossible to determine. This is often the case in economic evaluation of plant equipment. The program calculates the present or uniform annual after-tax cost which can be directly compared with income to find profit. The cost can be derived from the general formula for the net cash flow in any Year i:

$$\mathrm{NCF}(i) = [R(i) - \mathrm{OC}(i) - \mathrm{DEP}(i)](1 - \mathrm{TAX}) + \mathrm{DEP}(i)$$

where

NCF(i) = Net after-tax cash flow in Year i
R(i) = Revenue in Year i
OC(i) = Operating cost in Year i
DEP(i) = Depreciation in Year i
TAX = Income tax rate, decimal

If NCF(i) is *divided by* (1 − TAX), the before-tax profit can be found. Then a new form of this formula can be written as follows:

$$\frac{\mathrm{NCF}(i)}{(1 - \mathrm{TAX})} = \mathrm{R}(i) - \mathrm{OC}(i) + \frac{\mathrm{TAX}}{(1 - \mathrm{TAX})} * \mathrm{DEP}(i)$$

Therefore, the total after-tax cost in Year i will be:

$$\mathrm{C}(i) = \mathrm{OC}(i) - \frac{\mathrm{TAX}}{(1 - \mathrm{TAX})} * \mathrm{DEP}(i)$$

Using the above formula, the present (worth) value of costs (CPV0) is calculated by the program. The equivalent uniform annual cost is defined by the standard capital recovery formula:

$$\text{CAY} = \text{CPV0} * \frac{S*(1 + S)**i}{(1 + S)**(i - 1)}$$

Also, the average cost per hour is calculated by the program.

Description Of Variables

Notes:

>	means	Optional data; default values if 0. Input small value if input is 0.
==>	means	Data required in the data file.

==>	AI	= Capitalized cost of equipment, present value, k$
>	AIPR	= Percent of capital cost paid in cash, % (default: 0%)
	BKV	= Book value at the end of equipment life (NT), k$
	CAH	= Average cost per hour, in Year-0 dollar value, $/hour
	CPV0	= Cost present value (worth), in Year NYEAR0(0), k$
	DEP	= Depreciation, k$/yr
>	DEPR	= Depreciation calculation system (default: DEPR = 1) DEPR = 1—given in the data file DEPR = 2—straight line, with salvage value DEPR = 3—sum-of-years digits DEPR = 4—DDB—straight line DEPR = 5—MACRS—5 years, half-year convention DEPR = 6—MACRS—7 years, half-year convention DEPR = 7—MACRS—5 years, straight line option DEPR = 8—MACRS—10 years, straight line option
>	DR(i)	= Depreciation rates, %, must be given if DEPR = 1. It is assumed that the depreciation rate will be constant after the last value given through the end of equipment life [Sum DR(i) = 100%]
>	ER	= Cost inflation rate, % (default: 0.0%)

> LPS = Loan repayment system: (default: LPS = 1)
LPS = 1, uniform (principal + interest)
LPS = 2, constant principal + due interest
> INSR = Insurance rate, % (default = 0.0)
> MCH = Maintenance cost per hour, $/h
==> NT = Equipment operation period in years
> NMI = Total project life (default = NT)
> NYEAR = Calendar year, present; cost escalation starts (default = 1990)
> NYEAR0 = Calendar year, equipment purchased (Year 0) (default = 1991)
> NP = Loan repayment period (default = NT, must be NP $\leq$ NT
> OCH = Operating cost, $/h
==> OT = Operating time, h/yr
> PTAX = Property tax, % (default = 0.0)
> S = Effective before-tax return on investment, % (default = 0%)
> SI = Salvage value, k$ (default = 0)
> SL = Interest on loan, % (default = 0%)
> TAX = Federal/State tax rate, % (default = 34%)
> TCR = Investment tax credit rate, % (default = 0%)

If the project life or the equipment operation period is greater than six years, you will see the question: "Do you have a wide printer?" Answer **Y** if you have a wide printer and 15-inch-wide paper installed. In a contrary case, enter **N** and the table will be printed with narrow characters (only IBM compatible printers are supported).

EXAMPLE 19-8

A firm is considering purchasing a piece of machinery with an end-of-1990 cost of $650.000 and a five-year analysis period. It would be operated 4000 hours per year. If purchased at the end of 1990 (that is, December 31, 1990), half the cost would be paid in cash with the balance obtained from a five-year loan at 12% annual interest, with annual payments of one-fifth of the loan, plus interest each year.

The machinery would be depreciated over a seven-year period by the modified accelerated cost recovery system (MACRS) method. The firm's combined federal and state income tax rate is 34%. The annual property tax on the equipment is 1% of the purchase price, and the annual insurance cost is 2% of the purchase price. The operating cost is constant at $50 per hour (in 1990 dollars), while maintenance is $20 per hour (in 1990 dollars) during the first year of operation and will increase $2, $3, $5, and $10 (also in 1990 dollars) in each of the subsequent years. The estimated actual salvage value (in 1990 dollars) at the end of each year of operation is \$300,00, \$200,000, \$150,000, \$120,000, and \$100,000, respectively.

The estimated inflation rate from the end of 1990 on is $2\frac{1}{2}\%$. The firm sets 10% as the minimum attractive rate of return. Compute the equivalent uniform annual cost (stated as an hourly rate in 1990 dollars) to own and operate the equipment for the five-year period. Use EQANiPC.

Solution: Select EQANiPC from the main menu. Select "Run" from the next menu. Answer **N** to the question: "Do you want to use your data file?" A window will be displayed with a blank form for entering the project data. Press *Enter* after entering each item. If you want to leave any data entry blank, use the down arrow key to move to the next position. Entering data for the operating and maintenance costs is simplified. You need to enter only the first year value of 50 for the operating cost. The program automatically assumes the same value through the end of the equipment life.

If you save your data in a data file (for example EQ1.DAT), you can use it again later or review it or print it from the menu.

Data file:

```
*
*****************************************************
EQUIPMENT ECONOMIC ANALYSIS          PROGRAM: EQAN3PC
DATA FILE: EQ1.DAT
*****************************************************
(  > Optional data, input small value if input is 0.
 ==> Nonzero data required.)
GENERAL DATA OF THE PROJECT
(45 columns of text, I5 and F8.1 format for numbers):
*****************************************************
PRESENT YEAR                                : 1990
PROJECT STARTS IN YEAR                      : 1991
PROJECT DURATION PERIOD, years              :
EQUIPMENT OPERATION PERIOD, years        ==>:     5
LOAN REPAYMENT PERIOD, years                :
LOAN PAYMENT SCHEDULE                       :     2
DEPRECIATION SYSTEM                         :     6
EQUIPMENT OPERATION TIME, hours/year     ==>:  4000.0
PRESENT EQUIPMENT COST, k$               ==>:   650.0
PERCENT PAID CASH FOR EQUIPMENT, %          :    50.0
INTEREST ON LOAN, %/year                    :    12.0
FEDERAL/STATE TAX RATE, %                  >:
INVESTMENT TAX CREDIT RATE, %              >:
INSURANCE COST AS % OF EQUIPMENT PRICE, %   :     2.0
PROPERTY TAX AS % OF EQUIPMENT PRICE, %     :     1.0
EFFECTIVE BEFORE-TAX RETURN ON INVESTMENT,%:     10.0
COST INFLATION RATE, %                      :     2.5

DATA, VARIABLE IN TIME, PRESENT COST VALUES:
(22 columns of text, F8.1 format for the numbers, also beyond year 6)
*************************************************************************
OPERATING YEARS       :     0       1        2        3        4        5       6
                      :
=========================================================================
SALVAGE VALUE         :           300.0    200.0    150.0    120.0    100.0 *
DEPRECIATION, %       :   *
OPER. COST, $/hour    :            50.0 *
MAINT. COST, $/hour   :            20.0     22.0     25.0     30.0     40.0 *
=========================================================================
```

```
FOOTNOTES:

1. Loan payment schedule:
   1 = constant end-of-period payments
   2 = constant principal plus due interest.

2. Depreciation system DEPR:
   1 = depreciation rates provided by user
   2 = straight line, with salvage value
   3 = sum-of-years digits
   4 = double declining balance/straight line
   5 = MACRS, 5 years, half-year convention
   6 = MACRS, 7 years, half-year convention
   7 = MACRS, 5 years, straight line option
   8 = MACRS, 7 years, straight line option

3. Leave blanks after last data if not given for all years of operating
   period NT. Interpretation will be as follows:
   Salvage value: 0 assumed beyond last value given.
   Depreciation: needed only if DEPR = 1, it is assumed that the
    depreciation rate is constant after the last given value through
    the end of equipment life NT.
   Operating cost: last given value is used through the end of equipment
    life NT.
   Maintenance cost: the same as above.
```

Output:

Answer **Y** to the question "Do you want the output table?" The output will be displayed on the screen. Press *Esc* to get to the menu. Then the EQAN.OUT file can be printed:

```
*****************************************************
EQUIPMENT ECONOMIC ANALYSIS          PROGRAM: EQAN3PC
INPUT FILE:      EQ1.DAT ;  OUTPUT FILE: EQAN.OUT
*****************************************************

GENERAL DATA OF THE PROJECT

PRESENT YEAR                                    : 1990
PROJECT STARTS IN YEAR                          : 1991
PROJECT DURATION PERIOD, years                  :    5
EQUIPMENT OPERATION PERIOD, years            ==>:    5
LOAN REPAYMENT PERIOD, years                    :    5
LOAN PAYMENT SCHEDULE                           :    2
DEPRECIATION SYSTEM                             :    6
EQUIPMENT OPERATION TIME, hours/year         ==>:  4000.0
PRESENT EQUIPMENT COST, k$                   ==>:   650.0
PERCENT PAID CASH FOR EQUIPMENT, %              :    50.0
INTEREST ON LOAN, %/year                        :    12.0
FEDERAL/STATE TAX RATE, %                      >:    34.0
INVESTMENT TAX CREDIT RATE, %                  >:     0.0
INSURANCE COST AS % OF EQUIPMENT PRICE, %       :     2.0
PROPERTY TAX AS % OF EQUIPMENT PRICE, %         :     1.0
EFFECTIVE BEFORE-TAX RETURN ON INVESTMENT,%:         10.0
COST INFLATION RATE, %                          :     2.5

RESULTS:

COST PRESENT VALUE (YEAR 0)       (k$)      :  1842.0

EQUIVALENT O & M COST PER HOUR ($/hour) :      77.4
EQUIVALENT OWNERSHIP COST PER HOUR      :      44.1
EQUIVALENT TOTAL COST PER HOUR          :     121.5  ◀
====================================================
```

OUTPUT, EQUIPMENT COST ANALYSIS TABLE:
(All values in 1000 $ (k$) unless otherwise noted)

OPERATING YEARS	1991	1992	1993	1994	1995	1996
	0	1	2	3	4	5
DATA, VARIABLE IN TIME, PRESENT COST VALUES:						
SALVAGE VALUE		300.0	200.0	150.0	120.0	100.0
DEPRECIATION, %	14.29	24.49	17.49	12.49	8.93	8.92
OPER. COST, $/hour		50.0	50.0	50.0	50.0	50.0
MAINT. COST, $/hour		20.0	22.0	25.0	30.0	40.0
COST POSITIONS NOT SUBJECT TO TAX:						
EQUIPMENT COST, k$	666.2					
PAID CASH	333.1					
LOAN	333.1					
LOAN PAYMENT,PRINC.		66.6	66.6	66.6	66.6	66.6
INVESTM. TAX CREDIT						
ITC RECAPTURE						
DEPRECIATION	-95.2	-163.2	-116.5	-83.2	-59.5	-29.7
END BOOK VALUE						-118.9
TOTAL PER YEAR	237.9	-96.5	-49.9	-16.6	7.1	-82.0
PER HOUR ($/h)	59.5	-24.1	-12.5	-4.1	1.8	-20.5
COST POSITIONS SUBJECT TO TAX:						
LOAN REPAYMENT,INT.		40.0	32.0	24.0	16.0	8.0
DEPRECIATION	95.2	163.2	116.5	83.2	59.5	29.7
END BOOK VALUE						118.9
END SALVAGE VALUE						-116.0
OPERATING COST		210.1	215.4	220.8	226.3	231.9
MAINTENANCE COST		84.0	94.8	110.4	135.8	185.6
INS.COST + PROP.TAX		20.0	20.0	20.0	20.0	20.0
TOTAL PER YEAR	95.2	517.3	478.6	458.3	457.5	478.1
PER HOUR ($/h)	23.8	129.3	119.7	114.6	114.4	119.5
TOTAL COST PER YEAR	455.7	371.0	403.0	433.2	468.3	353.9
PER HOUR ($/h)	113.9	92.8	100.8	108.3	117.1	88.5
DISCOUNTED COST	444.6	321.0	309.3	294.9	282.7	189.5

EXAMPLE 19-9

This is Example 19-8 modified as follows:

- Six-year analysis period;
- 20% of the cost is paid in cash;
- Constant annual payments of the loan for the six-year analysis period;
- Depreciation is 10%, 20%, 20%, 20%, and 15%, respectively;
- Maintenance cost is $20 per hour during the first year of operation and will increase $5, $10, $15, $20, and $30 in each of the subsequent years.

Compute the equivalent uniform annual cost (stated as an hourly rate in 1990 dollars) to own and operate the equipment for the six-year period. Use EQANiPC.

Solution: Select EQANiPC from the main menu. Select *Run* from the next menu. Select EQAN1.DAT from the list of data files and press *Enter*. A window will be displayed with data of the project. Normally all data should match Example 19-9, unless some values were previously modified. Check them carefully and make corrections, if necessary. Press *Enter* after reviewing or editing each of two data windows. Answer **Y** to the question "Do you want the output table?"

The output will be displayed on the screen. Press *Esc* to get to the menu. Then the EQAN.OUT file can be printed.

```
*****************************************************
EQUIPMENT ECONOMIC ANALYSIS          PROGRAM: EQAN3PC
INPUT FILE:     EQAN1.DAT ;   OUTPUT FILE: EQAN.OUT
*****************************************************

GENERAL DATA OF THE PROJECT

PRESENT YEAR                                 : 1990
PROJECT STARTS IN YEAR                       : 1991
PROJECT DURATION PERIOD, years               :    6
EQUIPMENT OPERATION PERIOD, years         ==>:    6
LOAN REPAYMENT PERIOD, years                 :    6
LOAN PAYMENT SCHEDULE                        :    1
DEPRECIATION SYSTEM                          :    1
EQUIPMENT OPERATION TIME, hours/year      ==>:  4000.0
PRESENT EQUIPMENT COST, k$                ==>:   650.0
PERCENT PAID CASH FOR EQUIPMENT, %           :    20.0
INTEREST ON LOAN, %/year                     :    12.0
FEDERAL/STATE TAX RATE, %                   >:    34.0
INVESTMENT TAX CREDIT RATE, %               >:     0.0
INSURANCE COST AS % OF EQUIPMENT PRICE, %    :     2.0
PROPERTY TAX AS % OF EQUIPMENT PRICE, %      :     1.0
EFFECTIVE BEFORE-TAX RETURN ON INVESTMENT,%:      10.0
COST INFLATION RATE, %                       :     2.5

RESULTS:

COST PRESENT VALUE (YEAR 0)     (k$)      :  2319.4

EQUIVALENT O & M COST PER HOUR ($/hour) :    100.0
EQUIVALENT OWNERSHIP COST PER HOUR      :     33.1
EQUIVALENT TOTAL COST PER HOUR          :    133.1  ◄
===================================================

OUTPUT, EQUIPMENT COST ANALYSIS TABLE:
(All values in 1000 $ (k$) unless otherwise noted)
```

OPERATING YEARS	:	1991	1992	1993	1994	1995	1996	1997
	:	0	1	2	3	4	5	6
DATA, VARIABLE IN TIME, PRESENT COST VALUES:								
SALVAGE VALUE	:		300.0	200.0	150.0	120.0	100.0	100.0
DEPRECIATION, %	:	10.00	20.00	20.00	20.00	15.00	7.50	7.50
OPER. COST, $/hour	:		50.0	50.0	50.0	50.0	50.0	50.0
MAINT. COST, $/hour	:		20.0	25.0	35.0	50.0	70.0	100.0

```
COST POSITIONS NOT SUBJECT TO TAX:

EQUIPMENT COST, k$  :   666.2
         PAID CASH  :   133.2
              LOAN  :   533.0
```

LOAN PAYMENT,PRINC. :		65.7	73.6	82.4	92.3	103.3	115.7
INVESTM. TAX CREDIT :							
ITC RECAPTURE :							
DEPRECIATION :	-66.6	-133.2	-133.2	-133.2	-99.9	-50.0	-50.0
END BOOK VALUE :							
TOTAL PER YEAR :	66.6	-67.6	-59.7	-50.9	-7.7	53.4	65.8
PER HOUR ($/h):	16.7	-16.9	-14.9	-12.7	-1.9	13.3	16.4
COST POSITIONS SUBJECT TO TAX:							
LOAN REPAYMENT,INT. :		64.0	56.1	47.3	37.4	26.3	13.9
DEPRECIATION :	66.6	133.2	133.2	133.2	99.9	50.0	50.0
END BOOK VALUE :							
END SALVAGE VALUE :							-118.9
OPERATING COST :		210.1	215.4	220.8	226.3	231.9	237.7
MAINTENANCE COST :		84.0	107.7	154.5	226.3	324.7	475.5
INS.COST + PROP.TAX :		20.0	20.0	20.0	20.0	20.0	20.0
TOTAL PER YEAR :	66.6	511.4	532.4	575.8	609.9	652.9	678.2
PER HOUR ($/h):	16.7	127.8	133.1	143.9	152.5	163.2	169.5
TOTAL COST PER YEAR :	167.6	409.0	441.9	498.7	598.2	733.8	777.9
PER HOUR ($/h) :	41.9	102.2	110.5	124.7	149.6	183.4	194.5
DISCOUNTED COST :	163.5	353.9	339.2	339.5	361.1	392.9	369.4

Program For Capital Expenditure Analysis By Simulation

A program CEAPiPC (the number "i" indicates the current version of the program) was written by D. G. Newnan and modified and adapted to the PC by Dr. Jan Wolski. The program uses the simulation method. The useful life, capital cost, salvage value, expensed costs, and benefits can be defined in the form of probability distributions of various kinds. The result is a histogram of rate-of-return vs. relative frequency.

CEAPiPC Program Section-by-Section Discussion

Section 1. The logic of the capital expenditure analysis model is essentially linear and proceeds from Section 1 to HIST. In Section 1, four lines of required data are read in. They are required for every program, along with the associated descriptive and blank lines.

Section 2. This section is used to initially load all vectors with zeros. This may or may not be necessary.

Section 3. The four data lines in Section 1 provided an input of some control digits, so that the user might indicate whether the parameters are to be determined internally by built-in distributions, or whether exogeneous data are to be loaded at this time.

If established by the control digits, the following are read:

LCD	Cumulative life distribution
TP	Investment timing distribution
BRS	Loan repayment schedule
DEP	Depreciation schedule
CSV	Salvage value schedule
CCD	Cumulative expensed cost distribution
CBS	Cumulative annual benefit distribution

At this point in the program, the economic analysis problem has been completely defined.

Section 4. This section serves the single purpose of echoing the input data. The problem has been described in terms of eleven elements and, in Section 4, these elements are output.

Section 5. In this section, the main processing DO loop begins. The first task in Section 5 is to re-initialize all of those variables which are used in each individual iteration.

Section 6. The analysis period is determined in this section. A useful life control digit is analyzed by means of a computed GO TO. Alternatives available are:

a. useful life is fixed;
b. useful life is normally distributed;
c. useful life has a log normal distribution;
d. the useful life uses Erlang; (if the exponential distribution is desired, then it is obtained from Erlang simply by setting the value of K, here called KERL1, = 1); and,
e. cumulative life distribution LCD is used and was input in Section 3.

The value of useful life N has now been determined. The value is compared to a pre-set maximum useful life NMAX and, in the event that N exceeds NMAX, it is reset equal to NMAX. Thus, the analysis period = Asset useful life = N.

Section 7. In this section, the capitalized investment is determined. It follows the same general pattern followed in Section 6, that is, the capital control investment digit IFC is analyzed by means of a computed GO TO with the following options provided:

a. capitalized investment is fixed;
b. capitalized investment normally distributed;
c. log normal distribution; and,
d. Erlang.

Section 8. The salvage or terminal value of the asset (SV) is determined. Here the options are:

a. no salvage value;
b. salvage value fixed;
c. salvage value is a linear decline from salvage value SVMAX to salvage value SVLOW;
d. salvage value is determined from salvage value distribution CSV; and,
e. salvage value given for each year of project life from 1 through NMAX ($\leq$ 25); it is assumed as zero if not given. Values provided in CSV vector.

Section 9. This section addresses itself to the relatively complex problem of investment timing. The options provided are:

a. total investment is assumed to occur at the beginning of Year 1, which we describe in this program as Year 0;
b. investment is equally distributed over TN years.

Here the first investment is assumed to take place at the beginning of Year 1 (Year 0) and the last investment occurs at the beginning of Year TN. Since the program uses the end-of-year convention, the last investment is at the end of Year TN−1. In the event the useful life of the asset is less than TN years, the investment is re-distributed to occur over N years rather than TN years.

The last option is:

c. investment timing is provided by means of the investment timing schedule TP, which was input in Section 3.

The investment timing schedule permits the investment to be distributed over ten years. It is again possible that the investment useful life is less than 10, so the investment timing schedule TN may need to be altered; if this is necessary, it is accomplished in the vector REBTP (rebalanced TN).

Section 10. This section determines the deductible interest charge which results from borrowed money (if any) used in the project. The borrowed money control digit allows four options:

a. no borrowed money;
b. uniform repayment schedule;
c. equal principal payment schedule, plus annual interest; and,
d. repayment schedule is provided from Section 3.

The only function of this section is to develop the interest charge in the event the economic analysis is to be done on an after-tax basis. If this is true, then the interest paid in any year is assumed to be a tax deductible item. As in prior sections, there must be an examination here to ensure that the borrowed money repayment schedule does not exceed the useful life in any iteration. If it does, then the repayment schedule is accelerated so that the debt is extinguished by the end of the useful life N.

Section 11. This section computes the depreciation schedule for the asset based upon an established, depreciable life which is provided from the input data. Actual computation is done in SUBROUTINE DEPRC. By means of a control digit, the subroutine permits seven options:

a. straight line depreciation;
b. sum-of-years digits method;
c. double declining balance;
d. double declining balance with conversion to straight line;
e. declining balance at 150% of straight line;
f. depreciation schedule provided by Section 3; and,
g. depreciation is ignored.

This latter event would occur, presumably, where a before-tax computation is to be made.

In the event that the asset is not depreciated down to its salvage value, by the time the useful life N is reached in any iteration, the actual result for tax purposes would be a loss on disposal. This is assumed to occur at the end of the useful life if the book value of the asset is more than the salvage value. The program then makes the customary accounting computations. The tax adjustment itself is made in Section 14.

Section 12. In this section, the before-tax benefit schedule BENF is compiled. The benefit schedule control digit IBS provides for five options:

a. benefits are fixed;
b. benefits are normally distributed;
c. benefit distribution is log normal;
d. benefits follow Erlang distribution; and,
e. benefits are determined from the cumulative annual benefit distribution input in Section 3.

The distribution is sampled by means of the random number generator. Following the compilation of the benefit schedule, the program provides an option for introducing a trend into the benefit schedule. The benefit control digit ITB allows two options:

a. no trend; and,
b. trend equals plus-or-minus the decimal percentage per year.

The benefit schedule is adjusted for the benefit trend.

Section 13. The expensed cost schedule EXPC is computed in this section. There are six options:

a. no expensed annual cost;
b. expensed annual cost is fixed;

c. expensed cost is normally distributed;
d. expensed cost follows log normal distribution;
e. expensed cost is obtained from Erlang distribution; and,
f. expensed cost distribution was provided in Section 3 and is sampled using the random number generator.

Section 14. This section computes the income tax consequences schedule TXS. Taxable income is equal to annual benefit *minus* the depreciation *minus* the interest cost for the year *minus* the expensed cost for the year. The income tax control digit provides for three options:

a. computation is to be a before-tax computation (hence, income taxes are ignored);
b. uniform tax rate is applied; and,
c. uniform tax rate is used together with an investment tax credit allowance.

It is assumed that the investment tax benefit occurs at the time of the investment; thus, an investment at Year 0 would similarly have its investment tax credit during that year and would, therefore, be recorded as a saving in Year 0.

In the event the asset is retired prior to the end of its depreciable life, there may be a loss on disposal to take into account. The loss would be the depreciated book value *minus* the salvage value. If necessary, this adjustment is made.

Section 15. The after-tax cash flow schedule is computed in this section. Here, after-tax cash flow is defined as benefits *minus* expensed costs *minus* taxes *minus* loan payment *minus* the year's investment. If the parameter ISWCH is set to something other than zero, the entire cash flow table for each iteration is printed at this point. If ISWCH is equal to zero, printing is suppressed, and the section is concluded.

Section 16. This section does the preliminary calculations necessary to compute a rate of return for the after-tax cash flow, ATCH. The number of sign changes in the cash flow is computed, along with the sum of the cash flow.

For no sign changes in the cash flow	
and sum is negative or zero:	Rate of return set equal to 0%
and sum is positive:	Rate of return set equal to 100% (The true rate of return is ∞.)
For one sign change in the cash flow	
and sum is negative or zero:	Rate of return set equal to 0%
and sum is positive:	Rate of return computed by subroutine RORIE.

For more than one sign change in the cash flow, the rate of return is computed by subroutine RORIE.

Section 17. This section prints an output heading and then prints one line of summary statistics at the end of each iteration. As a result there will be ITER lines of output. This output is not controlled by ISWCH and always appears whenever ITER > 1.

Section 18. This section takes the trial rate of return for the iteration and assigns it to one cell of the vector RR, which will be used to print the rate of return histogram. At this point, the range of the main iteration DO loop, labeled 610, is reached. Control is returned to the beginning of Section 5. When the main iteration loop is satisfied, Section 18 calls the rate of return plot routine HIST. Following this, the program terminates.

***Subroutine* RORIE.** Printing in this subroutine is controlled by ISWCH.

If ISWCH = 0,	ISW = 1	printing is suppressed
ISWCH = 1,	ISW = 2	the tabulation and results are printed

The algebraic sum of the cash flow, ASUM, is computed. The ***original book*** method of computing an approximate rate of return is:

$$\text{Rate of return} = \frac{\text{Average annual profit after depreciation}}{\text{Original investment}}$$

Thus, an estimate of the rate of return is roughly:

$$\text{TRIAL} = \frac{\text{SUM/N}}{\text{CASHO}} = \frac{\text{SUM}}{\text{CASHO} * \text{N}}$$

In the program, TRIAL is purposely set 10% higher in an attempt to make TRIAL slightly higher than the answer being sought.

The search area is limited to the range 0% to 100%. If trial is computed to be outside the search area, it is set to either 5% or 80%. The search begins with the lower boundary TLOW equal to 0% and the upper boundary THI equal to 100%.

The trial internal rate of return equals TRIAL. The external interest rate depends on the method of computation MTH specified.

MTH	*External interest rate*
1	TRIAL
2	EXT

The subroutine computes the period-by-period results for the cash flow. If there is net internal investment, then in the next period compound interest will increase it by (1 + TRIAL). If there is an external investment, it will be increased by (1 + EXT) or (1 + TRIAL) depending on MTH.

If the TRIAL internal rate of return is correct, the net investment will equal zero at the end of the Nth year. If, instead, there is an unrecovered investment (a negative sum), it means that TRIAL is too high. THI is increased to TRIAL and a new trial halfway between the last TRIAL and TLOW is used.

$$\text{THI} = \text{TRIAL}$$
$$\text{TRIAL} = (\text{TLOW} + \text{THI})/2$$

A new period-by-period computation is done.

If the TRIAL internal rate of return had, instead, produced a positive sum at the end of the Nth year, we know that TRIAL is too low. This time TLOW would be increased to TRIAL and the binary search continued in the diminished interval.

$$\text{TLOW} = \text{TRIAL}$$
$$\text{TRIAL} = (\text{TLOW} + \text{THI})/2$$

When the interval between TLOW and TRIAL is less than or equal to 0.001, TRIAL is considered sufficiently close to the exact solution, and the computations are halted. The search should be completed within ten trials. If it is not, the internal rate of return is probably not within the search area. After twenty trials, the search is stopped.

***Subroutine* RANG.** To minimize the difficulty of implementing the program, a simple pseudo-random number generator is provided by subroutine RANG. The random numbers produced are greater than 0.0000 and less than or equal to 0.9999.

***Subroutine* NORML.** This subroutine generates random normal numbers, with mean equal to zero and standard deviation equal to 1.

***Subroutine* DEPRC.** This subroutine is called in Section 11. It provides for computation of the depreciation schedule by several methods.

***Subroutine* HIST.** HIST is arranged to compute a rate of return plot. The rate of return obtained in each iteration was loaded into the vector RR in Section 18. The graph is then output. The appropriate labeling on the axes of the table and the summary statistics (mean, variance, standard deviation, and N) are printed. The value N, in this case, indicates the number of iterations used in compiling the data on the graph.

CEAPiPC Variable List For Required Data

Data Line 1.

JOB — ***Job number*** or other identifying digits

NMAX — ***Maximum useful life*** (analysis period) NMAX $\leq$ 100

IL — ***Useful life control digit***

1 Useful life is fixed = NMAX

2 Normally distributed useful life

AVEN — Mean life
SDN — Standard deviation of life

3 Log normal distribution α = AVEN β = SDN

4 Erlang distribution

Mean of underlying exponential distribution = AVEN
For Erlang: Resulting mean = KERL1 * AVEN
KERL1 — K for Erlang. When $K = 1$ the exponential distribution is obtained

5 Cumulative useful life distribution is to be provided (LCD)

ISWCH — ***Cash flow table switch***

1 With ISWCH = 1 a complete cash flow table is printed each iteration with the result that ITER tables will be printed

0 Printing of cash flow tables is suppressed

IFC — ***Capitalized investment control digit***

1 Capitalized investment fixed

Fixed Investment = AVEI

2 Normally distributed capitalized investment

AVEI — Mean capitalized investment
SDI — Standard deviation of capitalized investment

3 Log normal distribution α = AVEI β = SDI

4 Erlang distribution

Mean of underlying exponential distribution = AVEI
Resulting mean = KERL2 * AVEI
KERL2 — K for Erlang. When $K = 1$ the exponential distribution is obtained

Data Line 2.

IT — ***Investment timing control digit***

1 Total investment made at beginning of Year 1 (Year 0)

2 Investment equally distributed over lesser of TN or N years

TN Years over which investment distributed

3 Percentage distribution of investment timing to be provided (TP)

IB ***Borrowed money control digit***

1 No borrowed money

2 Uniform repayment schedule

3 Equal principal payments *plus* annual interest

4 Repayment schedule to be provided (BRS)

BI ***Effective interest rate per annum charged on borrowed money, stated as a decimal*** (for example, 5% = 0.0500)

BN ***Year in which last portion of borrowed money is repaid***; that is, the term of the debt which began at beginning of Year 1 (Year 0), BN ≤ 99

BPC ***Portion of Year-0 investment that is borrowed, stated as a decimal*** (for example, 50% = 0.500); BPC may be greater than 1.000

ID ***Depreciation schedule control digit***

1 Straight line depreciation

2 Sum-of-years digits depreciation

3 Double declining balance depreciation

4 Double declining balance depreciation with conversion to straight line

5 Declining balance depreciation at 150% of straight line rate

6 Depreciation schedule to be provided (DEP)

7 Depreciation schedule not computed (this option would be used when a before-income-tax analysis is to be made)

ND ***Life of capital asset*** (years) to be used in calculation of depreciation schedule. If the value of ND exceeds the analysis period (N) in any iteration, the depreciation schedule is computed using N years rather than ND years as the depreciable life

ISV ***Salvage (Terminal) value control digit***

1 Salvage value = 0

AVESV 2 Salvage value is fixed = AVESV

3 Salvage value declines uniformly from SVMAX to SVLOW

SVMAX At Year 0 SV = SVMAX

SVLOW At Year NMAX SV = SVMIN

Linear interpolation between 0 and NMAX

4 Cumulative salvage value distribution is to be provided (CSV)

5 Salvage value given for each year of project life from 1 through NMAX (≤ 25); it is assumed as zero if not given (CSV)

Data Line 3.

IC	***Expensed cost control digit***
	1 No expensed annual cost
AVEC	2 Expensed cost is fixed = AVEC
	3 Normally distributed expensed cost
	Mean = AVEC
SDC	Standard deviation = SDC
	4 Log normal distribution α = AVEC β = SDC
	5 Erlang distribution
	Mean of underlying exponential distribution = AVEC
	Resulting mean = KERL3 * AVEC
KERL3	*K* for Erlang; when *K* = 1 the exponential distribution is obtained
	6 Cumulative expensed annual cost distribution to be provided (CCD)
ITX	***Income tax control digit***
	1 Income taxes not computed
TAX	2 Uniform income tax rate (TAX) stated as a decimal (for example, 39% = 0.39)
TXCD	3 Uniform income tax rate (TAX) and investment tax credit (TXCD); tax credit stated as a decimal (for example, 7% = 0.07)
EXT	***External investment interest rate (EXT)***
	Stated as a decimal (for example, 6% = 0.060)

Data Line 4.

IBS	***Benefit schedule (before taxes) control digit***
AVEB	1 Before-tax benefits fixed = AVEB
	2 Normally distributed before-tax benefits
	Mean = AVEB
SDB	Standard deviation = SDB
	3 Log normal distribution α = AVEB β = SDB
	4 Erlang distribution
	Mean of underlying exponential distribution = AVEB
KERL4	Resulting mean = KERL4 * AVEB
	K for Erlang; when *K* = 1 the exponential distribution is obtained
	5 Cumulative annual benefits schedule (before taxes) distribution to be provided (CBS)

ITB *Trend of benefit schedule control digit*

1 No trend adjustment of before-tax benefit schedule

2 Before-tax benefit schedule to be adjusted by a arithmetic gradient

TPC Annual percentage change per year (TPC) is stated as a signed decimal (a 4% declining gradient = −0.04)

ITER *Number of iterations of the simulation model desired*

Schedules Specified in Data Lines.

Listed in order of entry.

1. LCD *Cumulative life distribution*

25 integer values of useful life with a cumulative discrete probability of 0.04(I); I = 1,25

Example: NMAX
1 5 8 10 12 16 20 24 · · · xxx

2. TP *Investment timing distribution*

Ten percentage values (stated in decimal form) as distribution of investment over initial ten-year period; sum = 1.00

Example:
0.02 0.25 0.30 0.25 0.10 0.08 0.00 0.00 0.00 0.00

3. BRS *Borrowed money repayment schedule*

Dollar amount of each year's repayment provided for the term of the debt; number of elements = BN (= IBN) $BN_{max} = 99$

Example:
5000. 10000. 20000. · · · and so forth

4. DEP *Depreciation schedule*

Dollar amount of each year's depreciation provided for the depreciable life of the asset: ND years

Example:
2000. 52000. 46000. · · ·

5. CSV *Salvage value distribution*

When ISV = 4: 25 values of salvage value with a cumulative discrete probability of 0.04(I), I = 1,25; dollar amounts are input

When ISV = 5: Salvage value given for each year of project life from 1 through NMAX (≤ 25); it is assumed as zero if not given

Example:
3800. 3000. 2800. 2300. 2000. · · ·

6. CCD ***Cumulative expensed cost distribution***

25 values of expensed cost with a cumulative discrete probability of 0.04(I), I = 1,25; dollar amounts are input

Example:
2000. 2100. 2300. 2450. 2600. . . .

7. CBS ***Cumulative annual benefit distribution***

25 values of annual benefit with a cumulative discrete probability of 0.04 * I, I = 1,25; dollar amounts are input

Example:
2500. 3000. 3500. 4000. 4500. . . .

CEAPiPC Program Examples

The structure of the input file is shown in Example 19-10. Two lines of the text (title) are followed by four lines of data, and a number of optional data lines. Data files CEAP1.DAT and CEAP2.DAT from the diskette are used in the examples below.

EXAMPLE 19-10

This CEAPiPC problem uses essentially the same data as Example 19-8—the EQAN program. The problem is solved in a single iteration (ITER =1). Additional information needed for CEAP:

Maximum equipment life (NMAX) = 5 years
Constant benefit (AVEB) = $450/year

Solution: Select CEAPiPC from the main menu. Select "Run" from the next menu. Select CEAP1.DAT from the list of data files and press *Enter*. A window will be displayed with data of the project. Normally all data should match Example 19-8, unless some values were previously modified. Check them carefully and make corrections, if necessary.

Data file:

```
     EXAMPLE #1 TO TEST PROGRAM CEAP3PC
(See detailed documentation for description of variables)
JOB  NMAX  IL  AVEN  SDN  KERL1  ISWCH  IFC  AVEI  SDI  KERL2
1     6    1   0.    0.    0       1     1   650.  0.    0

IT   TN    IB  BI     BN   BPC   ID   ND  ISV  AVESV  SVMAX  SVMIN
1    0.    2  0.12   6.   0.75   6    6    5    0.     0.    0.

IC  AVEC    SDC    KERL3  ITX  TAX    TXCD    EXT
2   280.    0.       0    3   0.34    0.0    0.10

IBS  AVEB    SDB    KERL4   ITB  TPC   ITER
 1   450.    0.     0       1    0.    1
```

```
ID=6, DEPRECIATION SCHEDULE: 10,20,20,20,15,15 %, DEP(I):
65.  130.  130.  130.  97.5  97.5  *

ISV=5, SALVAGE  VALUE IN FOLLOWING YEARS, CSV(I):
300.  200.  150.  100.  80.  50.  50.  50.  *
```

Press *Enter* after reviewing or editing all data windows.

The output will be displayed on the screen. Press *Esc* to get to the menu. Then the CEAP.OUT output file can be printed.

Output file:

```
          CAPITAL EXPENDITURE ANALYSIS PROGRAM
      FROM NEWNAN - ENGINEERING ECONOMIC ANALYSIS

EXAMPLE #1 TO TEST PROGRAM CEAP3PC
PROJECT NMBR:     1      INPUT FILE: CEAP1.DAT   OUTPUT FILE: CEAP.OUT

INVESTMENT
     INVESTMENT FIXED AT       650.

INVESTMENT TIMING
     TOTAL INVESTMENT MADE AT BEGINNING OF YEAR 1 (YEAR 0)

ANALYSIS PERIOD
     MAX USEFUL LIFE =   6
     LIFE FIXED AT   6 YEARS

DEPRECIATION METHOD
     DEPRECIABLE LIFE =    6 YEARS
     DEPRECIATION SCHEDULE INPUT
               YEAR  DEPREC FOR YEAR
                 1          65.
                 2         130.
                 3         130.
                 4         130.
                 5          98.
                 6          98.

END-OF-LIFE SALVAGE VALUE
               YEAR   SALVAGE VALUE
                 1          300.
                 2          200.
                 3          150.
                 4          100.
                 5           80.
                 6           50.

BORROWED MONEY
     UNIFORM REPAYMENT SCHEDULE
     TERM OF LOAN =    6. YEARS
     PROPORTION OF YEAR 0 INVESTMENT BORROWED =    0.750
     INTEREST RATE PER ANNUM = 12.00 %

EXPENSED COSTS
     EXPENSED COSTS FIXED =     280.

INCOME TAX
     UNIFORM TAX RATE AND ANY INVESTMENT TAX CREDIT
     TAX CREDIT =    0.0 %
     TAX RATE    =  34.0 %

BENEFIT SCHEDULE
     UNIFORM BENEFIT =     450.
```

```
BENEFIT TREND
     NO TREND

EXTERNAL INVESTMENT
     INTEREST RATE EARNED ON ANY EXTERNAL INVESTMENT = 10.0 %
```

NET CASH FLOW CALCULATION TABLE

YR	CUMULATIVE INVESTMENT	ANNUAL INVESTMENT	DEPREC SCHEDULE	BORROWED MONEY	LOAN PAYMT PRINCIPAL	LOAN INTEREST
0	650.	650.		488.		
1	650.	0.	65.		60.	58.
2	650.	0.	130.		67.	51.
3	650.	0.	130.		75.	43.
4	650.	0.	130.		84.	34.
5	650.	0.	98.		95.	24.
6	650.	0.	98.		106.	13.

NET CASH FLOW CALCULATION TABLE - CONTINUED

YR	-INVESTMENT (NET)	- LOAN PAYMT PRIN + INT	+ PROJECT BENEFITS	- EXPENSED COSTS	- INCOME TAX	= NET CASH FLOW
0	162.					-162.
1	0.	119.	450.	280.	16.	36.
2	0.	119.	450.	280.	-4.	55.
3	0.	119.	450.	280.	-1.	53.
4	0.	119.	450.	280.	2.	49.
5	0.	119.	450.	280.	16.	35.
6	0.	119.	450.	280.	20.	81.
6	SALV. VAL.	50.				

```
TRIAL NUMBER  1 : INTERNAL RATE OF RETURN =  25.02 %
                   EXTERNAL INTEREST RATE =  10.00 %
```

YEAR	CASH FLOW	COMPUTATION			INTERNAL INVESTM.	EXTERNAL INVESTM.
0	-162.50				-162.50	0.00
1	35.62	-162.50*(1+0.250)+	35.62 =	-167.53	35.62	0.00
2	55.27	-167.53*(1+0.250)+	55.27 =	-154.18	55.27	0.00
3	52.52	-154.18*(1+0.250)+	52.52 =	-140.22	52.52	0.00
4	49.45	-140.22*(1+0.250)+	49.45 =	-125.85	49.45	0.00
5	34.95	-125.85*(1+0.250)+	34.95 =	-122.38	34.95	0.00
6	81.10	-122.38*(1+0.250)+	81.10 =	-71.90	81.10	0.00

```
TRIAL NUMBER  9 : INTERNAL RATE OF RETURN =  20.23 %
                   EXTERNAL INTEREST RATE =  10.00 %
```

YEAR	CASH FLOW	COMPUTATION			INTERNAL INVESTM.	EXTERNAL INVESTM.
0	-162.50				-162.50	0.00
1	35.62	-162.50*(1+0.202)+	35.62 =	-159.75	35.62	0.00
2	55.27	-159.75*(1+0.202)+	55.27 =	-136.80	55.27	0.00
3	52.52	-136.80*(1+0.202)+	52.52 =	-111.95	52.52	0.00
4	49.45	-111.95*(1+0.202)+	49.45 =	-85.15	49.45	0.00
5	34.95	-85.15*(1+0.202)+	34.95 =	-67.42	34.95	0.00
6	81.10	-67.42*(1+0.202)+	81.10 =	0.04	81.05	0.04

```
AFTER  9 TRIALS, THE TRUE INTERNAL RATE OF RETURN WAS FOUND
TO BE VERY CLOSE TO  20.228 %. THIS VALUE IS CONSIDERED THE ANSWER.
```

EXAMPLE 19-11

This problem uses Example 19-8 data, but with the following adjustments:

- Randomized equipment life, capital cost, and operating cost.
- Maximum equipment life (NMAX) = 8 years
- Standard deviation from average equipment life (SDN) = 0.5 year
- Equipment capital cost normally distributed with standard deviation (SDI) = $25
- Operating cost normally distributed with standard deviation (SDC) = $14 per year
- Constant benefit (AVEB) = $450/year
- Number of iterations (ITER) = 100

Data file:

```
      EXAMPLE #2 TO TEST PROGRAM CEAP3PC
 (See detailed documentation for description of variables)
 JOB  NMAX   IL  AVEN    SDN  KERL1   ISWCH   IFC      AVEI      SDI  KERL2
 4     8      2  6.00   0.50    0       0      2    650.00   25.00      0

 IT    TN    IB     BI    BN     BPC   ID   ND   ISV  AVESV  SVMAX  SVMIN
 1     0.     2   0.12  6.00   0.75    6    6    5    0.00    0.00   0.00

 IC        AVEC    SDC   KERL3  ITX    TAX    TXCD     EXT
 3       280.00  14.00     0     3    0.34    0.00    0.10

 IBS       AVEB    SDB   KERL4  ITB    TPC    ITER
  1      450.00   0.00     0     1    0.00    100

 ID=6, DEPRECIATION SCHEDULE: 10,20,20,20,15,15 %, DEP(I):
   65.00  130.00  130.00  130.00   97.50   97.50  *

 ISV=5, SALVAGE  VALUE IN FOLLOWING YEARS, CSV(I):
  300.00  200.00  150.00  100.00   80.00   50.00   50.00   50.00  *
```

Output file:

```
           CAPITAL EXPENDITURE ANALYSIS PROGRAM
      FROM NEWNAN - ENGINEERING ECONOMIC ANALYSIS

 EXAMPLE #2 TO TEST PROGRAM CEAP3PC
 PROJECT NMBR:    4      INPUT FILE: CEAP2.DAT    OUTPUT FILE: CEAP.OUT

 INVESTMENT
      NORMAL DISTRIBUTION WITH MEAN =       650.  HAND STD DEV =      25.

 INVESTMENT TIMING
      TOTAL INVESTMENT MADE AT BEGINNING OF YEAR 1 (YEAR 0)

 ANALYSIS PERIOD
      MAX USEFUL LIFE =   8
      NORMAL DISTRIBUTION WITH MEAN =  6.00  AND STD DEV =  0.50

 DEPRECIATION METHOD
      DEPRECIABLE LIFE =   6 YEARS
      DEPRECIATION SCHEDULE INPUT
                YEAR  DEPREC FOR YEAR
                  1         65.
                  2        130.
                  3        130.
                  4        130.
                  5         98.
                  6         98.
```

END-OF-LIFE SALVAGE VALUE

YEAR	SALVAGE VALUE
1	300.
2	200.
3	150.
4	100.
5	80.
6	50.
7	50.
8	50.

BORROWED MONEY
UNIFORM REPAYMENT SCHEDULE
TERM OF LOAN = 6. YEARS
PROPORTION OF YEAR 0 INVESTMENT BORROWED = 0.750
INTEREST RATE PER ANNUM = 12.00 %

EXPENSED COSTS
NORMAL DISTRIBUTION WITH MEAN = 280. AND STD DEV = 14.

INCOME TAX
UNIFORM TAX RATE AND ANY INVESTMENT TAX CREDIT
TAX CREDIT = 0.0 %
TAX RATE = 34.0 %

BENEFIT SCHEDULE
UNIFORM BENEFIT = 450.

BENEFIT TREND
NO TREND

EXTERNAL INVESTMENT
INTEREST RATE EARNED ON ANY EXTERNAL INVESTMENT = 10.0 %

MODEL ITERATIONS = 100

YR	ITER NO	CAPITALIZED INVESTMENT	TOTAL .BENEFITS	TOTAL EXP COST	TOTAL TAXES	SUM NET CASH FLOW	RATE OF RETURN
5	1	669.	2250.	1430.	-8.	45.	0.061
5	2	662.	2250.	1412.	-1.	65.	0.086
5	3	689.	2250.	1394.	2.	45.	0.068
7	5	681.	3150.	1900.	137.	284.	0.212
6	6	674.	2700.	1609.	83.	188.	0.193
7	7	686.	3150.	2041.	88.	185.	0.144
6	8	671.	2700.	1736.	41.	107.	0.112
7	9	629.	3150.	1918.	136.	334.	0.267
6	10	601.	2700.	1678.	67.	229.	0.234
6	11	662.	2700.	1627.	78.	190.	0.197
6	12	615.	2700.	1638.	79.	239.	0.263
6	13	689.	2700.	1676.	59.	126.	0.119
6	14	621.	2700.	1718.	52.	179.	0.186
6	15	694.	2700.	1691.	54.	111.	0.109
5	16	648.	2250.	1456.	-15.	53.	0.073
6	17	689.	2700.	1668.	62.	131.	0.132
6	18	695.	2700.	1679.	57.	117.	0.115
7	19	612.	3150.	1949.	127.	335.	0.260
5	20	694.	2250.	1329.	24.	82.	0.104
6	21	705.	2700.	1710.	46.	84.	0.078
6	23	627.	2700.	1560.	105.	277.	0.289
5	24	632.	2250.	1347.	24.	144.	0.200
6	25	682.	2700.	1647.	70.	154.	0.150
6	26	649.	2700.	1684.	60.	168.	0.155
6	27	687.	2700.	1668.	62.	134.	0.139
6	28	702.	2700.	1707.	47.	90.	0.083
6	29	633.	2700.	1644.	76.	214.	0.209
.	.	.	.	.	.	.	.
.	.	.	.	.	.	.	.
.	.	.	.	.	.	.	.

5	79	651.	2250.	1377.	12.	102.	0.141
5	80	680.	2250.	1403.	0.	50.	0.069
5	81	599.	2250.	1344.	28.	184.	0.260
5	82	684.	2250.	1395.	2.	50.	0.061
5	83	628.	2250.	1427.	-3.	96.	0.127
5	84	699.	2250.	1403.	-2.	27.	0.040
5	85	674.	2250.	1425.	-7.	43.	0.058
7	86	689.	3150.	1874.	145.	292.	0.221
7	87	701.	3150.	1853.	151.	292.	0.228
7	88	707.	3150.	2081.	73.	134.	0.099
7	89	604.	3150.	1996.	112.	312.	0.248
6	90	652.	2700.	1646.	73.	189.	0.183
5	91	683.	2250.	1408.	-2.	43.	0.056
7	92	698.	3150.	1885.	140.	274.	0.211
6	93	608.	2700.	1647.	77.	241.	0.257
7	94	674.	3150.	1969.	114.	247.	0.179
6	95	673.	2700.	1747.	37.	98.	0.095
7	96	655.	3150.	2008.	103.	245.	0.204
7	97	692.	3150.	1918.	130.	259.	0.191
6	98	708.	2700.	1710.	46.	81.	0.071
7	99	691.	3150.	1957.	116.	235.	0.177
5	100	646.	2250.	1410.	1.	85.	0.120

HISTOGRAM OF RATE OF RETURN VS. REL FREQUENCY

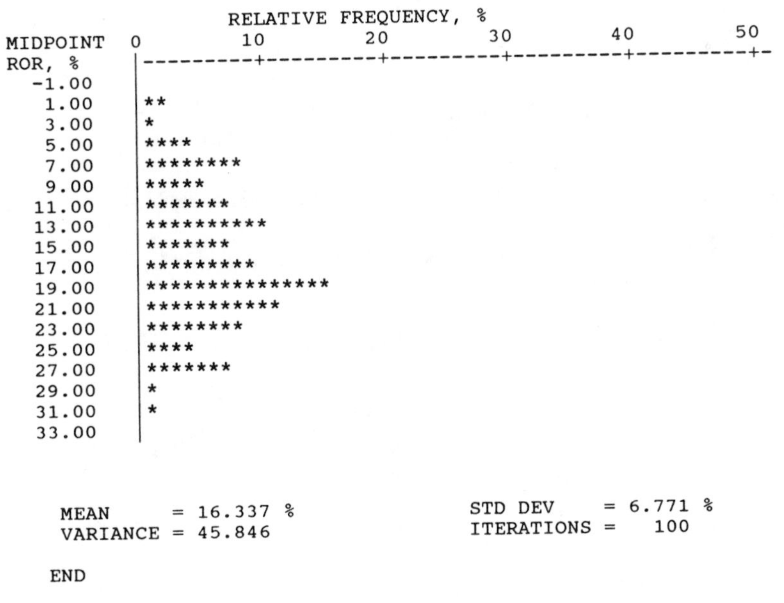

```
                        RELATIVE FREQUENCY, %
MIDPOINT  0          10         20         30         40         50
ROR, %    |---------+----------+----------+----------+----------+-
  -1.00   |
   1.00   |**
   3.00   |*
   5.00   |****
   7.00   |********
   9.00   |*****
  11.00   |*******
  13.00   |**********
  15.00   |*******
  17.00   |*********
  19.00   |***************
  21.00   |***********
  23.00   |********
  25.00   |****
  27.00   |*******
  29.00   |*
  31.00   |*
  33.00   |


     MEAN     = 16.337 %                STD DEV    = 6.771 %
     VARIANCE = 45.846                  ITERATIONS =   100

   END
```

EXAMPLE 19-12

Solve Example 14-12 using CEAPiPC and 200 iterations: If a more accurate scale is installed on a production line, it will reduce the error in computing postage charges and save $250 per year. The useful life of the scale is believed to be uniformly distributed and ranges from 12 to 16 years. The initial cost of the scale is estimated to be normally distributed with a mean of $1500 and a standard deviation of $150. Construct a graph of rate of return vs. frequency.

Solution: The problem is prepared by entering the parameters into a preformatted input data file. The completed file for this problem is:

Data file:

```
      EXAMPLE 14-12 Scales for Production Line
 (See detailed documentation for description of variables)
 JOB  NMAX   IL  AVEN     SDN  KERL1  ISWCH  IFC     AVEI     SDI  KERL2
 14    16     5  0.00    0.00     0       0    2  1500.00  150.00     0

 IT    TN    IB    BI      BN    BPC  ID   ND  ISV  AVESV  SVMAX  SVMIN
 1     0.00   1  0.00    1.00   0.00   7    0    1   0.00   0.00   0.00

 IC        AVEC    SDC   KERL3   ITX    TAX    TXCD     EXT
 1         0.00   0.00      0      1   0.00    0.00    0.06

 IBS       AVEB    SDB   KERL4   ITB    TPC    ITER
  1      250.00   0.00      0      1   0.00     200

 Cummulative life distribution LCD(i):
    12    12    12    12    12    13    13    13    13    13
    14    14    14    14    14    15    15    15    15    15
    16    16    16    16    16  *
```

Output file:

```
         CAPITAL EXPENDITURE ANALYSIS PROGRAM
      FROM NEWNAN - ENGINEERING ECONOMIC ANALYSIS

EXAMPLE 14-12 Scales for Production Line
PROJECT NMBR:    14       INPUT FILE: CEAP3.DAT    OUTPUT FILE: CEAP.OUT

INVESTMENT
     NORMAL DISTRIBUTION WITH MEAN =      1500.   HAND STD DEV =     150.

INVESTMENT TIMING
     TOTAL INVESTMENT MADE AT BEGINNING OF YEAR 1 (YEAR 0)
ANALYSIS PERIOD
     MAX USEFUL LIFE =  16
     CUMULATIVE DISTRIBUTION
               USEFUL LIFE CUMULATIVE PROBABILITY
                    12             0.04
                    12             0.08
                    12             0.12
                    12             0.16
                    12             0.20
                    13             0.24
                    13             0.28
                    13             0.32
                    13             0.36
                    13             0.40
                    14             0.44
                    14             0.48
                    14             0.52
                    14             0.56
                    14             0.60
                    15             0.64
                    15             0.68
                    15             0.72
                    15             0.76
                    15             0.80
                    16             0.84
                    16             0.88
                    16             0.92
                    16             0.96
                    16             1.00
```

```
DEPRECIATION METHOD
     DEPRECIABLE LIFE =   0 YEARS
     BEFORE TAX DEPRECIATION NOT COMPUTED

END-OF-LIFE SALVAGE VALUE
     NO SALVAGE VALUE

BORROWED MONEY
     NO BORROWED MONEY

EXPENSED COSTS
     NO EXPENSED ANNUAL COST

INCOME TAX
     INCOME TAXES IGNORED

BENEFIT SCHEDULE
     UNIFORM BENEFIT =     250.

BENEFIT TREND
     NO TREND

EXTERNAL INVESTMENT
     INTEREST RATE EARNED ON ANY EXTERNAL INVESTMENT =  6.0 %



HISTOGRAM OF RATE OF RETURN VS. REL FREQUENCY

                          RELATIVE FREQUENCY, %
  MIDPOINT  0         10        20        30        40        50
  ROR, %    |---------+---------+---------+---------+---------+-
     7.50   |
     8.50   |*
     9.50   |****
    10.50   |******
    11.50   |***************
    12.50   |*************
    13.50   |*********
    14.50   |*********
    15.50   |********
    16.50   |**********
    17.50   |********
    18.50   |**********
    19.50   |***
    20.50   |


      MEAN      = 14.240 %                 STD DEV     = 2.885 %
      VARIANCE =  8.326                    ITERATIONS =   200

     END
```

Problems

19-1 Solve Problem 6-12 using EAiPC.

19-2 Solve Problem 5-40 using EAiPC.

19-3 A 20-year home mortgage for $100,000 is to be repaid by 240 equal monthly payments. The nominal interest rate is 10% per year, compounded monthly.

a. What is the monthly payment?

b. How many months will it take to repay half the loan?

c. When will half or more of the monthly payment be applied to the loan principal?

19-4 Vincent van Gogh's painting "Irises" was sold at an art auction in 1987 for $53.9 million. The seller received $49 million, and Sotheby's, the art auctioneer, received a 10% commission, or $4.9 million. The painting was bought by the seller in 1947 for $84,000.

a. What rate of return did the seller receive from his investment in the painting?

b. If the new purchaser also holds the painting for 40 years, what must be the selling price in 2027 to produce the same rate of return? Assume a 10% sales commission at that time. (*Answers:* **a.** 17.26% ***b.*** $34.6 billion)

19-5 Solve Example 19-12 assuming the useful life of the scales is normally distributed with a mean of 15 years, and a standard deviation of 0.8 years.

19-6 Solve Example 19-8 assuming no annual cost inflation and a zero percent before-tax return on investment.

19-7 Solve Problem 9-6b. (*Answer:* 59 months)

19-8 Refer to Example 5-9. Compute the rate of return for the investment project, assuming if necessary a 10% interest rate on external investments.

19-9 Refer to Example 7A-2. Compute the rate of return on the Interair plane contract.

19-10 Solve Problem 11-5a using CEAPiPC.

19-11 Solve Problem 7A-5 using RORiPC. Why is the rate of return different from the answer given in Chapter 7A?

CHAPTER 20

Additional Homework Problems

1-12 (20-1) Company A has fixed expenses of \$15,000 per year and each unit of product has a \$0.002 variable cost. Company B has fixed expenses of \$5000 per year and can produce the same product at a \$0.05 variable cost. At what number of units of annual production will Company A have the same overall cost as Company B?

4-86 The United States recently purchased \$1 billion of 30-year zero-coupon bonds from a struggling foreign nation. The bonds yield 4½% per year interest. The zero-coupon bonds pay no interest during their 30-year life. Instead, at the end of 30 years, the U. S. Government is to receive back its \$1 billion together with interest at 4½% per year. A U. S. Senator objected to the purchase, claiming that the correct interest rate for bonds like this is 5¼%. The result, he said, was a multimillion dollar gift to the foreign country without the approval of the U. S. Congress. Assuming the Senator's 5¼% interest rate is correct, how much will the foreign country have saved in interest when they repay the bonds at the end of 30 years?

4-87 (20-9) In 1990 Mrs. John Hay Whitney sold her painting by Renoir, *Au Moulin de la Galette*, depicting an open-air Parisian dance hall, for \$71 million. The buyer also had to pay the auction house commission of 10%, or a total of \$78.1 million. Mrs. Whitney purchased the painting in 1929 for \$165,000.

a. What rate of return did she receive on her investment?

b. Was the rate of return really as high as you computed in ***a***? Explain.

4-88 (20-10) Derive an equation to find the end of year future sum F that is equivalent to a series of n *beginning-of-year* payments B at interest rate i. Then use the equation to determine the future sum F equivalent to six B payments of \$100 at 8% interest. (*Answer:* F = \$792.28)

4-89 (20-11) A woman made ten annual end-of-year purchases of $1000 worth of common stock. The stock paid no dividends. Then for four years she held the stock. At the end of the four years she sold all the stock for $28,000. What interest rate did she obtain on her investment?

4-90 (20-12) One thousand dollars is borrowed for one year at an interest rate of 1% per month compounded monthly. If this same sum of money could be borrowed for the same period at an interest rate of 12% per year compounded annually, how much could be saved in interest charges?

4-91 (20-13) For some interest rate i and some number of interest periods n, the uniform series capital recovery factor is 0.1728 and the sinking fund factor is 0.0378. What is the interest rate?

4-92 (20-14) What interest rate, compounded quarterly, is equivalent to a 9.31% effective interest rate?

4-93 (20-15) A contractor wishes to set up a special fund by making uniform semiannual end-of-period deposits for 20 years. The fund is to provide $10,000 at the end of each of the last five years of the 20-year period. If interest is 8%, compounded semiannually, what is the required semiannual deposit?

4-94 (20-16) How long will it take for $10,000, invested at 5% per year, compounded continuously, to triple in value?

4-95 (20-17) If $200 is deposited in a savings account at the beginning of each of 15 years, and the account draws interest at 7% per year, how much will be in the account at the end of 15 years?

4-96 (20-18) An automobile may be purchased with a $3000 downpayment now and 60 monthly payments of $280. If the interest rate is 12% compounded monthly, what is the price of the automobile?

4-97 (20-19) If the nominal annual interest rate is 12% compounded quarterly, what is the effective annual interest rate?

4-98 (20-20) A man is purchasing a small garden tractor. There will be no maintenance cost the first two years as the tractor is sold with two years free maintenance. The third year the maintenance is estimated at $20. In subsequent years the maintenance cost will increase by $20 per year (that is, fourth year maintenance will be $60; fifth year $80, and so on). How much would need to be set aside now at 8% interest to pay the maintenance costs on the tractor for the first six years of ownership?

4-99 (20-21) How many months, at an interest rate of one percent per month, does money have to be invested before it will double in value?

4-100 (20-22) Given a sum of money Q that will be received six years from now. At 5% annual interest the present worth now of Q is $60. At this same interest rate, what would be the value of Q ten years from now?

4-101 (20-23) A company deposits $2000 at the end of every year for ten years in a bank. The company makes no deposits during the subsequent five years. If the bank pays 8% interest, how much would be in the account at end of 15 years?

4-102 (20-24) A bank pays 10% nominal annual interest on special three-year certificates. What is the effective annual interest rate if interest is compounded:

a. every three months?

b. daily?

c. continuously?

5-44 (20-25) What amount of money deposited 50 years ago at 8% interest would provide a perpetual payment of $10,000 a year beginning this year?

5-45 (20-26) Annual maintenance costs for a particular section of highway pavement are $2000. The placement of a new surface would reduce the annual maintenance cost to $500 per year for the first five years and to $1000 per year for the next five years. After ten years the annual maintenance would again be $2000. If maintenance costs are the only saving, what investment can be justified for the new surface. Assume interest at 4%.

5-46 (20-27) A small dam was constructed for $2 million. The annual maintenance cost is $15,000. If interest is 5%, compute the capitalized cost of the dam, including maintenance.

5-47 (20-28) Twenty-five thousand dollars is deposited in a savings account that pays 5% interest, compounded semi-annually. Equal annual withdrawals are to be made from the account, beginning one year from now and continuing forever. What is the maximum equal annual withdrawal?

5-48 (20-29) Two alternative courses of action have the following schedules of disbursements:

Year	*A*	*B*
0	−$1300	
1	0	−$100
2	0	−200
3	0	−300
4	0	−400
5	0	−500
	−$1300	−$1500

Based on a 6% interest rate, which alternative should be selected?

5-49 (20-30) An investor is considering buying a 20-year corporate bond. The bond has a face value of $1000 and pays 6% interest per year in two semi-annual payments. Thus the purchaser of the bond will receive $30 every six months and, in addition, he/she will receive $1000 at the end of 20 years, along with the last $30 interest payment. If the investor thinks he/she should receive 8% interest, compounded semi-annually, how much would the investor be willing to pay for the bond?

5-50 (20-31) A trust fund is to be established for three purposes: (1) provide \$750,000 for the construction and \$250,000 for the initial equipment of a small engineering laboratory; (2) pay the \$150,000 per year laboratory operating cost; and (3) pay for \$100,000 of replacement equipment every four years, beginning four years from now.

At 6% interest, how much money is required in the trust fund to provide for the laboratory and equipment and its perpetual operation and equipment replacement?

5-51 (20-32) A city has developed a plan which will provide for future municipal water needs. The plan proposes an aqueduct which passes through 500 feet of tunnel in a nearby mountain. Two alternatives are being considered. The first proposes to build a full-capacity tunnel now for \$556,000. The second proposes to build a half-capacity tunnel now (cost = \$402,000) which should be adequate for 20 years, and then to build a second parallel half-capacity tunnel. The maintenance cost of the tunnel lining for the full-capacity tunnel is \$40,000 every 10 years, and for each half-capacity tunnel it is \$32,000 every 10 years.

The friction losses in the half-capacity tunnel will be greater than if the full-capacity tunnel were built. The estimated additional pumping costs in the single half-capacity tunnel will be \$2000 per year, and for the two half-capacity tunnels it will be \$4000 per year. Based on capitalized cost and a 7% interest rate, which alternative should be selected?

5-52 (20-33) A road building contractor has received a major highway construction contract that will require 50,000 m^3 of crushed stone each year for five years. The needed stone can be obtained from a quarry for \$5.80/m^3. As an alternative the contractor has decided to try and purchase the quarry. He believes if he owned the quarry the stone would only cost him \$4.30/m^3. He thinks he could resell the quarry at the end of five years for \$40,000. If the contractor uses a 10% interest rate, how much would he be willing to pay for the quarry?

5-53 (20-34) A new office building was constructed five years ago by a consulting engineering firm. At that time the firm obtained a bank loan for \$100,000 with a 12% annual interest rate, compounded quarterly. The terms of the loan call for equal quarterly payments to repay the loan in 10 years. The loan also allows for its prepayment at any time without penalty.

Due to internal changes in the firm, it is now proposed to refinance the loan through an insurance company. The new loan would be for a 20-year term with an interest rate of 8% per year, compounded quarterly. The new equal quarterly payments would repay the loan in the 20-year period. The insurance company requires the payment of a 5% loan initiation charge (often described as a "five-point loan fee") which will be added to the new loan.

a. What is the balance due on the original mortgage if 20 payments have been made in the last five years?

b. What is the difference between the equal quarterly payments on the present bank loan and the proposed insurance company loan?

6-42 (20-36) The town of Dry Gulch needs an additional supply of water from Pine Creek. The town engineer has selected two plans for comparison. *Gravity plan:* Divert water at a point ten miles up Pine Creek and carry it through a pipeline by gravity to the town. *Pumping plan*: Divert water at a point closer to the town and pump it to the town. The pumping plant would be built in two stages, with one-half capacity installed initially and the other half installed ten years later.

An analysis will assume a 40-year life, 10% interest and no salvage value. Costs are as follows:

	Gravity	*Pumping*
Initial investment	$2,800,000	$1,400,000
Additional investment in 10th year	None	200,000
Operation and maintenance	10,000/year	25,000/year
Power cost		
Average first 10 years	None	50,000/year
Average next 30 years	None	100,000/year

Determine the more economical plan.

7-47 (20-39) An investment of $5000 in Biotech common stock proved to be very profitable. At the end of three years the stock was sold for $25,000. What was the rate of return on the investment?

7-48 (20-40) A mine is for sale for $240,000. It is believed the mine will produce a profit of $65,000 the first year, but the profit will decline $5000 a year after that until it becomes zero and the mine is worthless. What rate of return would this produce for the purchaser of the mine?

7-49 (20-41) Two mutually exclusive alternatives are being considered.

Year	*A*	*B*
0	−$2500	−$6000
1	+746	+1664
2	+746	+1664
3	+746	+1664
4	+746	+1664
5	+746	+1664

If the minimum attractive rate of return is 8%, which alternative should be selected? Solve the problem by

a. present worth analysis.

b. annual cash flow analysis.

c. rate of return analysis.

11-44 Refer to Problem 11-28. To help pay for the pickup truck the Ogi Corp. obtained a $10,000 loan from the truck dealer at 10% interest, payable in four end-of-year payments of $2500 plus interest.

a. Compute the after-tax rate of return for the truck together with the loan. Note that the interest on the loan in tax deductible, but the $2500 principal payments are not.

b. Why is the after-tax rate of return computed in part ***a*** so much different from the 12.5% obtained in Problem 11-28?

11-45 (20-57) A firm has invested $14,000 in machinery with a seven-year useful life. The machinery will have no salvage value, as the cost to remove it will equal its scrap value. The uniform annual benefits from the machinery are $3600. For a 47% income tax rate, and sum-of-years digits depreciation, compute the after-tax rate of return.

Appendices

∞

$A = F(A/F, i, n)$

$P = C + \frac{A}{i}$

$i_{eff} = \left(1 + \frac{r}{m}\right)^m - 1$

APPENDIX **A**

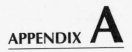

Compound Interest Tables

Discrete End-Of-Period Compounding

Tables for:

¼%	½%	¾%	1%	1¼%	1½%
1¾%	2%	2½%	3%	3½%	4%
4½%	5%	6%	7%	8%	9%
10%	12%	15%	18%	20%	25%
30%	35%	40%	45%	50%	60%

Values Of Interest Factors When N Equals Infinity

Single Payment:

$(F/P,i,\infty) = \infty$

$(P/F,i,\infty) = 0$

Arithmetic Gradient Series:

$(A/G,i,\infty) = 1/i$

$(P/G,i,\infty) = 1/i^2$

Uniform Payment Series:

$(A/F,i,\infty) = 0$

$(A/P,i,\infty) = i$

$(F/A,i,\infty) = \infty$

$(P/A,i,\infty) = 1/i$

	Single Payment		Uniform Payment Series				Arithmetic Gradient		
	Compound Amount Factor	Present Worth Factor	Sinking Fund Factor	Capital Recovery Factor	Compound Amount Factor	Present Worth Factor	Gradient Uniform Series	Gradient Present Worth	
n	Find F Given P F/P	Find P Given F P/F	Find A Given F A/F	Find A Given P A/P	Find F Given A F/A	Find P Given A P/A	Find A Given G A/G	Find P Given G P/G	n
1	1.003	.9975	1.0000	1.0025	1.000	0.998	0	0	**1**
2	1.005	.9950	.4994	.5019	2.003	1.993	0.504	1.005	**2**
3	1.008	.9925	.3325	.3350	3.008	2.985	1.005	2.999	**3**
4	1.010	.9901	.2491	.2516	4.015	3.975	1.501	5.966	**4**
5	1.013	.9876	.1990	.2015	5.025	4.963	1.998	9.916	**5**
6	1.015	.9851	.1656	.1681	6.038	5.948	2.498	14.861	**6**
7	1.018	.9827	.1418	.1443	7.053	6.931	2.995	20.755	**7**
8	1.020	.9802	.1239	.1264	8.070	7.911	3.490	27.611	**8**
9	1.023	.9778	.1100	.1125	9.091	8.889	3.987	35.440	**9**
10	1.025	.9753	.0989	.1014	10.113	9.864	4.483	44.216	**10**
11	1.028	.9729	.0898	.0923	11.139	10.837	4.978	53.950	**11**
12	1.030	.9705	.0822	.0847	12.167	11.807	5.474	64.634	**12**
13	1.033	.9681	.0758	.0783	13.197	12.775	5.968	76.244	**13**
14	1.036	.9656	.0703	.0728	14.230	13.741	6.464	88.826	**14**
15	1.038	.9632	.0655	.0680	15.266	14.704	6.957	102.301	**15**
16	1.041	.9608	.0613	.0638	16.304	15.665	7.451	116.716	**16**
17	1.043	.9584	.0577	.0602	17.344	16.624	7.944	132.063	**17**
18	1.046	.9561	.0544	.0569	18.388	17.580	8.437	148.319	**18**
19	1.049	.9537	.0515	.0540	19.434	18.533	8.929	165.492	**19**
20	1.051	.9513	.0488	.0513	20.482	19.485	9.421	183.559	**20**
21	1.054	.9489	.0464	.0489	21.534	20.434	9.912	202.531	**21**
22	1.056	.9465	.0443	.0468	22.587	21.380	10.404	222.435	**22**
23	1.059	.9442	.0423	.0448	23.644	22.324	10.894	243.212	**23**
24	1.062	.9418	.0405	.0430	24.703	23.266	11.384	264.854	**24**
25	1.064	.9395	.0388	.0413	25.765	24.206	11.874	287.407	**25**
26	1.067	.9371	.0373	.0398	26.829	25.143	12.363	310.848	**26**
27	1.070	.9348	.0358	.0383	27.896	26.078	12.852	335.150	**27**
28	1.072	.9325	.0345	.0370	28.966	27.010	13.341	360.343	**28**
29	1.075	.9301	.0333	.0358	30.038	27.940	13.828	386.366	**29**
30	1.078	.9278	.0321	.0346	31.114	28.868	14.317	413.302	**30**
36	1.094	.9140	.0266	.0291	37.621	34.387	17.234	592.632	**36**
40	1.105	.9049	.0238	.0263	42.014	38.020	19.171	728.882	**40**
48	1.127	.8871	.0196	.0221	50.932	45.179	23.025	1 040.22	**48**
50	1.133	.8826	.0188	.0213	53.189	46.947	23.984	1 125.96	**50**
52	1.139	.8782	.0180	.0205	55.458	48.705	24.941	1 214.76	**52**
60	1.162	.8609	.0155	.0180	64.647	55.653	28.755	1 600.31	**60**
70	1.191	.8396	.0131	.0156	76.395	64.144	33.485	2 147.87	**70**
72	1.197	.8355	.0127	.0152	78.780	65.817	34.426	2 265.81	**72**
80	1.221	.8189	.0113	.0138	88.440	72.427	38.173	2 764.74	**80**
84	1.233	.8108	.0107	.0132	93.343	75.682	40.037	3 030.06	**84**
90	1.252	.7987	.00992	.0124	100.789	80.504	42.820	3 447.19	**90**
96	1.271	.7869	.00923	.0117	108.349	85.255	45.588	3 886.62	**96**
100	1.284	.7790	.00881	.0113	113.451	88.383	47.425	4 191.60	**100**
104	1.297	.7713	.00843	.0109	118.605	91.480	49.256	4 505.93	**104**
120	1.349	.7411	.00716	.00966	139.743	103.563	56.512	5 852.52	**120**
240	1.821	.5492	.00305	.00555	328.306	180.312	107.590	19 399.75	**240**
360	2.457	.4070	.00172	.00422	582.745	237.191	152.894	36 264.96	**360**
480	3.315	.3016	.00108	.00358	926.074	279.343	192.673	53 821.93	**480**

Compound Interest Factors

	Single Payment		Uniform Payment Series				Arithmetic Gradient		
	Compound Amount Factor	Present Worth Factor	Sinking Fund Factor	Capital Recovery Factor	Compound Amount Factor	Present Worth Factor	Gradient Uniform Series	Gradient Present Worth	
n	Find *F* Given *P* *F/P*	Find *P* Given *F* *P/F*	Find *A* Given *F* *A/F*	Find *A* Given *P* *A/P*	Find *F* Given *A* *F/A*	Find *P* Given *A* *P/A*	Find *A* Given *G* *A/G*	Find *P* Given *G* *P/G*	*n*
1	1.005	.9950	1.0000	1.0050	1.000	0.995	0	0	**1**
2	1.010	.9901	.4988	.5038	2.005	1.985	0.499	0.991	**2**
3	1.015	.9851	.3317	.3367	3.015	2.970	0.996	2.959	**3**
4	1.020	.9802	.2481	.2531	4.030	3.951	1.494	5.903	**4**
5	1.025	.9754	.1980	.2030	5.050	4.926	1.990	9.803	**5**
6	1.030	.9705	.1646	.1696	6.076	5.896	2.486	14.660	**6**
7	1.036	.9657	.1407	.1457	7.106	6.862	2.980	20.448	**7**
8	1.041	.9609	.1228	.1278	8.141	7.823	3.474	27.178	**8**
9	1.046	.9561	.1089	.1139	9.182	8.779	3.967	34.825	**9**
10	1.051	.9513	.0978	.1028	10.228	9.730	4.459	43.389	**10**
11	1.056	.9466	.0887	.0937	11.279	10.677	4.950	52.855	**11**
12	1.062	.9419	.0811	.0861	12.336	11.619	5.441	63.218	**12**
13	1.067	.9372	.0746	.0796	13.397	12.556	5.931	74.465	**13**
14	1.072	.9326	.0691	.0741	14.464	13.489	6.419	86.590	**14**
15	1.078	.9279	.0644	.0694	15.537	14.417	6.907	99.574	**15**
16	1.083	.9233	.0602	.0652	16.614	15.340	7.394	113.427	**16**
17	1.088	.9187	.0565	.0615	17.697	16.259	7.880	128.125	**17**
18	1.094	.9141	.0532	.0582	18.786	17.173	8.366	143.668	**18**
19	1.099	.9096	.0503	.0553	19.880	18.082	8.850	160.037	**19**
20	1.105	.9051	.0477	.0527	20.979	18.987	9.334	177.237	**20**
21	1.110	.9006	.0453	.0503	22.084	19.888	9.817	195.245	**21**
22	1.116	.8961	.0431	.0481	23.194	20.784	10.300	214.070	**22**
23	1.122	.8916	.0411	.0461	24.310	21.676	10.781	233.680	**23**
24	1.127	.8872	.0393	.0443	25.432	22.563	11.261	254.088	**24**
25	1.133	.8828	.0377	.0427	26.559	23.446	11.741	275.273	**25**
26	1.138	.8784	.0361	.0411	27.692	24.324	12.220	297.233	**26**
27	1.144	.8740	.0347	.0397	28.830	25.198	12.698	319.955	**27**
28	1.150	.8697	.0334	.0384	29.975	26.068	13.175	343.439	**28**
29	1.156	.8653	.0321	.0371	31.124	26.933	13.651	367.672	**29**
30	1.161	.8610	.0310	.0360	32.280	27.794	14.127	392.640	**30**
36	1.197	.8356	.0254	.0304	39.336	32.871	16.962	557.564	**36**
40	1.221	.8191	.0226	.0276	44.159	36.172	18.836	681.341	**40**
48	1.270	.7871	.0185	.0235	54.098	42.580	22.544	959.928	**48**
50	1.283	.7793	.0177	.0227	56.645	44.143	23.463	1 035.70	**50**
52	1.296	.7716	.0169	.0219	59.218	45.690	24.378	1 113.82	**52**
60	1.349	.7414	.0143	.0193	69.770	51.726	28.007	1 448.65	**60**
70	1.418	.7053	.0120	.0170	83.566	58.939	32.468	1 913.65	**70**
72	1.432	.6983	.0116	.0166	86.409	60.340	33.351	2 012.35	**72**
80	1.490	.6710	.0102	.0152	98.068	65.802	36.848	2 424.65	**80**
84	1.520	.6577	.00961	.0146	104.074	68.453	38.576	2 640.67	**84**
90	1.567	.6383	.00883	.0138	113.311	72.331	41.145	2 976.08	**90**
96	1.614	.6195	.00814	.0131	122.829	76.095	43.685	3 324.19	**96**
100	1.647	.6073	.00773	.0127	129.334	78.543	45.361	3 562.80	**100**
104	1.680	.5953	.00735	.0124	135.970	80.942	47.025	3 806.29	**104**
120	1.819	.5496	.00610	.0111	163.880	90.074	53.551	4 823.52	**120**
240	3.310	.3021	.00216	.00716	462.041	139.581	96.113	13 415.56	**240**
360	6.023	.1660	.00100	.00600	1 004.5	166.792	128.324	21 403.32	**360**
480	10.957	.0913	.00050	.00550	1 991.5	181.748	151.795	27 588.37	**480**

Compound Interest Factors

3/4%

	Single Payment		Uniform Payment Series				Arithmetic Gradient		
	Compound Amount Factor	Present Worth Factor	Sinking Fund Factor	Capital Recovery Factor	Compound Amount Factor	Present Worth Factor	Gradient Uniform Series	Gradient Present Worth	
n	Find *F* Given *P* F/P	Find *P* Given *F* P/F	Find *A* Given *F* A/F	Find *A* Given *P* A/P	Find *F* Given *A* F/A	Find *P* Given *A* P/A	Find *A* Given *G* A/G	Find *P* Given *G* P/G	**n**
1	1.008	.9926	1.0000	1.0075	1.000	0.993	0	0	**1**
2	1.015	.9852	.4981	.5056	2.008	1.978	0.499	0.987	**2**
3	1.023	.9778	.3308	.3383	3.023	2.956	0.996	2.943	**3**
4	1.030	.9706	.2472	.2547	4.045	3.926	1.492	5.857	**4**
5	1.038	.9633	.1970	.2045	5.076	4.889	1.986	9.712	**5**
6	1.046	.9562	.1636	.1711	6.114	5.846	2.479	14.494	**6**
7	1.054	.9490	.1397	.1472	7.160	6.795	2.971	20.187	**7**
8	1.062	.9420	.1218	.1293	8.213	7.737	3.462	26.785	**8**
9	1.070	.9350	.1078	.1153	9.275	8.672	3.951	34.265	**9**
10	1.078	.9280	.0967	.1042	10.344	9.600	4.440	42.619	**10**
11	1.086	.9211	.0876	.0951	11.422	10.521	4.927	51.831	**11**
12	1.094	.9142	.0800	.0875	12.508	11.435	5.412	61.889	**12**
13	1.102	.9074	.0735	.0810	13.602	12.342	5.897	72.779	**13**
14	1.110	.9007	.0680	.0755	14.704	13.243	6.380	84.491	**14**
15	1.119	.8940	.0632	.0707	15.814	14.137	6.862	97.005	**15**
16	1.127	.8873	.0591	.0666	16.932	15.024	7.343	110.318	**16**
17	1.135	.8807	.0554	.0629	18.059	15.905	7.822	124.410	**17**
18	1.144	.8742	.0521	.0596	19.195	16.779	8.300	139.273	**18**
19	1.153	.8676	.0492	.0567	20.339	17.647	8.777	154.891	**19**
20	1.161	.8612	.0465	.0540	21.491	18.508	9.253	171.254	**20**
21	1.170	.8548	.0441	.0516	22.653	19.363	9.727	188.352	**21**
22	1.179	.8484	.0420	.0495	23.823	20.211	10.201	206.170	**22**
23	1.188	.8421	.0400	.0475	25.001	21.053	10.673	224.695	**23**
24	1.196	.8358	.0382	.0457	26.189	21.889	11.143	243.924	**24**
25	1.205	.8296	.0365	.0440	27.385	22.719	11.613	263.834	**25**
26	1.214	.8234	.0350	.0425	28.591	23.542	12.081	284.421	**26**
27	1.224	.8173	.0336	.0411	29.805	24.360	12.548	305.672	**27**
28	1.233	.8112	.0322	.0397	31.029	25.171	13.014	327.576	**28**
29	1.242	.8052	.0310	.0385	32.261	25.976	13.479	350.122	**29**
30	1.251	.7992	.0298	.0373	33.503	26.775	13.942	373.302	**30**
36	1.309	.7641	.0243	.0318	41.153	31.447	16.696	525.038	**36**
40	1.348	.7416	.0215	.0290	46.447	34.447	18.507	637.519	**40**
48	1.431	.6986	.0174	.0249	57.521	40.185	22.070	886.899	**48**
50	1.453	.6882	.0166	.0241	60.395	41.567	22.949	953.911	**50**
52	1.475	.6780	.0158	.0233	63.312	42.928	23.822	1 022.64	**52**
60	1.566	.6387	.0133	.0208	75.425	48.174	27.268	1 313.59	**60**
70	1.687	.5927	.0109	.0184	91.621	54.305	31.465	1 708.68	**70**
72	1.713	.5839	.0105	.0180	95.008	55.477	32.289	1 791.33	**72**
80	1.818	.5500	.00917	.0167	109.074	59.995	35.540	2 132.23	**80**
84	1.873	.5338	.00859	.0161	116.428	62.154	37.137	2 308.22	**84**
90	1.959	.5104	.00782	.0153	127.881	65.275	39.496	2 578.09	**90**
96	2.049	.4881	.00715	.0147	139.858	68.259	41.812	2 854.04	**96**
100	2.111	.4737	.00675	.0143	148.147	70.175	43.332	3 040.85	**100**
104	2.175	.4597	.00638	.0139	156.687	72.035	44.834	3 229.60	**104**
120	2.451	.4079	.00517	.0127	193.517	78.942	50.653	3 998.68	**120**
240	6.009	.1664	.00150	.00900	667.901	111.145	85.422	9 494.26	**240**
360	14.731	.0679	.00055	.00805	1 830.8	124.282	107.115	13 312.50	**360**
480	36.111	.0277	.00021	.00771	4 681.5	129.641	119 662	15 513.16	**480**

	Single Payment		Uniform Payment Series				Arithmetic Gradient		
	Compound Amount Factor	Present Worth Factor	Sinking Fund Factor	Capital Recovery Factor	Compound Amount Factor	Present Worth Factor	Gradient Uniform Series	Gradient Present Worth	
n	Find *F* Given *P* *F/P*	Find *P* Given *F* *P/F*	Find *A* Given *F* *A/F*	Find *A* Given *P* *A/P*	Find *F* Given *A* *F/A*	Find *P* Given *A* *P/A*	Find *A* Given *G* *A/G*	Find *P* Given *G* *P/G*	*n*
1	1.010	.9901	1.0000	1.0100	1.000	0.990	0	0	**1**
2	1.020	.9803	.4975	.5075	2.010	1.970	0.498	0.980	**2**
3	1.030	.9706	.3300	.3400	3.030	2.941	0.993	2.921	**3**
4	1.041	.9610	.2463	.2563	4.060	3.902	1.488	5.804	**4**
5	1.051	.9515	.1960	.2060	5.101	4.853	1.980	9.610	**5**
6	1.062	.9420	.1625	.1725	6.152	5.795	2.471	14.320	**6**
7	1.072	.9327	.1386	.1486	7.214	6.728	2.960	19.917	**7**
8	1.083	.9235	.1207	.1307	8.286	7.652	3.448	26.381	**8**
9	1.094	.9143	.1067	.1167	9.369	8.566	3.934	33.695	**9**
10	1.105	.9053	.0956	.1056	10.462	9.471	4.418	41.843	**10**
11	1.116	.8963	.0865	.0965	11.567	10.368	4.900	50.806	**11**
12	1.127	.8874	.0788	.0888	12.682	11.255	5.381	60.568	**12**
13	1.138	.8787	.0724	.0824	13.809	12.134	5.861	71.112	**13**
14	1.149	.8700	.0669	.0769	14.947	13.004	6.338	82.422	**14**
15	1.161	.8613	.0621	.0721	16.097	13.865	6.814	94.481	**15**
16	1.173	.8528	.0579	.0679	17.258	14.718	7.289	107.273	**16**
17	1.184	.8444	.0543	.0643	18.430	15.562	7.761	120.783	**17**
18	1.196	.8360	.0510	.0610	19.615	16.398	8.232	134.995	**18**
19	1.208	.8277	.0481	.0581	20.811	17.226	8.702	149.895	**19**
20	1.220	.8195	.0454	.0554	22.019	18.046	9.169	165.465	**20**
21	1.232	.8114	.0430	.0530	23.239	18.857	9.635	181.694	**21**
22	1.245	.8034	.0409	.0509	24.472	19.660	10.100	198.565	**22**
23	1.257	.7954	.0389	.0489	25.716	20.456	10.563	216.065	**23**
24	1.270	.7876	.0371	.0471	26.973	21.243	11.024	234.179	**24**
25	1.282	.7798	.0354	.0454	28.243	22.023	11.483	252.892	**25**
26	1.295	.7720	.0339	.0439	29.526	22.795	11.941	272.195	**26**
27	1.308	.7644	.0324	.0424	30.821	23.560	12.397	292.069	**27**
28	1.321	.7568	.0311	.0411	32.129	24.316	12.852	312.504	**28**
29	1.335	.7493	.0299	.0399	33.450	25.066	13.304	333.486	**29**
30	1.348	.7419	.0287	.0387	34.785	25.808	13.756	355.001	**30**
36	1.431	.6989	.0232	.0332	43.077	30.107	16.428	494.620	**36**
40	1.489	.6717	.0205	.0305	48.886	32.835	18.178	596.854	**40**
48	1.612	.6203	.0163	.0263	61.223	37.974	21.598	820.144	**48**
50	1.645	.6080	.0155	.0255	64.463	39.196	22.436	879.417	**50**
52	1.678	.5961	.0148	.0248	67.769	40.394	23.269	939.916	**52**
60	1.817	.5504	.0122	.0222	81.670	44.955	26.533	1 192.80	**60**
70	2.007	.4983	.00993	.0199	100.676	50.168	30.470	1 528.64	**70**
72	2.047	.4885	.00955	.0196	104.710	51.150	31.239	1 597.86	**72**
80	2.217	.4511	.00822	.0182	121.671	54.888	34.249	1 879.87	**80**
84	2.307	.4335	.00765	.0177	130.672	56.648	35.717	2 023.31	**84**
90	2.449	.4084	.00690	.0169	144.863	59.161	37.872	2 240.56	**90**
96	2.599	.3847	.00625	.0163	159.927	61.528	39.973	2 459.42	**96**
100	2.705	.3697	.00587	.0159	170.481	63.029	41.343	2 605.77	**100**
104	2.815	.3553	.00551	.0155	181.464	64.471	42.688	2 752.17	**104**
120	3.300	.3030	.00435	.0143	230.039	69.701	47.835	3 334.11	**120**
240	10.893	.0918	.00101	.0110	989.254	90.819	75.739	6 878.59	**240**
360	35.950	.0278	.00029	.0103	3 495.0	97.218	89.699	8 720.43	**360**
480	118.648	.00843	.00008	.0101	11 764.8	99.157	95.920	9 511.15	**480**

	Single Payment		Uniform Payment Series				Arithmetic Gradient		
	Compound Amount Factor	Present Worth Factor	Sinking Fund Factor	Capital Recovery Factor	Compound Amount Factor	Present Worth Factor	Gradient Uniform Series	Gradient Present Worth	
n	Find *F* Given *P* *F/P*	Find *P* Given *F* *P/F*	Find *A* Given *F* *A/F*	Find *A* Given *P* *A/P*	Find *F* Given *A* *F/A*	Find *P* Given *A* *P/A*	Find *A* Given *G* *A/G*	Find *P* Given *G* *P/G*	*n*
1	1.013	.9877	1.0000	1.0125	1.000	0.988	0	0	**1**
2	1.025	.9755	.4969	.5094	2.013	1.963	0.497	0.976	**2**
3	1.038	.9634	.3292	.3417	3.038	2.927	0.992	2.904	**3**
4	1.051	.9515	.2454	.2579	4.076	3.878	1.485	5.759	**4**
5	1.064	.9398	.1951	.2076	5.127	4.818	1.976	9.518	**5**
6	1.077	.9282	.1615	.1740	6.191	5.746	2.464	14.160	**6**
7	1.091	.9167	.1376	.1501	7.268	6.663	2.951	19.660	**7**
8	1.104	.9054	.1196	.1321	8.359	7.568	3.435	25.998	**8**
9	1.118	.8942	.1057	.1182	9.463	8.462	3.918	33.152	**9**
10	1.132	.8832	.0945	.1070	10.582	9.346	4.398	41.101	**10**
11	1.146	.8723	.0854	.0979	11.714	10.218	4.876	49.825	**11**
12	1.161	.8615	.0778	.0903	12.860	11.079	5.352	59.302	**12**
13	1.175	.8509	.0713	.0838	14.021	11.930	5.827	69.513	**13**
14	1.190	.8404	.0658	.0783	15.196	12.771	6.299	80.438	**14**
15	1.205	.8300	.0610	.0735	16.386	13.601	6.769	92.058	**15**
16	1.220	.8197	.0568	.0693	17.591	14.420	7.237	104.355	**16**
17	1.235	.8096	.0532	.0657	18.811	15.230	7.702	117.309	**17**
18	1.251	.7996	.0499	.0624	20.046	16.030	8.166	130.903	**18**
19	1.266	.7898	.0470	.0595	21.297	16.819	8.628	145.119	**19**
20	1.282	.7800	.0443	.0568	22.563	17.599	9.088	159.940	**20**
21	1.298	.7704	.0419	.0544	23.845	18.370	9.545	175.348	**21**
22	1.314	.7609	.0398	.0523	25.143	19.131	10.001	191.327	**22**
23	1.331	.7515	.0378	.0503	26.458	19.882	10.455	207.859	**23**
24	1.347	.7422	.0360	.0485	27.788	20.624	10.906	224.930	**24**
25	1.364	.7330	.0343	.0468	29.136	21.357	11.355	242.523	**25**
26	1.381	.7240	.0328	.0453	30.500	22.081	11.803	260.623	**26**
27	1.399	.7150	.0314	.0439	31.881	22.796	12.248	279.215	**27**
28	1.416	.7062	.0300	.0425	33.280	23.503	12.691	298.284	**28**
29	1.434	.6975	.0288	.0413	34.696	24.200	13.133	317.814	**29**
30	1.452	.6889	.0277	.0402	36.129	24.889	13.572	337.792	**30**
36	1.564	.6394	.0222	.0347	45.116	28.847	16.164	466.297	**36**
40	1.644	.6084	.0194	.0319	51.490	31.327	17.852	559.247	**40**
48	1.815	.5509	.0153	.0278	65.229	35.932	21.130	759.248	**48**
50	1.861	.5373	.0145	.0270	68.882	37.013	21.930	811.692	**50**
52	1.908	.5242	.0138	.0263	72.628	38.068	22.722	864.960	**52**
60	2.107	.4746	.0113	.0238	88.575	42.035	25.809	1 084.86	**60**
70	2.386	.4191	.00902	.0215	110.873	46.470	29.492	1 370.47	**70**
72	2.446	.4088	.00864	.0211	115.675	47.293	30.205	1 428.48	**72**
80	2.701	.3702	.00735	.0198	136.120	50.387	32.983	1 661.89	**80**
84	2.839	.3522	.00680	.0193	147.130	51.822	34.326	1 778.86	**84**
90	3.059	.3269	.00607	.0186	164.706	53.846	36.286	1 953.85	**90**
96	3.296	.3034	.00545	.0179	183.643	55.725	38.180	2 127.55	**96**
100	3.463	.2887	.00507	.0176	197.074	56.901	39.406	2 242.26	**100**
104	3.640	.2747	.00474	.0172	211.190	58.021	40.604	2 355.90	**104**
120	4.440	.2252	.00363	.0161	275.220	61.983	45.119	2 796.59	**120**
240	19.716	.0507	.00067	.0132	1 497.3	75.942	67.177	5 101.55	**240**
360	87.543	.0114	.00014	.0126	6 923.4	79.086	75.840	5 997.91	**360**
480	388.713	.00257	.00003	.0125	31 017.1	79.794	78.762	6 284.74	**480**

1½% Compound Interest Factors 1½%

	Single Payment		Uniform Payment Series				Arithmetic Gradient		
	Compound Amount Factor	Present Worth Factor	Sinking Fund Factor	Capital Recovery Factor	Compound Amount Factor	Present Worth Factor	Gradient Uniform Series	Gradient Present Worth	
n	Find F Given P F/P	Find P Given F P/F	Find A Given F A/F	Find A Given P A/P	Find F Given A F/A	Find P Given A P/A	Find A Given G A/G	Find P Given G P/G	n
1	1.015	.9852	1.0000	1.0150	1.000	0.985	0	0	**1**
2	1.030	.9707	.4963	.5113	2.015	1.956	0.496	0.970	**2**
3	1.046	.9563	.3284	.3434	3.045	2.912	0.990	2.883	**3**
4	1.061	.9422	.2444	.2594	4.091	3.854	1.481	5.709	**4**
5	1.077	.9283	.1941	.2091	5.152	4.783	1.970	9.422	**5**
6	1.093	.9145	.1605	.1755	6.230	5.697	2.456	13.994	**6**
7	1.110	.9010	.1366	.1516	7.323	6.598	2.940	19.400	**7**
8	1.126	.8877	.1186	.1336	8.433	7.486	3.422	25.614	**8**
9	1.143	.8746	.1046	.1196	9.559	8.360	3.901	32.610	**9**
10	1.161	.8617	.0934	.1084	10.703	9.222	4.377	40.365	**10**
11	1.178	.8489	.0843	.0993	11.863	10.071	4.851	48.855	**11**
12	1.196	.8364	.0767	.0917	13.041	10.907	5.322	58.054	**12**
13	1.214	.8240	.0702	.0852	14.237	11.731	5.791	67.943	**13**
14	1.232	.8118	.0647	.0797	15.450	12.543	6.258	78.496	**14**
15	1.250	.7999	.0599	.0749	16.682	13.343	6.722	89.694	**15**
16	1.269	.7880	.0558	.0708	17.932	14.131	7.184	101.514	**16**
17	1.288	.7764	.0521	.0671	19.201	14.908	7.643	113.937	**17**
18	1.307	.7649	.0488	.0638	20.489	15.673	8.100	126.940	**18**
19	1.327	.7536	.0459	.0609	21.797	16.426	8.554	140.505	**19**
20	1.347	.7425	.0432	.0582	23.124	17.169	9.005	154.611	**20**
21	1.367	.7315	.0409	.0559	24.470	17.900	9.455	169.241	**21**
22	1.388	.7207	.0387	.0537	25.837	18.621	9.902	184.375	**22**
23	1.408	.7100	.0367	.0517	27.225	19.331	10.346	199.996	**23**
24	1.430	.6995	.0349	.0499	28.633	20.030	10.788	216.085	**24**
25	1.451	.6892	.0333	.0483	30.063	20.720	11.227	232.626	**25**
26	1.473	.6790	.0317	.0467	31.514	21.399	11.664	249.601	**26**
27	1.495	.6690	.0303	.0453	32.987	22.068	12.099	266.995	**27**
28	1.517	.6591	.0290	.0440	34.481	22.727	12.531	284.790	**28**
29	1.540	.6494	.0278	.0428	35.999	23.376	12.961	302.972	**29**
30	1.563	.6398	.0266	.0416	37.539	24.016	13.388	321.525	**30**
36	1.709	.5851	.0212	.0362	47.276	27.661	15.901	439.823	**36**
40	1.814	.5513	.0184	.0334	54.268	29.916	17.528	524.349	**40**
48	2.043	.4894	.0144	.0294	69.565	34.042	20.666	703.537	**48**
50	2.105	.4750	.0136	.0286	73.682	35.000	21.428	749.955	**50**
52	2.169	.4611	.0128	.0278	77.925	35.929	22.179	796.868	**52**
60	2.443	.4093	.0104	.0254	96.214	39.380	25.093	988.157	**60**
70	2.835	.3527	.00817	.0232	122.363	43.155	28.529	1 231.15	**70**
72	2.921	.3423	.00781	.0228	128.076	43.845	29.189	1 279.78	**72**
80	3.291	.3039	.00655	.0215	152.710	46.407	31.742	1 473.06	**80**
84	3.493	.2863	.00602	.0210	166.172	47.579	32.967	1 568.50	**84**
90	3.819	.2619	.00532	.0203	187.929	49.210	34.740	1 709.53	**90**
96	4.176	.2395	.00472	.0197	211.719	50.702	36.438	1 847.46	**96**
100	4.432	.2256	.00437	.0194	228.802	51.625	37.529	1 937.43	**100**
104	4.704	.2126	.00405	.0190	246.932	52.494	38.589	2 025.69	**104**
120	5.969	.1675	.00302	.0180	331.286	55.498	42.518	2 359.69	**120**
240	35.632	.0281	.00043	.0154	2 308.8	64.796	59.737	3 870.68	**240**
360	212.700	.00470	.00007	.0151	14 113.3	66.353	64.966	4 310.71	**360**
480	1 269.7	.00079	.00001	.0150	84 577.8	66.614	66.288	4 415.74	**480**

1¾% Compound Interest Factors 1¾%

	Single Payment		Uniform Payment Series				Arithmetic Gradient		
	Compound Amount Factor	Present Worth Factor	Sinking Fund Factor	Capital Recovery Factor	Compound Amount Factor	Present Worth Factor	Gradient Uniform Series	Gradient Present Worth	
n	Find F Given P F/P	Find P Given F P/F	Find A Given F A/F	Find A Given P A/P	Find F Given A F/A	Find P Given A P/A	Find A Given G A/G	Find P Given G P/G	n
1	1.018	.9828	1.0000	1.0175	1.000	0.983	0	0	**1**
2	1.035	.9659	.4957	.5132	2.018	1.949	0.496	0.966	**2**
3	1.053	.9493	.3276	.3451	3.053	2.898	0.989	2.865	**3**
4	1.072	.9330	.2435	.2610	4.106	3.831	1.478	5.664	**4**
5	1.091	.9169	.1931	.2106	5.178	4.748	1.965	9.332	**5**
6	1.110	.9011	.1595	.1770	6.269	5.649	2.450	13.837	**6**
7	1.129	.8856	.1355	.1530	7.378	6.535	2.931	19.152	**7**
8	1.149	.8704	.1175	.1350	8.508	7.405	3.409	25.245	**8**
9	1.169	.8554	.1036	.1211	9.656	8.261	3.885	32.088	**9**
10	1.189	.8407	.0924	.1099	10.825	9.101	4.357	39.655	**10**
11	1.210	.8263	.0832	.1007	12.015	9.928	4.827	47.918	**11**
12	1.231	.8121	.0756	.0931	13.225	10.740	5.294	56.851	**12**
13	1.253	.7981	.0692	.0867	14.457	11.538	5.758	66.428	**13**
14	1.275	.7844	.0637	.0812	15.710	12.322	6.219	76.625	**14**
15	1.297	.7709	.0589	.0764	16.985	13.093	6.677	87.417	**15**
16	1.320	.7576	.0547	.0722	18.282	13.851	7.132	98.782	**16**
17	1.343	.7446	.0510	.0685	19.602	14.595	7.584	110.695	**17**
18	1.367	.7318	.0477	.0652	20.945	15.327	8.034	123.136	**18**
19	1.390	.7192	.0448	.0623	22.311	16.046	8.481	136.081	**19**
20	1.415	.7068	.0422	.0597	23.702	16.753	8.924	149.511	**20**
21	1.440	.6947	.0398	.0573	25.116	17.448	9.365	163.405	**21**
22	1.465	.6827	.0377	.0552	26.556	18.130	9.804	177.742	**22**
23	1.490	.6710	.0357	.0532	28.021	18.801	10.239	192.503	**23**
24	1.516	.6594	.0339	.0514	29.511	19.461	10.671	207.671	**24**
25	1.543	.6481	.0322	.0497	31.028	20.109	11.101	223.225	**25**
26	1.570	.6369	.0307	.0482	32.571	20.746	11.528	239.149	**26**
27	1.597	.6260	.0293	.0468	34.141	21.372	11.952	255.425	**27**
28	1.625	.6152	.0280	.0455	35.738	21.987	12.373	272.036	**28**
29	1.654	.6046	.0268	.0443	37.363	22.592	12.791	288.967	**29**
30	1.683	.5942	.0256	.0431	39.017	23.186	13.206	306.200	**30**
36	1.867	.5355	.0202	.0377	49.566	26.543	15.640	415.130	**36**
40	2.002	.4996	.0175	.0350	57.234	28.594	17.207	492.017	**40**
48	2.300	.4349	.0135	.0310	74.263	32.294	20.209	652.612	**48**
50	2.381	.4200	.0127	.0302	78.903	33.141	20.932	693.708	**50**
52	2.465	.4057	.0119	.0294	83.706	33.960	21.644	735.039	**52**
60	2.832	.3531	.00955	.0271	104.676	36.964	24.389	901.503	**60**
70	3.368	.2969	.00739	.0249	135.331	40.178	27.586	1 108.34	**70**
72	3.487	.2868	.00704	.0245	142.127	40.757	28.195	1 149.12	**72**
80	4.006	.2496	.00582	.0233	171.795	42.880	30.533	1 309.25	**80**
84	4.294	.2329	.00531	.0228	188.246	43.836	31.644	1 387.16	**84**
90	4.765	.2098	.00465	.0221	215.166	45.152	33.241	1 500.88	**90**
96	5.288	.1891	.00408	.0216	245.039	46.337	34.756	1 610.48	**96**
100	5.668	.1764	.00375	.0212	266.753	47.062	35.721	1 681.09	**100**
104	6.075	.1646	.00345	.0209	290.028	47.737	36.652	1 749.68	**104**
120	8.019	.1247	.00249	.0200	401.099	50.017	40.047	2 003.03	**120**
240	64.308	.0156	.00028	.0178	3 617.6	56.254	53.352	3 001.27	**240**
360	515.702	.00194	.00003	.0175	29 411.5	57.032	56.443	3 219.08	**360**
480	4 135.5	.00024		.0175	236 259.0	57.129	57.027	3 257.88	**480**

2% Compound Interest Factors 2%

	Single Payment		Uniform Payment Series				Arithmetic Gradient		
	Compound Amount Factor	Present Worth Factor	Sinking Fund Factor	Capital Recovery Factor	Compound Amount Factor	Present Worth Factor	Gradient Uniform Series	Gradient Present Worth	
n	Find *F* Given *P* *F/P*	Find *P* Given *F* *P/F*	Find *A* Given *F* *A/F*	Find *A* Given *P* *A/P*	Find *F* Given *A* *F/A*	Find *P* Given *A* *P/A*	Find *A* Given *G* *A/G*	Find *P* Given *G* *P/G*	*n*
1	1.020	.9804	1.0000	1.0200	1.000	0.980	0	0	**1**
2	1.040	.9612	.4951	.5151	2.020	1.942	0.495	0.961	**2**
3	1.061	.9423	.3268	.3468	3.060	2.884	0.987	2.846	**3**
4	1.082	.9238	.2426	.2626	4.122	3.808	1.475	5.617	**4**
5	1.104	.9057	.1922	.2122	5.204	4.713	1.960	9.240	**5**
6	1.126	.8880	.1585	.1785	6.308	5.601	2.442	13.679	**6**
7	1.149	.8706	.1345	.1545	7.434	6.472	2.921	18.903	**7**
8	1.172	.8535	.1165	.1365	8.583	7.325	3.396	24.877	**8**
9	1.195	.8368	.1025	.1225	9.755	8.162	3.868	31.571	**9**
10	1.219	.8203	.0913	.1113	10.950	8.983	4.337	38.954	**10**
11	1.243	.8043	.0822	.1022	12.169	9.787	4.802	46.996	**11**
12	1.268	.7885	.0746	.0946	13.412	10.575	5.264	55.669	**12**
13	1.294	.7730	.0681	.0881	14.680	11.348	5.723	64.946	**13**
14	1.319	.7579	.0626	.0826	15.974	12.106	6.178	74.798	**14**
15	1.346	.7430	.0578	.0778	17.293	12.849	6.631	85.200	**15**
16	1.373	.7284	.0537	.0737	18.639	13.578	7.080	96.127	**16**
17	1.400	.7142	.0500	.0700	20.012	14.292	7.526	107.553	**17**
18	1.428	.7002	.0467	.0667	21.412	14.992	7.968	119.456	**18**
19	1.457	.6864	.0438	.0638	22.840	15.678	8.407	131.812	**19**
20	1.486	.6730	.0412	.0612	24.297	16.351	8.843	144.598	**20**
21	1.516	.6598	.0388	.0588	25.783	17.011	9.276	157.793	**21**
22	1.546	.6468	.0366	.0566	27.299	17.658	9.705	171.377	**22**
23	1.577	.6342	.0347	.0547	28.845	18.292	10.132	185.328	**23**
24	1.608	.6217	.0329	.0529	30.422	18.914	10.555	199.628	**24**
25	1.641	.6095	.0312	.0512	32.030	19.523	10.974	214.256	**25**
26	1.673	.5976	.0297	.0497	33.671	20.121	11.391	229.196	**26**
27	1.707	.5859	.0283	.0483	35.344	20.707	11.804	244.428	**27**
28	1.741	.5744	.0270	.0470	37.051	21.281	12.214	259.936	**28**
29	1.776	.5631	.0258	.0458	38.792	21.844	12.621	275.703	**29**
30	1.811	.5521	.0247	.0447	40.568	22.396	13.025	291.713	**30**
36	2.040	.4902	.0192	.0392	51.994	25.489	15.381	392.036	**36**
40	2.208	.4529	.0166	.0366	60.402	27.355	16.888	461.989	**40**
48	2.587	.3865	.0126	.0326	79.353	30.673	19.755	605.961	**48**
50	2.692	.3715	.0118	.0318	84.579	31.424	20.442	642.355	**50**
52	2.800	.3571	.0111	.0311	90.016	32.145	21.116	678.779	**52**
60	3.281	.3048	.00877	.0288	114.051	34.761	23.696	823.692	**60**
70	4.000	.2500	.00667	.0267	149.977	37.499	26.663	999.829	**70**
72	4.161	.2403	.00633	.0263	158.056	37.984	27.223	1 034.050	**72**
80	4.875	.2051	.00516	.0252	193.771	39.744	29.357	1 166.781	**80**
84	5.277	.1895	.00468	.0247	213.865	40.525	30.361	1 230.413	**84**
90	5.943	.1683	.00405	.0240	247.155	41.587	31.793	1 322.164	**90**
96	6.693	.1494	.00351	.0235	284.645	42.529	33.137	1 409.291	**96**
100	7.245	.1380	.00320	.0232	312.230	43.098	33.986	1 464.747	**100**
104	7.842	.1275	.00292	.0229	342.090	43.624	34.799	1 518.082	**104**
120	10.765	.0929	.00205	.0220	488.255	45.355	37.711	1 710.411	**120**
240	115.887	.00863	.00017	.0202	5 744.4	49.569	47.911	2 374.878	**240**
360	1 247.5	.00080	.00002	.0200	62 326.8	49.960	49.711	2 483.567	**360**
480	13 429.8	.00007		.0200	671 442.0	49.996	49.964	2 498.027	**480**

2½% Compound Interest Factors 2½%

	Single Payment		Uniform Payment Series				Arithmetic Gradient		
	Compound Amount Factor	Present Worth Factor	Sinking Fund Factor	Capital Recovery Factor	Compound Amount Factor	Present Worth Factor	Gradient Uniform Series	Gradient Present Worth	
n	Find *F* Given *P* *F/P*	Find *P* Given *F* *P/F*	Find *A* Given *F* *A/F*	Find *A* Given *P* *A/P*	Find *F* Given *A* *F/A*	Find *P* Given *A* *P/A*	Find *A* Given *G* *A/G*	Find *P* Given *G* *P/G*	*n*
1	1.025	.9756	1.0000	1.0250	1.000	0.976	0	0	**1**
2	1.051	.9518	.4938	.5188	2.025	1.927	0.494	0.952	**2**
3	1.077	.9286	.3251	.3501	3.076	2.856	0.984	2.809	**3**
4	1.104	.9060	.2408	.2658	4.153	3.762	1.469	5.527	**4**
5	1.131	.8839	.1902	.2152	5.256	4.646	1.951	9.062	**5**
6	1.160	.8623	.1566	.1816	6.388	5.508	2.428	13.374	**6**
7	1.189	.8413	.1325	.1575	7.547	6.349	2.901	18.421	**7**
8	1.218	.8207	.1145	.1395	8.736	7.170	3.370	24.166	**8**
9	1.249	.8007	.1005	.1255	9.955	7.971	3.835	30.572	**9**
10	1.280	.7812	.0893	.1143	11.203	8.752	4.296	37.603	**10**
11	1.312	.7621	.0801	.1051	12.483	9.514	4.753	45.224	**11**
12	1.345	.7436	.0725	.0975	13.796	10.258	5.206	53.403	**12**
13	1.379	.7254	.0660	.0910	15.140	10.983	5.655	62.108	**13**
14	1.413	.7077	.0605	.0855	16.519	11.691	6.100	71.309	**14**
15	1.448	.6905	.0558	.0808	17.932	12.381	6.540	80.975	**15**
16	1.485	.6736	.0516	.0766	19.380	13.055	6.977	91.080	**16**
17	1.522	.6572	.0479	.0729	20.865	13.712	7.409	101.595	**17**
18	1.560	.6412	.0447	.0697	22.386	14.353	7.838	112.495	**18**
19	1.599	.6255	.0418	.0668	23.946	14.979	8.262	123.754	**19**
20	1.639	.6103	.0391	.0641	25.545	15.589	8.682	135.349	**20**
21	1.680	.5954	.0368	.0618	27.183	16.185	9.099	147.257	**21**
22	1.722	.5809	.0346	.0596	28.863	16.765	9.511	159.455	**22**
23	1.765	.5667	.0327	.0577	30.584	17.332	9.919	171.922	**23**
24	1.809	.5529	.0309	.0559	32.349	17.885	10.324	184.638	**24**
25	1.854	.5394	.0293	.0543	34.158	18.424	10.724	197.584	**25**
26	1.900	.5262	.0278	.0528	36.012	18.951	11.120	210.740	**26**
27	1.948	.5134	.0264	.0514	37.912	19.464	11.513	224.088	**27**
28	1.996	.5009	.0251	.0501	39.860	19.965	11.901	237.612	**28**
29	2.046	.4887	.0239	.0489	41.856	20.454	12.286	251.294	**29**
30	2.098	.4767	.0228	.0478	43.903	20.930	12.667	265.120	**30**
31	2.150	.4651	.0217	.0467	46.000	21.395	13.044	279.073	**31**
32	2.204	.4538	.0208	.0458	48.150	21.849	13.417	293.140	**32**
33	2.259	.4427	.0199	.0449	50.354	22.292	13.786	307.306	**33**
34	2.315	.4319	.0190	.0440	52.613	22.724	14.151	321.559	**34**
35	2.373	.4214	.0182	.0432	54.928	23.145	14.512	335.886	**35**
40	2.685	.3724	.0148	.0398	67.402	25.103	16.262	408.221	**40**
45	3.038	.3292	.0123	.0373	81.516	26.833	17.918	480.806	**45**
50	3.437	.2909	.0103	.0353	97.484	28.362	19.484	552.607	**50**
55	3.889	.2572	.00865	.0337	115.551	29.714	20.961	622.827	**55**
60	4.400	.2273	.00735	.0324	135.991	30.909	22.352	690.865	**60**
65	4.978	.2009	.00628	.0313	159.118	31.965	23.660	756.280	**65**
70	5.632	.1776	.00540	.0304	185.284	32.898	24.888	818.763	**70**
75	6.372	.1569	.00465	.0297	214.888	33.723	26.039	878.114	**75**
80	7.210	.1387	.00403	.0290	248.382	34.452	27.117	934.217	**80**
85	8.157	.1226	.00349	.0285	286.278	35.096	28.123	987.026	**85**
90	9.229	.1084	.00304	.0280	329.154	35.666	29.063	1 036.54	**90**
95	10.442	.0958	.00265	.0276	377.663	36.169	29.938	1 082.83	**95**
100	11.814	.0846	.00231	.0273	432.548	36.614	30.752	1 125.97	**100**

3% Compound Interest Factors 3%

	Single Payment		Uniform Payment Series				Arithmetic Gradient		
	Compound Amount Factor	Present Worth Factor	Sinking Fund Factor	Capital Recovery Factor	Compound Amount Factor	Present Worth Factor	Gradient Uniform Series	Gradient Present Worth	
n	Find F Given P F/P	Find P Given F P/F	Find A Given F A/F	Find A Given P A/P	Find F Given A F/A	Find P Given A P/A	Find A Given G A/G	Find P Given G P/G	**n**
1	1.030	.9709	1.0000	1.0300	1.000	0.971	0	0	**1**
2	1.061	.9426	.4926	.5226	2.030	1.913	0.493	0.943	**2**
3	1.093	.9151	.3235	.3535	3.091	2.829	0.980	2.773	**3**
4	1.126	.8885	.2390	.2690	4.184	3.717	1.463	5.438	**4**
5	1.159	.8626	.1884	.2184	5.309	4.580	1.941	8.889	**5**
6	1.194	.8375	.1546	.1846	6.468	5.417	2.414	13.076	**6**
7	1.230	.8131	.1305	.1605	7.662	6.230	2.882	17.955	**7**
8	1.267	.7894	.1125	.1425	8.892	7.020	3.345	23.481	**8**
9	1.305	.7664	.0984	.1284	10.159	7.786	3.803	29.612	**9**
10	1.344	.7441	.0872	.1172	11.464	8.530	4.256	36.309	**10**
11	1.384	.7224	.0781	.1081	12.808	9.253	4.705	43.533	**11**
12	1.426	.7014	.0705	.1005	14.192	9.954	5.148	51.248	**12**
13	1.469	.6810	.0640	.0940	15.618	10.635	5.587	59.419	**13**
14	1.513	.6611	.0585	.0885	17.086	11.296	6.021	68.014	**14**
15	1.558	.6419	.0538	.0838	18.599	11.938	6.450	77.000	**15**
16	1.605	.6232	.0496	.0796	20.157	12.561	6.874	86.348	**16**
17	1.653	.6050	.0460	.0760	21.762	13.166	7.294	96.028	**17**
18	1.702	.5874	.0427	.0727	23.414	13.754	7.708	106.014	**18**
19	1.754	.5703	.0398	.0698	25.117	14.324	8.118	116.279	**19**
20	1.806	.5537	.0372	.0672	26.870	14.877	8.523	126.799	**20**
21	1.860	.5375	.0349	.0649	28.676	15.415	8.923	137.549	**21**
22	1.916	.5219	.0327	.0627	30.537	15.937	9.319	148.509	**22**
23	1.974	.5067	.0308	.0608	32.453	16.444	9.709	159.656	**23**
24	2.033	.4919	.0290	.0590	34.426	16.936	10.095	170.971	**24**
25	2.094	.4776	.0274	.0574	36.459	17.413	10.477	182.433	**25**
26	2.157	.4637	.0259	.0559	38.553	17.877	10.853	194.026	**26**
27	2.221	.4502	.0246	.0546	40.710	18.327	11.226	205.731	**27**
28	2.288	.4371	.0233	.0533	42.931	18.764	11.593	217.532	**28**
29	2.357	.4243	.0221	.0521	45.219	19.188	11.956	229.413	**29**
30	2.427	.4120	.0210	.0510	47.575	19.600	12.314	241.361	**30**
31	2.500	.4000	.0200	.0500	50.003	20.000	12.668	253.361	**31**
32	2.575	.3883	.0190	.0490	52.503	20.389	13.017	265.399	**32**
33	2.652	.3770	.0182	.0482	55.078	20.766	13.362	277.464	**33**
34	2.732	.3660	.0173	.0473	57.730	21.132	13.702	289.544	**34**
35	2.814	.3554	.0165	.0465	60.462	21.487	14.037	301.627	**35**
40	3.262	.3066	.0133	.0433	75.401	23.115	15.650	361.750	**40**
45	3.782	.2644	.0108	.0408	92.720	24.519	17.156	420.632	**45**
50	4.384	.2281	.00887	.0389	112.797	25.730	18.558	477.480	**50**
55	5.082	.1968	.00735	.0373	136.072	26.774	19.860	531.741	**55**
60	5.892	.1697	.00613	.0361	163.053	27.676	21.067	583.052	**60**
65	6.830	.1464	.00515	.0351	194.333	28.453	22.184	631.201	**65**
70	7.918	.1263	.00434	.0343	230.594	29.123	23.215	676.087	**70**
75	9.179	.1089	.00367	.0337	272.631	29.702	24.163	717.698	**75**
80	10.641	.0940	.00311	.0331	321.363	30.201	25.035	756.086	**80**
85	12.336	.0811	.00265	.0326	377.857	30.631	25.835	791.353	**85**
90	14.300	.0699	.00226	.0323	443.349	31.002	26.567	823.630	**90**
95	16.578	.0603	.00193	.0319	519.272	31.323	27.235	853.074	**95**
100	19.219	.0520	.00165	.0316	607.287	31.599	27.844	879.854	**100**

3½% Compound Interest Factors 3½%

	Single Payment		Uniform Payment Series				Arithmetic Gradient		
	Compound Amount Factor	Present Worth Factor	Sinking Fund Factor	Capital Recovery Factor	Compound Amount Factor	Present Worth Factor	Gradient Uniform Series	Gradient Present Worth	
n	Find F Given P F/P	Find P Given F P/F	Find A Given F A/F	Find A Given P A/P	Find F Given A F/A	Find P Given A P/A	Find A Given G A/G	Find P Given G P/G	n
1	1.035	.9662	1.0000	1.0350	1.000	0.966	0	0	**1**
2	1.071	.9335	.4914	.5264	2.035	1.900	0.491	0.933	**2**
3	1.109	.9019	.3219	.3569	3.106	2.802	0.977	2.737	**3**
4	1.148	.8714	.2373	.2723	4.215	3.673	1.457	5.352	**4**
5	1.188	.8420	.1865	.2215	5.362	4.515	1.931	8.719	**5**
6	1.229	.8135	.1527	.1877	6.550	5.329	2.400	12.787	**6**
7	1.272	.7860	.1285	.1635	7.779	6.115	2.862	17.503	**7**
8	1.317	.7594	.1105	.1455	9.052	6.874	3.320	22.819	**8**
9	1.363	.7337	.0964	.1314	10.368	7.608	3.771	28.688	**9**
10	1.411	.7089	.0852	.1202	11.731	8.317	4.217	35.069	**10**
11	1.460	.6849	.0761	.1111	13.142	9.002	4.657	41.918	**11**
12	1.511	.6618	.0685	.1035	14.602	9.663	5.091	49.198	**12**
13	1.564	.6394	.0621	.0971	16.113	10.303	5.520	56.871	**13**
14	1.619	.6178	.0566	.0916	17.677	10.921	5.943	64.902	**14**
15	1.675	.5969	.0518	.0868	19.296	11.517	6.361	73.258	**15**
16	1.734	.5767	.0477	.0827	20.971	12.094	6.773	81.909	**16**
17	1.795	.5572	.0440	.0790	22.705	12.651	7.179	90.824	**17**
18	1.857	.5384	.0408	.0758	24.500	13.190	7.580	99.976	**18**
19	1.922	.5202	.0379	.0729	26.357	13.710	7.975	109.339	**19**
20	1.990	.5026	.0354	.0704	28.280	14.212	8.365	118.888	**20**
21	2.059	.4856	.0330	.0680	30.269	14.698	8.749	128.599	**21**
22	2.132	.4692	.0309	.0659	32.329	15.167	9.128	138.451	**22**
23	2.206	.4533	.0290	.0640	34.460	15.620	9.502	148.423	**23**
24	2.283	.4380	.0273	.0623	36.666	16.058	9.870	158.496	**24**
25	2.363	.4231	.0257	.0607	38.950	16.482	10.233	168.652	**25**
26	2.446	.4088	.0242	.0592	41.313	16.890	10.590	178.873	**26**
27	2.532	.3950	.0229	.0579	43.759	17.285	10.942	189.143	**27**
28	2.620	.3817	.0216	.0566	46.291	17.667	11.289	199.448	**28**
29	2.712	.3687	.0204	.0554	48.911	18.036	11.631	209.773	**29**
30	2.807	.3563	.0194	.0544	51.623	18.392	11.967	220.105	**30**
31	2.905	.3442	.0184	.0534	54.429	18.736	12.299	230.432	**31**
32	3.007	.3326	.0174	.0524	57.334	19.069	12.625	240.742	**32**
33	3.112	.3213	.0166	.0516	60.341	19.390	12.946	251.025	**33**
34	3.221	.3105	.0158	.0508	63.453	19.701	13.262	261.271	**34**
35	3.334	.3000	.0150	.0500	66.674	20.001	13.573	271.470	**35**
40	3.959	.2526	.0118	.0468	84.550	21.355	15.055	321.490	**40**
45	4.702	.2127	.00945	.0445	105.781	22.495	16.417	369.307	**45**
50	5.585	.1791	.00763	.0426	130.998	23.456	17.666	414.369	**50**
55	6.633	.1508	.00621	.0412	160.946	24.264	18.808	456.352	**55**
60	7.878	.1269	.00509	.0401	196.516	24.945	19.848	495.104	**60**
65	9.357	.1069	.00419	.0392	238.762	25.518	20.793	530.598	**65**
70	11.113	.0900	.00346	.0385	288.937	26.000	21.650	562.895	**70**
75	13.199	.0758	.00287	.0379	348.529	26.407	22.423	592.121	**75**
80	15.676	.0638	.00238	.0374	419.305	26.749	23.120	618.438	**80**
85	18.618	.0537	.00199	.0370	503.365	27.037	23.747	642.036	**85**
90	22.112	.0452	.00166	.0367	603.202	27.279	24.308	663.118	**90**
95	26.262	.0381	.00139	.0364	721.778	27.483	24.811	681.890	**95**
100	31.191	.0321	.00116	.0362	862.608	27.655	25.259	698.554	**100**

4% Compound Interest Factors 4%

	Single Payment		Uniform Payment Series				Arithmetic Gradient		
	Compound Amount Factor	Present Worth Factor	Sinking Fund Factor	Capital Recovery Factor	Compound Amount Factor	Present Worth Factor	Gradient Uniform Series	Gradient Present Worth	
n	Find *F* Given *P* *F/P*	Find *P* Given *F* *P/F*	Find *A* Given *F* *A/F*	Find *A* Given *P* *A/P*	Find *F* Given *A* *F/A*	Find *P* Given *A* *P/A*	Find *A* Given *G* *A/G*	Find *P* Given *G* *P/G*	*n*
1	1.040	.9615	1.0000	1.0400	1.000	0.962	0	0	**1**
2	1.082	.9246	.4902	.5302	2.040	1.886	0.490	0.925	**2**
3	1.125	.8890	.3203	.3603	3.122	2.775	0.974	2.702	**3**
4	1.170	.8548	.2355	.2755	4.246	3.630	1.451	5.267	**4**
5	1.217	.8219	.1846	.2246	5.416	4.452	1.922	8.555	**5**
6	1.265	.7903	.1508	.1908	6.633	5.242	2.386	12.506	**6**
7	1.316	.7599	.1266	.1666	7.898	6.002	2.843	17.066	**7**
8	1.369	.7307	.1085	.1485	9.214	6.733	3.294	22.180	**8**
9	1.423	.7026	.0945	.1345	10.583	7.435	3.739	27.801	**9**
10	1.480	.6756	.0833	.1233	12.006	8.111	4.177	33.881	**10**
11	1.539	.6496	.0741	.1141	13.486	8.760	4.609	40.377	**11**
12	1.601	.6246	.0666	.1066	15.026	9.385	5.034	47.248	**12**
13	1.665	.6006	.0601	.1001	16.627	9.986	5.453	54.454	**13**
14	1.732	.5775	.0547	.0947	18.292	10.563	5.866	61.962	**14**
15	1.801	.5553	.0499	.0899	20.024	11.118	6.272	69.735	**15**
16	1.873	.5339	.0458	.0858	21.825	11.652	6.672	77.744	**16**
17	1.948	.5134	.0422	.0822	23.697	12.166	7.066	85.958	**17**
18	2.026	.4936	.0390	.0790	25.645	12.659	7.453	94.350	**18**
19	2.107	.4746	.0361	.0761	27.671	13.134	7.834	102.893	**19**
20	2.191	.4564	.0336	.0736	29.778	13.590	8.209	111.564	**20**
21	2.279	.4388	.0313	.0713	31.969	14.029	8.578	120.341	**21**
22	2.370	.4220	.0292	.0692	34.248	14.451	8.941	129.202	**22**
23	2.465	.4057	.0273	.0673	36.618	14.857	9.297	138.128	**23**
24	2.563	.3901	.0256	.0656	39.083	15.247	9.648	147.101	**24**
25	2.666	.3751	.0240	.0640	41.646	15.622	9.993	156.104	**25**
26	2.772	.3607	.0226	.0626	44.312	15.983	10.331	165.121	**26**
27	2.883	.3468	.0212	.0612	47.084	16.330	10.664	174.138	**27**
28	2.999	.3335	.0200	.0600	49.968	16.663	10.991	183.142	**28**
29	3.119	.3207	.0189	.0589	52.966	16.984	11.312	192.120	**29**
30	3.243	.3083	.0178	.0578	56.085	17.292	11.627	201.062	**30**
31	3.373	.2965	.0169	.0569	59.328	17.588	11.937	209.955	**31**
32	3.508	.2851	.0159	.0559	62.701	17.874	12.241	218.792	**32**
33	3.648	.2741	.0151	.0551	66.209	18.148	12.540	227.563	**33**
34	3.794	.2636	.0143	.0543	69.858	18.411	12.832	236.260	**34**
35	3.946	.2534	.0136	.0536	73.652	18.665	13.120	244.876	**35**
40	4.801	.2083	.0105	.0505	95.025	19.793	14.476	286.530	**40**
45	5.841	.1712	.00826	.0483	121.029	20.720	15.705	325.402	**45**
50	7.107	.1407	.00655	.0466	152.667	21.482	16.812	361.163	**50**
55	8.646	.1157	.00523	.0452	191.159	22.109	17.807	393.689	**55**
60	10.520	.0951	.00420	.0442	237.990	22.623	18.697	422.996	**60**
65	12.799	.0781	.00339	.0434	294.968	23.047	19.491	449.201	**65**
70	15.572	.0642	.00275	.0427	364.290	23.395	20.196	472.479	**70**
75	18.945	.0528	.00223	.0422	448.630	23.680	20.821	493.041	**75**
80	23.050	.0434	.00181	.0418	551.244	23.915	21.372	511.116	**80**
85	28.044	.0357	.00148	.0415	676.089	24.109	21.857	526.938	**85**
90	34.119	.0293	.00121	.0412	827.981	24.267	22.283	540.737	**90**
95	41.511	.0241	.00099	.0410	1 012.8	24.398	22.655	552.730	**95**
100	50.505	.0198	.00081	.0408	1 237.6	24.505	22.980	563.125	**100**

4½% Compound Interest Factors 4½%

	Single Payment		Uniform Payment Series				Arithmetic Gradient		
	Compound Amount Factor	Present Worth Factor	Sinking Fund Factor	Capital Recovery Factor	Compound Amount Factor	Present Worth Factor	Gradient Uniform Series	Gradient Present Worth	
n	Find *F* Given *P* *F/P*	Find *P* Given *F* *P/F*	Find *A* Given *F* *A/F*	Find *A* Given *P* *A/P*	Find *F* Given *A* *F/A*	Find *P* Given *A* *P/A*	Find *A* Given *G* *A/G*	Find *P* Given *G* *P/G*	*n*
1	1.045	.9569	1.0000	1.0450	1.000	0.957	0	0	**1**
2	1.092	.9157	.4890	.5340	2.045	1.873	0.489	0.916	**2**
3	1.141	.8763	.3188	.3638	3.137	2.749	0.971	2.668	**3**
4	1.193	.8386	.2337	.2787	4.278	3.588	1.445	5.184	**4**
5	1.246	.8025	.1828	.2278	5.471	4.390	1.912	8.394	**5**
6	1.302	.7679	.1489	.1939	6.717	5.158	2.372	12.233	**6**
7	1.361	.7348	.1247	.1697	8.019	5.893	2.824	16.642	**7**
8	1.422	.7032	.1066	.1516	9.380	6.596	3.269	21.564	**8**
9	1.486	.6729	.0926	.1376	10.802	7.269	3.707	26.948	**9**
10	1.553	.6439	.0814	.1264	12.288	7.913	4.138	32.743	**10**
11	1.623	.6162	.0722	.1172	13.841	8.529	4.562	38.905	**11**
12	1.696	.5897	.0647	.1097	15.464	9.119	4.978	45.391	**12**
13	1.772	.5643	.0583	.1033	17.160	9.683	5.387	52.163	**13**
14	1.852	.5400	.0528	.0978	18.932	10.223	5.789	59.182	**14**
15	1.935	.5167	.0481	.0931	20.784	10.740	6.184	66.416	**15**
16	2.022	.4945	.0440	.0890	22.719	11.234	6.572	73.833	**16**
17	2.113	.4732	.0404	.0854	24.742	11.707	6.953	81.404	**17**
18	2.208	.4528	.0372	.0822	26.855	12.160	7.327	89.102	**18**
19	2.308	.4333	.0344	.0794	29.064	12.593	7.695	96.901	**19**
20	2.412	.4146	.0319	.0769	31.371	13.008	8.055	104.779	**20**
21	2.520	.3968	.0296	.0746	33.783	13.405	8.409	112.715	**21**
22	2.634	.3797	.0275	.0725	36.303	13.784	8.755	120.689	**22**
23	2.752	.3634	.0257	.0707	38.937	14.148	9.096	128.682	**23**
24	2.876	.3477	.0240	.0690	41.689	14.495	9.429	136.680	**24**
25	3.005	.3327	.0224	.0674	44.565	14.828	9.756	144.665	**25**
26	3.141	.3184	.0210	.0660	47.571	15.147	10.077	152.625	**26**
27	3.282	.3047	.0197	.0647	50.711	15.451	10.391	160.547	**27**
28	3.430	.2916	.0185	.0635	53.993	15.743	10.698	168.420	**28**
29	3.584	.2790	.0174	.0624	57.423	16.022	10.999	176.232	**29**
30	3.745	.2670	.0164	.0614	61.007	16.289	11.295	183.975	**30**
31	3.914	.2555	.0154	.0604	64.752	16.544	11.583	191.640	**31**
32	4.090	.2445	.0146	.0596	68.666	16.789	11.866	199.220	**32**
33	4.274	.2340	.0137	.0587	72.756	17.023	12.143	206.707	**33**
34	4.466	.2239	.0130	.0580	77.030	17.247	12.414	214.095	**34**
35	4.667	.2143	.0123	.0573	81.497	17.461	12.679	221.380	**35**
40	5.816	.1719	.00934	.0543	107.030	18.402	13.917	256.098	**40**
45	7.248	.1380	.00720	.0522	138.850	19.156	15.020	287.732	**45**
50	9.033	.1107	.00560	.0506	178.503	19.762	15.998	316.145	**50**
55	11.256	.0888	.00439	.0494	227.918	20.248	16.860	341.375	**55**
60	14.027	.0713	.00345	.0485	289.497	20.638	17.617	363.571	**60**
65	17.481	.0572	.00273	.0477	366.237	20.951	18.278	382.946	**65**
70	21.784	.0459	.00217	.0472	461.869	21.202	18.854	399.750	**70**
75	27.147	.0368	.00172	.0467	581.043	21.404	19.354	414.242	**75**
80	33.830	.0296	.00137	.0464	729.556	21.565	19.785	426.680	**80**
85	42.158	.0237	.00109	.0461	914.630	21.695	20.157	437.309	**85**
90	52.537	.0190	.00087	.0459	1 145.3	21.799	20.476	446.359	**90**
95	65.471	.0153	.00070	.0457	1 432.7	21.883	20.749	454.039	**95**
100	81.588	.0123	.00056	.0456	1 790.9	21.950	20.981	460.537	**100**

5% Compound Interest Factors 5%

	Single Payment		Uniform Payment Series				Arithmetic Gradient		
	Compound Amount Factor	Present Worth Factor	Sinking Fund Factor	Capital Recovery Factor	Compound Amount Factor	Present Worth Factor	Gradient Uniform Series	Gradient Present Worth	
n	Find *F* Given *P* *F/P*	Find *P* Given *F* *P/F*	Find *A* Given *F* *A/F*	Find *A* Given *P* *A/P*	Find *F* Given *A* *F/A*	Find *P* Given *A* *P/A*	Find *A* Given *G* *A/G*	Find *P* Given *G* *P/G*	*n*
1	1.050	.9524	1.0000	1.0500	1.000	0.952	0	0	**1**
2	1.102	.9070	.4878	.5378	2.050	1.859	0.488	0.907	**2**
3	1.158	.8638	.3172	.3672	3.152	2.723	0.967	2.635	**3**
4	1.216	.8227	.2320	.2820	4.310	3.546	1.439	5.103	**4**
5	1.276	.7835	.1810	.2310	5.526	4.329	1.902	8.237	**5**
6	1.340	.7462	.1470	.1970	6.802	5.076	2.358	11.968	**6**
7	1.407	.7107	.1228	.1728	8.142	5.786	2.805	16.232	**7**
8	1.477	.6768	.1047	.1547	9.549	6.463	3.244	20.970	**8**
9	1.551	.6446	.0907	.1407	11.027	7.108	3.676	26.127	**9**
10	1.629	.6139	.0795	.1295	12.578	7.722	4.099	31.652	**10**
11	1.710	.5847	.0704	.1204	14.207	8.306	4.514	37.499	**11**
12	1.796	.5568	.0628	.1128	15.917	8.863	4.922	43.624	**12**
13	1.886	.5303	.0565	.1065	17.713	9.394	5.321	49.988	**13**
14	1.980	.5051	.0510	.1010	19.599	9.899	5.713	56.553	**14**
15	2.079	.4810	.0463	.0963	21.579	10.380	6.097	63.288	**15**
16	2.183	.4581	.0423	.0923	23.657	10.838	6.474	70.159	**16**
17	2.292	.4363	.0387	.0887	25.840	11.274	6.842	77.140	**17**
18	2.407	.4155	.0355	.0855	28.132	11.690	7.203	84.204	**18**
19	2.527	.3957	.0327	.0827	30.539	12.085	7.557	91.327	**19**
20	2.653	.3769	.0302	.0802	33.066	12.462	7.903	98.488	**20**
21	2.786	.3589	.0280	.0780	35.719	12.821	8.242	105.667	**21**
22	2.925	.3419	.0260	.0760	38.505	13.163	8.573	112.846	**22**
23	3.072	.3256	.0241	.0741	41.430	13.489	8.897	120.008	**23**
24	3.225	.3101	.0225	.0725	44.502	13.799	9.214	127.140	**24**
25	3.386	.2953	.0210	.0710	47.727	14.094	9.524	134.227	**25**
26	3.556	.2812	.0196	.0696	51.113	14.375	9.827	141.258	**26**
27	3.733	.2678	.0183	.0683	54.669	14.643	10.122	148.222	**27**
28	3.920	.2551	.0171	.0671	58.402	14.898	10.411	155.110	**28**
29	4.116	.2429	.0160	.0660	62.323	15.141	10.694	161.912	**29**
30	4.322	.2314	.0151	.0651	66.439	15.372	10.969	168.622	**30**
31	4.538	.2204	.0141	.0641	70.761	15.593	11.238	175.233	**31**
32	4.765	.2099	.0133	.0633	75.299	15.803	11.501	181.739	**32**
33	5.003	.1999	.0125	.0625	80.063	16.003	11.757	188.135	**33**
34	5.253	.1904	.0118	.0618	85.067	16.193	12.006	194.416	**34**
35	5.516	.1813	.0111	.0611	90.320	16.374	12.250	200.580	**35**
40	7.040	.1420	.00828	.0583	120.799	17.159	13.377	229.545	**40**
45	8.985	.1113	.00626	.0563	159.699	17.774	14.364	255.314	**45**
50	11.467	.0872	.00478	.0548	209.347	18.256	15.223	277.914	**50**
55	14.636	.0683	.00367	.0537	272.711	18.633	15.966	297.510	**55**
60	18.679	.0535	.00283	.0528	353.582	18.929	16.606	314.343	**60**
65	23.840	.0419	.00219	.0522	456.795	19.161	17.154	328.691	**65**
70	30.426	.0329	.00170	.0517	588.525	19.343	17.621	340.841	**70**
75	38.832	.0258	.00132	.0513	756.649	19.485	18.018	351.072	**75**
80	49.561	.0202	.00103	.0510	971.222	19.596	18.353	359.646	**80**
85	63.254	.0158	.00080	.0508	1 245.1	19.684	18.635	366.800	**85**
90	80.730	.0124	.00063	.0506	1 594.6	19.752	18.871	372.749	**90**
95	103.034	.00971	.00049	.0505	2 040.7	19.806	19.069	377.677	**95**
100	131.500	.00760	.00038	.0504	2 610.0	19.848	19.234	381.749	**100**

	Single Payment		Uniform Payment Series				Arithmetic Gradient		
	Compound Amount Factor	Present Worth Factor	Sinking Fund Factor	Capital Recovery Factor	Compound Amount Factor	Present Worth Factor	Gradient Uniform Series	Gradient Present Worth	
n	Find *F* Given *P* *F/P*	Find *P* Given *F* *P/F*	Find *A* Given *F* *A/F*	Find *A* Given *P* *A/P*	Find *F* Given *A* *F/A*	Find *P* Given *A* *P/A*	Find *A* Given *G* *A/G*	Find *P* Given *G* *P/G*	**n**
1	1.060	.9434	1.0000	1.0600	1.000	0.943	0	0	**1**
2	1.124	.8900	.4854	.5454	2.060	1.833	0.485	0.890	**2**
3	1.191	.8396	.3141	.3741	3.184	2.673	0.961	2.569	**3**
4	1.262	.7921	.2286	.2886	4.375	3.465	1.427	4.945	**4**
5	1.338	.7473	.1774	.2374	5.637	4.212	1.884	7.934	**5**
6	1.419	.7050	.1434	.2034	6.975	4.917	2.330	11.459	**6**
7	1.504	.6651	.1191	.1791	8.394	5.582	2.768	15.450	**7**
8	1.594	.6274	.1010	.1610	9.897	6.210	3.195	19.841	**8**
9	1.689	.5919	.0870	.1470	11.491	6.802	3.613	24.577	**9**
10	1.791	.5584	.0759	.1359	13.181	7.360	4.022	29.602	**10**
11	1.898	.5268	.0668	.1268	14.972	7.887	4.421	34.870	**11**
12	2.012	.4970	.0593	.1193	16.870	8.384	4.811	40.337	**12**
13	2.133	.4688	.0530	.1130	18.882	8.853	5.192	45.963	**13**
14	2.261	.4423	.0476	.1076	21.015	9.295	5.564	51.713	**14**
15	2.397	.4173	.0430	.1030	23.276	9.712	5.926	57.554	**15**
16	2.540	.3936	.0390	.0990	25.672	10.106	6.279	63.459	**16**
17	2.693	.3714	.0354	.0954	28.213	10.477	6.624	69.401	**17**
18	2.854	.3503	.0324	.0924	30.906	10.828	6.960	75.357	**18**
19	3.026	.3305	.0296	.0896	33.760	11.158	7.287	81.306	**19**
20	3.207	.3118	.0272	.0872	36.786	11.470	7.605	87.230	**20**
21	3.400	.2942	.0250	.0850	39.993	11.764	7.915	93.113	**21**
22	3.604	.2775	.0230	.0830	43.392	12.042	8.217	98.941	**22**
23	3.820	.2618	.0213	.0813	46.996	12.303	8.510	104.700	**23**
24	4.049	.2470	.0197	.0797	50.815	12.550	8.795	110.381	**24**
25	4.292	.2330	.0182	.0782	54.864	12.783	9.072	115.973	**25**
26	4.549	.2198	.0169	.0769	59.156	13.003	9.341	121.468	**26**
27	4.822	.2074	.0157	.0757	63.706	13.211	9.603	126.860	**27**
28	5.112	.1956	.0146	.0746	68.528	13.406	9.857	132.142	**28**
29	5.418	.1846	.0136	.0736	73.640	13.591	10.103	137.309	**29**
30	5.743	.1741	.0126	.0726	79.058	13.765	10.342	142.359	**30**
31	6.088	.1643	.0118	.0718	84.801	13.929	10.574	147.286	**31**
32	6.453	.1550	.0110	.0710	90.890	14.084	10.799	152.090	**32**
33	6.841	.1462	.0103	.0703	97.343	14.230	11.017	156.768	**33**
34	7.251	.1379	.00960	.0696	104.184	14.368	11.228	161.319	**34**
35	7.686	.1301	.00897	.0690	111.435	14.498	11.432	165.743	**35**
40	10.286	.0972	.00646	.0665	154.762	15.046	12.359	185.957	**40**
45	13.765	.0727	.00470	.0647	212.743	15.456	13.141	203.109	**45**
50	18.420	.0543	.00344	.0634	290.335	15.762	13.796	217.457	**50**
55	24.650	.0406	.00254	.0625	394.171	15.991	14.341	229.322	**55**
60	32.988	.0303	.00188	.0619	533.126	16.161	14.791	239.043	**60**
65	44.145	.0227	.00139	.0614	719.080	16.289	15.160	246.945	**65**
70	59.076	.0169	.00103	.0610	967.928	16.385	15.461	253.327	**70**
75	79.057	.0126	.00077	.0608	1 300.9	16.456	15.706	258.453	**75**
80	105.796	.00945	.00057	.0606	1 746.6	16.509	15.903	262.549	**80**
85	141.578	.00706	.00043	.0604	2 343.0	16.549	16.062	265.810	**85**
90	189.464	.00528	.00032	.0603	3 141.1	16.579	16.189	268.395	**90**
95	253.545	.00394	.00024	.0602	4 209.1	16.601	16.290	270.437	**95**
100	339.300	.00295	.00018	.0602	5 638.3	16.618	16.371	272.047	**100**

300 + 300(A/F, i, n)

Maint - Previous

Compound Interest Factors

	Single Payment		Uniform Payment Series				Arithmetic Gradient		
	Compound Amount Factor	Present Worth Factor	Sinking Fund Factor	Capital Recovery Factor	Compound Amount Factor	Present Worth Factor	Gradient Uniform Series	Gradient Present Worth	
n	Find *F* Given *P* *F/P*	Find *P* Given *F* *P/F*	Find *A* Given *F* *A/F*	Find *A* Given *P* *A/P*	Find *F* Given *A* *F/A*	Find *P* Given *A* *P/A*	Find *A* Given *G* *A/G*	Find *P* Given *G* *P/G*	*n*
1	1.070	.9346	1.0000	1.0700	1.000	0.935	0	0	**1**
2	1.145	.8734	.4831	.5531	2.070	1.808	0.483	0.873	**2**
3	1.225	.8163	.3111	.3811	3.215	2.624	0.955	2.506	**3**
4	1.311	.7629	.2252	.2952	4.440	3.387	1.416	4.795	**4**
5	1.403	.7130	.1739	.2439	5.751	4.100	1.865	7.647	**5**
6	1.501	.6663	.1398	.2098	7.153	4.767	2.303	10.978	**6**
7	1.606	.6227	.1156	.1856	8.654	5.389	2.730	14.715	**7**
8	1.718	.5820	.0975	.1675	10.260	5.971	3.147	18.789	**8**
9	1.838	.5439	.0835	.1535	11.978	6.515	3.552	23.140	**9**
10	1.967	.5083	.0724	.1424	13.816	7.024	3.946	27.716	**10**
11	2.105	.4751	.0634	.1334	15.784	7.499	4.330	32.467	**11**
12	2.252	.4440	.0559	.1259	17.888	7.943	4.703	37.351	**12**
13	2.410	.4150	.0497	.1197	20.141	8.358	5.065	42.330	**13**
14	2.579	.3878	.0443	.1143	22.551	8.745	5.417	47.372	**14**
15	2.759	.3624	.0398	.1098	25.129	9.108	5.758	52.446	**15**
16	2.952	.3387	.0359	.1059	27.888	9.447	6.090	57.527	**16**
17	3.159	.3166	.0324	.1024	30.840	9.763	6.411	62.592	**17**
18	3.380	.2959	.0294	.0994	33.999	10.059	6.722	67.622	**18**
19	3.617	.2765	.0268	.0968	37.379	10.336	7.024	72.599	**19**
20	3.870	.2584	.0244	.0944	40.996	10.594	7.316	77.509	**20**
21	4.141	.2415	.0223	.0923	44.865	10.836	7.599	82.339	**21**
22	4.430	.2257	.0204	.0904	49.006	11.061	7.872	87.079	**22**
23	4.741	.2109	.0187	.0887	53.436	11.272	8.137	91.720	**23**
24	5.072	.1971	.0172	.0872	58.177	11.469	8.392	96.255	**24**
25	5.427	.1842	.0158	.0858	63.249	11.654	8.639	100.677	**25**
26	5.807	.1722	.0146	.0846	68.677	11.826	8.877	104.981	**26**
27	6.214	.1609	.0134	.0834	74.484	11.987	9.107	109.166	**27**
28	6.649	.1504	.0124	.0824	80.698	12.137	9.329	113.227	**28**
29	7.114	.1406	.0114	.0814	87.347	12.278	9.543	117.162	**29**
30	7.612	.1314	.0106	.0806	94.461	12.409	9.749	120.972	**30**
31	8.145	.1228	.00980	.0798	102.073	12.532	9.947	124.655	**31**
32	8.715	.1147	.00907	.0791	110.218	12.647	10.138	128.212	**32**
33	9.325	.1072	.00841	.0784	118.934	12.754	10.322	131.644	**33**
34	9.978	.1002	.00780	.0778	128.259	12.854	10.499	134.951	**34**
35	10.677	.0937	.00723	.0772	138.237	12.948	10.669	138.135	**35**
40	14.974	.0668	.00501	.0750	199.636	13.332	11.423	152.293	**40**
45	21.002	.0476	.00350	.0735	285.750	13.606	12.036	163.756	**45**
50	29.457	.0339	.00246	.0725	406.530	13.801	12.529	172.905	**50**
55	41.315	.0242	.00174	.0717	575.930	13.940	12.921	180.124	**55**
60	57.947	.0173	.00123	.0712	813.523	14.039	13.232	185.768	**60**
65	81.273	.0123	.00087	.0709	1 146.8	14.110	13.476	190.145	**65**
70	113.990	.00877	.00062	.0706	1 614.1	14.160	13.666	193.519	**70**
75	159.877	.00625	.00044	.0704	2 269.7	14.196	13.814	196.104	**75**
80	224.235	.00446	.00031	.0703	3 189.1	14.222	13.927	198.075	**80**
85	314.502	.00318	.00022	.0702	4 478.6	14.240	14.015	199.572	**85**
90	441.105	.00227	.00016	.0702	6 287.2	14.253	14.081	200.704	**90**
95	618.673	.00162	.00011	.0701	8 823.9	14.263	14.132	201.558	**95**
100	867.720	.00115	.00008	.0701	12 381.7	14.269	14.170	202.200	**100**

	Single Payment		Uniform Payment Series				Arithmetic Gradient		
	Compound Amount Factor	Present Worth Factor	Sinking Fund Factor	Capital Recovery Factor	Compound Amount Factor	Present Worth Factor	Gradient Uniform Series	Gradient Present Worth	
n	Find *F* Given *P* *F/P*	Find *P* Given *F* *P/F*	Find *A* Given *F* *A/F*	Find *A* Given *P* *A/P*	Find *F* Given *A* *F/A*	Find *P* Given *A* *P/A*	Find *A* Given *G* *A/G*	Find *P* Given *G* *P/G*	***n***
1	1.080	.9259	1.0000	1.0800	1.000	0.926	0	0	**1**
2	1.166	.8573	.4808	.5608	2.080	1.783	0.481	0.857	**2**
3	1.260	.7938	.3080	.3880	3.246	2.577	0.949	2.445	**3**
4	1.360	.7350	.2219	.3019	4.506	3.312	1.404	4.650	**4**
5	1.469	.6806	.1705	.2505	5.867	3.993	1.846	7.372	**5**
6	1.587	.6302	.1363	.2163	7.336	4.623	2.276	10.523	**6**
7	1.714	.5835	.1121	.1921	8.923	5.206	2.694	14.024	**7**
8	1.851	.5403	.0940	.1740	10.637	5.747	3.099	17.806	**8**
9	1.999	.5002	.0801	.1601	12.488	6.247	3.491	21.808	**9**
10	2.159	.4632	.0690	.1490	14.487	6.710	3.871	25.977	**10**
11	2.332	.4289	.0601	.1401	16.645	7.139	4.240	30.266	**11**
12	2.518	.3971	.0527	.1327	18.977	7.536	4.596	34.634	**12**
13	2.720	.3677	.0465	.1265	21.495	7.904	4.940	39.046	**13**
14	2.937	.3405	.0413	.1213	24.215	8.244	5.273	43.472	**14**
15	3.172	.3152	.0368	.1168	27.152	8.559	5.594	47.886	**15**
16	3.426	.2919	.0330	.1130	30.324	8.851	5.905	52.264	**16**
17	3.700	.2703	.0296	.1096	33.750	9.122	6.204	56.588	**17**
18	3.996	.2502	.0267	.1067	37.450	9.372	6.492	60.843	**18**
19	4.316	.2317	.0241	.1041	41.446	9.604	6.770	65.013	**19**
20	4.661	.2145	.0219	.1019	45.762	9.818	7.037	69.090	**20**
21	5.034	.1987	.0198	.0998	50.423	10.017	7.294	73.063	**21**
22	5.437	.1839	.0180	.0980	55.457	10.201	7.541	76.926	**22**
23	5.871	.1703	.0164	.0964	60.893	10.371	7.779	80.673	**23**
24	6.341	.1577	.0150	.0950	66.765	10.529	8.007	84.300	**24**
25	6.848	.1460	.0137	.0937	73.106	10.675	8.225	87.804	**25**
26	7.396	.1352	.0125	.0925	79.954	10.810	8.435	91.184	**26**
27	7.988	.1252	.0114	.0914	87.351	10.935	8.636	94.439	**27**
28	8.627	.1159	.0105	.0905	95.339	11.051	8.829	97.569	**28**
29	9.317	.1073	.00962	.0896	103.966	11.158	9.013	100.574	**29**
30	10.063	.0994	.00883	.0888	113.283	11.258	9.190	103.456	**30**
31	10.868	.0920	.00811	.0881	123.346	11.350	9.358	106.216	**31**
32	11.737	.0852	.00745	.0875	134.214	11.435	9.520	108.858	**32**
33	12.676	.0789	.00685	.0869	145.951	11.514	9.674	111.382	**33**
34	13.690	.0730	.00630	.0863	158.627	11.587	9.821	113.792	**34**
35	14.785	.0676	.00580	.0858	172.317	11.655	9.961	116.092	**35**
40	21.725	.0460	.00386	.0839	259.057	11.925	10.570	126.042	**40**
45	31.920	.0313	.00259	.0826	386.506	12.108	11.045	133.733	**45**
50	46.902	.0213	.00174	.0817	573.771	12.233	11.411	139.593	**50**
55	68.914	.0145	.00118	.0812	848.925	12.319	11.690	144.006	**55**
60	101.257	.00988	.00080	.0808	1 253.2	12.377	11.902	147.300	**60**
65	148.780	.00672	.00054	.0805	1 847.3	12.416	12.060	149.739	**65**
70	218.607	.00457	.00037	.0804	2 720.1	12.443	12.178	151.533	**70**
75	321.205	.00311	.00025	.0802	4 002.6	12.461	12.266	152.845	**75**
80	471.956	.00212	.00017	.0802	5 887.0	12.474	12.330	153.800	**80**
85	693.458	.00144	.00012	.0801	8 655.7	12.482	12.377	154.492	**85**
90	1 018.9	.00098	.00008	.0801	12 724.0	12.488	12.412	154.993	**90**
95	1 497.1	.00067	.00005	.0801	18 701.6	12.492	12.437	155.352	**95**
100	2 199.8	.00045	.00004	.0800	27 484.6	12.494	12.455	155.611	**100**

(P-S)(A/P, i, n) + Si

SOYD = N/sum (P-S)   2/N (P - sum)

	Single Payment		Uniform Payment Series				Arithmetic Gradient		
	Compound Amount Factor	Present Worth Factor	Sinking Fund Factor	Capital Recovery Factor	Compound Amount Factor	Present Worth Factor	Gradient Uniform Series	Gradient Present Worth	
n	Find *F* Given *P* *F/P*	Find *P* Given *F* *P/F*	Find *A* Given *F* *A/F*	Find *A* Given *P* *A/P*	Find *F* Given *A* *F/A*	Find *P* Given *A* *P/A*	Find *A* Given *G* *A/G*	Find *P* Given *G* *P/G*	*n*
1	1.090	.9174	1.0000	1.0900	1.000	0.917	0	0	**1**
2	1.188	.8417	.4785	.5685	2.090	1.759	0.478	0.842	**2**
3	1.295	.7722	.3051	.3951	3.278	2.531	0.943	2.386	**3**
4	1.412	.7084	.2187	.3087	4.573	3.240	1.393	4.511	**4**
5	1.539	.6499	.1671	.2571	5.985	3.890	1.828	7.111	**5**
6	1.677	.5963	.1329	.2229	7.523	4.486	2.250	10.092	**6**
7	1.828	.5470	.1087	.1987	9.200	5.033	2.657	13.375	**7**
8	1.993	.5019	.0907	.1807	11.028	5.535	3.051	16.888	**8**
9	2.172	.4604	.0768	.1668	13.021	5.995	3.431	20.571	**9**
10	2.367	.4224	.0658	.1558	15.193	6.418	3.798	24.373	**10**
11	2.580	.3875	.0569	.1469	17.560	6.805	4.151	28.248	**11**
12	2.813	.3555	.0497	.1397	20.141	7.161	4.491	32.159	**12**
13	3.066	.3262	.0436	.1336	22.953	7.487	4.818	36.073	**13**
14	3.342	.2992	.0384	.1284	26.019	7.786	5.133	39.963	**14**
15	3.642	.2745	.0341	.1241	29.361	8.061	5.435	43.807	**15**
16	3.970	.2519	.0303	.1203	33.003	8.313	5.724	47.585	**16**
17	4.328	.2311	.0270	.1170	36.974	8.544	6.002	51.282	**17**
18	4.717	.2120	.0242	.1142	41.301	8.756	6.269	54.886	**18**
19	5.142	.1945	.0217	.1117	46.019	8.950	6.524	58.387	**19**
20	5.604	.1784	.0195	.1095	51.160	9.129	6.767	61.777	**20**
21	6.109	.1637	.0176	.1076	56.765	9.292	7.001	65.051	**21**
22	6.659	.1502	.0159	.1059	62.873	9.442	7.223	68.205	**22**
23	7.258	.1378	.0144	.1044	69.532	9.580	7.436	71.236	**23**
24	7.911	.1264	.0130	.1030	76.790	9.707	7.638	74.143	**24**
25	8.623	.1160	.0118	.1018	84.701	9.823	7.832	76.927	**25**
26	9.399	.1064	.0107	.1007	93.324	9.929	8.016	79.586	**26**
27	10.245	.0976	.00973	.0997	102.723	10.027	8.191	82.124	**27**
28	11.167	.0895	.00885	.0989	112.968	10.116	8.357	84.542	**28**
29	12.172	.0822	.00806	.0981	124.136	10.198	8.515	86.842	**29**
30	13.268	.0754	.00734	.0973	136.308	10.274	8.666	89.028	**30**
31	14.462	.0691	.00669	.0967	149.575	10.343	8.808	91.102	**31**
32	15.763	.0634	.00610	.0961	164.037	10.406	8.944	93.069	**32**
33	17.182	.0582	.00556	.0956	179.801	10.464	9.072	94.931	**33**
34	18.728	.0534	.00508	.0951	196.983	10.518	9.193	96.693	**34**
35	20.414	.0490	.00464	.0946	215.711	10.567	9.308	98.359	**35**
40	31.409	.0318	.00296	.0930	337.883	10.757	9.796	105.376	**40**
45	48.327	.0207	.00190	.0919	525.860	10.881	10.160	110.556	**45**
50	74.358	.0134	.00123	.0912	815.085	10.962	10.430	114.325	**50**
55	114.409	.00874	.00079	.0908	1 260.1	11.014	10.626	117.036	**55**
60	176.032	.00568	.00051	.0905	1 944.8	11.048	10.768	118.968	**60**
65	270.847	.00369	.00033	.0903	2 998.3	11.070	10.870	120.334	**65**
70	416.731	.00240	.00022	.0902	4 619.2	11.084	10.943	121.294	**70**
75	641.193	.00156	.00014	.0901	7 113.3	11.094	10.994	121.965	**75**
80	986.555	.00101	.00009	.0901	10 950.6	11.100	11.030	122.431	**80**
85	1 517.9	.00066	.00006	.0901	16 854.9	11.104	11.055	122.753	**85**
90	2 335.5	.00043	.00004	.0900	25 939.3	11.106	11.073	122.976	**90**
95	3 593.5	.00028	.00003	.0900	39 916.8	11.108	11.085	123.129	**95**
100	5 529.1	.00018	.00002	.0900	61 422.9	11.109	11.093	123.233	**100**

Compound Interest Factors

10%

	Single Payment		Uniform Payment Series				Arithmetic Gradient		
	Compound Amount Factor	Present Worth Factor	Sinking Fund Factor	Capital Recovery Factor	Compound Amount Factor	Present Worth Factor	Gradient Uniform Series	Gradient Present Worth	
n	Find *F* Given *P* *F/P*	Find *P* Given *F* *P/F*	Find *A* Given *F* *A/F*	Find *A* Given *P* *A/P*	Find *F* Given *A* *F/A*	Find *P* Given *A* *P/A*	Find *A* Given *G* *A/G*	Find *P* Given *G* *P/G*	*n*
1	1.100	.9091	1.0000	1.1000	1.000	0.909	0	0	**1**
2	1.210	.8264	.4762	.5762	2.100	1.736	0.476	0.826	**2**
3	1.331	.7513	.3021	.4021	3.310	2.487	0.937	2.329	**3**
4	1.464	.6830	.2155	.3155	4.641	3.170	1.381	4.378	**4**
5	1.611	.6209	.1638	.2638	6.105	3.791	1.810	6.862	**5**
6	1.772	.5645	.1296	.2296	7.716	4.355	2.224	9.684	**6**
7	1.949	.5132	.1054	.2054	9.487	4.868	2.622	12.763	**7**
8	2.144	.4665	.0874	.1874	11.436	5.335	3.004	16.029	**8**
9	2.358	.4241	.0736	.1736	13.579	5.759	3.372	19.421	**9**
10	2.594	.3855	.0627	.1627	15.937	6.145	3.725	22.891	**10**
11	2.853	.3505	.0540	.1540	18.531	6.495	4.064	26.396	**11**
12	3.138	.3186	.0468	.1468	21.384	6.814	4.388	29.901	**12**
13	3.452	.2897	.0408	.1408	24.523	7.103	4.699	33.377	**13**
14	3.797	.2633	.0357	.1357	27.975	7.367	4.996	36.801	**14**
15	4.177	.2394	.0315	.1315	31.772	7.606	5.279	40.152	**15**
16	4.595	.2176	.0278	.1278	35.950	7.824	5.549	43.416	**16**
17	5.054	.1978	.0247	.1247	40.545	8.022	5.807	46.582	**17**
18	5.560	.1799	.0219	.1219	45.599	8.201	6.053	49.640	**18**
19	6.116	.1635	.0195	.1195	51.159	8.365	6.286	52.583	**19**
20	6.728	.1486	.0175	.1175	57.275	8.514	6.508	55.407	**20**
21	7.400	.1351	.0156	.1156	64.003	8.649	6.719	58.110	**21**
22	8.140	.1228	.0140	.1140	71.403	8.772	6.919	60.689	**22**
23	8.954	.1117	.0126	.1126	79.543	8.883	7.108	63.146	**23**
24	9.850	.1015	.0113	.1113	88.497	8.985	7.288	65.481	**24**
25	10.835	.0923	.0102	.1102	98.347	9.077	7.458	67.696	**25**
26	11.918	.0839	.00916	.1092	109.182	9.161	7.619	69.794	**26**
27	13.110	.0763	.00826	.1083	121.100	9.237	7.770	71.777	**27**
28	14.421	.0693	.00745	.1075	134.210	9.307	7.914	73.650	**28**
29	15.863	.0630	.00673	.1067	148.631	9.370	8.049	75.415	**29**
30	17.449	.0573	.00608	.1061	164.494	9.427	8.176	77.077	**30**
31	19.194	.0521	.00550	.1055	181.944	9.479	8.296	78.640	**31**
32	21.114	.0474	.00497	.1050	201.138	9.526	8.409	80.108	**32**
33	23.225	.0431	.00450	.1045	222.252	9.569	8.515	81.486	**33**
34	25.548	.0391	.00407	.1041	245.477	9.609	8.615	82.777	**34**
35	28.102	.0356	.00369	.1037	271.025	9.644	8.709	83.987	**35**
40	45.259	.0221	.00226	.1023	442.593	9.779	9.096	88.953	**40**
45	72.891	.0137	.00139	.1014	718.905	9.863	9.374	92.454	**45**
50	117.391	.00852	.00086	.1009	1 163.9	9.915	9.570	94.889	**50**
55	189.059	.00529	.00053	.1005	1 880.6	9.947	9.708	96.562	**55**
60	304.482	.00328	.00033	.1003	3 034.8	9.967	9.802	97.701	**60**
65	490.371	.00204	.00020	.1002	4 893.7	9.980	9.867	98.471	**65**
70	789.748	.00127	.00013	.1001	7 887.5	9.987	9.911	98.987	**70**
75	1 271.9	.00079	.00008	.1001	12 709.0	9.992	9.941	99.332	**75**
80	2 048.4	.00049	.00005	.1000	20 474.0	9.995	9.961	99.561	**80**
85	3 299.0	.00030	.00003	.1000	32 979.7	9.997	9.974	99.712	**85**
90	5 313.0	.00019	.00002	.1000	53 120.3	9.998	9.983	99.812	**90**
95	8 556.7	.00012	.00001	.1000	85 556.9	9.999	9.989	99.877	**95**
100	13 780.6	.00007	.00001	.1000	137 796.3	9.999	9.993	99.920	**100**

12% Compound Interest Factors 12%

n	Single Payment: Compound Amount Factor	Single Payment: Present Worth Factor	Uniform Payment Series: Sinking Fund Factor	Uniform Payment Series: Capital Recovery Factor	Uniform Payment Series: Compound Amount Factor	Uniform Payment Series: Present Worth Factor	Arithmetic Gradient: Gradient Uniform Series	Arithmetic Gradient: Gradient Present Worth	n
	Find F Given P F/P	Find P Given F P/F	Find A Given F A/F	Find A Given P A/P	Find F Given A F/A	Find P Given A P/A	Find A Given G A/G	Find P Given G P/G	
1	1.120	.8929	1.0000	1.1200	1.000	0.893	0	0	**1**
2	1.254	.7972	.4717	.5917	2.120	1.690	0.472	0.797	**2**
3	1.405	.7118	.2963	.4163	3.374	2.402	0.925	2.221	**3**
4	1.574	.6355	.2092	.3292	4.779	3.037	1.359	4.127	**4**
5	1.762	.5674	.1574	.2774	6.353	3.605	1.775	6.397	**5**
6	1.974	.5066	.1232	.2432	8.115	4.111	2.172	8.930	**6**
7	2.211	.4523	.0991	.2191	10.089	4.564	2.551	11.644	**7**
8	2.476	.4039	.0813	.2013	12.300	4.968	2.913	14.471	**8**
9	2.773	.3606	.0677	.1877	14.776	5.328	3.257	17.356	**9**
10	3.106	.3220	.0570	.1770	17.549	5.650	3.585	20.254	**10**
11	3.479	.2875	.0484	.1684	20.655	5.938	3.895	23.129	**11**
12	3.896	.2567	.0414	.1614	24.133	6.194	4.190	25.952	**12**
13	4.363	.2292	.0357	.1557	28.029	6.424	4.468	28.702	**13**
14	4.887	.2046	.0309	.1509	32.393	6.628	4.732	31.362	**14**
15	5.474	.1827	.0268	.1468	37.280	6.811	4.980	33.920	**15**
16	6.130	.1631	.0234	.1434	42.753	6.974	5.215	36.367	**16**
17	6.866	.1456	.0205	.1405	48.884	7.120	5.435	38.697	**17**
18	7.690	.1300	.0179	.1379	55.750	7.250	5.643	40.908	**18**
19	8.613	.1161	.0158	.1358	63.440	7.366	5.838	42.998	**19**
20	9.646	.1037	.0139	.1339	72.052	7.469	6.020	44.968	**20**
21	10.804	.0926	.0122	.1322	81.699	7.562	6.191	46.819	**21**
22	12.100	.0826	.0108	.1308	92.503	7.645	6.351	48.554	**22**
23	13.552	.0738	.00956	.1296	104.603	7.718	6.501	50.178	**23**
24	15.179	.0659	.00846	.1285	118.155	7.784	6.641	51.693	**24**
25	17.000	.0588	.00750	.1275	133.334	7.843	6.771	53.105	**25**
26	19.040	.0525	.00665	.1267	150.334	7.896	6.892	54.418	**26**
27	21.325	.0469	.00590	.1259	169.374	7.943	7.005	55.637	**27**
28	23.884	.0419	.00524	.1252	190.699	7.984	7.110	56.767	**28**
29	26.750	.0374	.00466	.1247	214.583	8.022	7.207	57.814	**29**
30	29.960	.0334	.00414	.1241	241.333	8.055	7.297	58.782	**30**
31	33.555	.0298	.00369	.1237	271.293	8.085	7.381	59.676	**31**
32	37.582	.0266	.00328	.1233	304.848	8.112	7.459	60.501	**32**
33	42.092	.0238	.00292	.1229	342.429	8.135	7.530	61.261	**33**
34	47.143	.0212	.00260	.1226	384.521	8.157	7.596	61.961	**34**
35	52.800	.0189	.00232	.1223	431.663	8.176	7.658	62.605	**35**
40	93.051	.0107	.00130	.1213	767.091	8.244	7.899	65.116	**40**
45	163.988	.00610	.00074	.1207	1 358.2	8.283	8.057	66.734	**45**
50	289.002	.00346	.00042	.1204	2 400.0	8.304	8.160	67.762	**50**
55	509.321	.00196	.00024	.1202	4 236.0	8.317	8.225	68.408	**55**
60	897.597	.00111	.00013	.1201	7 471.6	8.324	8.266	68.810	**60**
65	1 581.9	.00063	.00008	.1201	13 173.9	8.328	8.292	69.058	**65**
70	2 787.8	.00036	.00004	.1200	23 223.3	8.330	8.308	69.210	**70**
75	4 913.1	.00020	.00002	.1200	40 933.8	8.332	8.318	69.303	**75**
80	8 658.5	.00012	.00001	.1200	72 145.7	8.332	8.324	69.359	**80**
85	15 259.2	.00007	.00001	.1200	127 151.7	8.333	8.328	69.393	**85**
90	26 891.9	.00004		.1200	224 091.1	8.333	8.330	69.414	**90**
95	47 392.8	.00002		.1200	394 931.4	8.333	8.331	69.426	**95**
100	83 522.3	.00001		.1200	696 010.5	8.333	8.332	69.434	**100**

15% Compound Interest Factors 15%

	Single Payment		Uniform Payment Series				Arithmetic Gradient		
	Compound Amount Factor	Present Worth Factor	Sinking Fund Factor	Capital Recovery Factor	Compound Amount Factor	Present Worth Factor	Gradient Uniform Series	Gradient Present Worth	
n	Find *F* Given *P* *F/P*	Find *P* Given *F* *P/F*	Find *A* Given *F* *A/F*	Find *A* Given *P* *A/P*	Find *F* Given *A* *F/A*	Find *P* Given *A* *P/A*	Find *A* Given *G* *A/G*	Find *P* Given *G* *P/G*	*n*
1	1.150	.8696	1.0000	1.1500	1.000	0.870	0	0	**1**
2	1.322	.7561	.4651	.6151	2.150	1.626	0.465	0.756	**2**
3	1.521	.6575	.2880	.4380	3.472	2.283	0.907	2.071	**3**
4	1.749	.5718	.2003	.3503	4.993	2.855	1.326	3.786	**4**
5	2.011	.4972	.1483	.2983	6.742	3.352	1.723	5.775	**5**
6	2.313	.4323	.1142	.2642	8.754	3.784	2.097	7.937	**6**
7	2.660	.3759	.0904	.2404	11.067	4.160	2.450	10.192	**7**
8	3.059	.3269	.0729	.2229	13.727	4.487	2.781	12.481	**8**
9	3.518	.2843	.0596	.2096	16.786	4.772	3.092	14.755	**9**
10	4.046	.2472	.0493	.1993	20.304	5.019	3.383	16.979	**10**
11	4.652	.2149	.0411	.1911	24.349	5.234	3.655	19.129	**11**
12	5.350	.1869	.0345	.1845	29.002	5.421	3.908	21.185	**12**
13	6.153	.1625	.0291	.1791	34.352	5.583	4.144	23.135	**13**
14	7.076	.1413	.0247	.1747	40.505	5.724	4.362	24.972	**14**
15	8.137	.1229	.0210	.1710	47.580	5.847	4.565	26.693	**15**
16	9.358	.1069	.0179	.1679	55.717	5.954	4.752	28.296	**16**
17	10.761	.0929	.0154	.1654	65.075	6.047	4.925	29.783	**17**
18	12.375	.0808	.0132	.1632	75.836	6.128	5.084	31.156	**18**
19	14.232	.0703	.0113	.1613	88.212	6.198	5.231	32.421	**19**
20	16.367	.0611	.00976	.1598	102.444	6.259	5.365	33.582	**20**
21	18.822	.0531	.00842	.1584	118.810	6.312	5.488	34.645	**21**
22	21.645	.0462	.00727	.1573	137.632	6.359	5.601	35.615	**22**
23	24.891	.0402	.00628	.1563	159.276	6.399	5.704	36.499	**23**
24	28.625	.0349	.00543	.1554	184.168	6.434	5.798	37.302	**24**
25	32.919	.0304	.00470	.1547	212.793	6.464	5.883	38.031	**25**
26	37.857	.0264	.00407	.1541	245.712	6.491	5.961	38.692	**26**
27	43.535	.0230	.00353	.1535	283.569	6.514	6.032	39.289	**27**
28	50.066	.0200	.00306	.1531	327.104	6.534	6.096	39.828	**28**
29	57.575	.0174	.00265	.1527	377.170	6.551	6.154	40.315	**29**
30	66.212	.0151	.00230	.1523	434.745	6.566	6.207	40.753	**30**
31	76.144	.0131	.00200	.1520	500.957	6.579	6.254	41.147	**31**
32	87.565	.0114	.00173	.1517	577.100	6.591	6.297	41.501	**32**
33	100.700	.00993	.00150	.1515	664.666	6.600	6.336	41.818	**33**
34	115.805	.00864	.00131	.1513	765.365	6.609	6.371	42.103	**34**
35	133.176	.00751	.00113	.1511	881.170	6.617	6.402	42.359	**35**
40	267.864	.00373	.00056	.1506	1 779.1	6.642	6.517	43.283	**40**
45	538.769	.00186	.00028	.1503	3 585.1	6.654	6.583	43.805	**45**
50	1 083.7	.00092	.00014	.1501	7 217.7	6.661	6.620	44.096	**50**
55	2 179.6	.00046	.00007	.1501	14 524.1	6.664	6.641	44.256	**55**
60	4 384.0	.00023	.00003	.1500	29 220.0	6.665	6.653	44.343	**60**
65	8 817.8	.00011	.00002	.1500	58 778.6	6.666	6.659	44.390	**65**
70	17 735.7	.00006	.00001	.1500	118 231.5	6.666	6.663	44.416	**70**
75	35 672.9	.00003		.1500	237 812.5	6.666	6.665	44.429	**75**
80	71 750.9	.00001		.1500	478 332.6	6.667	6.666	44.436	**80**
85	144 316.7	.00001		.1500	962 104.4	6.667	6.666	44.440	**85**

ODB SUM EOY SL

EOY−S / n−1

18% Compound Interest Factors 18%

	Single Payment		Uniform Payment Series				Arithmetic Gradient		
	Compound Amount Factor	Present Worth Factor	Sinking Fund Factor	Capital Recovery Factor	Compound Amount Factor	Present Worth Factor	Gradient Uniform Series	Gradient Present Worth	
n	Find *F* Given *P* *F/P*	Find *P* Given *F* *P/F*	Find *A* Given *F* *A/F*	Find *A* Given *P* *A/P*	Find *F* Given *A* *F/A*	Find *P* Given *A* *P/A*	Find *A* Given *G* *A/G*	Find *P* Given *G* *P/G*	**n**
1	1.180	.8475	1.0000	1.1800	1.000	0.847	0	0	**1**
2	1.392	.7182	.4587	.6387	2.180	1.566	0.459	0.718	**2**
3	1.643	.6086	.2799	.4599	3.572	2.174	0.890	1.935	**3**
4	1.939	.5158	.1917	.3717	5.215	2.690	1.295	3.483	**4**
5	2.288	.4371	.1398	.3198	7.154	3.127	1.673	5.231	**5**
6	2.700	.3704	.1059	.2859	9.442	3.498	2.025	7.083	**6**
7	3.185	.3139	.0824	.2624	12.142	3.812	2.353	8.967	**7**
8	3.759	.2660	.0652	.2452	15.327	4.078	2.656	10.829	**8**
9	4.435	.2255	.0524	.2324	19.086	4.303	2.936	12.633	**9**
10	5.234	.1911	.0425	.2225	23.521	4.494	3.194	14.352	**10**
11	6.176	.1619	.0348	.2148	28.755	4.656	3.430	15.972	**11**
12	7.288	.1372	.0286	.2086	34.931	4.793	3.647	17.481	**12**
13	8.599	.1163	.0237	.2037	42.219	4.910	3.845	18.877	**13**
14	10.147	.0985	.0197	.1997	50.818	5.008	4.025	20.158	**14**
15	11.974	.0835	.0164	.1964	60.965	5.092	4.189	21.327	**15**
16	14.129	.0708	.0137	.1937	72.939	5.162	4.337	22.389	**16**
17	16.672	.0600	.0115	.1915	87.068	5.222	4.471	23.348	**17**
18	19.673	.0508	.00964	.1896	103.740	5.273	4.592	24.212	**18**
19	23.214	.0431	.00810	.1881	123.413	5.316	4.700	24.988	**19**
20	27.393	.0365	.00682	.1868	146.628	5.353	4.798	25.681	**20**
21	32.324	.0309	.00575	.1857	174.021	5.384	4.885	26.300	**21**
22	38.142	.0262	.00485	.1848	206.345	5.410	4.963	26.851	**22**
23	45.008	.0222	.00409	.1841	244.487	5.432	5.033	27.339	**23**
24	53.109	.0188	.00345	.1835	289.494	5.451	5.095	27.772	**24**
25	62.669	.0160	.00292	.1829	342.603	5.467	5.150	28.155	**25**
26	73.949	.0135	.00247	.1825	405.272	5.480	5.199	28.494	**26**
27	87.260	.0115	.00209	.1821	479.221	5.492	5.243	28.791	**27**
28	102.966	.00971	.00177	.1818	566.480	5.502	5.281	29.054	**28**
29	121.500	.00823	.00149	.1815	669.447	5.510	5.315	29.284	**29**
30	143.370	.00697	.00126	.1813	790.947	5.517	5.345	29.486	**30**
31	169.177	.00591	.00107	.1811	934.317	5.523	5.371	29.664	**31**
32	199.629	.00501	.00091	.1809	1 103.5	5.528	5.394	29.819	**32**
33	235.562	.00425	.00077	.1808	1 303.1	5.532	5.415	29.955	**33**
34	277.963	.00360	.00065	.1806	1 538.7	5.536	5.433	30.074	**34**
35	327.997	.00305	.00055	.1806	1 816.6	5.539	5.449	30.177	**35**
40	750.377	.00133	.00024	.1802	4 163.2	5.548	5.502	30.527	**40**
45	1 716.7	.00058	.00010	.1801	9 531.6	5.552	5.529	30.701	**45**
50	3 927.3	.00025	.00005	.1800	21 813.0	5.554	5.543	30.786	**50**
55	8 984.8	.00011	.00002	.1800	49 910.1	5.555	5.549	30.827	**55**
60	20 555.1	.00005	.00001	.1800	114 189.4	5.555	5.553	30.846	**60**
65	47 025.1	.00002		.1800	261 244.7	5.555	5.554	30.856	**65**
70	107 581.9	.00001		.1800	597.671.7	5.556	5.555	30.860	**70**

20% Compound Interest Factors 20%

	Single Payment		Uniform Payment Series				Arithmetic Gradient		
	Compound Amount Factor	Present Worth Factor	Sinking Fund Factor	Capital Recovery Factor	Compound Amount Factor	Present Worth Factor	Gradient Uniform Series	Gradient Present Worth	
n	Find *F* Given *P* *F/P*	Find *P* Given *F* *P/F*	Find *A* Given *F* *A/F*	Find *A* Given *P* *A/P*	Find *F* Given *A* *F/A*	Find *P* Given *A* *P/A*	Find *A* Given *G* *A/G*	Find *P* Given *G* *P/G*	**n**
1	1.200	.8333	1.0000	1.2000	1.000	0.833	0	0	**1**
2	1.440	.6944	.4545	.6545	2.200	1.528	0.455	0.694	**2**
3	1.728	.5787	.2747	.4747	3.640	2.106	0.879	1.852	**3**
4	2.074	.4823	.1863	.3863	5.368	2.589	1.274	3.299	**4**
5	2.488	.4019	.1344	.3344	7.442	2.991	1.641	4.906	**5**
6	2.986	.3349	.1007	.3007	9.930	3.326	1.979	6.581	**6**
7	3.583	.2791	.0774	.2774	12.916	3.605	2.290	8.255	**7**
8	4.300	.2326	.0606	.2606	16.499	3.837	2.576	9.883	**8**
9	5.160	.1938	.0481	.2481	20.799	4.031	2.836	11.434	**9**
10	6.192	.1615	.0385	.2385	25.959	4.192	3.074	12.887	**10**
11	7.430	.1346	.0311	.2311	32.150	4.327	3.289	14.233	**11**
12	8.916	.1122	.0253	.2253	39.581	4.439	3.484	15.467	**12**
13	10.699	.0935	.0206	.2206	48.497	4.533	3.660	16.588	**13**
14	12.839	.0779	.0169	.2169	59.196	4.611	3.817	17.601	**14**
15	15.407	.0649	.0139	.2139	72.035	4.675	3.959	18.509	**15**
16	18.488	.0541	.0114	.2114	87.442	4.730	4.085	19.321	**16**
17	22.186	.0451	.00944	.2094	105.931	4.775	4.198	20.042	**17**
18	26.623	.0376	.00781	.2078	128.117	4.812	4.298	20.680	**18**
19	31.948	.0313	.00646	.2065	154.740	4.843	4.386	21.244	**19**
20	38.338	.0261	.00536	.2054	186.688	4.870	4.464	21.739	**20**
21	46.005	.0217	.00444	.2044	225.026	4.891	4.533	22.174	**21**
22	55.206	.0181	.00369	.2037	271.031	4.909	4.594	22.555	**22**
23	66.247	.0151	.00307	.2031	326.237	4.925	4.647	22.887	**23**
24	79.497	.0126	.00255	.2025	392.484	4.937	4.694	23.176	**24**
25	95.396	.0105	.00212	.2021	471.981	4.948	4.735	23.428	**25**
26	114.475	.00874	.00176	.2018	567.377	4.956	4.771	23.646	**26**
27	137.371	.00728	.00147	.2015	681.853	4.964	4.802	23.835	**27**
28	164.845	.00607	.00122	.2012	819.223	4.970	4.829	23.999	**28**
29	197.814	.00506	.00102	.2010	984.068	4.975	4.853	24.141	**29**
30	237.376	.00421	.00085	.2008	1 181.9	4.979	4.873	24.263	**30**
31	284.852	.00351	.00070	.2007	1 419.3	4.982	4.891	24.368	**31**
32	341.822	.00293	.00059	.2006	1 704.1	4.985	4.906	24.459	**32**
33	410.186	.00244	.00049	.2005	2 045.9	4.988	4.919	24.537	**33**
34	492.224	.00203	.00041	.2004	2 456.1	4.990	4.931	24.604	**34**
35	590.668	.00169	.00034	.2003	2 948.3	4.992	4.941	24.661	**35**
40	1 469.8	.00068	.00014	.2001	7 343.9	4.997	4.973	24.847	**40**
45	3 657.3	.00027	.00005	.2001	18 281.3	4.999	4.988	24.932	**45**
50	9 100.4	.00011	.00002	.2000	45 497.2	4.999	4.995	24.970	**50**
55	22 644.8	.00004	.00001	.2000	113 219.0	5.000	4.998	24.987	**55**
60	56 347.5	.00002		.2000	281 732.6	5.000	4.999	24.994	**60**

Compound Interest Factors

25%

	Single Payment		Uniform Payment Series				Arithmetic Gradient		
	Compound Amount Factor	Present Worth Factor	Sinking Fund Factor	Capital Recovery Factor	Compound Amount Factor	Present Worth Factor	Gradient Uniform Series	Gradient Present Worth	
n	Find *F* Given *P* *F/P*	Find *P* Given *F* *P/F*	Find *A* Given *F* *A/F*	Find *A* Given *P* *A/P*	Find *F* Given *A* *F/A*	Find *P* Given *A* *P/A*	Find *A* Given *G* *A/G*	Find *P* Given *G* *P/G*	*n*
1	1.250	.8000	1.0000	1.2500	1.000	0.800	0	0	**1**
2	1.563	.6400	.4444	.6944	2.250	1.440	0.444	0.640	**2**
3	1.953	.5120	.2623	.5123	3.813	1.952	0.852	1.664	**3**
4	2.441	.4096	.1734	.4234	5.766	2.362	1.225	2.893	**4**
5	3.052	.3277	.1218	.3718	8.207	2.689	1.563	4.204	**5**
6	3.815	.2621	.0888	.3388	11.259	2.951	1.868	5.514	**6**
7	4.768	.2097	.0663	.3163	15.073	3.161	2.142	6.773	**7**
8	5.960	.1678	.0504	.3004	19.842	3.329	2.387	7.947	**8**
9	7.451	.1342	.0388	.2888	25.802	3.463	2.605	9.021	**9**
10	9.313	.1074	.0301	.2801	33.253	3.571	2.797	9.987	**10**
11	11.642	.0859	.0235	.2735	42.566	3.656	2.966	10.846	**11**
12	14.552	.0687	.0184	.2684	54.208	3.725	3.115	11.602	**12**
13	18.190	.0550	.0145	.2645	68.760	3.780	3.244	12.262	**13**
14	22.737	.0440	.0115	.2615	86.949	3.824	3.356	12.833	**14**
15	28.422	.0352	.00912	.2591	109.687	3.859	3.453	13.326	**15**
16	35.527	.0281	.00724	.2572	138.109	3.887	3.537	13.748	**16**
17	44.409	.0225	.00576	.2558	173.636	3.910	3.608	14.108	**17**
18	55.511	.0180	.00459	.2546	218.045	3.928	3.670	14.415	**18**
19	69.389	.0144	.00366	.2537	273.556	3.942	3.722	14.674	**19**
20	86.736	.0115	.00292	.2529	342.945	3.954	3.767	14.893	**20**
21	108.420	.00922	.00233	.2523	429.681	3.963	3.805	15.078	**21**
22	135.525	.00738	.00186	.2519	538.101	3.970	3.836	15.233	**22**
23	169.407	.00590	.00148	.2515	673.626	3.976	3.863	15.362	**23**
24	211.758	.00472	.00119	.2512	843.033	3.981	3.886	15.471	**24**
25	264.698	.00378	.00095	.2509	1 054.8	3.985	3.905	15.562	**25**
26	330.872	.00302	.00076	.2508	1 319.5	3.988	3.921	15.637	**26**
27	413.590	.00242	.00061	.2506	1 650.4	3.990	3.935	15.700	**27**
28	516.988	.00193	.00048	.2505	2 064.0	3.992	3.946	15.752	**28**
29	646.235	.00155	.00039	.2504	2 580.9	3.994	3.955	15.796	**29**
30	807.794	.00124	.00031	.2503	3 227.2	3.995	3.963	15.832	**30**
31	1 009.7	.00099	.00025	.2502	4 035.0	3.996	3.969	15.861	**31**
32	1 262.2	.00079	.00020	.2502	5 044.7	3.997	3.975	15.886	**32**
33	1 577.7	.00063	.00016	.2502	6 306.9	3.997	3.979	15.906	**33**
34	1 972.2	.00051	.00013	.2501	7 884.6	3.998	3.983	15.923	**34**
35	2 465.2	.00041	.00010	.2501	9 856.8	3.998	3.986	15.937	**35**
40	7 523.2	.00013	.00003	.2500	30 088.7	3.999	3.995	15.977	**40**
45	22 958.9	.00004	.00001	.2500	91 831.5	4.000	3.998	15.991	**45**
50	70 064.9	.00001		.2500	280 255.7	4.000	3.999	15.997	**50**
55	213 821.2	.00000		.2500	855 280.7	4.000	4.000	15.999	**55**

30% Compound Interest Factors 30%

	Single Payment		Uniform Payment Series				Arithmetic Gradient		
	Compound Amount Factor	Present Worth Factor	Sinking Fund Factor	Capital Recovery Factor	Compound Amount Factor	Present Worth Factor	Gradient Uniform Series	Gradient Present Worth	
n	Find *F* Given *P* *F/P*	Find *P* Given *F* *P/F*	Find *A* Given *F* *A/F*	Find *A* Given *P* *A/P*	Find *F* Given *A* *F/A*	Find *P* Given *A* *P/A*	Find *A* Given *G* *A/G*	Find *P* Given *G* *P/G*	*n*
1	1.300	.7692	1.0000	1.3000	1.000	0.769	0	0	**1**
2	1.690	.5917	.4348	.7348	2.300	1.361	0.435	0.592	**2**
3	2.197	.4552	.2506	.5506	3.990	1.816	0.827	1.502	**3**
4	2.856	.3501	.1616	.4616	6.187	2.166	1.178	2.552	**4**
5	3.713	.2693	.1106	.4106	9.043	2.436	1.490	3.630	**5**
6	4.827	.2072	.0784	.3784	12.756	2.643	1.765	4.666	**6**
7	6.275	.1594	.0569	.3569	17.583	2.802	2.006	5.622	**7**
8	8.157	.1226	.0419	.3419	23.858	2.925	2.216	6.480	**8**
9	10.604	.0943	.0312	.3312	32.015	3.019	2.396	7.234	**9**
10	13.786	.0725	.0235	.3235	42.619	3.092	2.551	7.887	**10**
11	17.922	.0558	.0177	.3177	56.405	3.147	2.683	8.445	**11**
12	23.298	.0429	.0135	.3135	74.327	3.190	2.795	8.917	**12**
13	30.287	.0330	.0102	.3102	97.625	3.223	2.889	9.314	**13**
14	39.374	.0254	.00782	.3078	127.912	3.249	2.969	9.644	**14**
15	51.186	.0195	.00598	.3060	167.286	3.268	3.034	9.917	**15**
16	66.542	.0150	.00458	.3046	218.472	3.283	3.089	10.143	**16**
17	86.504	.0116	.00351	.3035	285.014	3.295	3.135	10.328	**17**
18	112.455	.00889	.00269	.3027	371.518	3.304	3.172	10.479	**18**
19	146.192	.00684	.00207	.3021	483.973	3.311	3.202	10.602	**19**
20	190.049	.00526	.00159	.3016	630.165	3.316	3.228	10.702	**20**
21	247.064	.00405	.00122	.3012	820.214	3.320	3.248	10.783	**21**
22	321.184	.00311	.00094	.3009	1 067.3	3.323	3.265	10.848	**22**
23	417.539	.00239	.00072	.3007	1 388.5	3.325	3.278	10.901	**23**
24	542.800	.00184	.00055	.3006	1 806.0	3.327	3.289	10.943	**24**
25	705.640	.00142	.00043	.3004	2 348.8	3.329	3.298	10.977	**25**
26	917.332	.00109	.00033	.3003	3 054.4	3.330	3.305	11.005	**26**
27	1 192.5	.00084	.00025	.3003	3 971.8	3.331	3.311	11.026	**27**
28	1 550.3	.00065	.00019	.3002	5 164.3	3.331	3.315	11.044	**28**
29	2 015.4	.00050	.00015	.3001	6 714.6	3.332	3.319	11.058	**29**
30	2 620.0	.00038	.00011	.3001	8 730.0	3.332	3.322	11.069	**30**
31	3 406.0	.00029	.00009	.3001	11 350.0	3.332	3.324	11.078	**31**
32	4 427.8	.00023	.00007	.3001	14 756.0	3.333	3.326	11.085	**32**
33	5 756.1	.00017	.00005	.3001	19 183.7	3.333	3.328	11.090	**33**
34	7 483.0	.00013	.00004	.3000	24 939.9	3.333	3.329	11.094	**34**
35	9 727.8	.00010	.00003	.3000	32 422.8	3.333	3.330	11.098	**35**
40	36 118.8	.00003	.00001	.3000	120 392.6	3.333	3.332	11.107	**40**
45	134 106.5	.00001		.3000	447 018.3	3.333	3.333	11.110	**45**

	Single Payment		Uniform Payment Series				Arithmetic Gradient		
	Compound Amount Factor	Present Worth Factor	Sinking Fund Factor	Capital Recovery Factor	Compound Amount Factor	Present Worth Factor	Gradient Uniform Series	Gradient Present Worth	
n	Find *F* Given *P* *F/P*	Find *P* Given *F* *P/F*	Find *A* Given *F* *A/F*	Find *A* Given *P* *A/P*	Find *F* Given *A* *F/A*	Find *P* Given *A* *P/A*	Find *A* Given *G* *A/G*	Find *P* Given *G* *P/G*	*n*
1	1.350	.7407	1.0000	1.3500	1.000	0.741	0	0	**1**
2	1.822	.5487	.4255	.7755	2.350	1.289	0.426	0.549	**2**
3	2.460	.4064	.2397	.5897	4.173	1.696	0.803	1.362	**3**
4	3.322	.3011	.1508	.5008	6.633	1.997	1.134	2.265	**4**
5	4.484	.2230	.1005	.4505	9.954	2.220	1.422	3.157	**5**
6	6.053	.1652	.0693	.4193	14.438	2.385	1.670	3.983	**6**
7	8.172	.1224	.0488	.3988	20.492	2.508	1.881	4.717	**7**
8	11.032	.0906	.0349	.3849	28.664	2.598	2.060	5.352	**8**
9	14.894	.0671	.0252	.3752	39.696	2.665	2.209	5.889	**9**
10	20.107	.0497	.0183	.3683	54.590	2.715	2.334	6.336	**10**
11	27.144	.0368	.0134	.3634	74.697	2.752	2.436	6.705	**11**
12	36.644	.0273	.00982	.3598	101.841	2.779	2.520	7.005	**12**
13	49.470	.0202	.00722	.3572	138.485	2.799	2.589	7.247	**13**
14	66.784	.0150	.00532	.3553	187.954	2.814	2.644	7.442	**14**
15	90.158	.0111	.00393	.3539	254.739	2.825	2.689	7.597	**15**
16	121.714	.00822	.00290	.3529	344.897	2.834	2.725	7.721	**16**
17	164.314	.00609	.00214	.3521	466.611	2.840	2.753	7.818	**17**
18	221.824	.00451	.00158	.3516	630.925	2.844	2.776	7.895	**18**
19	299.462	.00334	.00117	.3512	852.748	2.848	2.793	7.955	**19**
20	404.274	.00247	.00087	.3509	1 152.2	2.850	2.808	8.002	**20**
21	545.769	.00183	.00064	.3506	1 556.5	2.852	2.819	8.038	**21**
22	736.789	.00136	.00048	.3505	2 102.3	2.853	2.827	8.067	**22**
23	994.665	.00101	.00035	.3504	2 839.0	2.854	2.834	8.089	**23**
24	1 342.8	.00074	.00026	.3503	3 833.7	2.855	2.839	8.106	**24**
25	1 812.8	.00055	.00019	.3502	5 176.5	2.856	2.843	8.119	**25**
26	2 447.2	.00041	.00014	.3501	6 989.3	2.856	2.847	8.130	**26**
27	3 303.8	.00030	.00011	.3501	9 436.5	2.856	2.849	8.137	**27**
28	4 460.1	.00022	.00008	.3501	12 740.3	2.857	2.851	8.143	**28**
29	6 021.1	.00017	.00006	.3501	17 200.4	2.857	2.852	8.148	**29**
30	8 128.5	.00012	.00004	.3500	23 221.6	2.857	2.853	8.152	**30**
31	10 973.5	.00009	.00003	.3500	31 350.1	2.857	2.854	8.154	**31**
32	14 814.3	.00007	.00002	.3500	42 323.7	2.857	2.855	8.157	**32**
33	19 999.3	.00005	.00002	.3500	57 137.9	2.857	2.855	8.158	**33**
34	26 999.0	.00004	.00001	.3500	77 137.2	2.857	2.856	8.159	**34**
35	36 448.7	.00003	.00001	.3500	104 136.3	2.857	2.856	8.160	**35**

Compound Interest Factors

	Single Payment		Uniform Payment Series				Arithmetic Gradient		
	Compound Amount Factor	Present Worth Factor	Sinking Fund Factor	Capital Recovery Factor	Compound Amount Factor	Present Worth Factor	Gradient Uniform Series	Gradient Present Worth	
n	Find F Given P F/P	Find P Given F P/F	Find A Given F A/F	Find A Given P A/P	Find F Given A F/A	Find P Given A P/A	Find A Given G A/G	Find P Given G P/G	n
1	1.400	.7143	1.0000	1.4000	1.000	0.714	0	0	1
2	1.960	.5102	.4167	.8167	2.400	1.224	0.417	0.510	2
3	2.744	.3644	.2294	.6294	4.360	1.589	0.780	1.239	3
4	3.842	.2603	.1408	.5408	7.104	1.849	1.092	2.020	4
5	5.378	.1859	.0914	.4914	10.946	2.035	1.358	2.764	5
6	7.530	.1328	.0613	.4613	16.324	2.168	1.581	3.428	6
7	10.541	.0949	.0419	.4419	23.853	2.263	1.766	3.997	7
8	14.758	.0678	.0291	.4291	34.395	2.331	1.919	4.471	8
9	20.661	.0484	.0203	.4203	49.153	2.379	2.042	4.858	9
10	28.925	.0346	.0143	.4143	69.814	2.414	2.142	5.170	10
11	40.496	.0247	.0101	.4101	98.739	2.438	2.221	5.417	11
12	56.694	.0176	.00718	.4072	139.235	2.456	2.285	5.611	12
13	79.371	.0126	.00510	.4051	195.929	2.469	2.334	5.762	13
14	111.120	.00900	.00363	.4036	275.300	2.478	2.373	5.879	14
15	155.568	.00643	.00259	.4026	386.420	2.484	2.403	5.969	15
16	217.795	.00459	.00185	.4018	541.988	2.489	2.426	6.038	16
17	304.913	.00328	.00132	.4013	759.783	2.492	2.444	6.090	17
18	426.879	.00234	.00094	.4009	1 064.7	2.494	2.458	6.130	18
19	597.630	.00167	.00067	.4007	1 419.6	2.496	2.468	6.160	19
20	836.682	.00120	.00048	.4005	2 089.2	2.497	2.476	6.183	20
21	1 171.4	.00085	.00034	.4003	2 925.9	2.498	2.482	6.200	21
22	1 639.9	.00061	.00024	.4002	4 097.2	2.498	2.487	6.213	22
23	2 295.9	.00044	.00017	.4002	5 737.1	2.499	2.490	6.222	23
24	3 214.2	.00031	.00012	.4001	8 033.0	2.499	2.493	6.229	24
25	4 499.9	.00022	.00009	.4001	11 247.2	2.499	2.494	6.235	25
26	6 299.8	.00016	.00006	.4001	15 747.1	2.500	2.496	6.239	26
27	8 819.8	.00011	.00005	.4000	22 046.9	2.500	2.497	6.242	27
28	12 347.7	.00008	.00003	.4000	30 866.7	2.500	2.498	6.244	28
29	17 286.7	.00006	.00002	.4000	43 214.3	2.500	2.498	6.245	29
30	24 201.4	.00004	.00002	.4000	60 501.0	2.500	2.499	6.247	30
31	33 882.0	.00003	.00001	.4000	84 702.5	2.500	2.499	6.248	31
32	47 434.8	.00002	.00001	.4000	118 584.4	2.500	2.499	6.248	32
33	66 408.7	.00002	.00001	.4000	166 019.2	2.500	2.500	6.249	33
34	92 972.1	.00001		.4000	232 427.9	2.500	2.500	6.249	34
35	130 161.0	.00001		.4000	325 400.0	2.500	2.500	6.249	35

n	Single Payment		Uniform Payment Series				Arithmetic Gradient		n
	Compound Amount Factor	Present Worth Factor	Sinking Fund Factor	Capital Recovery Factor	Compound Amount Factor	Present Worth Factor	Gradient Uniform Series	Gradient Present Worth	
	Find *F* Given *P* *F/P*	Find *P* Given *F* *P/F*	Find *A* Given *F* *A/F*	Find *A* Given *P* *A/P*	Find *F* Given *A* *F/A*	Find *P* Given *A* *P/A*	Find *A* Given *G* *A/G*	Find *P* Given *G* *P/G*	n
1	1.450	.6897	1.0000	1.4500	1.000	0.690	0	0	**1**
2	2.103	.4756	.4082	.8582	2.450	1.165	0.408	0.476	**2**
3	3.049	.3280	.2197	.6697	4.553	1.493	0.758	1.132	**3**
4	4.421	.2262	.1316	.5816	7.601	1.720	1.053	1.810	**4**
5	6.410	.1560	.0832	.5332	12.022	1.876	1.298	2.434	**5**
6	9.294	.1076	.0543	.5043	18.431	1.983	1.499	2.972	**6**
7	13.476	.0742	.0361	.4861	27.725	2.057	1.661	3.418	**7**
8	19.541	.0512	.0243	.4743	41.202	2.109	1.791	3.776	**8**
9	28.334	.0353	.0165	.4665	60.743	2.144	1.893	4.058	**9**
10	41.085	.0243	.0112	.4612	89.077	2.168	1.973	4.277	**10**
11	59.573	.0168	.00768	.4577	130.162	2.185	2.034	4.445	**11**
12	86.381	.0116	.00527	.4553	189.735	2.196	2.082	4.572	**12**
13	125.252	.00798	.00362	.4536	276.115	2.204	2.118	4.668	**13**
14	181.615	.00551	.00249	.4525	401.367	2.210	2.145	4.740	**14**
15	263.342	.00380	.00172	.4517	582.982	2.214	2.165	4.793	**15**
16	381.846	.00262	.00118	.4512	846.325	2.216	2.180	4.832	**16**
17	553.677	.00181	.00081	.4508	1 228.2	2.218	2.191	4.861	**17**
18	802.831	.00125	.00056	.4506	1 781.8	2.219	2.200	4.882	**18**
19	1 164.1	.00086	.00039	.4504	2 584.7	2.220	2.206	4.898	**19**
20	1 688.0	.00059	.00027	.4503	3 748.8	2.221	2.210	4.909	**20**
21	2 447.5	.00041	.00018	.4502	5 436.7	2.221	2.214	4.917	**21**
22	3 548.9	.00028	.00013	.4501	7 884.3	2.222	2.216	4.923	**22**
23	5 145.9	.00019	.00009	.4501	11 433.2	2.222	2.218	4.927	**23**
24	7 461.6	.00013	.00006	.4501	16 579.1	2.222	2.219	4.930	**24**
25	10 819.3	.00009	.00004	.4500	24 040.7	2.222	2.220	4.933	**25**
26	15 688.0	.00006	.00003	.4500	34 860.1	2.222	2.221	4.934	**26**
27	22 747.7	.00004	.00002	.4500	50 548.1	2.222	2.221	4.935	**27**
28	32 984.1	.00003	.00001	.4500	73 295.8	2.222	2.221	4.936	**28**
29	47 826.9	.00002	.00001	.4500	106 279.9	2.222	2.222	4.937	**29**
30	69 349.1	.00001	.00001	.4500	154 106.8	2.222	2.222	4.937	**30**
31	100 556.1	.00001		.4500	223 455.9	2.222	2.222	4.938	**31**
32	145 806.4	.00001		.4500	324 012.0	2.222	2.222	4.938	**32**
33	211 419.3			.4500	469 818.5	2.222	2.222	4.938	**33**
34	306 558.0			.4500	681 237.8	2.222	2.222	4.938	**34**
35	444 509.2			.4500	987 795.9	2.222	2.222	4.938	**35**

	Single Payment		Uniform Payment Series				Arithmetic Gradient		
	Compound Amount Factor	Present Worth Factor	Sinking Fund Factor	Capital Recovery Factor	Compound Amount Factor	Present Worth Factor	Gradient Uniform Series	Gradient Present Worth	
n	Find *F* Given *P* *F/P*	Find *P* Given *F* *P/F*	Find *A* Given *F* *A/F*	Find *A* Given *P* *A/P*	Find *F* Given *A* *F/A*	Find *P* Given *A* *P/A*	Find *A* Given *G* *A/G*	Find *P* Given *G* *P/G*	*n*
1	1.500	.6667	1.0000	1.5000	1.000	0.667	0	0	**1**
2	2.250	.4444	.4000	.9000	2.500	1.111	0.400	0.444	**2**
3	3.375	.2963	.2105	.7105	4.750	1.407	0.737	1.037	**3**
4	5.063	.1975	.1231	.6231	8.125	1.605	1.015	1.630	**4**
5	7.594	.1317	.0758	.5758	13.188	1.737	1.242	2.156	**5**
6	11.391	.0878	.0481	.5481	20.781	1.824	1.423	2.595	**6**
7	17.086	.0585	.0311	.5311	32.172	1.883	1.565	2.947	**7**
8	25.629	.0390	.0203	.5203	49.258	1.922	1.675	3.220	**8**
9	38.443	.0260	.0134	.5134	74.887	1.948	1.760	3.428	**9**
10	57.665	.0173	.00882	.5088	113.330	1.965	1.824	3.584	**10**
11	86.498	.0116	.00585	.5058	170.995	1.977	1.871	3.699	**11**
12	129.746	.00771	.00388	.5039	257.493	1.985	1.907	3.784	**12**
13	194.620	.00514	.00258	.5026	387.239	1.990	1.933	3.846	**13**
14	291.929	.00343	.00172	.5017	581.859	1.993	1.952	3.890	**14**
15	437.894	.00228	.00114	.5011	873.788	1.995	1.966	3.922	**15**
16	656.841	.00152	.00076	.5008	1 311.7	1.997	1.976	3.945	**16**
17	985.261	.00101	.00051	.5005	1 968.5	1.998	1.983	3.961	**17**
18	1 477.9	.00068	.00034	.5003	2 953.8	1.999	1.988	3.973	**18**
19	2 216.8	.00045	.00023	.5002	4 431.7	1.999	1.991	3.981	**19**
20	3 325.3	.00030	.00015	.5002	6 648.5	1.999	1.994	3.987	**20**
21	4 987.9	.00020	.00010	.5001	9 973.8	2.000	1.996	3.991	**21**
22	7 481.8	.00013	.00007	.5001	14 961.7	2.000	1.997	3.994	**22**
23	11 222.7	.00009	.00004	.5000	22 443.5	2.000	1.998	3.996	**23**
24	16 834.1	.00006	.00003	.5000	33 666.2	2.000	1.999	3.997	**24**
25	25 251.2	.00004	.00002	.5000	50 500.3	2.000	1.999	3.998	**25**
26	37 876.8	.00003	.00001	.5000	75 751.5	2.000	1.999	3.999	**26**
27	56 815.1	.00002	.00001	.5000	113 628.3	2.000	2.000	3.999	**27**
28	85 222.7	.00001	.00001	.5000	170 443.4	2.000	2.000	3.999	**28**
29	127 834.0	.00001		.5000	255 666.1	2.000	2.000	4.000	**29**
30	191 751.1	.00001		.5000	383 500.1	2.000	2.000	4.000	**30**
31	287 626.6			.5000	575 251.2	2.000	2.000	4.000	**31**
32	431 439.9			.5000	862 877.8	2.000	2.000	4.000	**32**

	Single Payment		Uniform Payment Series				Arithmetic Gradient		
	Compound Amount Factor	Present Worth Factor	Sinking Fund Factor	Capital Recovery Factor	Compound Amount Factor	Present Worth Factor	Gradient Uniform Series	Gradient Present Worth	
n	Find *F* Given *P* *F/P*	Find *P* Given *F* *P/F*	Find *A* Given *F* *A/F*	Find *A* Given *P* *A/P*	Find *F* Given *A* *F/A*	Find *P* Given *A* *P/A*	Find *A* Given *G* *A/G*	Find *P* Given *G* *P/G*	**n**
1	1.600	.6250	1.0000	1.6000	1.000	0.625	0	0	**1**
2	2.560	.3906	.3846	.9846	2.600	1.016	0.385	0.391	**2**
3	4.096	.2441	.1938	.7938	5.160	1.260	0.698	0.879	**3**
4	6.554	.1526	.1080	.7080	9.256	1.412	0.946	1.337	**4**
5	10.486	.0954	.0633	.6633	15.810	1.508	1.140	1.718	**5**
6	16.777	.0596	.0380	.6380	26.295	1.567	1.286	2.016	**6**
7	26.844	.0373	.0232	.6232	43.073	1.605	1.396	2.240	**7**
8	42.950	.0233	.0143	.6143	69.916	1.628	1.476	2.403	**8**
9	68.719	.0146	.00886	.6089	112.866	1.642	1.534	2.519	**9**
10	109.951	.00909	.00551	.6055	181.585	1.652	1.575	2.601	**10**
11	175.922	.00568	.00343	.6034	291.536	1.657	1.604	2.658	**11**
12	281.475	.00355	.00214	.6021	467.458	1.661	1.624	2.697	**12**
13	450.360	.00222	.00134	.6013	748.933	1.663	1.638	2.724	**13**
14	720.576	.00139	.00083	.6008	1 199.3	1.664	1.647	2.742	**14**
15	1 152.9	.00087	.00052	.6005	1 919.9	1.665	1.654	2.754	**15**
16	1 844.7	.00054	.00033	.6003	3 072.8	1.666	1.658	2.762	**16**
17	2 951.5	.00034	.00020	.6002	4 917.5	1.666	1.661	2.767	**17**
18	4 722.4	.00021	.00013	.6001	7 868.9	1.666	1.663	2.771	**18**
19	7 555.8	.00013	.00008	.6011	12 591.3	1.666	1.664	2.773	**19**
20	12 089.3	.00008	.00005	.6000	20 147.1	1.667	1.665	2.775	**20**
21	19 342.8	.00005	.00003	.6000	32 236.3	1.667	1.666	2.776	**21**
22	30 948.5	.00003	.00002	.6000	51 579.2	1.667	1.666	2.777	**22**
23	49 517.6	.00002	.00001	.6000	82 527.6	1.667	1.666	2.777	**23**
24	79 228.1	.00001	.00001	.6000	132 045.2	1.667	1.666	2.777	**24**
25	126 765.0	.00001		.6000	211 273.4	1.667	1.666	2.777	**25**
26	202 824.0			.6000	338 038.4	1.667	1.667	2.778	**26**
27	324 518.4			.6000	540 862.4	1.667	1.667	2.778	**27**
28	519 229.5			.6000	865 380.9	1.667	1.667	2.778	**28**

Continuous Compounding—Single Payment Factors

rn	Compound Amount Factor e^{rn} Find F Given P F/P	Present Worth Factor e^{-rn} Find P Given F P/F
.01	1.0101	.9900
.02	1.0202	.9802
.03	1.0305	.9704
.04	1.0408	.9608
.05	1.0513	.9512
.06	1.0618	.9418
.07	1.0725	.9324
.08	1.0833	.9231
.09	1.0942	.9139
.10	1.1052	.9048
.11	1.1163	.8958
.12	1.1275	.8869
.13	1.1388	.8781
.14	1.1503	.8694
.15	1.1618	.8607
.16	1.1735	.8521
.17	1.1853	.8437
.18	1.1972	.8353
.19	1.2092	.8270
.20	1.2214	.8187
.21	1.2337	.8106
.22	1.2461	.8025
.23	1.2586	.7945
.24	1.2712	.7866
.25	1.2840	.7788
.26	1.2969	.7711
.27	1.3100	.7634
.28	1.3231	.7558
.29	1.3364	.7483
.30	1.3499	.7408
.31	1.3634	.7334
.32	1.3771	.7261
.33	1.3910	.7189
.34	1.4049	.7118
.35	1.4191	.7047
.36	1.4333	.6977
.37	1.4477	.6907
.38	1.4623	.6839
.39	1.4770	.6771
.40	1.4918	.6703
.41	1.5068	.6637
.42	1.5220	.6570
.43	1.5373	.6505
.44	1.5527	.6440
.45	1.5683	.6376
.46	1.5841	.6313
.47	1.6000	.6250
.48	1.6161	.6188
.49	1.6323	.6126
.50	1.6487	.6065

rn	Compound Amount Factor e^{rn} Find F Given P F/P	Present Worth Factor e^{-rn} Find P Given F P/F
.51	1.6653	.6005
.52	1.6820	.5945
.53	1.6989	.5886
.54	1.7160	.5827
.55	1.7333	.5769
.56	1.7507	.5712
.57	1.7683	.5655
.58	1.7860	.5599
.59	1.8040	.5543
.60	1.8221	.5488
.61	1.8404	.5434
.62	1.8589	.5379
.63	1.8776	.5326
.64	1.8965	.5273
.65	1.9155	.5220
.66	1.9348	.5169
.67	1.9542	.5117
.68	1.9739	.5066
.69	1.9937	.5016
.70	2.0138	.4966
.71	2.0340	.4916
.72	2.0544	.4868
.73	2.0751	.4819
.74	2.0959	.4771
.75	2.1170	.4724
.76	2.1383	.4677
.77	2.1598	.4630
.78	2.1815	.4584
.79	2.2034	.4538
.80	2.2255	.4493
.81	2.2479	.4449
.82	2.2705	.4404
.83	2.2933	.4360
.84	2.3164	.4317
.85	2.3396	.4274
.86	2.3632	.4232
.87	2.3869	.4190
.88	2.4109	.4148
.89	2.4351	.4107
.90	2.4596	.4066
.91	2.4843	.4025
.92	2.5093	.3985
.93	2.5345	.3946
.94	2.5600	.3906
.95	2.5857	.3867
.96	2.6117	.3829
.97	2.6379	.3791
.98	2.6645	.3753
.99	2.6912	.3716
1.00	2.7183	.3679

Appendix **B**

Approximate Calculations Using Depreciation

When asked the question, "What uniform annual payment would be required to repay $5000 in five years with interest at 8%?" we have calculated

$$A = P(A/P,8\%,5 \text{ years}) = \$5000(0.2505) = \$1252$$

But what does the $1252 represent? A little thought reveals that there are two components:

1. A portion of the $5000 principal sum must be repaid each year by the borrower.
2. The borrower must also pay 8% interest on the amount of the unpaid debt each year.

The details of this repayment plan were described in Plan 3 of Table 4-1. We saw that $1252 was the uniform annual payment that was *equivalent* to a present sum of $5000 for the stated situation.

Suppose one wished to estimate the uniform annual payment without using Compound Interest Tables. The only practical thing to do would be to make a reasonable estimate of the annual principal payment and of the annual interest payment. The sum of these two components would be an estimate of the uniform annual payment. One method is described in the next section.

Annual Cost By Straight Line Depreciation Plus Average Interest

If we assume that a quantity of money is repaid in equal annual payments, then the annual payment is identical to straight line depreciation with zero salvage value. The resulting plot of the remaining debt *vs.* time is shown in Fig. B-1.

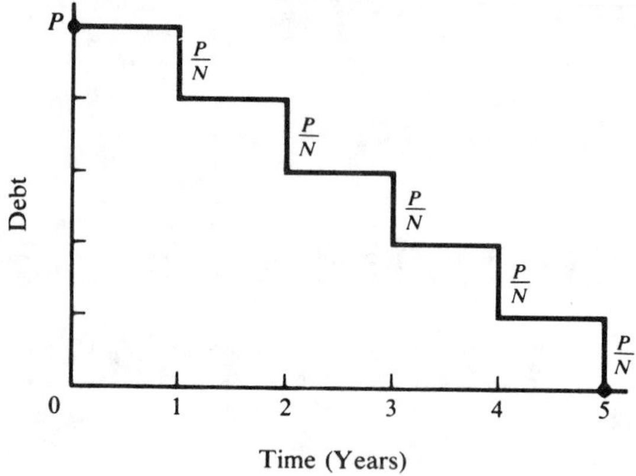

Figure B-1 Debt repayment with equal principal payments.

In addition, there is an annual interest payment. At the end of the first year, it is Pi; in the last year, it is Pi/N. Since the principal repayment plan is uniform, the annual interest would decline uniformly from the initial Pi to Pi/N in the last year. We may compute the average annual interest as:

$$\text{Average interest} = \frac{\text{First-year interest} + \text{Last-year interest}}{2}$$

$$= \frac{Pi + (Pi/N)}{2} = \frac{Pi}{2}\left(1 + \frac{1}{N}\right)$$

$$= \frac{Pi}{2}\left(\frac{N + 1}{N}\right)$$

Combining the components, where $S = 0$:

$$\text{Average annual cost} = \frac{P}{N} + \frac{Pi}{2}\left(\frac{N + 1}{N}\right)$$

EXAMPLE B-1

What will be the average annual payment to repay a debt of $5000 in five years with interest at 8%?

Solution: Using the method of straight line depreciation plus average interest:

$$\text{Average annual cost} = \frac{P}{N} + \frac{Pi}{2}\left(\frac{N+1}{N}\right)$$

$$= \frac{5000}{5} + \frac{5000(0.08)}{2}\left(\frac{6}{5}\right) = 1000 + 240$$

$$= \$1240 \blacktriangleleft$$

What is the difference between the \$1240 calculated in Example B-1 and the previously calculated \$1252? We have seen that the \$1252 represents the *equivalent* uniform annual cost while the Example B-1 calculation produces the *average* uniform annual cost.

When the straight line depreciation plus average interest method is to be used in computing the *average* annual cost where there is a salvage value, the equation must be adjusted. Where there is a salvage value S, as in Figure B-2,

$$\text{Straight line depreciation} = \frac{P-S}{N}$$

$$\text{Average interest} = \frac{\text{First-year interest} + \text{Last-year interest}}{2}$$

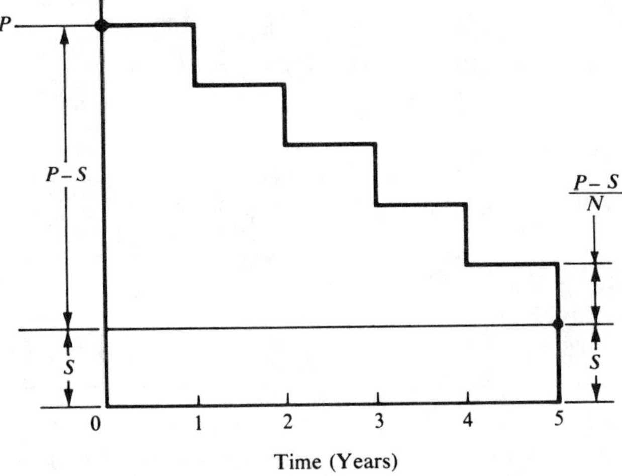

Figure B-2

$$\text{Average interest} = \frac{(P - S)i + Si + [(P - S)i/N] + Si}{2}$$

$$= (P - S)\left(\frac{i}{2}\right)\left(1 + \frac{1}{N}\right) + Si$$

$$= (P - S)\left(\frac{i}{2}\right)\left(\frac{N + 1}{N}\right) + Si \quad ◀$$

Thus the general equation for straight line depreciation plus average interest is:

$$\textbf{\textit{Average annual cost}} = \frac{P - S}{N} + (P - S)\left(\frac{i}{2}\right)\left(\frac{N + 1}{N}\right) + Si$$

EXAMPLE B-2

A firm buys equipment with an initial cost of $8000, and an estimated salvage value of $2000 at the end of an eight-year useful life. If interest is 6%, what is:

a. the equivalent uniform annual cost?

b. the average uniform annual cost based on straight line depreciation plus average interest?

Solution:

a. $\text{EUAC} = (8000 - 2000)(A/P,6\%,8 \text{ years}) + 2000(0.06)$
$= 6000(0.1610) + 2000(0.06) = 966 + 120$
$= \$1086$

b. Average uniform annual cost

$$= \frac{8000 - 2000}{8} + (8000 - 2000)\left(\frac{0.06}{2}\right)\left(\frac{9}{8}\right) + 2000(0.06)$$

$$= 750.00 + 202.50 + 120.00$$

$$= \$1072.50$$

EXAMPLE B-3

Consider Problem 6-10. An electronics firm invested $60,000 in a precision inspection device. It cost $4000 to operate and maintain in the first year, and $3000 in each of the subsequent years. At the end of four years, the firm changed their inspection procedure, eliminating the need for the device. The purchasing agent was very fortunate in being able to sell the inspection device for the $60,000 that had originally been paid for it. Assume interest at 10% per year.

a. Compute the average uniform annual cost.

b. The equivalent uniform annual cost is $9287. Why is this different from the answer in part **a**?

Solution:

a. Average uniform annual cost

$$= \frac{60{,}000 - 60{,}000}{4} + (60{,}000 - 60{,}000)\left(\frac{0.10}{2}\right)\left(\frac{5}{4}\right)$$
$$+ 60{,}000(0.10) + \frac{4000 + 3(3000)}{4}$$
$$= 60{,}000(0.10) + \frac{13{,}000}{4} = \$9250$$

b. In this problem the salvage value equals the cost. Straight line depreciation plus average interest is 60,000(0.10) = $6000. This is the same as computing 60,000(A/P,10%,4) – 60,000(A/F,10%,4) = $6000. The EUAC of operation and maintenance is

$$3000 + 1000(P/F,10\%,1)(A/P,10\%,4) = \$3287$$

The average uniform annual cost of operation and maintenance is

$$\frac{4000 + 3(3000)}{4} = \$3250$$ ◀

This explains the $37 difference.

The approximate method usually gives values that are smaller than the exact capital recovery calculations. The error becomes larger as N and i increase. Since the exact computation can be readily done if Compound Interest Tables are available, there seems little reason to resort to an approximate calculation like straight line depreciation plus average interest.

References

AASHTO. *A Manual on User Benefit Analysis of Highway and Bus–Transit Improvements*. Washington, D.C.: American Association of State Highway and Transportation Officials, 1978.

American Telephone and Telegraph Co. *Engineering Economy*, 3rd ed. New York: McGraw-Hill, 1977.

Au, T., and Au, P. *Engineering Economics for Capital Investment Analysis*. Boston: Allyn and Bacon, 1983.

Barish, N. N., and Kaplan, S. *Economic Analysis for Engineering and Managerial Decision Making*, 2nd ed. New York: McGraw-Hill, 1978.

Bernhard, R. H. "A Comprehensive Comparison and Critique of Discounting Indices Proposed for Capital Investment Evaluation," *The Engineering Economist*. Vol. 16, No. 3, pp. 157–186.

Blank, L. T., and Tarquin, A. J. *Engineering Economy*, 3rd ed. New York: McGraw-Hill, 1989.

Bussey, L. E. *The Economic Analysis of Industrial Projects*. Englewood Cliffs, New Jersey: Prentice-Hall, 1978.

Cassimatis, P. *A Concise Introduction to Engineering Economics*. Boston: Unwin Hyman, 1988.

Collier, C. A., and Ledbetter, W. B. *Engineering Economic and Cost Analysis*, 2nd ed. New York: Harper and Row, 1988.

DeGarmo, E. P., Sullivan, W. G., and Bontadelli, J. A. *Engineering Economy*, 8th ed. New York: Macmillan, 1988.

Engineering Economist, The. A quarterly journal of the Engineering Economy Divisions of ASEE and IIE. Norcross, Georgia: Institute of Industrial Engineers.

Eschenbach, T. *Cases in Engineering Economy*. New York: John Wiley & Sons, 1989.

Fleischer, G. A. *Engineering Economy: Capital Allocation Theory*. Boston: PWS Engineering, 1984.

Grant, E. L., Ireson, W. G., and Leavenworth, R. S. *Principles of Engineering Economy*, 8th ed. New York: John Wiley & Sons, 1990.

Jones, B. W. *Inflation in Engineering Economic Analysis*. New York: John Wiley & Sons, 1982.

Kleinfeld, I. *Engineering and Managerial Economics*. New York: Holt, Rinehart and Winston, 1986.

Lorie, J. H., and Savage, L. J. "Three Problems in Rationing Capital," *The Journal of Business*. Vol. 28, No. 4, pp. 229–239.

Mallik, A. K. *Engineering Economy with Computer Applications*. Mahomet, Illinois: Engineering Technology, 1979.

Merrett, A. J., and Sykes, A. *The Finance and Analysis of Capital Projects*, 2nd ed. London: Longman, 1973.

Newnan, D. G., editor. *Engineering Economy Exam File*. San Jose, California: Engineering Press, 1984.

———, "Determining Rate of Return by Means of Payback Period and Useful Life," *The Engineering Economist*. Vol. 15, No. 1, pp. 29–39.

Oglesby, C. H., and Hicks, R. G. *Highway Engineering*, 4th ed. New York: John Wiley & Sons, 1982.

Park, W. R., and Jackson, D. E. *Cost Engineering Analysis*, 2nd ed. New York: John Wiley & Sons, 1984.

Riggs, J. L., and West, T. M. *Engineering Economics*, 3rd ed. New York: McGraw-Hill, 1986.

Smith, G. W. *Engineering Economy: Analysis of Capital Expenditures*, 4th ed. Ames, Iowa: The Iowa State University Press, 1987.

Steiner, H. M. *Public and Private Investments: Socioeconomic Analysis*. New York: John Wiley & Sons, 1980.

Stevens, G. T. *Economic and Financial Analysis of Capital Investments*. New York: John Wiley & Sons, 1979.

Swalm, R. O., and Lopez-Leautaud, J. L. *Engineering Economic Analysis: A Future Wealth Approach*. New York: John Wiley & Sons, 1984.

Terborgh, G. *Business Investment Management*. Washington, D.C.: Machinery and Allied Products Institute, 1967.

Thuesen, H. G., Fabrycky, W. J., and Thuesen, G. J. *Engineering Economy*, 7th ed. Englewood Cliffs, New Jersey: Prentice-Hall, 1988.

Wellington, A. M. *The Economic Theory of Railway Location*. New York: John Wiley & Sons, 1887.

White, J. A., Agee, M. H., and Case, K. E. *Principles of Engineering Economic Analysis*, 3rd ed. New York: John Wiley & Sons, 1989.

Index

A

B

C

D

E

F

G

N

O

P

R

S

T

U

V

Exam Files

Professors around the country have opened their exam files and revealed their examination problems and solutions. These are actual exam problems with the complete solutions prepared by the same professors who wrote the problems. **EXAM FILES** are currently available for these topics:

Calculus I
Calculus II
Calculus III
Chemistry
Circuit Analysis
College Algebra
Differential Equations
Dynamics
Fluid Mechanics
Linear Algebra
Materials Science
Mechanics of Materials
Organic Chemistry
Physics I—Mechanics
Physics II—Heat, Light, and Sound
Physics III—Electricity and Magnetism
Probability and Statistics
Statics
Thermodynamics

The **EXAM FILE** series also includes four engineering license review books:

Civil Engineering License Exam File
Electrical Engineering License Exam File
Engineer-In-Training Exam File
Mechanical Engineering License Exam File

The **EXAM FILES** for courses (about $14.45 to $21.45) and engineering licenses (about $22.45 to $36.45) may be ordered with a Visa or MasterCard by telephone.

Engineering Press/Maxway: (212) 947-6100

For a detailed description of the **EXAM FILES**, ask at your college or technical bookstore, or write to:

Engineering Press, Inc.
P.O. Box 1
San Jose, California 95103-0001

Economic Criteria

Method of Analysis	*Fixed Input*	*Fixed Output*	*Neither Input Nor Output Fixed*
PRESENT WORTH	Maximize PW of Benefits	Minimize PW of Costs	Maximize (PW of Benefits − PW of Costs), or Maximize Net Present Worth
ANNUAL CASH FLOW	Maximize Equivalent Uniform Annual Benefits (EUAB)	Minimize Equivalent Uniform Annual Cost (EUAC)	Maximize (EUAB − EUAC)
FUTURE WORTH	Maximize FW of Benefits	Minimize FW of Costs	Maximize (FW of Benefits − FW of Costs), or Maximize Net Future Worth
BENEFIT–COST RATIO	Maximize Benefit–Cost Ratio	Maximize Benefit–Cost Ratio	*Two Alternatives:* Compute the incremental Benefit–Cost ratio ($\Delta B/\Delta C$) on the increment of *investment* between the alternatives. If $\Delta B/\Delta C \geq 1$, choose higher-cost alternative; if not, choose lower-cost alternative. *Three or more Alternatives:* Incremental analysis is required (see Ch. 9).
RATE OF RETURN	*Two Alternatives:* Compute the incremental rate of return (ΔROR) on the increment of *investment* between the alternatives. If ΔROR $\geq$ minimum attractive rate of return, choose the higher-cost alternative; if not, choose lower-cost alternative. *Three or more Alternatives:* Incremental analysis is required (see Ch. 8).		